The Elements of
Real Analysis

The Elements of Real Analysis

Second Edition

Robert G. Bartle

Professor of Mathematics
University of Illinois
Urbana-Champaign

JOHN WILEY & SONS

New York • Chichester • Brisbane • Toronto • Singapore

Library of Congress Cataloging in Publication Data:

Bartle, Robert Gardner, 1927–
The elements of real analysis.

Bibliography: p.
Includes index.
1. Mathematical analysis. II. Title.
QA300.B29 1975   515'.8   75-15979
ISBN 0 471 05464 X

Printed in the United States of America

21 22 23 24 25 26 27 28 29 30

To my parents

Preface

At one time an undergraduate student of advanced mathematics was expected to develop technique in solving problems that involved considerable computation; however, he was not expected to master "theoretical subtleties" such as uniform convergence or uniform continuity. He was required to be able to use the Implicit Function Theorem, but not to know its hypotheses. This situation has changed; it is now considered important that all advanced students of mathematics—future mathematicians, computer scientists, physicists, engineers, or economists—grasp the basic theoretical nature of the subject. They will then understand both the power and the limitation of the general theory more fully.

This textbook developed from my experience in teaching real analysis at the University of Illinois since 1955. My audience often ranges from unusually well-prepared freshmen to graduate students. Most of them are usually not mathematics majors, but they have studied at least the equivalent of three semesters of (nonrigorous) calculus, including partial derivatives, multiple integrals, line integrals, and infinite series. It is desirable for all of the students to have a semester of linear or modern algebra to prepare the way for this course in which analytic theorems are proved. However, since many of the students I encounter do not possess this background, I begin the study of analysis with a few algebraic proofs to start them on their way.

In this edition, I introduce the algebraic and order properties of the real number system in Sections 4 and 5 in a simpler manner than was used in the first edition. In addition, I introduce the definitions of a vector space and a normed space in Section 8, since these notions occur so frequently in modern mathematics. I also shortened several sections to make the material more readily available and to provide additional flexibility in using this book as a text. I added many new exercises and projects but tried to keep the book at the same level of sophistication as the first edition. There have been only minor changes in the first part of the book. However, since experience has shown that the discussion of differentiation and integration in $\mathbf{R}^p$ was too brief in the first edition, I assembled the theory of functions of one variable into a single chapter, and expanded considerably the treatment of functions of several variables.

In Sections 1 to 3, I present the set-theoretic terminology and notation that is employed subsequently and introduce a few basic concepts. However, these sections do *not* give a systematic presentation of set theory. (Such a presentation is not needed or desirable at this stage.) These sections should be examined briefly and consulted later if necessary. The text really starts with Section 4, and Section 6 introduces "analysis." It is possible to cover Sections 4 to 12, 14 to 17, 20 to 24.1, and most of 27 to 31 in one semester. I would exercise the instructor's prerogative and briefly introduce certain other topics (such as series) at the expense of soft-pedaling (or even omitting) various results that are not essential to the later material. Since the entire text provides a little more material than can usually be covered in one year at this level, the instructor probably will limit discussion of some sections. However, it will be useful to the student to have the additional material available for future reference. Most of the topics generally associated with courses in "advanced calculus" are dealt with here. The main exception is the subject line and surface integrals and Stokes's Theorem; this topic is not discussed, since an intuitive treatment is properly a part of calculus, and a rigorous treatment requires a rather extensive discussion in order to be useful.

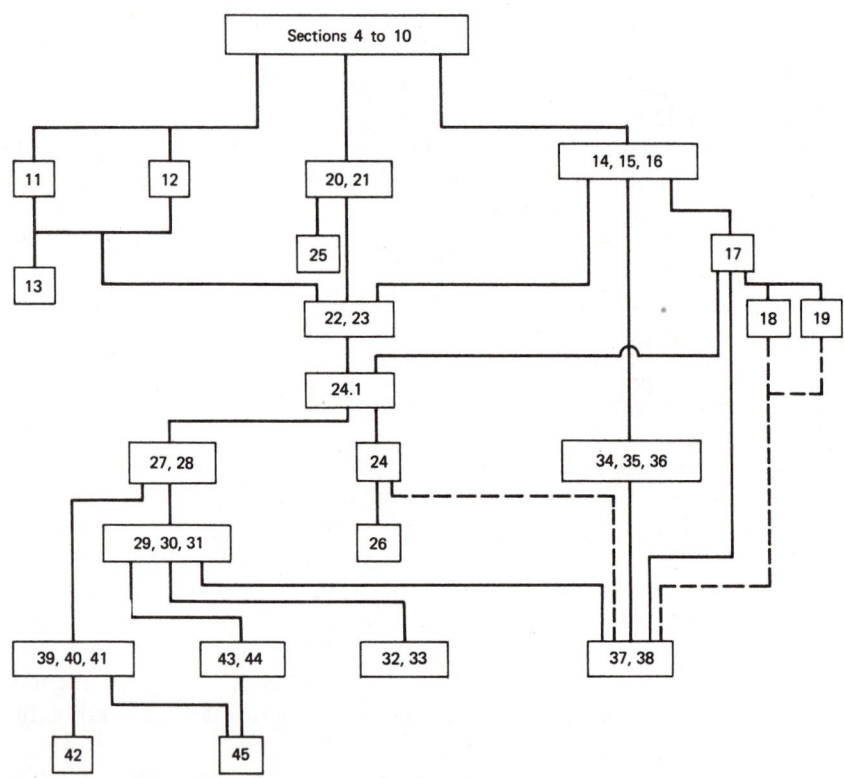

The logical dependence of the various sections of this textbook is indicated by the adjoining diagram. A solid line in this diagram indicates a dependence on the preceding section; a dotted line indicates a slight dependence. All definitions, theorems, corollaries, and lemmas, for instance, are numbered consecutively according to the section number. I assigned names to the more important theorems whenever a name seemed appropriate. The proofs are set off from the text by the head PROOF and end Q.E.D.

It is not possible to over-emphasize the importance of the exercises and projects; only by applying serious and concerted efforts to their solution can one hope to master the material in this book. The projects develop a specific topic in a connected sequence of exercises; we believe they convey to the student at least a taste of the pleasure (and torment!) of doing research in mathematics. I hope that no student will fail to try his hand on several of these projects, for I believe that they are a particularly valuable feature of this book.

In writing this book, I have drawn from my classroom experience and have been influenced by many sources. I benefited by discussions with students and colleagues and, since the publication of the first edition, I have had an extensive correspondence with students and teachers at other institutions. I thank everyone who has made comments and suggestions. Their interest in improving the book encouraged me to undertake this revision. Professors K. W. Anderson, W. G. Bade, and A. L. Peressini read the manuscript of th first edition and offered useful suggestions. I particularly thank my colleague, Professor B. C. Berndt, for his extensive and incisive comments and corrections. I am also grateful to Carolyn J. Bloemker for her patience and painstaking typing of the revised manuscript under a variety of circumstances. Finally, I appreciate the assistance and cooperation of the Wiley staff.

23 June 1975
Urbana-Champaign, Illinois

Robert G. Bartle

Chapter Summaries

Introduction: A Glimpse at Set Theory 1

I. The Real Numbers 27

II. The Topology of Cartesian Spaces 52

III. Convergence 90

IV. Continuous Functions 136

V. Functions of One Variable 193

VI. Infinite Series 286

VII. Differentiation in R^p 346

VIII. Integration in R^p 412

INTRODUCTION: A GLIMPSE AT SET THEORY

The idea of a set is basic to all of mathematics, and all mathematical objects and constructions ultimately go back to set theory. In view of the fundamental importance of set theory, we shall present here a brief resumé of the set-theoretic notions that will be used frequently in this text. However, since the aim of this book is to present the *elements* (rather than the *foundations*) of real analysis, we adopt a rather pragmatic and naïve point of view. We shall be content with an informal discussion and shall regard the word "set" as understood and synonymous with the words "class," "collection," "aggregate," and "ensemble." No attempt will be made to define these terms or to present a list of axioms for set theory. A reader who is sophisticated enough to be troubled by our informal development should consult the references on set theory that are given at the end of this text. There he will learn how this material can be put on an axiomatic basis. He will find this axiomatization to be an interesting development in the foundations of mathematics. However, since we regard it to be outside the subject area of the present book, we shall not go through the details here.

The reader is strongly urged to read this introduction quickly to absorb the notations we shall employ. Unlike the later chapters, which must be *studied*, this introduction is to be considered background material. One should not spend much time on it.

Section 1   The Algebra of Sets

If A denotes a set and if x is an element, it is often convenient to write

$$x \in A$$

as an abbreviation for the statement that x is an **element** of A, or that x is a

member of the set A, or that the set A **contains** the element x, or that x **is in** A. We shall not examine the nature of this property of being an element of a set any further. For most purposes it is possible to employ the naïve meaning of "membership," and an axiomatic characterization of this relation is not necessary.

If A is a set and x is an element which does *not* belong to A, we shall often write

$$x \notin A.$$

In accordance with our naïve conception of a set, we shall require that exactly one of the two possibilities

$$x \in A, \qquad x \notin A,$$

holds for an element x and a set A.

If A and B are two sets and x is an element, then there are, in principle, four possibilities (see Figure 1.1):

(1) $x \in A$ and $x \in B$;   (2) $x \in A$ and $x \notin B$;

(3) $x \notin A$ and $x \in B$;   (4) $x \notin A$ and $x \notin B$.

If the second case cannot occur (that is, if every element of A is also an element of B), then we shall say that A is **contained** in B, or that B **contains** A, or that A is a **subset** of B, and we shall write

$$A \subseteq B \qquad \text{or} \qquad B \supseteq A.$$

If $A \subseteq B$ and there exists an element in B which is not in A, we say that A is a **proper subset** of B.

It should be noted that the statement that $A \subseteq B$ does not automatically preclude the possibility that A exhausts all of B. When this is true the sets A and B are "equal" in the sense we now define.

1.1 DEFINITION. Two sets are **equal** if they contain the same elements. If the sets A and B are equal, we write $A = B$.

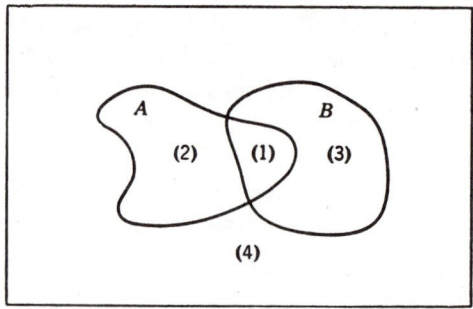

Figure 1.1

Thus in order to show that the sets A and B are equal we must show that the possibilities (2) and (3) mentioned above cannot occur. Equivalently, we must show that both $A \subseteq B$ and $B \subseteq A$,

The word "property" is not easy to define precisely. However, we shall not hesitate to use it in the usual (informal) fashion. If P denotes a property that is meaningful for a collection of elements, then we agree to write

$$\{x : P(x)\}$$

for the set of all elements x for which the property P holds. We usually read this as "the set of all x such that $P(x)$." It is often worthwhile to specify which elements we are testing for the property P. Hence we shall often write

$$\{x \in S : P(x)\}$$

for the subset of S for which the property P holds.

EXAMPLES. (a) If $N = \{1, 2, 3, \ldots\}$ denotes the set of natural numbers, then the set

$$\{x \in N : x^2 - 3x + 2 = 0\}$$

consists of those natural numbers satisfying the stated equation. Now the only solutions of the quadratic equation $x^2 - 3x + 2 = 0$ are $x = 1$ and $x = 2$. Hence, instead of writing the above expression (since we have detailed information concerning all of the elements in the set under examination) we shall ordinarily denote this set by $\{1, 2\}$ thereby listing the elements of the set.

(b) Sometimes a formula can be used to abbreviate the description of a set. For example, the set of all even natural numbers could be denoted by $\{2x : x \in N\}$, instead of the more cumbersome $\{y \in N : y = 2x, x \in N\}$.

(c) The set $\{x \in N : 6 < x < 9\}$ can be written explicitly as $\{7, 8\}$, thereby exhibiting the elements of the set. Of course, there are many other possible descriptions of this set. For example:

$$\{x \in N : 40 < x^2 < 80\},$$
$$\{x \in N : x^2 - 15x + 56 = 0\},$$
$$\{7 + x : x = 0 \quad \text{or} \quad x = 1\}.$$

(d) In addition to the set of **natural numbers** (consisting of the elements denoted by 1, 2, 3, ...) which we shall systematically denote by N, there are a few other sets for which we introduce a standard notation. The set of **integers** is

$$Z = \{0, 1, -1, 2, -2, 3, -3, \ldots\}.$$

The set of **rational numbers** is

$$Q = \{m/n : m, n \in Z \quad \text{and} \quad n \neq 0\}.$$

We shall treat the sets N, Z, and Q as if they are well understood and shall not re-examine their properties in much detail. Of basic importance for our later study is the set R of all real numbers which will be examined in Sections 4–6. A particular subset of R that will be useful is the **unit interval**

$$I = \{x \in R : 0 \leq x \leq 1\}.$$

Finally, we denote the set of **complex numbers** by C. A more detailed definition of C and a brief description of some of its properties will be given in Section 13.

Set Operations

We now introduce some methods of constructing new sets from given ones.

1.2 DEFINITION. If A and B are sets, then their **intersection** is the set of all elements that belong to both A and B. We shall denote the intersection of the sets A, B by the symbol $A \cap B$, which is read "A intersect B." (See Figure 1.2.)

1.3 DEFINITION. If A and B are sets, then their **union** is the set of all elements which belong either to A or to B or to both A and B. We shall denote the union of the sets A, B by the symbol $A \cup B$, which is read "A union B." (See Figure 1.2.)

We could also define $A \cap B$ and $A \cup B$ by

$$A \cap B = \{x : x \in A \quad \text{and} \quad x \in B\},$$
$$A \cup B = \{x : x \in A \quad \text{or} \quad x \in B\}.$$

In connection with the latter, it is important to realize that the word "or" is being used in the inclusive sense that is customary in mathematics and logic. In legal terminology this inclusive sense is sometimes indicated by "and/or."

We have tacitly assumed that the intersection and the union of two sets is again a set. Among other things this requires that there must exist a set which has no elements at all (for if A and B have no common elements, their intersection has no elements).

1.4 DEFINITION. The set which has no elements is called the **empty** or the **void** set and will be denoted by the symbol $\emptyset$. If A and B are sets with no common elements (that is, if $A \cap B = \emptyset$), then we say that A and B are **disjoint** or that they are **non-intersecting.**

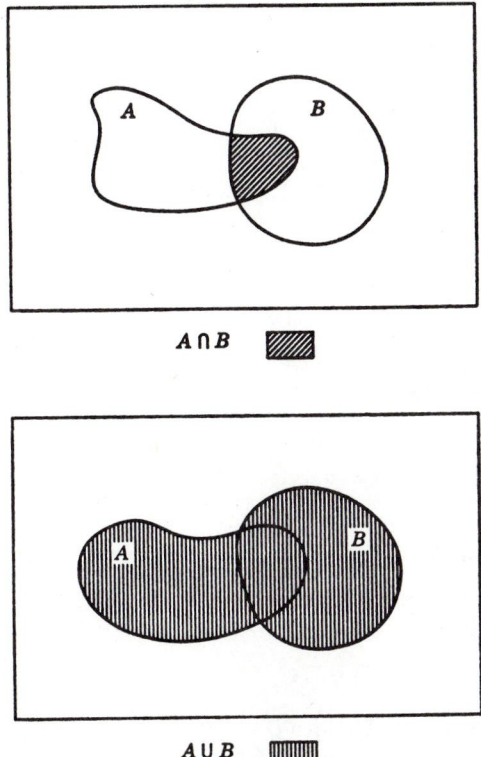

$A \cap B$ ▨

$A \cup B$ ▥

Figure 1.2.   The intersection and union of two sets.

The next result gives some of the algebraic properties of the operations on sets that we have just defined. Since the proofs of these assertions are routine, we shall leave most of them to the reader as exercises.

1.5   THEOREM.   *Let A, B, C, be any sets, then*
(a) $A \cap A = A, \quad A \cup A = A$;
(b) $A \cap B = B \cap A, \quad A \cup B = B \cup A$;
(c) $(A \cap B) \cap C = A \cap (B \cap C), \quad (A \cup B) \cup C = A \cup (B \cup C)$;
(d) $A \cap (B \cup C) = (A \cap B) \cup (A \cap C)$,
$\quad A \cup (B \cap C) = (A \cup B) \cap (A \cup C)$.

These equalities are sometimes referred to as the *idempotent*, the *commutative*, the *associative*, and the *distributive properties*, respectively, of the operations of intersection and union of sets.

In order to give a sample proof, we shall prove the first equation in (d). Let x be an element of $A \cap (B \cup C)$, then $x \in A$ and $x \in B \cup C$. This means that $x \in A$, and either

$x \in B$ or $x \in C$. Hence we either have (i) $x \in A$ and $x \in B$, or we have (ii) $x \in A$ and $x \in C$. Therefore, either $x \in A \cap B$ or $x \in A \cap C$, so $x \in (A \cap B) \cup (A \cap C)$. This shows that $A \cap (B \cup C)$ is a subset of $(A \cap B) \cup (A \cap C)$.

Conversely, let y be an element of $(A \cap B) \cup (A \cap C)$. Then, either (iii) $y \in A \cap B$, or (iv) $y \in A \cap C$. It follows that $y \in A$, and either $y \in B$ or $y \in C$. Therefore, $y \in A$ and $y \in B \cup C$ so that $y \in A \cap (B \cup C)$. Hence $(A \cap B) \cup (A \cap C)$ is a subset of $A \cap (B \cup C)$. In view of Definition 1.1, we conclude that the sets $A \cap (B \cup C)$ and $(A \cap B) \cup (A \cap C)$ are equal.

As an indication of an alternate method, we note that there are, in principle, a total of $8 (= 2^3)$ possibilities for an element x relative to three sets A, B, C (see Figure 1.3); namely:

(1) $x \in A$, $x \in B$, $x \in C$;   (2) $x \in A$, $x \in B$, $x \notin C$;

(3) $x \in A$, $x \notin B$, $x \in C$;   (4) $x \in A$, $x \notin B$, $x \notin C$;

(5) $x \notin A$, $x \in B$, $x \in C$;   (6) $x \notin A$, $x \in B$, $x \notin C$;

(7) $x \notin A$, $x \notin B$, $x \in C$;   (8) $x \notin A$, $x \notin B$, $x \notin C$.

The proof consists in showing that both sides of the first equation in (d) contain those and only those elements x belonging to the cases (1), (2), or (3).

In view of the relations in Theorem 1.5(c), we usually drop the parentheses and write merely

$$A \cap B \cap C, \quad A \cup B \cup C.$$

It is possible to show that if $\{A_1, A_2, \ldots, A_n\}$ is a collection of sets, then there is a uniquely defined set A consisting of all elements which belong to *at least one* of the sets A_j, $j = 1, 2, \ldots, n$; and there exists a uniquely defined set B consisting of all elements which belong to *all* of the sets A_j, $j = 1, 2, \ldots, n$. Dropping the use of parentheses, we write

$$A = A_1 \cup A_2 \cup \cdots \cup A_n, \quad B = A_1 \cap A_2 \cap \cdots \cap A_n.$$

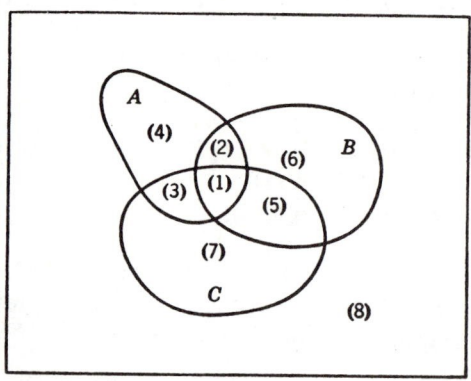

Figure 1.3

Sometimes, in order to save space, we mimic the notation used in calculus for sums and employ a more condensed notation, such as

$$A = \bigcup_{j=1}^{n} A_j = \bigcup \{A_j : j = 1, 2, \ldots, n\},$$

$$B = \bigcap_{j=1}^{n} A_j = \bigcap \{A_j : j = 1, 2, \ldots, n\}.$$

Similarly, if for each j in a set J there is a set A_j, then $\bigcup \{A_j : j \in J\}$ denotes the set of all elements which belong to *at least one* of the sets A_j. In the same way, $\bigcap \{A_j : j \in J\}$ denotes the set of all elements which belong to *all* of the sets A_j for $j \in J$.

We now introduce another method of constructing a new set from two given ones.

1.6. DEFINITION. If A and B are sets, then the **complement of B relative to** A is the set of all elements of A which do not belong to B. We shall denote this set by $A \setminus B$ (read "A minus B"), although the related notations $A - B$ and $A \sim B$ are sometimes used by other authors. (See Figure 1.4.)

In the notation introduced above, we have

$$A \setminus B = \{x \in A : x \notin B\}.$$

Sometimes the set A is understood and does not need to be mentioned explicitly. In this situation we refer simply to the *complement* of B and denote $A \setminus B$ by $\mathscr{C}(B)$.

Returning to Figure 1.1, we note that the elements x which satisfy (1) belong to $A \cap B$; those which satisfy (2) belong to $A \setminus B$; and those which satisfy (3) belong to $B \setminus A$. We shall now show that A is the union of the sets $A \cap B$ and $A \setminus B$.

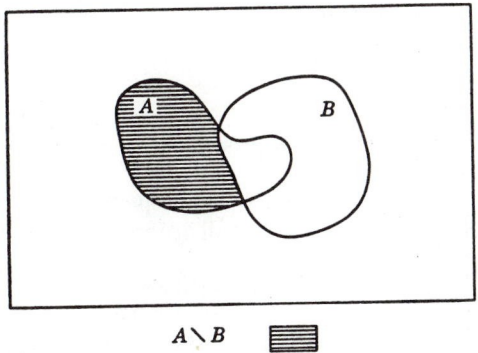

$A \setminus B$ ▤

Figure 1.4. The relative complement.

1.7 THEOREM. *The sets $A \cap B$ and $A \setminus B$ are non-intersecting and*

$$A = (A \cap B) \cup (A \setminus B).$$

PROOF. Suppose $x \in A \cap B$ and $x \in A \setminus B$. The latter asserts that $x \in A$ and $x \notin B$ which contradicts the relation $x \in A \cap B$. Hence the sets are disjoint.

If $x \in A$, then either $x \in B$ or $x \notin B$. In the former case $x \in A$ and $x \in B$ so that $x \in A \cap B$. In the latter situation, $x \in A$ and $x \notin B$ so that $x \in A \setminus B$. This shows that A is a subset of $(A \cap B) \cup (A \setminus B)$. Conversely, if $y \in (A \cap B) \cup (A \setminus B)$, then either $y \in A \cap B$, or $y \in A \setminus B$. In either case we have $y \in A$, showing that $(A \cap B) \cup (A \setminus B)$ is a subset of A.   Q.E.D.

We shall now state the *De Morgan†* laws for three sets; a more general formulation will be given in the exercises.

1.8 THEOREM. *If A, B, C, are any sets, then*

$$A \setminus (B \cup C) = (A \setminus B) \cap (A \setminus C),$$
$$A \setminus (B \cap C) = (A \setminus B) \cup (A \setminus C).$$

PROOF. We shall carry out a demonstration of the first relation, leaving the second one to the reader. To establish the equality of the sets, we show that every element in $A \setminus (B \cup C)$ is contained in both $(A \setminus B)$ and $(A \setminus C)$ and conversely.

If x is in $A \setminus (B \cup C)$, then x is in A but x is not in $B \cup C$. Hence x is in A, but x is neither in B nor in C. (Why?) Therefore, x is in A but not B, and x is in A but not C. That is, $x \in A \setminus B$ and $x \in A \setminus C$, showing that $x \in (A \setminus B) \cap (A \setminus C)$.

Conversely, if $x \in (A \setminus B) \cap (A \setminus C)$, then $x \in (A \setminus B)$ and $x \in (A \setminus C)$. Thus $x \in A$ and both $x \notin B$ and $x \notin C$. It follows that $x \in A$ and $x \notin (B \cup C)$, so that $x \in A \setminus (B \cup C)$.

Since the sets $(A \setminus B) \cap (A \setminus C)$ and $A \setminus (B \cup C)$ contain the same elements, they are equal by Definition 1.1.   Q.E.D.

Cartesian Product

We now define the Cartesian‡ product of two sets.

1.9 DEFINITION. If A and B are two non-void sets, then the **Cartesian product** $A \times B$ of A and B is the set of all ordered pairs (a, b) with $a \in A$ and $b \in B$. (See Figure 1.5.)

---

† AUGUSTUS DE MORGAN (1806–1873) taught at University College, London. He was a mathematician and logician and helped prepare the way for modern mathematical logic.
‡ RENÉ DESCARTES (1596–1650), the creator of analytic geometry, was a French gentleman, soldier, mathematician, and one of the greatest philosophers of all time.

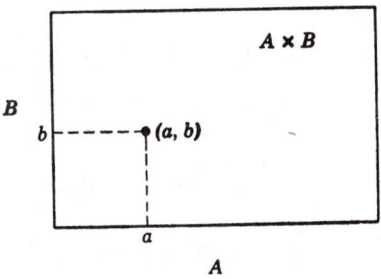

Figure 1.5. The Cartesian product.

(The definition just given is somewhat informal as we have not defined what is meant by an "ordered pair." We shall not examine the matter further except to mention that the ordered pair (a, b) could be defined to be the set whose sole elements are $\{a\}, \{a, b\}$. It can then be shown that the ordered pairs (a, b) and (a', b') are equal if and only if $a = a'$ and $b = b'$. This is the fundamental property of ordered pairs.)

Thus if $A = \{1, 2, 3\}$ and $B = \{4, 5\}$, then the set $A \times B$ is the set whose elements are the ordered pairs

$$(1, 4), (1, 5), (2, 4), (2, 5), (3, 4), (3, 5).$$

We may visualize the set $A \times B$ as the set of six points in the plane with the coordinates which we have just listed.

We often draw a diagram (such as Figure 1.5) to indicate the Cartesian product of two sets A, B. However, it should be realized that this diagram may be somewhat of a simplification. For example, if $A = \{x \in \mathbf{R} : 1 \le x \le 2\}$ and $B = \{x \in \mathbf{R} : 0 \le x \le 1 \text{ or } 2 \le x \le 3\}$, then instead of a rectangle, we should have a drawing like Figure 1.6.

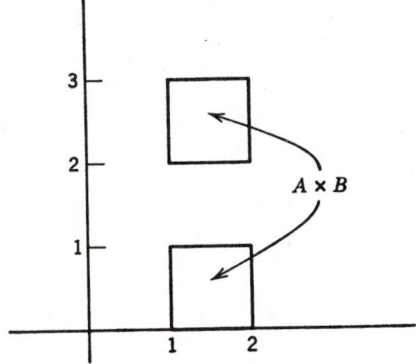

Figure 1.6. The Cartesian product.

Exercises

1.A. Draw a diagram to represent each of the sets mentioned in Theorem 1.5.

1.B. Prove part (c) of Theorem 1.5.

1.C. Prove the second part of (d) of Theorem 1.5.

1.D. Prove that $A \subseteq B$ if and only if $A \cap B = A$.

1.E. Show that the set D of all elements that belong either to A or to B but not to both is given by

$$D = (A \setminus B) \cup (B \setminus A).$$

This set D is often called the **symmetric difference** of A and B. Represent it by a diagram.

1.F. Show that the symmetric difference D, defined in the preceding exercise is also given by $D = (A \cup B) \setminus (A \cap B)$.

1.G. If $B \subseteq A$, show that $B = A \setminus (A \setminus B)$.

1.H. If A and B are any sets, show that $A \cap B = A \setminus (A \setminus B)$.

1.I. If $\{A_1, A_2, \ldots, A_n\}$ is a collection of sets, and if E is any set, show that

$$E \cap \bigcup_{j=1}^{n} A_j = \bigcup_{j=1}^{n} (E \cap A_j), \qquad E \cup \bigcup_{j=1}^{n} A_j = \bigcup_{j=1}^{n} (E \cup A_j).$$

1.J. If $\{A_1, A_2, \ldots, A_n\}$ is a collection of sets, and if E is any set, show that

$$E \cap \bigcap_{j=1}^{n} A_j = \bigcap_{j=1}^{n} (E \cap A_j), \qquad E \cup \bigcap_{j=1}^{n} A_j = \bigcap_{j=1}^{n} (E \cup A_j).$$

1.K. Let E be a set and $\{A_1, A_2, \ldots, A_n\}$ be a collection of sets. Establish the De Morgan laws:

$$E \setminus \bigcap_{j=1}^{n} A_j = \bigcup_{j=1}^{n} (E \setminus A_j), \qquad E \setminus \bigcup_{j=1}^{n} A_j = \bigcap_{j=1}^{n} (E \setminus A_j).$$

Note that if $E \setminus A_j$ is denoted by $\mathscr{C}(A_j)$, these relations take the form

$$\mathscr{C}\left(\bigcap_{j=1}^{n} A_j\right) = \bigcup_{j=1}^{n} \mathscr{C}(A_j), \qquad \mathscr{C}\left(\bigcup_{j=1}^{n} A_j\right) = \bigcap_{j=1}^{n} \mathscr{C}(A_j).$$

1.L. Let J be any set and, for each $j \in J$, let A_j be contained in X. Show that

$$\mathscr{C}(\bigcap\{A_j : j \in J\}) = \bigcup\{\mathscr{C}(A_j) : j \in J\},$$
$$\mathscr{C}(\bigcup\{A_j : j \in J\}) = \bigcap\{\mathscr{C}(A_j) : j \in J\}.$$

1.M. If B_1 and B_2 are subsets of B and if $B = B_1 \cup B_2$, then

$$A \times B = (A \times B_1) \cup (A \times B_2).$$

Section 2   Functions

We now turn to a discussion of the fundamental notion of a *function* or *mapping*. It will be seen that a function is a special kind of a set, although there are other visualizations which are often suggestive. All of the later sections will be concerned with various types of functions, but they will usually be of less abstract nature than considered in the present intoductory section.

To the mathematician of a century ago the word "function" ordinarily meant a definite formula, such as

$$f(x) = x^2 + 3x - 5,$$

which associates to each real number x another real number $f(x)$. The fact that certain formulas, such as

$$g(x) = \sqrt{x - 5},$$

do not give rise to real numbers for all real values of x was, of course, well-known but was not regarded as sufficient grounds to require an extension of the notion of function. Probably one could arouse controversy among those mathematicians as to whether the absolute value

$$h(x) = |x|$$

of a real number is an "honest function" or not. For, after all, the definition of $|x|$ is given "in pieces" by

$$|x| = \begin{cases} x, & \text{if} \quad x \ge 0, \\ -x, & \text{if} \quad x < 0. \end{cases}$$

As mathematics developed, it became increasingly clear that the requirement that a function be a formula was unduly restrictive and that a more general definition would be useful. It also became evident that it is important to make a clear distinction between the function itself and the values of the function. The reader probably finds himself in the position of the mathematician of a century ago in these two respects due to no fault of his own. We propose to bring him up to date with the current usage, but we shall do so in two steps. Our first revised definition of a function would be:

A function f from a set A to a set B is a rule of correspondence that assigns to each x in a certain subset D of A, a uniquely determined element $f(x)$ of B.

Certainly, the explicit formulas of the type mentioned above are included in this tentative definition. The proposed definition allows the possibility that the function might not be defined for certain elements of A and also allows the consideration of functions for which the sets A and B are not necessarily real numbers (but might even be desks and chairs—or cats and dogs).

However suggestive the proposed definition may be, it has a significant defect: it is not clear. There remains the difficulty of interpreting the phrase "rule of correspondence." Doubtless the reader can think of phrases that will satisfy him better than the above one, but it is not likely that he can dispel the fog entirely. The most satisfactory solution seems to be to define "function" entirely in terms of sets and the notions introduced in the preceding section. This has the disadvantage of being more artificial and losing some of the intuitive content of the earlier description, but the gain in clarity outweighs these disadvantages.

The key idea is to think of the graph of the function: that is, a collection of ordered pairs. We notice that an arbitrary collection of ordered pairs cannot be the graph of a function, for once the first member of the ordered pair is named, the second is uniquely determined.

2.1. DEFINITION. Let A and B be sets (which are not necessarily distinct). A **function from** A **to** B is a set f of ordered pairs in $A \times B$ with the property that if (a, b) and (a, b') are elements of f, then $b = b'$. The set of all elements of A that can occur as first members of elements in f is called the **domain** of f and will be denoted $D(f)$. The set of all elements of B that can occur as second members of elements f is called the **range** of f (or the **set of values** of f) and will be denoted by $R(f)$. In case $D(f) = A$, we often say that f **maps** A **into** B (or is a **mapping** of A into B) and write $f : A \to B$.

If (a, b) is an element of a function f, then it is customary to write

$$b = f(a) \qquad \text{or} \qquad f : a \mapsto b$$

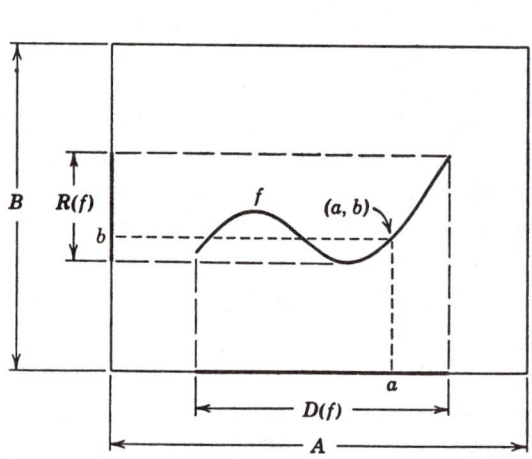

Figure 2.1.   A function as a graph.

instead of $(a, b) \in f$. We often refer to the element b as the **value** of f at the point a, or the **image under** f of the point a.

Tabular Representation

One way of visualizing a function is as a graph. Another way which is important and widely used is as a *table*. Consider Table 2.1, which might be found in the sports page of the *Foosland Bugle-Gazette*.

The domain of this free-throw function f consists of the nine players

$$D(f) = \{\text{Anderson, Bade, Bateman, Hochschild, Kakutani,}$$
$$\text{Kovalevsky, Osborn, Peressini, Rosenberg}\},$$

while the range of the function consists of the six numbers

$$R(f) = \{0, 1, 2, 4, 5, 8\}.$$

The actual elements of the function are ordered pairs

(Anderson, 2), (Bade, 0), (Bateman, 5),

(Hochschild, 1), (Kakutani, 4), (Kovalevsky, 8),

(Osborn, 0), (Peressini, 2), (Rosenberg, 4).

In such tabular representations, we ordinarily write down only the domain of the function in the left-hand column (for there is no need to mention the members of the team that did not play). We could say that the value of this free-throw function f at Anderson is 2 and write $f(\text{Anderson}) = 2$, or Anderson $\mapsto 2$, and so on.

We are all familiar with such use of tables to convey information. They are important examples of functions and are usually of a nature that would be difficult to express in terms of a formula.

TABLE 2.1	
Player	Free Throws Made
Anderson	2
Bade	0
Bateman	5
Hochschild	1
Kakutani	4
Kovalevsky	8
Osborn	0
Peressini	2
Rosenberg	4

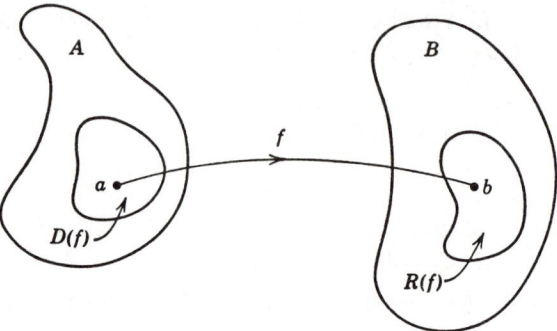

Figure 2.2.   A function as a transformation.

Transformations and Machines

There is another way of visualizing a function: as a *transformation* of part of the set A into part of B. In this phraseology, when $(a, b) \in f$, we think of f as taking the element a from the subset $D(f)$ of A and "transforming" or "mapping" it into an element $b = f(a)$ in the subset $R(f)$ of B. We often draw a diagram such as Figure 2.2. We frequently use this geometrical representation of a function even when the sets A and B are not subsets of the plane.

There is another way of visualizing a function: namely, as a *machine* which will accept elements of $D(f)$ as inputs and yield corresponding elements of $R(f)$ as outputs. If we take an element x from $D(f)$ and put it into f, then out comes the corresponding value $f(x)$. If we put a different element y of $D(f)$ into f, we get $f(y)$ (which may or may not differ from $f(x)$). If we try to insert something which does not belong to $D(f)$ into f, we find that it is not accepted, for f can operate only on elements belonging to $D(f)$. (See Figure 2.3.)

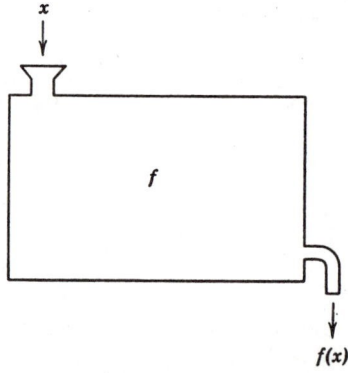

Figure 2.3.   A function as a machine.

This last visualization makes clear the distinction between f and $f(x)$: the first is the machine, the second is the output of the machine when we put x into it. Certainly it is useful to distinguish between a machine and its outputs. Only a fool would confuse a meat grinder with ground meat; however, enough people have confused functions with their values that it is worthwhile to make a modest effort to distinguish between them notationally.

Restrictions and Extensions of Functions

If f is a function with domain $D(f)$ and D_1 is a subset of $D(f)$, it is often useful to define a new function f_1 with domain D_1 by $f_1(x) = f(x)$ for all $x \in D_1$. This function f_1 is called the *restriction of f to the set D_1*. In terms of Definition 2.2, we have

$$f_1 = \{(a, b) \in f : a \in D_1\}.$$

Sometimes we write $f_1 = f \mid D_1$ to denote the restriction of the function f to the set D_1.

A similar construction (that appears less artificial) is the notion of an "extension." If g is a function with domain $D(g)$ and $D_2 \supseteq D(g)$, then any function g_2 with domain D_2 such that $g_2(x) = g(x)$ for all $x \in D(g)$ is called an *extension of g to the set D_2*.

Composition of Functions

We now want to "compose" two functions by first applying f to each x in $D(f)$ and then applying g to $f(x)$ whenever possible (that is, when $f(x)$ belongs to $D(g)$). In doing so, some care needs to be exercised concerning the domain of the resulting function. For example, if f is defined on **R** by $f(x) = x^3$ and if g is defined for $x \geq 0$ by $g(x) = \sqrt{x}$, then the composition $g \circ f$ can be defined only for $x \geq 0$, and for these real numbers it is to have the value $\sqrt{x^3}$.

2.2 DEFINITION. Let f be a function with domain $D(f)$ in A and range $R(f)$ in B and let g be a function with domain $D(g)$ in B and range $R(g)$ in C. (See Figure 2.4.) The **composition** $g \circ f$ (note the order!) is the function from A to C given by

$$g \circ f = \{(a, c) \in A \times C : \text{there exists an element } b \in B$$
$$\text{such that } (a, b) \in f \text{ and } (b, c) \in g\}.$$

2.3. THEOREM. *If f and g are functions, the composition $g \circ f$ is a function with*

$$D(g \circ f) = \{x \in D(f) : f(x) \in D(g)\},$$
$$R(g \circ f) = \{g(f(x)) : x \in D(g \circ f)\}.$$

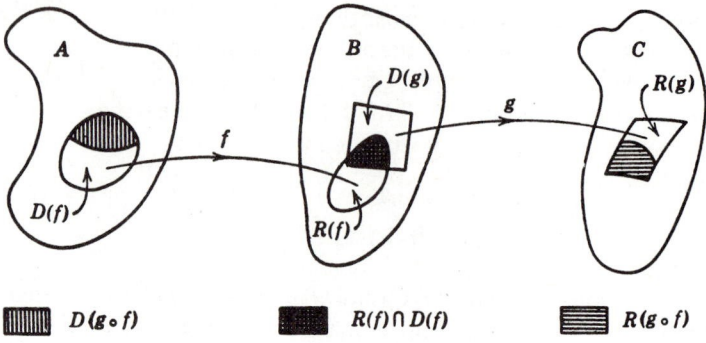

Figure 2.4. Composition of functions.

2.4 EXAMPLES. (a) Let f, g be functions whose values at the real number x are the real numbers given by†

$$f(x) = 2x, \qquad g(x) = 3x^2 - 1.$$

Since $D(g)$ is the set $\mathbf{R}$ of all real numbers and $R(f) \subseteq D(g)$, the domain $D(g \circ f)$ is also $\mathbf{R}$ and $g \circ f(x) = 3(2x)^2 - 1 = 12x^2 - 1$. On the other hand, $D(f \circ g) = \mathbf{R}$, but $f \circ g(x) = 2(3x^2 - 1) = 6x^2 - 2$.

(b) If h is the function with $D(h) = \{x \in \mathbf{R} : x \geq 1\}$ defined by

$$h(x) = \sqrt{x - 1},$$

and if f is as in part (a), then $D(h \circ f) = \{x \in \mathbf{R} : 2x \geq 1\} = \{x \in \mathbf{R} : x \geq \frac{1}{2}\}$ and $h \circ f(x) = \sqrt{2x - 1}$. Also $D(f \circ h) = \{x \in \mathbf{R} : x \geq 1\}$ and $f \circ h(x) = 2\sqrt{x - 1}$. If g is the function in part (a), then $D(h \circ g) = \{x \in \mathbf{R} : 3x^2 - 1 \geq 1\} = \{x \in \mathbf{R} : x \leq -\sqrt{\frac{2}{3}} \text{ or } x \geq \sqrt{\frac{2}{3}}\}$ and $h \circ g(x) = \sqrt{3x^2 - 2}$. Also $D(g \circ h) = \{x \in \mathbf{R} : x \geq 1\}$ and $g \circ h(x) = 3x - 4$. (Note that the formula expressing $g \circ h$ has meaning for values of x other than those in the domain of $g \circ h$.)

(c) Let F, G be the functions with domains $D(F) = \{x \in \mathbf{R} : x \geq 0\}$, and $D(G) = \mathbf{R}$, such that the values of F and G at a point x in their domains are

$$F(x) = \sqrt{x}, \qquad G(x) = -x^2 - 1.$$

Then $D(G \circ F) = \{x \in \mathbf{R} : x \geq 0\}$ and $G \circ F(x) = -x - 1$, whereas $D(F \circ G) = \{x \in D(G) : G(x) \in D(f)\}$. This last set is void as $G(x) < 0$ for all $x \in D(G)$. Hence the function $F \circ G$ is not defined at any point, so $F \circ G$ is the "void function."

---

† We also denote this by writing $f : x \mapsto 2x$ and $g : x \mapsto 3x^2 - 1$ for $x \in \mathbf{R}$.

Injective and Inverse Functions

We now give a way of constructing a new function from a given one in case the original function does not take on the same value twice.

2.5 DEFINITION. Let f be a function with domain $D(f)$ in A and range $R(f)$ in B. We say that f is **injective** or **one-one** if, whenever (a, b) and (a', b) are elements of f, then $a = a'$. If f is injective we may say that f is an **injection.**

In other words, f is injective if and only if the two relations $f(a) = b$, $f(a') = b$ imply that $a = a'$. Alternatively, f is injective if and only if when a, a' are in $D(f)$ and $a \neq a'$, then $f(a) \neq f(a')$.

We claim that if f is injective from A to B, then the set of ordered pairs in $B \times A$ obtained by interchanging the first and second members of ordered pairs in f yields a function g which is also injective.

We omit the proof of this assertion, leaving it as an exercise; it is a good test for the reader. The connections between f and g are:

$$D(g) = R(f), \qquad R(g) = D(f),$$
$$(a, b) \in f \qquad \text{if and only if} \qquad (b, a) \in g.$$

This last statement can be written in the more usual form:

$$b = f(a) \qquad \text{if and only if} \qquad a = g(b).$$

2.6 DEFINITION. Let f be an injection with domain $D(f)$ in A and range $R(f)$ in B. If $g = \{(b, a) \in B \times A : (a, b) \in f\}$, then g is an injection with domain $D(g) = R(f)$ in B and with range $R(g) = D(f)$ in A. The function g is called the function **inverse** to f and is denoted by f^{-1}.

The inverse function can be interpreted from the mapping point of view. (See Figure 2.5.) If f is injective, it maps distinct elements of $D(f)$ into distinct elements of $R(f)$. Thus, each element b of $R(f)$ is the image under f of a unique element a in $D(f)$. The inverse function f^{-1} maps the element b into this unique element a.

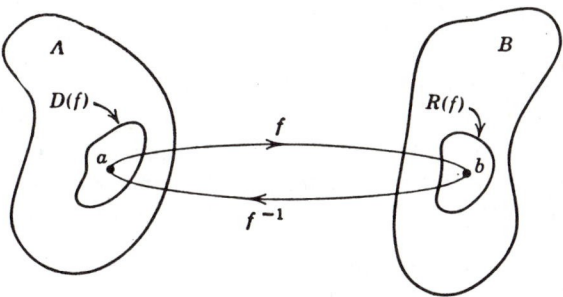

Figure 2.5. The inverse function.

2.7 EXAMPLES. (a) Let $F: x \mapsto x^2$ be the function with domain $D(F) = \textbf{R}$, the set of all real numbers, and range in $\textbf{R}$ such that the value of F at the real number x is $F(x) = x^2$. (In other words, F is the function $\{(x. x^2) : x \in \textbf{R}\}$.) It is readily seen that F is not one-one; in fact, the ordered pairs $(2, 4)$, $(-2, 4)$ both belong to F. Since F is not one-one, it does not have an inverse.

(b) Let f be the function with domain $D(f) = \{x \in \textbf{R} : x \geq 0\}$ and $R(f) = \textbf{R}$ whose value at x in $D(f)$ is $f(x) = x^2$. Note that f is the restriction to $D(f)$ of the function F in part (a). In terms of ordered pairs, $f = \{(x, x^2) : x \in \textbf{R}, x \geq 0\}$. Unlike the function F in part (a), f is injective, for if $x^2 = y^2$ with x, y in $D(f)$, then $x = y$. (Why?) Therefore, f has an inverse function g with $D(g) = R(f) = \{x \in \textbf{R} : x \geq 0\}$ and $R(g) = D(f) = \{x \in \textbf{R} : x \geq 0\}$. Furthermore, $y = x^2 = f(x)$ if and only if $x = g(y)$. This inverse function g is ordinarily called the **positive square root function** and is denoted by

$$g(y) = \sqrt{y}, \qquad y \in \textbf{R}, \quad y \geq 0.$$

(c) If f_1 is the function $\{(x, x^2) : x \in \textbf{R}, x \leq 0\}$, then as in (b), f_1 is one-one and has domain $D(f_1) = \{x \in \textbf{R} : x \leq 0\}$ and range $R(f_1) = \{x \in \textbf{R} : x \geq 0\}$. Note that f_1 is the restriction to $D(f_1)$ of the function F of part (a). The function g_1 inverse to f is called the **negative square root function** and is denoted by

$$g_1(y) = -\sqrt{y}, \qquad y \in \textbf{R}, \quad y \geq 0,$$

so that $g_1(y) \leq 0$.

(d) The sine function F introduced in trigonometry with $D(F) = \textbf{R}$ and $R(F) = \{y \in \textbf{R} : -1 \leq y \leq +1\}$ is well-known not to be injective (for example, $\sin 0 = \sin 2\pi = 0$). However, if we let f be its restriction to the set $D(f) = \{x \in \textbf{R} : -\pi/2 \leq x \leq +\pi/2\}$, then f is injective. It therefore has an inverse function g with $D(g) = R(f)$ and $R(g) = D(f)$. Also, $y = \sin x$ with $x \in D(f)$ if and only if $x = g(y)$. The function g is called the (principal branch) of the **inverse sine function** and is often denoted by

$$g(y) = \text{Arc sin } y \qquad \text{or} \qquad g(y) = \text{Sin}^{-1} y.$$

Surjective and Bijective Functions

2.8 DEFINITION. Let f be a function with domain $D(f) \subseteq A$ and range $R(f) \subseteq B$. We say that f is **surjective,** or that f **maps onto** B, in case the range $R(f) = B$. If f is surjective, we may say that f is a **surjection.**

In defining a function it is important to specify the domain of the function and the set in which the values are taken. Once this has been done it is possible to inquire whether or not the function is surjective.

2.9   DEFINITION.   A function f with domain $D(f) \subseteq A$ and range $R(f) \subseteq B$ is said to be **bijective** if (i) it is injective (that is, it is one-one), and (ii) it is surjective (that is, it maps $D(f)$ onto B).   If f is bijective, we may say that f is a **bijection.**

Direct and Inverse Images

Let f be an arbitrary function with domain $D(f)$ in A and range $R(f)$ in B. We do *not* assume that f is injective.

2.10   DEFINITION.   If E is a subset of A, then the **direct image** of E under f is the subset of $R(f)$ given by

$$\{f(x) : x \in E \cap D(f)\}.$$

We usually denote the direct image of a set E under f by the notation $f(E)$. (See Figure 2.6.)

It will be observed that if $E \cap D(f) = \emptyset$, then $f(E) = \emptyset$.   If E contains the single point p in $D(f)$, then the set $f(E)$ contains the single point $f(p)$.   Certain properties of sets are preserved under the direct image, as we now show.

2.11   THEOREM.   *Let f be a function with domain in A and range in B and let E, F be subsets of A.*
(a) *If $E \subseteq F$, then $f(E) \subseteq f(F)$.*   (b) $f(E \cap F) \subseteq f(E) \cap f(F)$.
(c) $f(E \cup F) = f(E) \cup f(F)$.   (d) $f(E \setminus F) \subseteq f(E)$.

PROOF.   (a) If $x \in E$, then $x \in F$ and hence $f(x) \in f(F)$.   Since this is true for all $x \in E$, we infer that $f(E) \subseteq f(F)$.

(b) Since $E \cap F \subseteq E$, it follows from part (a) that $f(E \cap F) \subseteq f(E)$; likewise, $f(E \cap F) \subseteq f(F)$.   Therefore, we conclude that $f(E \cap F) \subseteq f(E) \cap f(F)$.

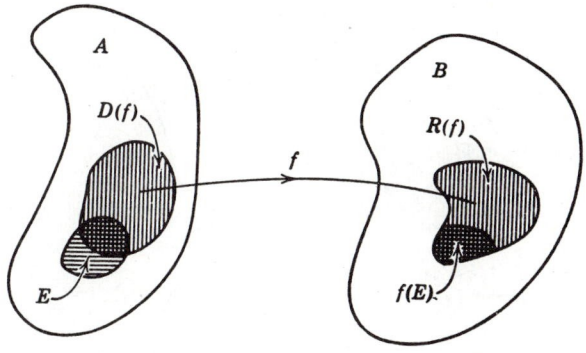

Figure 2.6.   Direct images.

(c) Since $E \subseteq E \cup F$ and $F \subseteq E \cup F$, it follows from part (a) that $f(E) \cup f(F) \subseteq f(E \cup F)$. Conversely, if $y \in f(E \cup F)$, then there exists an element $x \in E \cup F$ such that $y = f(x)$. Since $x \in E$ or $x \in F$, it follows that either $y = f(x) \in f(E)$ or that $y \in f(F)$. Therefore, we conclude that $f(E \cup F) \subseteq f(E) \cup f(F)$, which completes the proof of part (c).

(d) Part (d) follows immediately from (a).   Q.E.D.

It will be seen in Exercise 2.J that, in general, it is not possible to replace the inclusion sign in (b) by equality.

We now introduce the notion of the inverse image of a set under a function. Note that it is not required that the function be injective.

2.12 DEFINITION. If H is a subset of B, then the **inverse image** of H under f is the subset of $D(f)$ given by

$$\{x : f(x) \in H\}.$$

We usually denote the inverse image of a set H under f by the symbol $f^{-1}(H)$. (See Figure 2.7.)

Once again, we emphasize that f need not be injective so that the inverse function f^{-1} need not exist. (However, if f^{-1} does exist, then $f^{-1}(H)$ is the direct image of H under f^{-1}).

2.13 THEOREM. *Let f be a function with domain in A and range in B and let G, H be subsets of B.*
  (a) *If $G \subseteq H$, then $f^{-1}(G) \subseteq f^{-1}(H)$.* (b) $f^{-1}(G \cap H) = f^{-1}(G) \cap f^{-1}(H)$.
  (c) $f^{-1}(G \cup H) = f^{-1}(G) \cup f^{-1}(H)$.   (d) $f^{-1}(G \backslash H) = f^{-1}(G) \backslash f^{-1}(H)$.

PROOF. (a) Suppose that $x \in f^{-1}(G)$; then $f(x) \in G \subseteq H$ and hence $x \in f^{-1}(H)$.

(b) Since $G \cap H$ is a subset of G and H, it follows from part (a) that

$$f^{-1}(G \cap H) \subseteq f^{-1}(G) \cap f^{-1}(H).$$

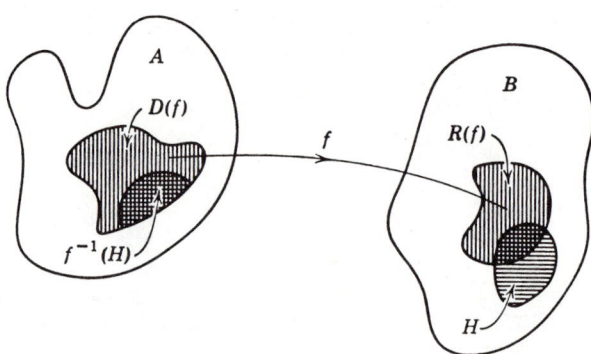

Figure 2.7.   Inverse Images.

Conversely, if $x \in f^{-1}(G) \cap f^{-1}(H)$, then $f(x) \in G$ and $f(x) \in H$. Therefore, $f(x) \in G \cap H$ and $x \in f^{-1}(G \cap H)$.

(c) Since G and H are subsets of $G \cup H$, it follows from part (a) that

$$f^{-1}(G \cup H) \supseteq f^{-1}(G) \cup f^{-1}(H).$$

Conversely, if $x \in f^{-1}(G \cup H)$, then $f(x) \in G \cup H$. It follows that either $f(x) \in G$ whence $x \in f^{-1}(G)$, or $f(x) \in H$ in which case $x \in f^{-1}(H)$. Hence

$$f^{-1}(G \cup H) \subseteq f^{-1}(G) \cup f^{-1}(H).$$

(d) If $x \in f^{-1}(G \setminus H)$, then $f(x) \in G \setminus H$. Therefore, $x \in f^{-1}(G)$ and $x \notin f^{-1}(H)$, whence it follows that

$$f^{-1}(G \setminus H) \subseteq f^{-1}(G) \setminus f^{-1}(H).$$

Conversely, if $w \in f^{-1}(G) \setminus f^{-1}(H)$, then $f(w) \in G$ and $f(w) \notin H$. Hence $f(w) \in G \setminus H$ and it follows that

$$f^{-1}(G) \setminus f^{-1}(H) \subseteq f^{-1}(G \setminus H).$$

Q.E.D.

Exercises

2.A. Prove that Definition 2.2 actually yields a function and not just a subset.

2.B. Let $A = B = \mathbf{R}$ and consider the subset $C = \{(x, y) : x^2 + y^2 = 1\}$ of $A \times B$. Is this set a function with domain in $\mathbf{R}$ and range in $\mathbf{R}$?

2.C. Consider the subset of $\mathbf{R} \times \mathbf{R}$ defined by $D = \{(x, y) : |x| + |y| = 1\}$. Describe this set in words. Is it a function?

2.D. Give an example of two functions f, g on $\mathbf{R}$ to $\mathbf{R}$ such that $f \neq g$, but such that $f \circ g = g \circ f$.

2.E. Prove that if f is an injection from A to B, then $f^{-1} = \{(b, a) : (a, b) \in f\}$ is a function. Then prove it is an injection.

2.F. Suppose f is an injection. Show that $f^{-1} \circ f(x) = x$ for all x in $D(f)$ and $f \circ f^{-1}(y) = y$ for all y in $R(f)$.

2.G. Let f and g be functions and suppose that $g \circ f(x) = x$ for all x in $D(f)$. Show that f is injection and that $R(f) \subseteq D(g)$ and $R(g) \supseteq D(f)$.

2.H. Let f, g be functions such that

$$g \circ f(x) = x \qquad \text{for all } x \text{ in } D(f),$$
$$f \circ g(y) = y \qquad \text{for all } y \text{ in } D(g).$$

Prove that $g = f^{-1}$.

2.I. Show that the direct image $f(E)$ is empty if and only if $E \cap D(f) = \emptyset$.

2.J. Let f be the function on $\mathbf{R}$ to $\mathbf{R}$ given by $f(x) = x^2$, and let $E = \{x \in \mathbf{R} : -1 \leq x \leq 0\}$ and $F = \{x \in \mathbf{R} : 0 \leq x \leq 1\}$. Then $E \cap F = \{0\}$ and $f(E \cap F) = \{0\}$ while $f(E) = f(F) = \{y \in \mathbf{R} : 0 \leq y \leq 1\}$. Hence $f(E \cap F)$ is a proper subset of $f(E) \cap f(F)$. Now delete 0 from E and F.

2.K. If f, E, F are as in Exercise 2.J, then $E \setminus F = \{x \in R : -1 \le x < 0\}$ and $f(E) \setminus f(F) = \emptyset$. Hence, it does *not* follow that

$$f(E \setminus F) \subseteq f(E) \setminus f(F).$$

2.L. Show that if f is an injection of $D(f)$ into $R(f)$ and if H is a subset of $R(f)$, then the inverse image of H under f coincides with the direct image of H under the inverse function f^{-1}.

2.M. If f and g are as in Definition 2.2, then $D(g \circ f) = f^{-1}(D(g))$.

Section 3   Finite and Infinite Sets

The purpose of this section is very restricted: it is to introduce the terms "finite," "countable," and "infinite." It provides a basis for the study of cardinal numbers, but it does not pursue this study. Although the theories of cardinal and ordinal numbers are fascinating in their own right, it turns out that very little exposure to these topics is really essential for the material in this text.†

We shall assume familiarity with the set of *natural numbers*. We shall denote this set by the symbol N; the elements of N are denoted by the familiar symbols

$$1, 2, 3, \ldots.$$

If $n, m \in N$, we all have an intuitive idea of what is meant by saying that n is less than or equal to m. We now borrow this notion, realizing that complete precision requires more analysis than we have given. We assume that *every non-empty subset of N has a least element*. This is an important property of N; we sometimes say that N is *well-ordered* meaning that N has this property. This Well-Ordering Property is equivalent to *mathematical induction*. We shall feel free to make use of arguments based on mathematical induction, which we suppose to be familiar to the reader.

By an **initial segment** of N is meant a set consisting of all of the natural numbers which are less than or equal to some fixed element of N. Thus an initial segment S_n of N determines and is determined by an element n of N as follows:

An element x of N belongs to S_n if and only if $x \le n$.

For example: the subset $S_2 = \{1, 2\}$ is the initial segment of N determined by the natural number 2; the subset $S_4 = \{1, 2, 3, 4\}$ is the initial segment of N determined by the natural number 4; but the subset $\{1, 3, 5\}$ of N is *not* an initial segment of N, since it contains 3 but not 2, and 5 but not 4.

---

† A reader wishing to learn about these topics should consult the book of Halmos cited in the References.

3.1 DEFINITION. A set B is **finite** if it is empty or if there is a bijection with domain B and range in an initial segment of N. If there is no such function, the set is **infinite.** If there is a bijection of B onto N, then the set B is **denumerable** (or **enumerable**). If a set is either finite or denumerable, it is said to be **countable.**

When there is an injective (or one-one) function with domain B and range C, we sometimes say that B can be put into *one-one correspondence* with C. By using this terminology, we rephrase Definition 3.1 and say that a set B is finite if it is empty or can be put into one-one correspondence with a subset of an initial segment of N. We say that B is denumerable if it can be put into one-one correspondence with all of N.

It will be noted that, by definition, a set B is either finite or infinite. However, it may be that, owing to the description of the set, it may not be a trivial matter to decide whether the given set B is finite or infinite.

The subsets of N denoted by $\{1, 3, 5\}, \{2, 4, 5, 8, 10\}, \{2, 3, \ldots, 100\}$, are finite since, although they are not initial segments of N, they are contained in initial segments of N and hence can be put into one-one correspondence with subsets of initial segments of N. The set E of even natural numbers

$$E = \{2, 4, 6, 8, \ldots\}$$

and the set O of odd natural numbers

$$O = \{1, 3, 5, 7, \ldots\}$$

are not initial segments of N. However, since they can be put into one-one correspondence with all of N (how?), they are both denumerable.

Even though the set Z of all integers

$$Z = \{\ldots, -2, -1, 0, 1, 2, \ldots\},$$

contains the set N, it may be seen that Z is a denumerable set. (How?)

We now state some theorems without proof. At first reading it is probably best to accept them without further examination; on a later reading, however, the reader will do well to attempt to provide proofs for these statements. In doing so, he will find the inductive property of the set N of natural numbers to be useful.[†]

3.2 THEOREM. *A set B is countable if and only if there is an injection with domain B and range in N.*

3.3 THEOREM. *Any subset of a finite set is finite. Any subset of a countable set is countable.*

3.4 THEOREM. *The union of a finite collection of finite sets is a finite set. The union of a countable collection of countable set is a countable set.*

† See the books of Halmos and Hamilton-Landin cited in the References.

It is a consequence of the second part of Theorem 3.4 that the set Q of all rational numbers forms a countable set. (We recall that a rational number is a fraction m/n, where m and n are integers and $n \neq 0$. To see that Q is a countable set we form the sets

$$A_0 = \{0\},$$
$$A_1 = \{\tfrac{1}{1}, -\tfrac{1}{1}, \tfrac{2}{1}, -\tfrac{2}{1}, \tfrac{3}{1}, -\tfrac{3}{1}, \ldots\},$$
$$A_2 = \{\tfrac{1}{2}, -\tfrac{1}{2}, \tfrac{2}{2}, -\tfrac{2}{2}, \tfrac{3}{2}, -\tfrac{3}{2}, \ldots\},$$

$$\ldots\ldots\ldots\ldots\ldots\ldots\ldots\ldots$$

$$A_n = \left\{\frac{1}{n}, -\frac{1}{n}, \frac{2}{n}, -\frac{2}{n}, \frac{3}{n}, -\frac{3}{n}, \ldots\right\},$$

$$\ldots\ldots\ldots\ldots\ldots\ldots\ldots\ldots$$

Note that each of the sets A_n is countable and that their union is all of Q. Hence Theorem 3.4 asserts that Q is countable. In fact, we can enumerate Q by the "diagonal procedure":

$$0, \tfrac{1}{1}, -\tfrac{1}{1}, \tfrac{1}{2}, \tfrac{2}{1}, -\tfrac{1}{2}, \tfrac{1}{3}, \ldots.$$

By using this type of argument, the reader should be able to construct a proof of Theorem 3.4. See also Exercise 3.K.

The Uncountability of R and I

Despite the fact that the set of rational numbers is countable, the entire set R of real numbers is not countable. In fact, the set I of real numbers x satisfying $0 \leq x \leq 1$ is not countable. To demonstrate this, we shall use the elegant "diagonal" argument of G. Cantor.† We assume it is known that every real number x with $0 \leq x \leq 1$ has a decimal representation in the form $x = 0.a_1 a_2 a_3 \cdots$, where each a_k denotes one of the digits 0, 1, 2, 3, 4, 5, 6, 7, 8, 9. It is to be realized that certain real numbers have two representations in this form (for example, the rational number $\tfrac{1}{10}$ has the two representations

$$0.1000 \cdots \quad \text{and} \quad 0.0999 \cdots).$$

We could decide in favor of one of these two representations, but it is not necessary to do so. Since there are infinitely many rational numbers in the interval $0 \leq x \leq 1$, (why?) the set I cannot be finite. We shall now show that it is not denumerable. Suppose that there is an enumeration $x_1, x_2, x_3, \ldots$ of all

---

† GEORG CANTOR (1845–1918) was born in St. Petersburg, studied in Berlin with Weierstrass, and taught at Halle. He is best known for his work on set theory, which he developed during the years 1874–1895.

real numbers satisfying $0 \le x \le 1$ given by

$$x_1 = 0.a_1a_2a_3 \cdots$$
$$x_2 = 0.b_1b_2b_3 \cdots$$
$$x_3 = 0.c_1c_2c_3 \cdots$$
$$\cdots\cdots\cdots\cdots\cdots$$

Now let y_1 be a digit different from 0, a_1, and 9; let y_2 be a digit different from 0, b_2, and 9; let y_3 be a digit different from 0, c_3, and 9, etc. Consider the number y with decimal representation

$$y = 0.y_1y_2y_3 \cdots$$

which clearly satisfies $0 \le y \le 1$. The number y is not one of the numbers with two decimal representations, since $y_n \neq 0, 9$. At the same time $y \neq x_n$ for any n (since the nth digits in the decimal representations for y and x_n are different). Therefore, any denumerable collection of real numbers in this interval will omit at least one real number belonging to this interval. Therefore, this interval is not a countable set.

Suppose that a set A is infinite; we shall *suppose* that there is a one-one correspondence with a subset of A and all of **N**. In other words, we *assume that every infinite set contains a denumerable subset*. This assertion is a weak form of the so-called "Axiom of Choice," which is one of the usual axioms of set theory. After the reader has digested the contents of this book, he may turn to an axiomatic treatment of the foundations which we have been discussing in a somewhat informal fashion. However, for the moment he would do well to take the above statement as a temporary axiom. It can be replaced later by a more far-reaching axiom of set theory.

Exercises

3.A. Exhibit a one-one correspondence between the set E of even natural numbers and **N**.

3.B. Exhibit a one-one correspondence between the set O of odd natural numbers and **N**.

3.C. Exhibit a one-one correspondence between **N** and a proper subset of **N**.

3.D. If A is contained in some initial segment of **N**, use the well-ordering property of **N** to define a bijection of A onto some initial segment of **N**.

3.E. Given an example of a countable collection of finite sets whose union is not finite.

3.F. Use the fact that every infinite set has a denumerable subset to show that every infinite set can be put into one-one correspondence with a proper subset of itself.

3.G. Show that if the set A can be put into one-one correspondence with a set B, then B can be put into one-one correspondence with A.

3.H. Show that if the set A can be put into one-one correspondence with a set B, and if B can be put into one-one correspondence with a set C, then A can be put into one-one correspondence with C.

3.I. Using induction on $n \in N$, show that the initial segment determined by n cannot be put into one-one correspondence with the initial segment determined by $m \in N$, if $m < n$.

3.J. Prove that N cannot be put into one-one correspondence with any initial segment of N.

3.K. For each $n \in N$ let $A_n = \{a_{nj} : j \in N\}$, and suppose that $A_n \cap A_m = \emptyset$ for $n \neq m$, $n, m \in N$. Show that the function $f(n, j) = \frac{1}{2}(n + j - 2)(n + j - 1) + n$ gives an enumeration of $\bigcup \{A_n : n \in N\}$.

I
THE REAL NUMBERS

In this chapter we shall discuss the properties of the real number system. Although it would be possible to construct this system from a more primitive set (such as the set N of the natural numbers or the set Q of rational numbers), we shall not do so. Instead, we shall exhibit a list of properties that are associated with the real number system and show how other properties can be deduced from the ones assumed.

For the sake of clarity we prefer not to state all the properties of the real number system at once. Instead, we shall introduce first, in Section 4, the "algebraic properties" based on the two operations of addition and multiplication and discuss briefly some of their consequences. Next, we introduce the "order properties." In Section 6, we make the final step by adding the "completeness property." There are several reasons for this somewhat piecemeal procedure. First, there are a number of properties to be considered, and it is well to take a few at a time. Furthermore, the proofs required in the preliminary algebraic stages are more natural at first than some of the later proofs. Finally, since there are several other interesting methods of adding the "completeness property," we wish to have it isolated from the other assumptions.

Part of the purpose of Sections 4 and 5 is to provide examples of proofs of elementary theorems which are derived from explicitly stated assumptions. It is our experience that students who have not had much exposure to rigorous proofs can grasp the arguments presented in these sections readily and can then proceed into Section 6. *However, students who are familiar with the axiomatic method and the technique of proofs can go into Section 6 after a cursory look at Sections 4 and 5.*

In Section 7, we introduce the notion of a cut in the real number system, and define various types of cells and intervals. The important Nested Cells Property of R is established, and the Cantor set is briefly discussed.

Section 4   The Algebraic Properties of R

In this section we shall give the "algebraic" structure of the real number system. Briefly expressed, the real numbers form a "field" in the sense of abstract algebra. We shall now explain what that means.

By a **binary operation** in a set F we mean a function B with domain $F \times F$ and range in F. Instead of using the notation $B(a, b)$ to denote the value of the binary operation B at the point (a, b) in $F \times F$, it is conventional to use a notation such as aBb, or $a + b$, or $a \cdot b$.

4.1   ALGEBRAIC PROPERTIES OF R.   In the set R of real numbers there are two binary operations (denoted by $+$ and $\cdot$ and called addition and multiplication, respectively) satisfying the following† properties:

(A1) $a + b = b + a$ for all a, b in R;

(A2) $(a + b) + c = a + (b + c)$ for all a, b, c in R;

(A3) there exists an element 0 in R with $0 + a = a$ and $a + 0 = a$ for all a in R;

(A4) for each element a in R there is an element $-a$ in R such that $a + (-a) = 0$ and $(-a) + a = 0$;

(M1) $a \cdot b = b \cdot a$ for all a, b in R;

(M2) $(a \cdot b) \cdot c = a \cdot (b \cdot c)$ for all a, b, c in R;

(M3) the element 1 in R is distinct from 0 and has the property that $1 \cdot a = a$ and $a \cdot 1 = a$ for all a in R;

(M4) for each element $a \neq 0$ in R there is an element $1/a$ in R such that $a \cdot (1/a) = 1$ and $(1/a) \cdot a = 1$;

(D) $a \cdot (b + c) = (a \cdot b) + (a \cdot c)$ and $(b + c) \cdot a = (b \cdot a) + (c \cdot a)$ for all a, b, c in R.

These properties are certainly familiar to the reader. We will now obtain a few easy (but important) consequences of them. First of all we shall prove that 0 is the only element of R that satisfies (A3), and 1 is the only element that satisfies (M3).

4.2   THEOREM.   (a) *If z and a are elements of R such that $z + a = a$, then $z = 0$.*

(b) *If w and $b \neq 0$ are elements of R such that $w \cdot b = b$, then $w = 1$.*

PROOF.   (a) The hypothesis is that $z + a = a$. Add $-a$ to both sides and use (A4), (A2), (A4), and (A3) to obtain

$$0 = a + (-a) = (z + a) + (-a) = z + (a + (-a))$$
$$= z + 0 = z.$$

† This list is not intended to be "minimal." Thus the second assertions in (A3) and (A4) follow from the first assertions by using (A1).

The proof of part (b) is left as an exercise. Note that it uses the hypothesis that $b \neq 0$. Q.E.D.

We now show that the elements $-a$ and $1/a$ (when $a \neq 0$) are uniquely determined by the properties given in (A4) and (M4).

4.3 THEOREM. (a) *If a and b are elements of **R** and $a + b = 0$, then $b = -a$.*
(b) *If $a \neq 0$ and b are elements of **R** and $a \cdot b = 1$, then $b = 1/a$.*

PROOF. (a) If $a + b = 0$, add $-a$ to both sides to obtain $(-a) + (a + b) = -a + 0$. Now use (A2) on the left and (A3) on the right side to obtain

$$((-a) + a) + b = -a.$$

If we use (A4) and (A3) on the left side, we obtain $b = -a$.
The proof of (b) is left as an exercise. Note that it uses the hypothesis that $a \neq 0$. Q.E.D.

Properties (A4) and (M4) guarantee the possibility of solving the equations

$$a + x = 0, \qquad a \cdot x = 1 \quad (a \neq 0),$$

for x, and Theorem 4.3 implies the uniqueness of the solutions. We now show that the right-hand sides of these equations can be arbitrary elements of **R**.

4.4 THEOREM. (a) *Let a, b be arbitrary elements of **R**. Then the equation $a + x = b$ has the unique solution $x = (-a) + b$.*
(b) *Let $a \neq 0$ and b be arbitrary elements of **R**. Then the equation $a \cdot x = b$ has the unique solution $x = (1/a) \cdot b$.*

PROOF. Since $a + ((-a) + b) = (a + (-a)) + b = 0 + b = b$, it is clear that $x = (-a) + b$ is a solution of the equation $a + x = b$. To establish that it is the only solution, let x_1 be any solution of this equation; hence

$$a + x_1 = b.$$

We add $-a$ to both sides to obtain

$$(-a) + (a + x_1) = (-a) + b.$$

If we employ (A3), (A4), and (A2), we get

$$x_1 = 0 + x_1 = (-a + a) + x_1$$

$$= (-a) + (a + x_1) = (-a) + b.$$

Hence $x_1 = (-a) + b$.
The proof of part (b) is left as an exercise. Q.E.D.

4.5   THEOREM.   *If a and b are any elements of **R**, then*
(a) $a \cdot 0 = 0$;   (b) $-a = (-1) \cdot a$;
(c) $-(a+b) = (-a) + (-b)$;   (d) $-(-a) = a$;
(e) $(-1) \cdot (-1) = 1$.

PROOF.   (a) From (M3), we know that $a \cdot 1 = a$. Hence

$$a + a \cdot 0 = a \cdot 1 + a \cdot 0 = a \cdot (1+0)$$
$$= a \cdot 1 = a.$$

If we apply Theorem 4.2(a), we infer that $a \cdot 0 = 0$.
    (b) It is seen that

$$a + (-1) \cdot a = 1 \cdot a + (-1) \cdot a = (1 + (-1)) \cdot a$$
$$= 0 \cdot a = 0.$$

It follows from Theorem 4.3(a) that $(-1) \cdot a = -a$.
    (c) We have

$$-(a+b) = (-1) \cdot (a+b) = (-1) \cdot a + (-1) \cdot b$$
$$= (-a) + (-b).$$

(d) By (A4) we have $(-a) + a = 0$.   According to the uniqueness assertion
of Theorem 4.3(a), it follows that $a = -(-a)$.
    (e) In part (b), substitute $a = -1$.   We have

$$-(-1) = (-1) \cdot (-1).$$

Hence the assertion follows from part (d) with $a = 1$.   Q.E.D.

4.6   THEOREM   (a) *If $a \in \mathbf{R}$ and $a \neq 0$, then $1/a \neq 0$ and $1/(1/a) = a$.*
(b) *If $a, b \in \mathbf{R}$ and $a \cdot b = 0$, then either $a = 0$ or $b = 0$.*
(c) *If $a, b \in \mathbf{R}$, then $(-a) \cdot (-b) = a \cdot b$.*
(d) *If $a \in \mathbf{R}$ and $a \neq 0$, then $1/(-a) = -(1/a)$.*

PROOF.   (a) If $a \neq 0$, then $1/a \neq 0$ for otherwise $1 = a \cdot (1/a) = a \cdot 0 = 0$,
contrary to (M3).   Since $(1/a) \cdot a = 1$, it follows from Theorem 4.3(b) that
$a = 1/(1/a)$.
    (b) Suppose that $a \cdot b = 0$ and that $a \neq 0$. If we multiply by $1/a$, we
obtain

$$b = 1 \cdot b = ((1/a) \cdot a) \cdot b = (1/a) \cdot (a \cdot b)$$
$$= (1/a) \cdot 0 = 0.$$

A similar argument holds if $b \neq 0$.

(c) From Theorem 4.5, we have $-a = (-1) \cdot a$, and $-b = (-1) \cdot b$; hence

$$(-a) \cdot (-b) = ((-1) \cdot a) \cdot ((-1) \cdot b)$$
$$= (a \cdot (-1)) \cdot ((-1) \cdot b)$$
$$= a \cdot ((-1) \cdot (-1)) \cdot b = a \cdot 1 \cdot b$$
$$= a \cdot b.$$

(d) If $a \neq 0$, then $1/a \neq 0$ and $-a \neq 0$. Since $a \cdot (1/a) = 1$, it follows from part (c) that $(-a) \cdot (-(1/a)) = 1$. If we apply Theorem 4.3(b), we deduce that $1/(-a) = -(1/a)$ as claimed.   Q.E.D.

Rational Numbers

From now on we shall generally drop the use of the dot to denote multiplication and write ab for $a \cdot b$. As usual we shall write a^2 for aa, a^3 for $aaa = (a^2)a$, and if $n \in \mathbf{N}$ we define $a^{n+1} = (a^n)a$. It follows by use of mathematical induction that if $m, n \in \mathbf{N}$, then

$$(*) \qquad\qquad a^{m+n} = a^m a^n$$

for any $a \in \mathbf{R}$. Similarly we shall write 2 for $1 + 1$,   3 for $2 + 1 = (1 + 1) + 1$, and so forth. In addition we shall generally write $b - a$ instead of $(-a) + b = b + (-a)$ and, if $a \neq 0$, we shall generally write

$$b/a \qquad \text{or} \qquad \frac{b}{a}$$

instead of $(1/a) \cdot b = b \cdot (1/a)$. We shall also write a^{-1} for $1/a$, and a^{-n} for $1/a^n$. It can then be shown that formula $(*)$ above holds for $m, n \in \mathbf{Z}$ when $a \neq 0$, provided we adopt the conventions $a^0 = 1$, $a^1 = a$.

Elements of $\mathbf{R}$ which are of the form

$$\frac{b}{a} \qquad \text{or} \qquad \frac{-b}{a}$$

for $a, b \in \mathbf{N}$, $a \neq 0$, are said to be **rational numbers,** and the set of all rational numbers in $\mathbf{R}$ will be denoted by the standard notation $\mathbf{Q}$. All of the elements of $\mathbf{R}$ which are *not* rational numbers are said to be **irrational numbers.** Although this terminology is unfortunate, it is also quite standard and we shall adopt it.

We shall close this section with a proof of the fact that there does not exist a rational number whose square is 2.

4.7 THEOREM. *There does not exist a rational number r such that* $r^2 = 2$.

PROOF. Suppose, on the contrary that $(p/q)^2 = 2$, where p and q are integers. We may, without loss of generality, suppose that p and q have no common integral factors. (Why?) Since $p^2 = 2q^2$, it follows that p must be an even integer (for if $p = 2k + 1$ is odd, then $p^2 = 4k^2 + 4k + 1 = 2(2k^2 + 2k) + 1$ is odd). Therefore $p = 2k$ for some integer k and hence $4k^2 = 2q^2$. It follows that $q^2 = 2k^2$, whence q must also be even. Therefore both p and q are divisible by 2, contrary to our hypothesis.   Q.E.D.

Exercises

4.A. Prove part (b) of Theorem 4.2.

4.B. Prove part (b) of Theorem 4.3.

4.C. Prove part (b) of Theorem 4.4.

4.D. Using mathematical induction, show that if $a \in \textbf{R}$ and $m, n \in \textbf{N}$, then $a^{m+n} = a^m a^n$.

4.E. Show that if $a \in \textbf{R}$, $a \neq 0$, and $m, n \in \textbf{Z}$, then $a^{m+n} = a^m a^n$.

4.F. Use the argument in Theorem 4.7 to show that there does not exist a rational number s such that $s^2 = 6$.

4.G. Modify the argument in Theorem 4.7 to show that there does not exist a rational number t such that $t^2 = 3$.

4.H. If $\xi \in \textbf{R}$ is irrational and $r \in \textbf{R}$, $r \neq 0$, is rational, show that $r + \xi$ and $r\xi$ are irrational.

Section 5   The Order Properties of *R*

The purpose of this section is to introduce the important "order" properties of *R*, which will play a very important role in subsequent sections. The simplest way to introduce the notion of order is to make use of the notion of "strict positivity," which we now explain.

5.1 THE ORDER PROPERTIES OF *R*. There is a non-empty subset P of *R*, called the set of **strictly positive real numbers,** satisfying the properties:

(i) If a, b belong to P, then $a + b$ belongs to P.

(ii) If a, b belong to P, then ab belongs to P.

(iii) If a belongs to R, then precisely one of the following relations holds: $a \in P$. $a = 0$. $-a \in P$.

Condition (iii) is sometimes called the **property of trichotomy.** It implies that the set $N = \{-a : a \in P\}$, sometimes called the set of **strictly negative real numbers,** has no elements in common with P. In fact the entire set *R* is the union of the three disjoint sets P, $\{0\}$, N.

5.2 DEFINITION. If $a \in P$, we say that a is a **strictly positive real number** and write $a > 0$. If a is either in P or is 0, we say that a is a

positive real number and write $a \ge 0$. If $-a \in P$, we say that a is a **strictly negative real number** and write $a < 0$. If $-a$ is either in P or is 0, we say that a is a **negative real number** and write $a \le 0$.

It should be noted that, according to the terminology just introduced, the number 0 is both positive and negative; it is the only number with this dual status. This terminology may seem a bit strange at first, but it will prove to be a convenience. Some authors reserve the term "positive" for the elements of the set P and use the term "non-negative" for the elements of $P \cup \{0\}$.

We now introduce the order relations.

5.3 DEFINITION. Let a, b be elements of **R**. If $a - b \in P$, then we write $a > b$. If $-(a - b) \in P$, then we write $a < b$. If $a - b \in P \cup \{0\}$, then we write $a \ge b$. If $-(a - b) \in P \cup \{0\}$, then we write $a \le b$.

As usual, it is often convenient to turn the signs around and write

$$b < a, \qquad b > a, \qquad b \le a, \qquad b \ge a,$$

respectively. In addition, if $a < b$ and $b < c$, then we often write

$$a < b < c \qquad \text{or} \qquad c > b > a.$$

If $a \le b$ and $b < c$, then we often write

$$a \le b < c \qquad \text{or} \qquad c > b \ge a.$$

Properties of the Order

We shall now establish the basic properties of the order relation in **R**. These are the familiar "laws" for inequalities which the reader has met in earlier courses. They will be frequently used in later sections and are of great importance.

5.4 THEOREM. *Let a, b, c be elements of* **R**.
(a) *If $a > b$ and $b > c$, then $a > c$.*
(b) *Exactly one of the following holds:* $a > b$, $a = b$, $a < b$.
(c) *If $a \ge b$ and $b \ge a$, then $a = b$.*

PROOF. (a) If $a - b$ and $b - c$ belong to P, then from 5.1(i) we infer that $a - c = (a - b) + (b - c)$ also belongs to P. Hence $a > c$.

(b) By 5.1(iii) exactly one of the following possibilities takes place: $a - b \in P$, $a - b = 0$, $b - a = -(a - b) \in P$.

(c) If $a \ne b$, then from part (b) we must have either $a - b$ or $b - a$ in P. Hence, either $a > b$ or $b > a$. In either case one of the hypotheses is contradicted.   Q.E.D.

5.5 Theorem. (a) *If $0 \neq a \in \mathbf{R}$, then $a^2 > 0$.*
(b) *$1 > 0$.*
(c) *If $n \in \mathbf{N}$, then $n > 0$.*

proof. (a) Either a or $-a$ belongs to P. If $a \in P$, then by 5.1(ii) we have $a^2 = aa \in P$. If $-a \in P$, then by Theorem 4.6(c) we have $a^2 = (-a)(-a) \in P$. Hence, in either case, $a^2 \in P$.

(b) Since $1 = (1)^2$, the conclusion follows from (a).

(c) We use mathematical induction. The validity of the assertion with $n = 1$ is part (b). If the assertion is true for the natural number k (that is, supposing $k \in P$), then since $1 \in P$, it follows from 5.1(i) that $k + 1 \in P$. Hence the assertion is true for all natural numbers.   Q.E.D.

The next properties are probably familiar to the reader.

5.6 Theorem. *Let a, b, c, d be elements of $\mathbf{R}$.*
(a) *If $a > b$, then $a + c > b + c$.*
(b) *If $a > b$ and $c > d$, then $a + c > b + d$.*
(c) *If $a > b$ and $c > 0$, then $ac > bc$.*
(c′) *If $a > b$ and $c < 0$, then $ac < bc$.*
(d) *If $a > 0$, then $1/a > 0$.*
(d′) *If $a < 0$, then $1/a < 0$.*

proof. (a) Observe that $(a + c) - (b + c) = a - b$.

(b) If $a - b$ and $c - d$ belong to P, then by 5.1(i) we conclude that $(a + c) - (b + d) = (a - b) + (c - d)$ also belongs to P.

(c) If $a - b$ and c belong to P, then by 5.1(ii) $ac - bc = (a - b)c$ also belongs to P.

(c′) If $a - b$ and $-c$ belong to P, then by 5.1(ii) $bc - ac = (a - b)(-c)$ also belongs to P.

(d) If $a > 0$, then by 5.1(iii) $a \neq 0$ so that the element $1/a$ exists. If $1/a = 0$, then $1 = a(1/a) = 0$, a contradiction. If $1/a < 0$, then part (c′) with $c = 1/a$ implies that $1 = a(1/a) < 0$, contradicting 5.5(b). Therefore we must have $1/a > 0$, since the other two possibilities have been excluded.

(d′) This can be proved by an argument similar to that in (d). Alternatively, we can invoke Theorem 4.6(d) and use (d) directly.   Q.E.D.

5.7 Theorem. *If $a > b$, then $a > \frac{1}{2}(a + b) > b$.*

proof. Since $a > b$ it follows from Theorem 5.6(a) with $c = a$ that $2a > a + b$ and from Theorem 5.6(a) with $c = b$ that $a + b > 2b$. By Theorem 5.5(c) we know that $2 > 0$ and from Theorem 5.6(d) that $\frac{1}{2} > 0$. Applying Theorem 5.6(c) with $c = \frac{1}{2}$, we deduce that $a > \frac{1}{2}(a + b)$ and $\frac{1}{2}(a + b) > b$. Hence $a > \frac{1}{2}(a + b) > b$, as claimed.   Q.E.D.

The theorem just proved (with $b = 0$) implies that given any strictly positive number a, there is another strictly smaller and strictly positive number (namely $\frac{1}{2}a$). Thus, *there is no smallest strictly positive real number.*

We have already seen that if $a > 0$ and $b > 0$, then $ab > 0$. Also if $a < 0$ and $b < 0$, then $ab > 0$. We now show that the converse is true.

5.8 THEOREM. *If $ab > 0$, then we either have $a > 0$ and $b > 0$, or we have $a < 0$ and $b < 0$.*

PROOF. If $ab > 0$, then $a \neq 0$ and $b \neq 0$. (Why?) If $a > 0$, then from Theorem 5.6(d) we infer that $1/a > 0$ and from 5.6(c) that $b = ((1/a)a)b = (1/a)(ab) > 0$. On the other hand, if $a < 0$, then from Theorem 5.6(d') we infer that $1/a < 0$ and from 5.6(c') that $b = ((1/a)a)b = (1/a)(ab) < 0$. Q.E.D.

5.9 COROLLARY. *If $ab < 0$, then we either have $a > 0$ and $b < 0$, or we have $a < 0$ and $b > 0$.*

The proof of this assertion is left an exercise.

Absolute Value

The trichotomy property 5.1(iii) assures that if $a \neq 0$, then one of the numbers a and $-a$ is strictly positive. The absolute value of $a \neq 0$ is defined to be the strictly positive one of the pair $\{a, -a\}$, and the absolute value of 0 is defined to be 0.

5.10 DEFINITION. If $a \in \mathbf{R}$, the **absolute value** of a is denoted by $|a|$ and is defined by

$$|a| = a \qquad \text{if} \quad a \geq 0,$$
$$ = -a \qquad \text{if} \quad a < 0.$$

Thus the domain of the absolute value function is all of **R**, its range is $P \cup \{0\}$, and it maps the elements a, $-a$, into the same element.

5.11 THEOREM. (a) $|a| = 0$ *if and only if $a = 0$.*
(b) $|-a| = |a|$ *for all $a \in \mathbf{R}$.*
(c) $|ab| = |a| \, |b|$ *for all $a, b \in \mathbf{R}$.*
(d) *If $c \geq 0$, then $|a| \leq c$ if and only if $-c \leq a \leq c$.*
(e) $-|a| \leq a \leq |a|$ *for all $a \in \mathbf{R}$.*

PROOF. (a) If $a = 0$, then by definition $|0| = 0$. If $a \neq 0$, then also $-a \neq 0$, so that $|a| \neq 0$.
(b) If $a = 0$, then $|0| = 0 = |-0|$. If $a > 0$, then $|a| = a = |-a|$. If $a < 0$, then $|a| = -a = |-a|$.
(c) If $a > 0$ and $b > 0$, then $ab > 0$ so that $|ab| = ab = |a| \, |b|$. If $a < 0$ and $b > 0$, then $ab < 0$ so that $|ab| = -(ab) = (-a)b = |a| \, |b|$. The other cases are handled similarly.

(d) If $|a| \le c$, then both $a \le c$ and $-a \le c$. (Why?) From the latter and Theorem 5.6(c') we infer that $-c \le a$ so that $-c \le a \le c$. Conversely, if this relation holds then both $a \le c$ and $-a \le c$, whence $|a| \le c$.

(e) Use part (d) with $c = |a| \ge 0$.   Q.E.D.

The next result will be used very frequently in the sequel. (Recall that $a \pm b$ means both $a + b$ and $a - b$.)

5.12   THE TRIANGLE INEQUALITY.   *If a, b are any real numbers, then*

$$\big| |a| - |b| \big| \le |a \pm b| \le |a| + |b|.$$

PROOF.   According to Theorem 5.11(e), we have $-|a| \le a \le |a|$ and $-|b| \le \pm b \le |b|$. Employing 5.6(b) we infer that

$$-(|a| + |b|) = -|a| - |b| \le a \pm b \le |a| + |b|.$$

From Theorem 5.11(d) it follows that $|a \pm b| \le |a| + |b|$, proving the second part of the inequality.

Since $|a| = |(a - b) + b| \le |a - b| + |b|$ (why?), it follows that $|a| - |b| \le |a - b|$. Similarly $|b| - |a| \le |a - b|$. (Why?) Combining these two inequalities, we deduce that $\big| |a| - |b| \big| \le |a - b|$, which is the first part of the inequality with the minus sign. To obtain the inequality with the plus sign, replace b by $-b$.   Q.E.D.

5.13   COROLLARY.   *If a_1, a_2, ..., a_n are any n real numbers, then*

$$|a_1 + a_2 + \cdots + a_n| \le |a_1| + |a_2| + \cdots + |a_n|.$$

PROOF.   If $n = 2$, the conclusion is precisely 5.12. If $n > 2$, we use mathematical induction and the fact that

$$|a_1 + a_2 + \cdots + a_k + a_{k+1}| = |(a_1 + a_2 + \cdots + a_k) + a_{k+1}|$$
$$\le |a_1 + a_2 + \cdots + a_k| + |a_{k+1}|. \quad \text{Q.E.D.}$$

Exercises

5.A. If a, $b \in R$ and $a^2 + b^2 = 0$, show that $a = b = 0$.

5.B. If $n \in N$, show that $n^2 \ge n$ and hence $1/n^2 \le 1/n$.

5.C. If $a > -1$, $a \in R$, show that $(1 + a)^n \ge 1 + na$ for all $n \in N$. This inequality is called **Bernoulli's Inequality.**† (Hint: use mathematical induction.)

5.D. If $c > 1$, $c \in R$, show that $c^n \ge c$ for all $n \in N$. (Hint: $c = 1 + a$ with $a > 0$).

5.E. If $c > 1$, $c \in R$, show that $c^m \ge c^n$ for $m \ge n$, m, $n \in N$.

5.F. Suppose that $0 < c < 1$. If $m \ge n$, m, $n \in N$, show that $0 < c^m \le c^n < 1$.

5.G. Show that $n < 2^n$ for all $n \in N$. Hence $1/2^n < 1/n$ for all $n \in N$.

5.H. If a and b are positive real numbers and $n \in N$, then $a^n < b^n$ if and only if $a < b$.

† JACOB BERNOULLI (1654–1705) was a member of a Swiss family that produced several mathematicians who played an important role in the development of calculus.

5.I. Show that if $a \le x \le b$ and $a \le y \le b$, then $|x - y| \le b - a$. Interpret this geometrically.

5.J. Let $\delta > 0$, $a \in \mathbf{R}$. Show that $a - \delta < x < a + \delta$ if and only if $|x - a| < \delta$. Similarly, $a - \delta \le x \le a + \delta$ if and only if $|x - a| \le \delta$.

5.K. If $a, b \in \mathbf{R}$ and $b \ne 0$, show that $|a/b| = |a|/|b|$.

5.L. Show if $a, b \in \mathbf{R}$, then $|a + b| = |a| + |b|$ if and only if $ab \ge 0$.

5.M. Sketch the points (x, y) in the plane $\mathbf{R} \times \mathbf{R}$ for which $|y| = |x|$.

5.N. Sketch the points (x, y) in the plane $\mathbf{R} \times \mathbf{R}$ for which $|x| + |y| = 1$.

5.O. If x, y, z belong to $\mathbf{R}$, then $x \le y \le z$ if and only if $|x - y| + |y - z| = |x - z|$.

5.P. If $0 < a < 1$, then $0 < a^2 < a < 1$, while if $1 < a$, then $1 < a < a^2$.

Section 6   The Completeness Property of **R**

In this section we shall present one more property of the real number system which is often called the "completeness property" since it guarantees the existence of elements in **R** when certain hypotheses are satisfied. There are various versions of this completeness property, but we choose to give here what is probably the most efficient method by assuming that bounded sets in **R** have a supremum.

Suprema and Infima

We now introduce the notion of an upper bound of a set of real numbers. This idea will be of utmost importance in later sections.

6.1 DEFINITION. Let S be a subset of **R**.

(a) An element $u \in \mathbf{R}$ is said to be an **upper bound** of S if $s \le u$ for all $s \in S$.

(b) An element $w \in \mathbf{R}$ is said to be a **lower bound** of S if $w \le s$ for all $s \in S$.

We note that a subset $S \subseteq \mathbf{R}$ may not have an upper bound (for example, take $S = \mathbf{R}$). However, if it has one upper bound, then it has infinitely many (for if u is an upper bound of S, then $u + n$ is also an upper bound of S for any $n \in \mathbf{N}$). Again, the set $S_1 = \{x \in \mathbf{R} : 0 < x < 1\}$ has 1 for an upper bound; in fact, any number $u \ge 1$ is an upper bound of S_1. Similarly the set $S_2 = \{x \in \mathbf{R} : 0 \le x \le 1\}$ has the same upper bounds as S_1. Note, however, that S_2 contains the upper bound 1, while S_1 does not contain any of its upper bounds. (Why can no number $c < 1$ be an upper bound of S_1?)

To show that a number $u \in \mathbf{R}$ is *not* an upper bound of $S \subseteq \mathbf{R}$ we must produce an element $s_0 \in S$ such that $u < s_0$. If $S = \emptyset$, the empty set, that cannot be done. Hence the empty set has the unusual property that *every* real number is an upper bound; also every real number is a lower bound of $\emptyset$. This may seem artificial, but it is a logical consequence of our definitions, so we must accept it.

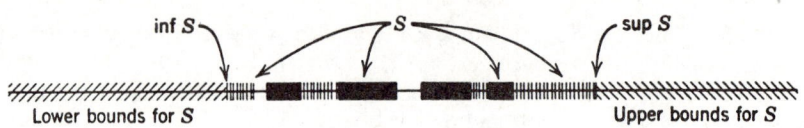

Figure 6.1.   Suprema and infima.

As a matter of terminology, when a set has an upper bound, we shall say that it is **bounded above,** and when a set has a lower bound, we shall say that it is **bounded below.** If a set has both an upper and a lower bound, we shall say that it is **bounded.** If a set lacks either an upper or a lower bound, we shall say that it is **unbounded.** Thus the sets S_1 and S_2 above are both bounded. However the subset $P = \{x \in \mathbf{R} : x > 0\}$ of $\mathbf{R}$ is unbounded since it does not have an upper bound. Similarly, the set $\mathbf{R}$ is unbounded since it does not have either an upper or a lower bound.

6.2   Definition.   Let S be a subset of $\mathbf{R}$.

(a) If S is bounded above, then an upper bound of S is said to be a **supremum** (or a **least upper bound**) of S if it is less than any other upper bound of S.

(b) If S is bounded below, then a lower bound of S is said to be an **infimum** (or a **greatest lower bound**) of S if it is greater than any other lower bound of S. (See Figure 6.1.)

Expressed differently, a number $u \in \mathbf{R}$ is a supremum of a subset S of $\mathbf{R}$ if it satisfies the two conditions:

(i) $s \le u$ for all $s \in S$;

(ii) if v is any number such that $s \le v$ for all $s \in S$, then $u \le v$.

Indeed, the condition (i) makes u an upper bound of S, and (ii) shows that u is less than any other upper bound of S.

It is apparent that there can be only one supremum for a given subset S of $\mathbf{R}$. For, if u_1 and u_2 are suprema of S, then they are both upper bounds of S. Since u_1 is a supremum of S and u_2 is an upper bound of S, we must have $u_1 \le u_2$. A similar argument shows that we must have $u_2 \le u_1$. Therefore $u_1 = u_2$. In a similar manner one shows that there can be only one infimum for a given subset S of $\mathbf{R}$. Where these numbers exist, we shall denote them by

$$\sup S \qquad \text{and} \qquad \inf S.$$

It is often convenient to have another characterization of the supremum of a subset of $\mathbf{R}$.

6.3   Lemma.   *A number $u \in \mathbf{R}$ is the supremum of a non-empty subset $S \subseteq \mathbf{R}$ if and only if it has the following properties:*

(i) *There are no elements $s \in S$ with $u < s$.*

(ii) *If $v < u$, then there is an element $s_v \in S$ such that $v < s_v$.*

PROOF. Suppose that u satisfies (i) and (ii). The condition (i) implies that u is an upper bound of S. If v is any number with $v < u$, then property (ii) shows that v cannot be an upper bound of S. Hence u is the supremum of S.

Conversely, let u be the supremum of S. Since u is an upper bound of S, condition (i) holds. If $v < u$, then v is not an upper bound of S. Therefore, there exists an element $s_v \in S$ such that $v < s_v$. Q.E.D.

The reader should convince himself that the number 1 is the supremum of both of the sets S_1 and S_2 which were defined after Definition 6.1. We note that S_2 contains its supremum, but that S_1 does not contain its supremum. Thus, *when we say that a set has a supremum, we are making no statement as to whether the set contains the supremum as an element or not.*

It is a deep and fundamental property of the real number system that every non-empty subset of **R** which is bounded above has a supremum. We shall make frequent and essential use of this property, which we take as our final assumption about **R**.

6.4 SUPREMUM PROPERTY. *Every non-empty set of real numbers which has an upper bound has a supremum.*

The analogous property of infima can be readily established from the Supremum Property.

6.5 INFIMUM PROPERTY. *Every non-empty set of real numbers which has a lower bound has an infimum.*

PROOF. Let S be bounded below and let $S_1 = \{-s : s \in S\}$ so that S_1 is bounded above. The Supremum Property assures that S_1 has a supremum u. We leave it to the reader to show that $-u$ is the infimum of S. Q.E.D.

The Archimedean† Property

One important consequence of the Supremum Property is that the subset **N** of natural numbers is not bounded above in **R**. In particular this means that given any real number x, there exists a natural number n_x which is greater than x (otherwise x would be an upper bound for **N**). We shall now prove this assertion.

6.6 ARCHIMEDEAN PROPERTY. *If $x \in \mathbf{R}$, there is a natural number $n_x \in \mathbf{N}$ such that $x < n_x$.*

PROOF. If the conclusion fails, then x is an upper bound for **N**.

† This property of **R** is named after Archimedes (287–212 B.C.), who has been called "the greatest intellect of antiquity," and was one of the founders of the scientific method.

Therefore, by the Supremum Property, N has a supremum u. Since x is an upper bound for N, it follows that $u \le x$. Since $u - 1 < u$, it follows from Lemma 6.3(ii) that there exists $n_1 \in N$ such that $u - 1 < n_1$. Therefore $u < n_1 + 1$, but since $n_1 + 1 \in N$ this contradicts the assumption that u is an upper bound of N.   Q.E.D.

6.7 COROLLARY.   *Let y and z be strictly positive real numbers.*
(a) *There is a natural number n such that $ny > z$.*
(b) *There is a natural number n such that $0 < 1/n < z$.*
(c) *There is a natural number n such that $n - 1 \le y < n$.*

PROOF.   (a) Since y and z are strictly positive, then $x = z/y$ is also strictly positive. Let $n \in N$ be such that $z/y = x < n$. Then $z < ny$, as claimed.

(b) Let $n \in N$ be such that $0 < 1/z < n$.   Then $0 < 1/n < z$.

(c) The Archimedean Property assures that there exist natural numbers m such that $y < m$. Let n be the least such natural number (see Section 3). Then $n - 1 \le y < n$.   Q.E.D.

We noted after Theorem 5.7 that there is no smallest strictly positive real number.   Corollary 6.7(b) shows that given any $z > 0$ there is a rational number of the form $1/n$ with $0 < 1/n < z$.   Sometimes one says "there are arbitrarily small rational numbers of the form $1/n$."

The Existence of $\sqrt{2}$

One important property of the Supremum Property is that, as we have said before, it assures the existence of certain real numbers.   We shall make use of it many times in this way.   At the moment we will show that it guarantees the existence of a positive real number x such that $x^2 = 2$; that is, a **positive square root** of 2.   This result complements Theorem 4.7.

6.8 THEOREM.   *There exists a positive number $x \in R$ such that $x^2 = 2$.*

PROOF.   Let $S = \{y \in R : 0 \le y, y^2 \le 2\}$.   The set S is bounded above by 2; for, if not, then there exists an element $s \in S$ such that $2 < s$ whence it follows that $4 < s^2 \le 2$, a contradiction.   By the Supremum Property the set S has a supremum and we let $x = \sup S$.   Clearly $x > 0$.

We claim that $x^2 = 2$.   If not, then either $x^2 < 2$ or $x^2 > 2$.   If $x^2 < 2$, let $n \in N$ be chosen such that $1/n < (2 - x^2)/(2x + 1)$.   In this case

$$\left(x + \frac{1}{n}\right)^2 = x^2 + \frac{2x}{n} + \frac{1}{n^2} \le x^2 + \frac{2x + 1}{n} < x^2 + (2 - x^2) = 2,$$

which means that $x + 1/n \in S$, contrary to the fact that x is an upper bound of S.

If $x^2 > 2$, we choose $m \in N$ such that $1/m < (x^2 - 2)/2x$.   Since $x = \sup S$, there exists an $s_0 \in S$ with $x - 1/m < s_0$. But this implies that

$$2 < x^2 - \frac{2x}{m} < x^2 - \frac{2x}{m} + \frac{1}{m^2} = \left(x - \frac{1}{m}\right)^2 < s_0^2.$$

Hence $s_0^2 > 2$, contrary to the fact that $s_0 \in S$.

Since we have excluded the possibilities that $x^2 < 2$ and $x^2 > 2$, we must have $x^2 = 2$. Q.E.D.

By modifying the argument in Theorem 6.8 very slightly, the reader can show that if $a \geq 0$, then there is a unique number $b \geq 0$ such that $b^2 = a$. We call b the **positive square root** of a and denote it by

$$b = \sqrt{a} \qquad \text{or} \qquad b = a^{1/2}.$$

We now know that there exists at least one irrational element, namely, $\sqrt{2}$ (the positive square root of 2). Actually there are "more" irrational numbers than rational numbers in the sense that (as we have seen in Section 3) the set of rational numbers is countable while the set of irrational numbers is not countable. We shall now show that there are arbitrarily small irrational numbers; this result complements Corollary 6.7.

6.9 COROLLARY. *Let $\xi > 0$ be an irrational number and let $z > 0$. Then there exists a natural number m such that the irrational number ξ/m satisfies $0 < \xi/m < z$.*

PROOF. Since $\xi > 0$, $z > 0$, it follows from Theorem 5.6(d) and 5.6(c) that $\xi/z > 0$. By the Archimedean Property there exists a natural number m such that $0 < \xi/z < m$. Therefore $0 < \xi/m < z$ and it is an exercise to show that ξ/m is irrational. Q.E.D.

We now show that between any two distinct real numbers there is a rational number and an irrational number. (In fact, there are infinitely many of both kinds!)

6.10 THEOREM. *Let x and y be real numbers with $x < y$.*

(a) *Then there is a rational number r such that $x < r < y$.*

(b) *If $\xi > 0$ is any irrational number, then there is a rational number s such that the irrational number $s\xi$ satisfies $x < s\xi < y$.*

PROOF. It is no loss of generality to assume that $0 < x$. (Why?)

(a) Since $y - x > 0$, it follows from Corollary 6.7(b) that there is a natural number m such that $0 < 1/m < y - x$. From Corollary 6.7(a) there is a natural number k such that

$$\frac{k}{m} = k\frac{1}{m} > x,$$

and we let n be the least such natural number. Therefore

$$\frac{n-1}{m} \leq x < \frac{n}{m}.$$

We must also have $n/m < y$, for otherwise

$$\frac{n-1}{m} \le x < y \le \frac{n}{m}$$

which implies that $y - x \le 1/m$ contrary to the choice of m. Therefore $x < n/m < y$.

(b) Supposing that $0 < x < y$ and $\xi > 0$, we have $x/\xi < y/\xi$. By part (a) there exists a rational number s such that $x/\xi < s < y/\xi$. Therefore $x < s\xi < y$. (Show that $s\xi$ is irrational.)   Q.E.D.

Exercises

6.A. Prove that a non-empty finite set of real numbers has a supremum and an infimum.

6.B. If a subset S of R contains an upper bound, then this upper bound is the supremum of S.

6.C. Give an example of a set of rational numbers which is bounded but does not have a rational supremum.

6.D. Give an example of a set of irrational numbers that has a rational supremum.

6.E. Prove that the union of two bounded sets is bounded.

6.F. Give an example of a countable collection of bounded sets whose union is bounded, and an example where the union is unbounded.

6.G. If S is a bounded set in R and if S_0 is a nonempty subset of S, then show that

$$\inf S \le \inf S_0 \le \sup S_0 \le \sup S.$$

Sometimes it is more convenient to express this in another way. Let $D \ne \emptyset$ and let $f : D \to R$ have bounded range. If D_0 is a non-empty subset of D, then

$$\inf \{f(x) : x \in D\} \le \inf \{f(x) : x \in D_0\} \le \sup \{f(x) : x \in D_0\} \le \sup \{f(x) : x \in D\}.$$

6.H. Let X and Y be non-empty sets and let $f : X \times Y \to R$ have bounded range in R. Let

$$f_1(x) = \sup \{f(x, y) : y \in Y\}, \qquad f_2(y) = \sup \{f(x, y) : x \in X\}.$$

Establish the **Principle of Iterated Suprema:**

$$\sup \{f(x, y) : x \in X, y \in Y\} = \sup \{f_1(x) : x \in X\}$$
$$= \sup \{f_2(y) : y \in Y\}.$$

We sometimes express this in symbols by:

$$\sup_{x, y} f(x, y) = \sup_{x} \sup_{y} f(x, y) = \sup_{y} \sup_{x} f(x, y).$$

6.I. Let f and f_1 be as in the preceding exercise and let

$$g_2(y) = \inf \{f(x, y) : x \in X\}$$

Prove that

$$\sup \{g_2(y) : y \in Y\} \le \inf \{f_1(x) : x \in X\}.$$

Show that strict inequality can hold. We sometimes express this inequality by

$$\sup_{y} \inf_{x} f(x, y) \le \inf_{x} \sup_{y} f(x, y).$$

6.J. Let X be a non-empty set and let $f : X \to \textbf{R}$ have bounded range in **R**. If $a \in \textbf{R}$, show that

$$\sup \{a + f(x) : x \in X\} = a + \sup \{f(x) : x \in X\},$$
$$\inf \{a + f(x) : x \in X\} = a + \inf \{f(x) : x \in X\}.$$

6.K. Let X be a non-empty set and let f and g be defined on X and have bounded ranges in **R**. Show that

$$
\begin{aligned}
\inf \{f(x) : x \in X\} + \inf \{g(x) : x \in X\} &\le \inf \{f(x) + g(x) : x \in X\} \\
&\le \inf \{f(x) : x \in X\} + \sup \{g(x) : x \in X\} \\
&\le \sup \{f(x) + g(x) : x \in X\} \\
&\le \sup \{f(x) : x \in X\} + \sup \{g(x) : x \in X\}.
\end{aligned}
$$

Give examples to show that each inequality can be strict.

6.L. If $z > 0$ show that there exists $n \in \textbf{N}$ such that $1/2^n < z$.

6.M. Modify the argument given in Theorem 6.8 to show that if $a > 0$, then the number

$$b = \sup \{y \in \textbf{R} : 0 \le y, \quad y^2 \le a\}$$

exists and has the property that $b^2 = a$. This number will be denoted by $\sqrt{a}$ or $a^{1/2}$, and is called the **positive square root** of a.

6.N. Use Exercise 5.P to show that if $0 < a < 1$, then $0 < a < \sqrt{a} < 1$, while if $1 < a$, then $1 < \sqrt{a} < a$.

Projects†

6.α. If a and b are strictly positive real numbers and if $n \in \textbf{N}$, we have defined a^n and b^n. It follows by mathematical induction that if $m, n \in \textbf{N}$, then

(i) $a^m a^n = a^{m+n}$;

(ii) $(a^m)^n = a^{mn}$;

(iii) $(ab)^n = a^n b^n$;

(iv) $a < b$ if and only if $a^n < b^n$.

† The projects are intended to be somewhat more challenging to the reader, but they differ considerably in difficulty. We have put these three (rather difficult) projects here because they belong here logically. The reader should return to them later after he has accumulated more experience with suprema.

We shall adopt the convention that $a^0 = 1$ and $a^{-n} = 1/a^n$. Thus we have defined a^x for x in Z and it is readily checked that properties (i)–(iii) remain valid.

We wish to define a^x for rational numbers x in such a way that (i)–(iii) hold. The following steps can be used as an outline. Throughout we shall assume that a and b are real numbers exceeding 1.

(a) If r is a rational number given by $r = m/n$, where m and n are integers and $n > 0$ we define $S_r(a) = \{x \in R : 0 \le x^n \le a^m\}$. Show that $S_r(a)$ is a bounded non-empty subset of R and define $a^r = \sup S_r(a)$.

(b) Prove that $z = a^r$ is the unique positive root of the equation $z^n = a^m$. (Hint: there is a constant K such that if $0 < \varepsilon < 1$, then $(1 + \varepsilon)^n < 1 + K\varepsilon$. Hence if $x^n < a^m < y^n$, there exists an $\varepsilon > 0$ such that

$$x^n(1+\varepsilon)^n < a^m < y^n/(1+\varepsilon)^n.)$$

(c) Show that the value of a^r given in part (a) does not depend on the representation of r in the form m/n. Also show that if r is an integer, then the new definition of a^r gives the same value as the old one.

(d) Show that if $r, s \in Q$, then $a^r a^s = a^{r+s}$ and $(a^r)^s = a^{rs}$.

(e) Show that $a^r b^r = (ab)^r$.

(f) If $r \in Q$, $r > 0$, then $a < b$ if and only if $a^r < b^r$.

(g) If $r, s \in Q$, then $r < s$ if and only if $a^r < a^s$.

(h) If c is a real number satisfying $0 < c < 1$, we define $c^r = (1/c)^{-r}$. Show that parts (d) and (e) hold and that a result similar to (g), but with the inequality reversed, holds.

6.β. Now that a^x has been defined for rational numbers x, we wish to define it for real x. In doing so, make free use of the results of the preceding project. As before, let a and b be real numbers exceeding 1. If $u \in R$, let

$$T_u(a) = \{a^r : r \in Q, r \le u\}.$$

Show that $T_u(a)$ is a bounded non-empty subset of R and define

$$a^u = \sup T_u(a).$$

Prove that this definition yields the same result as the previous one when u is rational. Establish the properties that correspond to the statements given in parts (d)–(g) of the preceding project. The very important function which has been defined on R in this project is called the **exponential function** (to the base a). Some alternative definitions will be given in later sections. Sometimes it is convenient to denote this function by the symbol

$$\exp_a$$

and denote its value at the real number u by $\exp_a(u)$ instead of a^u.

6.γ. Making use of the properties of the exponential function that were established in the preceding project, show that $\exp_a$ is an injective function with domain R and range $\{y \in R : y > 0\}$. Under our standing assumption that $a > 1$, this exponential function is strictly increasing in the sense that if $x < u$, then $\exp_a(x) < \exp_a(u)$. Therefore, the inverse function exists with domain $\{v \in R : v > 0\}$ and range R. We call this inverse function the **logarithm** (to the base a) and denote it by

$$\log_a.$$

Show that $\log_a$ is a strictly increasing function and that

$$\exp_a(\log_a(v)) = v \quad \text{for } v > 0, \qquad \log_a(\exp_a(u)) = u \quad \text{for } u \in R.$$

Also show that $\log_a(1) = 0$, $\log_a(a) = 1$, and that

$$\log_a(v) < 0 \quad \text{for } v < 1, \qquad \log_a(v) > 0 \quad \text{for } v > 1.$$

Prove that if v, $w > 0$, then

$$\log_a(vw) = \log_a(v) + \log_a(w).$$

Moreover, if $v > 0$ and $x \in R$, then

$$\log_a(v^x) = x \log_a(v).$$

Section 7   Cuts, Intervals, and the Cantor Set

Another method of completing the rational numbers to obtain R was devised by Dedekind†; it is based on the notion of a "cut."

7.1   DEFINITION.   An ordered pair (A, B) of non-void subsets of R is said to form a **cut** if $A \cap B = \emptyset$, $A \cup B = R$, and $a < b$ for all $a \in A$ and all $b \in B$.

A typical example of a cut in R is obtained for a fixed element $\xi \in R$ by defining

$$A = \{x \in R : x \leq \xi\}, \qquad B = \{x \in R : x > \xi\}.$$

Alternatively, we could take

$$A_1 = \{x \in R : x < \xi\}, \qquad B_1 = \{x \in R : x \geq \xi\}.$$

It is an important property of R, that every cut in R is determined by some real number.   We shall now establish this property.

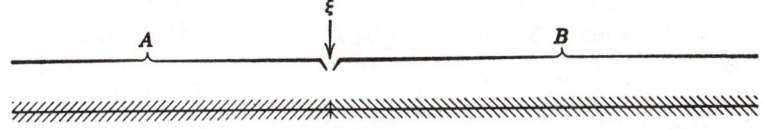

Figure 7.1.   A Dedekind cut.

† RICHARD DEDEKIND (1831–1916) was a student of Gauss. He contributed to number theory, but is best known for his work on the foundations of the real number system.

7.2 CUT PROPERTY. *If (A, B) is a cut in $\mathbf{R}$, then there exists a unique number $\xi \in \mathbf{R}$ such that $a \leq \xi$ for all $a \in A$ and $\xi \leq b$ for all $b \in B$.*

PROOF. By hypothesis, the sets A and B are non-void. Any element of B is an upper bound of A. Hence A has a supremum which we denote by ξ. Since ξ is an upper bound of A, then $a \leq \xi$ for all $a \in A$.

If $b \in B$, then from the definition of a cut $a \leq b$ for all $a \in A$. Hence b is an upper bound of A and so $\xi \leq b$. Thus the existence of a number with the stated properties is demonstrated.

To establish the uniqueness of ξ, let $\eta \in \mathbf{R}$ be such that $a \leq \eta$ for all $a \in A$ and $\eta \leq b$ for all $b \in B$. It follows that η is an upper bound of A; hence $\xi \leq \eta$. If $\xi < \eta$, then there exists a number $\zeta = (\xi + \eta)/2$ such that $\xi < \zeta < \eta$. Now either $\zeta \in A$ or $\zeta \in B$. If $\zeta \in A$, we have a contradiction of the fact that $a \leq \xi$ for all $a \in A$. If $\zeta \in B$, we have a contradiction of the fact that $\eta \leq b$ for all $b \in B$. Therefore we must have $\xi = \eta$.   Q.E.D.

Actually, what Dedekind did was, in essence, to define a real number to be a cut in the rational number system. This procedure enables one "to construct" the real number system $\mathbf{R}$ from the set $\mathbf{Q}$ of rational numbers.

Cells and Intervals

If $a \in \mathbf{R}$, then the sets

$$\{x \in \mathbf{R} : x < a\}, \qquad \{x \in \mathbf{R} : x > a\}$$

are called the **open rays** determined by a. Similarly, the sets

$$\{x \in \mathbf{R} : x \leq a\}, \qquad \{x \in \mathbf{R} : x \geq a\}$$

are called the **closed rays** determined by a. The point a is called the **end point** of these rays.   These sets are often denoted by the notations

$$(-\infty, a), \qquad (a, +\infty), \qquad (-\infty, a], \qquad [a, +\infty),$$

respectively; here $-\infty$ and $+\infty$ are merely symbols and are *not* to be considered elements of $\mathbf{R}$.

If $a, b \in \mathbf{R}$ and $a \leq b$, then the set

$$\{x \in \mathbf{R} : a < x < b\}$$

is called the **open cell** determined by a and b and is often denoted by (a, b). The set

$$\{x \in \mathbf{R} : a \leq x \leq b\}$$

is called the **closed cell** determined by a and b and is denoted by $[a, b]$. The sets

$$\{x \in \mathbf{R} : a \leq x < b\}, \qquad \{x \in \mathbf{R} : a < x \leq b\}$$

are called the **half-open** (or **half-closed**) **cells** determined by a and b and are denoted by

$$[a, b), \qquad (a, b],$$

respectively. The points a, b are called the **end points** of these cells.

By an **interval** in R, we mean either a ray, or a cell, or all of R. Thus there are ten different kinds of intervals in R; namely,

$$\emptyset, \qquad (-\infty, a), \quad (-\infty, a], \qquad [a, b], \quad [a, b),$$

$$(a, b], \qquad (a, b), \qquad [b, +\infty), \qquad (b, +\infty), \qquad R,$$

where a, $b \in R$ and $a < b$. Five of these intervals are bounded. Two are bounded above but not below, and two are bounded below but not above.

The **unit cell** (or the **unit interval**) is the set $[0, 1] = \{x \in R : 0 \le x \le 1\}$. It will be denoted by the standard notion I.

We shall say that a sequence of intervals I_n, $n \in N$, is **nested** in case the chain

$$I_1 \supseteq I_2 \supseteq I_3 \supseteq \cdots \supseteq I_n \supseteq I_{n+1} \supseteq \cdots$$

of inclusions holds. It is important to notice that a nested sequence of intervals does not need to have a common point. Indeed, it is an exercise to show that if $I_n = (n, +\infty)$, $n \in N$, then the sequence of intervals obtained is nested but has no common point. Similarly, if $J_n = (0, 1/n)$, $n \in N$, then the sequence is nested but has no common point.

However, it is a very important property of R that every nested sequence of *closed cells* has a common point. We shall now prove that fact.

7.3 NESTED CELLS PROPERTY. *If $n \in N$, let I_n be a non-void closed cell in R and suppose that this sequence is nested in the sense that*

$$I_1 \supseteq I_2 \supseteq \cdots \supseteq I_n \supseteq \cdots.$$

Then there exists an element which belongs to all of these cells.

PROOF. Suppose that $I_n = [a_n, b_n]$, where $a_n \le b_n$ for all $n \in N$. We note that $I_n \subseteq I_1$ for all n, hence $a_n \le b_1$ for all n. Hence the set $\{a_n : n \in N\}$ is bounded above. We let ξ be its supremum; hence $a_n \le \xi$ for all n.

We claim that $\xi \le b_n$ for all $n \in N$. If not, there exists some $m \in N$ such that $b_m < \xi$. Since ξ is the supremum of $\{a_n : n \in N\}$ there must exist a_p such that $b_m < a_p$. Now let q be the larger of the natural numbers m and p. Since $a_1 \le a_2 \le \cdots \le a_n \le \cdots$ and $b_1 \ge b_2 \ge \cdots \ge b_n \ge \cdots$ we infer that $b_q \le b_m < a_p \le a_q$. But this implies that $b_q < a_q$, contrary to the assumption that $I_q = [a_q, b_q]$ is a *non-void* closed cell. Therefore $\xi \le b_n$ for all $n \in N$. Since $a_n \le \xi \le b_n$, we infer that $\xi \in I_n = [a_n, b_n]$ for all $n \in N$.

Q.E.D.

We note that, under the hypotheses of 7.3, there may be more than one common element. In fact, if we let $\eta = \inf\{b_n : n \in N\}$, it is an exercise to show that

$$[\xi, \eta] = \bigcap_{n \in N} I_n.$$

The Cantor Set

We shall now introduce a subset of the unit cell I which is of considerable interest and is frequently useful in constructing examples and counter-examples. We shall denote this set by F and refer to it as the **Cantor set** (although it is also sometimes called **Cantor's ternary set** or the **Cantor discontinuum**).

One way of describing F is as the set of real numbers in I which have a ternary (= base 3) expansion using only the digits 0, 2. However, we choose to define it in different terms. In a sense that will be made more precise, F consists of those points in I that remain after "middle third" intervals have been successively removed.

To be more explicit: if we remove the open middle third of I, we obtain the set

$$F_1 = [0, \tfrac{1}{3}] \cup [\tfrac{2}{3}, 1].$$

If we remove the open middle third of each of the two closed intervals in F_1, we obtain the set

$$F_2 = [0, \tfrac{1}{9}] \cup [\tfrac{2}{9}, \tfrac{1}{3}] \cup [\tfrac{2}{3}, \tfrac{7}{9}] \cup [\tfrac{8}{9}, 1].$$

Hence F_2 is the union of $4(= 2^2)$ closed intervals all of which are of the form $[k/3^2, (k+1)/3^2]$. We now obtain the set F_3 by removing the open middle third of each of these sets. In general, if F_n has been constructed and consists of the union of 2^n intervals of the form $[k/3^n, (k+1)/3^n]$, then we obtain F_{n+1} by removing the open middle third of each of these intervals. The Cantor set is what remains after this process has been carried out for each n in N.

7.4 DEFINITION. The Cantor set F is the intersection of the sets F_n, $n \in N$, obtained by successive removal of open middle thirds.

At first glance, it may appear that every point is ultimately removed by this process. However, this is evidently not the case since the points $0, \tfrac{1}{3}, \tfrac{2}{3}, 1$ belong to all the sets F_n, $n \in N$, and hence to the Cantor set F. In fact, it is easily seen that there are an infinite number of points in F, even though F is relatively thin in some other respects. Indeed, it is not difficult to show that there are a non-denumerable number of elements of F and that the points of F can be put into one-one correspondence with the points of I. Hence the set F contains a large number of elements.

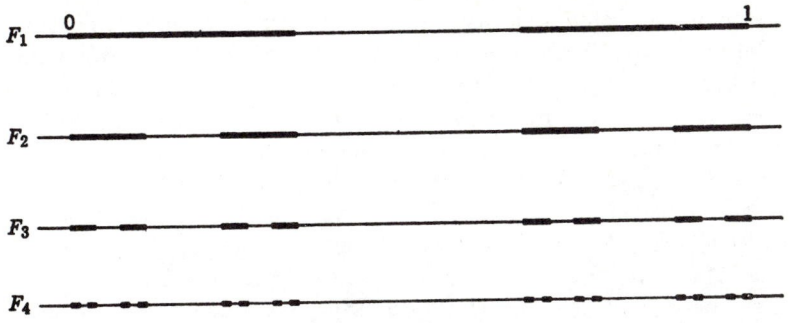

Figure 7.2. The Cantor set.

We now give two senses in which **F** is "thin." First we observe that **F** does not contain any non-void interval. For if x belongs to **F** and (a, b) is an open interval containing x, then (a, b) contains some middle thirds that were removed to obtain **F**. (Why?) Hence (a, b) is not a subset of the Cantor set, but contains infinitely many points in its complement $\mathscr{C}(\textbf{F})$.

A second sense in which **F** is thin refers to "length." While it is not possible to define length for arbitrary subsets of **R**, it is easy to convince oneself that **F** cannot have positive length. For, the length of F_1 is $\frac{2}{3}$, that of F_2 is $\frac{4}{9}$, and, in general, the length of F_n is $(\frac{2}{3})^n$. Since **F** is a subset of F_n, it cannot have length exceeding that of F_n. Since this must be true for each n in **N**, we conclude that **F**, although uncountable, cannot have positive length.

As strange as the Cantor set may seem, it is relatively well-behaved in many respects. It provides us with a bit of insight into how complicated subsets of **R** can be and how little our intuition guides us. It also serves as a test for the concepts that we will introduce in later sections and whose import are not fully grasped in terms of intervals and other very elementary subsets.

Models for **R**

In Sections 4–6, we have introduced **R** axiomatically in the sense that we have listed some properties that we assume it to have. This approach raises the question as to whether such a set actually exists and to what extent it is uniquely determined. While we shall not settle these questions, a few remarks about them is certainly appropriate.

The existence of a set which is a complete ordered field can be demonstrated by actual construction. If one feels sufficiently familiar with the rational field **Q**, one can define real numbers to be special subsets of **Q** and define addition, multiplication, and order relations between these subsets in such a way as to obtain a complete ordered field. There are two standard procedures that are used in doing this: one is Dedekind's method of "cuts" which is discussed in the book of Rudin that is cited in the References. The second way is Cantor's method of "Cauchy sequences" which is discussed in the book of Hamilton and Landin.

In the last paragraph we have asserted that it is possible to construct a model of R from Q (in at least two different ways). It is also possible to construct a model of R from the set N of natural numbers and this is often taken as the starting point by those who, like Kronecker,[†] regard the natural numbers as given by God. However, since even the set of natural numbers has its subtleties (such as the Well-ordering Property), we feel that the most satisfactory procedure is to go through the process of first constructing the set N from primitive set theoretic concepts, then developing the set Z of integers, next constructing the field Q of rationals, and finally the set R. This procedure is not particularly difficult to follow and it is edifying; however, it is rather lengthy. Since it is presented in detail in the book of Hamilton and Landin, it will not be given here.

From the remarks already made, it is clear that complete ordered fields can be constructed in different ways. Thus we cannot say that there is a *unique* complete ordered field. In a sense, all of the methods of construction suggested above lead to complete ordered fields that are "isomorphic." (This means that if R_1 and R_2 are complete ordered fields obtained by these constructions, then there exists a one-one mapping φ of R_1 onto R_2 such that (i) φ sends a rational element of R_1 into the corresponding rational element of R_2, (ii) φ sends $a + b$ into $\varphi(a) + \varphi(b)$, (iii) φ sends ab into $\varphi(a)\varphi(b)$, and (iv) φ sends a positive element of R_1 into a positive element of R_2.) Within naive set theory, we can provide an argument showing that any two complete ordered fields are isomorphic in the sense described. Whether this argument can be formalized within a given system of logic depends on the rules of inference employed in the system. Thus the question of the extent to which the real number system can be regarded as being uniquely determined is a rather delicate logical issue. However, for our purposes this uniqueness (or lack of it) is not important, for we can choose any particular complete ordered field as our model for the real number system.

Exercises

7.A. If (A, B) is a cut in R, show that sup $A = $ inf B.

7.B. If the cuts (A, B) and (A', B') determine the real numbers ξ and ξ', respectively, show that $\xi < \xi'$ implies that $A \subseteq A'$, $A \neq A'$.

7.C. Is the converse of the preceding exercise true?

7.D. Let $A = \{x \in R : x \le 0 \text{ or } x^2 \le 2\}$ and $B = \{x \in R : x > 0 \text{ and } x^2 > 2\}$. Show that (A, B) is a cut in R.

---

† LEOPOLD KRONECKER (1823–1891) studied with Dirichlet in Berlin and Kummer in Bonn. After making a fortune before he was thirty, he returned to mathematics. He is known for his work in algebra and number theory and for his personal opposition to the ideas of Cantor on set theory.

7.E. Let $I_n = (n, +\infty)$ for $n \in N$. Show that the sequence of intervals is nested, but that there is no common point.

7.F. Let $J_n = (0, 1/n)$ for $n \in N$. Show that this sequence of intervals is nested, but that there is no common point.

7.G. If $I_n = [a_n, b_n]$, $n \in N$, is a nested sequence of closed cells, show that

$$a_1 \leq a_2 \leq \cdots \leq a_n \leq \cdots \leq b_m \leq \cdots \leq b_2 \leq b_1.$$

If we put $\xi = \sup \{a_n : n \in N\}$ and $\eta = \inf \{b_m : m \in N\}$, show that $[\xi, \eta] = \bigcap_{n \in N} I_n$.

7.H. Show that every number in the Cantor set has a ternary (= base 3) expansion using only the digits 0, 2.

7.I. Show that the collection of "right hand" end points in F is denumerable. Show that if all these end points are deleted from F, then what remains can be put onto one-one correspondence with *all* of $[0, 1)$. Conclude that the set F is not countable.

7.J. Every open interval (a, b) which contains a point of F also contains an entire "middle third" set which belongs to $\mathscr{C}(F)$. Hence F does not contain any non-void open interval.

7.K. By removing sets with ever decreasing length, show that we can construct a "Cantor-like" set which has positive length. How large can we make the length of this set?

7.L. Show that F is not the union of a countable collection of closed intervals.

II
THE TOPOLOGY
OF CARTESIAN SPACES

The sections of Chapter I were devoted to developing the algebraic properties, the order properties, and the completeness property of the real number system. Considerable use of these properties will be made in this and later chapters.

Although it would be possible to turn immediately to a discussion of sequences of real numbers and continuous real functions, we prefer to delay the study of these topics a little longer. Indeed, we shall insert here the definitions of a vector space, a normed space, and an inner product space. We do so because these notions are easily grasped and because such spaces arise throughout all of analysis (to say nothing of its applications to geometry, physics, engineering, economics, etc.) Of course, the Cartesian spaces R^p will be of especial interest to us. Fortunately, our intuition for R^2 and R^3 usually carries over without much change to the space R^p, and a knowledge of these spaces is of help in analyzing more general spaces.

Section 8   Vector and Cartesian Spaces

A "vector space" is a set in which one can add two elements, and can multiply an element by a real number, in such a way that certain familiar properties hold. We shall now be more precise.

8.1   DEFINITION. A **vector space** is a set V (whose elements are called **vectors**) equipped with two binary operations, called **vector addition** and **scalar multiplication.**

If $x, y \in V$ there is an element $x + y$ in V, called the **vector sum** of x and y. This vector addition operation satisfies the following properties:

(A1) $x + y = y + x$ for all x, y in V;

(A2) $(x + y) + z = x + (y + z)$ for all x, y, z in V;

(A3) there exists an element 0 in V such that $0 + x = x$ and $x + 0 = x$ for all x in V;

(A4) given x in V there is an element $-x$ in V such that $x + (-x) = 0$ and $(-x) + x = 0$.

If $a \in R$ and $x \in V$ there is an element ax in V, called the **multiple** of a and x. This scalar multiplication operation satisfies the following properties:

(M1) $1x = x$ for all $x \in V$;

(M2) $a(bx) = (ab)x$ for all $a, b \in R$ and $x \in V$;

(D) $a(x + y) = ax + ay$ and $(a + b)x = ax + bx$ for all real $a, b \in R$ and $x, y \in V$.

We shall now give some elementary, but important, examples of vector spaces.

8.2 EXAMPLES. (a) The real number system is a vector space where the addition and scalar multiplication operations are the usual addition and multiplication of real numbers.

(b) Let R^2 denote the Cartesian product $R \times R$. Hence R^2 consists of all ordered pairs (x_1, x_2) of real numbers. If we define vector addition and scalar multiplication by

$$(x_1, x_2) + (y_1, y_2) = (x_1 + y_1, x_2 + y_2),$$

$$a(x_1, x_2) = (ax_1, ax_2),$$

then it can readily be checked that the properties in Definition 8.1 are satisfied. [Here $0 = (0, 0)$ and $-(x_1, x_2) = (-x_1, -x_2)$.] Hence R^2 is a vector space under these operations.

(c) Let $p \in N$ and let R^p denote the collection of all ordered "p-tuples"

$$(x_1, x_2, \ldots, x_p)$$

with $x_i \in R$ for $i = 1, \ldots, p$. If we define vector addition and scalar multiplication by

$$(x_1, x_2, \ldots, x_p) + (y_1, y_2, \ldots, y_p) = (x_1 + y_1, x_2 + y_2, \ldots, x_p + y_p)$$

$$a(x_1, x_2, \ldots, x_p) = (ax_1, ax_2, \ldots, ax_p),$$

then it can readily be checked that R^p is a vector space under these operations. [Here it is seen that $0 = (0, 0, \ldots, 0)$ and $-(x_1, x_2, \ldots, x_p) = (-x_1, -x_2, \ldots, -x_p)$.]

(d) Let S be any set and let R^S denote the collection of all functions u with domain S and range in R. (Hence R^S is the collection of all real-valued functions defined in S.) If we define $u + v$ and au by

$$(u + v)(s) = u(s) + u(s),$$

$$(au)(s) = au(s),$$

for all $s \in S$, then it can readily be checked that $\boldsymbol{R}^S$ is a vector space under these operations. [Here 0 is the function identically equal to zero, and $-u$ is the function whose value at $s \in S$ is $-u(s)$.]

In later sections, we shall encounter many other vector spaces.

Generally we shall write $x - y$ instead of $x + (-y)$.

Inner Products and Norms

The reader will note that the scalar multiplication in a vector space V is a function with domain $\boldsymbol{R} \times V$ and range V. Many vector spaces are also equipped with a function with domain $V \times V$ and range $\boldsymbol{R}$ that is of importance.

8.3 DEFINITION. If V is a vector space, then an **inner product** (or **dot product**) is a function on $V \times V$ to $\boldsymbol{R}$, denoted by $(x, y) \mapsto x \cdot y$, satisfying the properties:

(i) $x \cdot x \geq 0$ for all $x \in V$;

(ii) $x \cdot x = 0$ if and only if $x = 0$;

(iii) $x \cdot y = y \cdot x$ for all $x, y \in V$;

(iv) $x \cdot (y + z) = x \cdot y + x \cdot z$ and $(x + y) \cdot z = x \cdot z + y \cdot z$ for all x, y, $z \in V$;

(v) $(ax) \cdot y = a(x \cdot y) = x \cdot (ay)$ for all $a \in \boldsymbol{R}$, and $x, y \in V$.

A vector space in which an inner product has been defined is called an **inner product space.**

It is possible for different inner products to be defined in the same vector space (cf. Exercise 8.D).

8.4 EXAMPLES. (a) The ordinary multiplication in $\boldsymbol{R}$ satisfies the above properties, so $\boldsymbol{R}$ is an inner product space.

(b) In $\boldsymbol{R}^2$, we define

$$(x_1, x_2) \cdot (y_1, y_2) = x_1 y_1 + x_2 y_2.$$

It is easy to check that this defines an inner product on $\boldsymbol{R}^2$.

(c) In $\boldsymbol{R}^p$, we define

$$(x_1, x_2, \ldots, x_p) \cdot (y_1, y_2, \ldots, y_p) = x_1 y_1 + x_2 y_2 + \cdots + x_p y_p.$$

It is easy to check that this defines an inner product on $\boldsymbol{R}^p$.

8.5 DEFINITION. If V is a vector space, then a **norm** on V is a function on V to $\boldsymbol{R}$ denoted by $x \mapsto \|x\|$ satisfying the properties:

(i) $\|x\| \geq 0$ for all $x \in V$;

(ii) $\|x\| = 0$ if and only if $x = 0$;

(iii) $\|ax\| = |a| \|x\|$ for all $a \in \boldsymbol{R}$, $x \in V$;

(iv) $\|x + y\| \leq \|x\| + \|y\|$ for all $x, y \in V$.

A vector space in which a norm has been defined is called a **normed space.**

As we will see in the exercises, the same vector space can have several interesting norms.

8.6 EXAMPLES. (a) The absolute value function on $\mathbf{R}$ satisfies the properties in 8.5.

(b) In $\mathbf{R}^2$, we define

$$\|(x_1, x_2)\| = (x_1{}^2 + x_2{}^2)^{1/2}.$$

Properties (i), (ii), and (iii) are very easily checked. Property (iv) is a bit more complicated.

(c) In $\mathbf{R}^p$, we define

$$\|(x_1, x_2, \ldots, x_p)\| = (x_1{}^2 + x_2{}^2 + \cdots + x_p{}^2)^{1/2}.$$

Again, properties (i), (ii), and (iii) are easy.

We shall now give a theorem which asserts that an inner product can always be used to define a norm in a very natural way.

8.7 THEOREM. *Let V be an inner product space and define $\|x\|$ by*

$$\|x\| = \sqrt{x \cdot x} \qquad for \quad x \in V.$$

Then $x \mapsto \|x\|$ is a norm on V and satisfies the property that

(*) $$x \cdot y \leq \|x\| \, \|y\|.$$

Moreover, if x and y are non-zero, then the equality holds in (*) if and only if there is some strictly positive real number c such that $x = cy$.

PROOF   Since $x \cdot x \geq 0$ for all $x \in V$, then the square root of $x \cdot x$ exists, so $\|x\|$ is well-defined. The first three properties of the norm are direct consequences of 8.3(i), (ii), and (v). To prove (*), let $a, b \in \mathbf{R}$, $x, y \in V$, and let $z = ax - by$. If we use properties 8.3(i), (iii), (iv), and (v), we get

$$0 \leq z \cdot z = a^2 \, x \cdot x - 2ab \, x \cdot y + b^2 \, y \cdot y.$$

Now take $a = \|y\|$ and $b = \|x\|$, to get

$$0 \leq \|y\|^2 \, \|x\|^2 - 2 \, \|y\| \, \|x\| \, x \cdot y + \|x\|^2 \, \|y\|^2$$
$$= 2 \, \|x\| \, \|y\| \, (\|x\| \, \|y\| - x \cdot y).$$

Hence the inequality (*) holds.

If $x = cy$ with $c > 0$, then $\|x\| = c \, \|y\|$ and so

$$x \cdot y = (cy) \cdot y = c(y \cdot y) = c \, \|y\|^2$$
$$= \|x\| \, \|y\|$$

so that equality holds in (*). Conversely, if $x \cdot y = \|x\| \, \|y\|$ the calculation in the preceding paragraph shows that $z = \|y\| \, x - \|x\| \, y$ has the property that

$z \cdot z = 0$. Therefore $z = 0$ and, since x and y are non-zero vectors, we can take $c = \|x\|/\|y\|$. To establish 8.5(iv), we use $(*)$ to show that

$$
\begin{aligned}
\|x + y\|^2 &= (x + y) \cdot (x + y) \\
&= x \cdot x + x \cdot y + y \cdot x + y \cdot y \\
&= \|x\|^2 + 2(x \cdot y) + \|y\|^2 \\
&\le \|x\|^2 + 2\,\|x\|\,\|y\| + \|y\|^2 \\
&\le (\|x\| + \|y\|)^2,
\end{aligned}
$$

whence it follows that $\|x + y\| \le \|x\| + \|y\|$ for all $x, y \in V$.   Q.E.D.

We leave the proof of the following corollary as an exercise.

8.8   COROLLARY.   *If x, y are elements of V, then*

$(**)$ $|x \cdot y| \le \|x\|\,\|y\|$.

Moreover, if $y \ne 0$ then the equality can hold in $(**)$ if and only if there is a real number c such that $x = cy$.

Both of the inequalities $(*)$ and $(**)$ are called the **Schwarz Inequality,** or the **Cauchy-Bunyakovskii-Schwarz Inequality.**[†] They will be frequently used. The inequality 8.5(iv) is called the **Triangle Inequality.** We leave it to the reader to show that

$$
\big|\, \|x\| - \|y\| \,\big| \le \|x \pm y\| \le \|x\| + \|y\|,
$$

for any x, y in a normed space.

The Cartesian Space R^p

By the **p-dimensional real Cartesian space** we mean the set R^p equipped with the vector addition and scalar multiplication defined in Example 8.2(c), and the inner product defined in Example 8.4(c). As we have seen, this inner product induces the norm

$$
\|(x_1, x_2, \ldots, x_p)\| = \sqrt{x_1^{\,2} + x_2^{\,2} + \cdots + x_p^{\,2}}.
$$

[†] AUGUSTIN-LOUIS CAUCHY (1789–1857) was the founder of modern analysis but also made profound contributions to other mathematical areas. He served as an engineer under Napoleon, followed Charles X into self-imposed exile, and was excluded from his position at the Collège de France during the years of the July monarchy because he would not take a loyalty oath. Despite his political and religious activities, he found time to write 789 mathematical papers.

VICTOR BUNYAKOVSKIĬ (1804–1889), a professor at St. Petersburg, established a generalization of the Cauchy Inequality for integrals in 1859. His contribution was overlooked by western writers and was later discovered independently by Schwarz.

HERMANN AMANDUS SCHWARZ (1843–1921) was a student and successor of Weierstrass at Berlin. He made numerous contributions, especially to complex analysis.

The real numbers $x_1, x_2, \ldots, x_p$ are called the **first, second, ..., p-th coordinates** (or **components**) of the vector $x = (x_1, x_2, \ldots, x_p)$.

In $\mathbf{R}^p$, the real number $\|x\|$ can be thought of either as the "length" of x or as the distance from x to 0. More generally, we think of $\|x - y\|$ as the distance from x to y. With this interpretation, property 8.5(ii) asserts that the distance from x to y is zero if and only if $x = y$. Property 8.5(iii) with $a = -1$ asserts that $\|x - y\| = \|y - x\|$, which means that the distance from x to y is equal to the distance from y to x. The Triangle Inequality implies that

$$\|x - y\| \le \|x - z\| + \|z - y\|,$$

which means that the distance from x to y is no greater than the sum of the distance from x to z and the distance from z to y.

8.9 DEFINITION. Let $x \in \mathbf{R}^p$ and let $r > 0$. Then the set $\{y \in \mathbf{R}^p : \|x - y\| < r\}$ is called the **open ball** with **center** x and **radius** r. The set $\{y \in \mathbf{R}^p : \|x - y\| \le r\}$ is called the **closed ball** with **center** x and **radius** r. The set $\{y \in \mathbf{R}^p : \|x - y\| = r\}$ is called the **sphere** with **center** x and **radius** r.

The notion of a ball depends on the norm. It will be seen in the exercises that some balls are not very "round."

It is often convenient to have relations between the norm of a vector in $\mathbf{R}^p$ and the magnitude of its components.

8.10 THEOREM. *If $x = (x_1, x_2, \ldots, x_p)$ is any element of $\mathbf{R}^p$, then*

$$|x_i| \le \|x\| \le \sqrt{p} \sup \{|x_1|, |x_2|, \ldots, |x_p|\}.$$

PROOF. Since $\|x\|^2 = x_1^2 + x_2^2 + \cdots + x_p^2$, it is plain that $|x_i| \le \|x\|$ for all i. Similarly, if $M = \sup \{|x_1|, |x_2|, \ldots, |x_p|\}$, then $\|x\|^2 \le pM^2$, so that $\|x\| \le \sqrt{p} M$. Q.E.D.

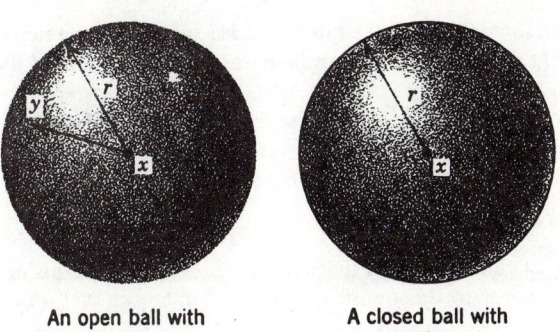

An open ball with
center x.

A closed ball with
center x.

Figure 8.1

The inequality just established asserts, in a quantitative fashion, that if the norm of x is small, then the lengths of its components are small, and conversely.

Exercises

8.A. If V is a vector space and if $x + z = x$ for some x and z in V, show that $z = 0$. Hence the zero element in V is unique.

8.B. If $x + y = 0$ for some x and y in V, show that $y = -x$.

8.C. Let $S = \{1, 2, \ldots, p\}$, for some $p \in N$. Show that the vector space R^S is "essentially the same" as the space R^p.

8.D. If w_1 and w_2 are strictly positive, show that the definition

$$(x_1, x_2) \cdot (y_1, y_2) = x_1 y_1 w_1 + x_2 y_2 w_2,$$

yields an inner product on R^2. Generalize this to R^p.

8.E. The definition

$$(x_1, x_2) \cdot (y_1, y_2) = x_1 y_1$$

is *not* an inner product on R^2. Why?

8.F. If $x = (x_1, x_2, \ldots, x_p) \in R^p$, define $\|x\|_1$ by

$$\|x\|_1 = |x_1| + |x_2| + \cdots + |x_p|.$$

Prove that $x \mapsto \|x\|_1$ is a norm on R^p.

8.G. If $x = (x_1, x_2, \ldots, x_p) \in R^p$, define $\|x\|_\infty$ by

$$\|x\|_\infty = \sup \{|x_1|, |x_2|, \ldots, |x_p|\}.$$

Prove that $x \mapsto \|x\|_\infty$ is a norm on R^p.

8.H. In the set R^2, describe the sets

$$S_1 = \{x \in R^2 : \|x\|_1 < 1\}, \qquad S_\infty = \{x \in R^2 : \|x\|_\infty < 1\}.$$

8.I. If $x, y \in R^p$, the norm defined in 8.4(c) satisfies the **Parallelogram Identity:**

$$\|x + y\|^2 + \|x - y\|^2 = 2(\|x\|^2 + \|y\|^2).$$

Prove this and show that it can be interpreted as saying that the sum of the squares of the lengths of the four sides of a parallelogram equals the sum of the squares of the diagonals.

8.J. Show that the norms defined in Exercises 8.F and 8.G do not satisfy the Parallelogram Identity.

8.K. Show that there exist positive constants a, b such that

$$a \|x\|_1 \leq \|x\| \leq b \|x\|_1 \qquad \text{for all} \quad x \in R^p.$$

Find the largest constant a and the smallest constant b with this property.

8.L. Show that there exist positive constants a, b such that

$$a \|x\|_1 \leq \|x\|_\infty \leq b \|x\|_1 \qquad \text{for all} \quad x \in R^p.$$

Find the largest constant a and the smallest constant b with this property.

8.M. If x, y belong to $\boldsymbol{R}^p$, is it true that

$$|x \cdot y| \leq \|x\|_1 \|y\|_1 \quad \text{and} \quad |x \cdot y| \leq \|x\|_\infty \|y\|_\infty \, ?$$

8.N. If x, y belong to $\boldsymbol{R}^p$, then is it true that the relation

$$\|x + y\| = \|x\| + \|y\|$$

holds if and only if $x = cy$ or $y = cx$ with $c \geq 0$?

8.O. Let x, y belong to $\boldsymbol{R}^p$, then is it true that the relation

$$\|x + y\|_\infty = \|x\|_\infty + \|y\|_\infty$$

holds if and only if $x = cy$ or $y = cx$ with $c \geq 0$?

8.P. If x, y belongs to $\boldsymbol{R}^p$, then

$$\|x + y\|^2 = \|x\|^2 + \|y\|^2$$

holds if and only if $x \cdot y = 0$. In this case, one says that x and y are **orthogonal** or **perpendicular.**

8.Q. A subset K of $\boldsymbol{R}^p$ is said to be **convex** if, whenever x, y belong to K and t is a real number such that $0 \leq t \leq 1$, then the point

$$(1 - t)x + ty = x + t(y - x)$$

also belongs to K. Interpret this condition geometrically and show that the subsets

$$K_1 = \{x \in \boldsymbol{R}^2 : \|x\| \leq 1\},$$
$$K_2 = \{(\xi, \eta) \in \boldsymbol{R}^2 : 0 < \xi < \eta\},$$
$$K_3 = \{(\xi, \eta) \in \boldsymbol{R}^2 : 0 \leq \eta \leq \xi \leq 1\},$$

are convex but that the subset

$$K_4 = \{x \in \boldsymbol{R}^2 : \|x\| = 1\}$$

is not convex.

8.R. The intersection of any collection of convex subsets of $\boldsymbol{R}^p$ is convex. The union of two convex subsets of $\boldsymbol{R}^p$ may not be convex.

8.S. If M is any set, then a function $d : M \times M \to \boldsymbol{R}$ is called a **metric** on M if it satisfies:

(i) $d(x, y) \geq 0$ for all x, y in M;
(ii) $d(x, y) = 0$ if and only if $x = y$;
(iii) $d(x, y) = d(y, x)$ for all x, y in M;
(iv) $d(x, y) \leq d(x, z) + d(z, y)$ for all x, y, z in M.

Show that if $x \mapsto \|x\|$ is any norm on a vector space V and if we define d by $d(x, y) = \|x - y\|$, for x, $y \in V$, then d is a metric on V.

8.T. Suppose that d is a metric on a set M. By using Definition 8.9 as a model, define an open ball with center $x \in M$ and radius r. Interpret the sets S_1 and S_∞ in Exercise 8.H as open balls in R^2 with respect to two different metrics. Interpret Exercise 8.K as saying that a ball with center 0 relative to the metric d_2 (derived from the norm in 8.6(b)) contains and is contained in balls with center 0 relative to the metric d_1 derived from $\| \ \|_1$. Make similar interpretations of Exercise 8.L and Theorem 8.10.

8.U. Let M be any set and let d be defined on $M \times M$ by the requirement that

$$d(x, y) = \begin{cases} 0 & \text{if} & x = y, \\ 1 & \text{if} & x \neq y. \end{cases}$$

Show that d gives a metric on M in the sense defined in Exercise 8.S. If x is any point in M, then the open ball with center x and radius 1 (relative to the metric d) consists of precisely one point. However, the open ball with center x and radius 2 (relative to d) consists of all of M. This metric d, is sometimes called the **discrete metric** on the set M.

Projects

8.α. In this project we develop some important inequalities.

(a) Let a and b be positive real numbers. Show that

$$ab \leq (a^2 + b^2)/2,$$

and that the equality holds if and only if $a = b$. (Hint: consider $(a - b)^2$.)

(b) Let a_1 and a_2 be positive real numbers. Show that

$$\sqrt{a_1 a_2} \leq (a_1 + a_2)/2$$

and that the equality holds if and only if $a_1 = a_2$.

(c) Let $a_1, a_2, \ldots, a_m$ be $m = 2^n$ positive real numbers. Show that

$$(*) \qquad (a_1 a_2 \cdots a_m)^{1/m} \leq (a_1 + a_2 + \cdots + a_m)/m$$

and that the equality holds if and only if $a_1 = \cdots = a_m$.

(d) Show that the inequality $(*)$ between the geometric mean and the arithmetic mean holds even when m is not a power of 2. (Hint: if $2^{n-1} < m < 2^n$, let $b_j = a_j$ for $j = 1, \ldots, m$ and let

$$b_j = (a_1 + a_2 + \cdots + a_m)/m$$

for $j = m + 1, \ldots, 2^n$. Now apply part (c) to the numbers $b_1, b_2, \ldots, b_{2^n}$.)

(e) Let $a_1, a_2, \ldots, a_n$ and $b_1, b_2, \ldots, b_n$ be two sets of real numbers. Prove **Lagrange's Identity**†

$$\left\{ \sum_{j=1}^{n} a_j b_j \right\}^2 = \left\{ \sum_{j=1}^{n} a_j^2 \right\}\left\{ \sum_{k=1}^{n} b_k^2 \right\} - \frac{1}{2} \sum_{j,k=1}^{n} (a_j b_k - a_k b_j)^2.$$

(Hint: experiment with the cases $n = 2$ and $n = 3$ first.)

(f) Use part (e) to establish **Cauchy's Inequality**

$$\left\{ \sum_{j=1}^{n} a_j b_j \right\}^2 \leq \left\{ \sum_{j=1}^{n} a_j^2 \right\}\left\{ \sum_{k=1}^{n} b_k^2 \right\}.$$

Show that the equality holds if and only if the ordered sets $(a_1, a_2, \ldots, a_n)$ and $(b_1, b_2, \ldots, b_n)$ are proportional.

† JOSEPH-LOUIS LAGRANGE (1736–1813) was born in Turin, where he become professor at the age of nineteen. He later went to Berlin for twenty years as successor to Euler and then to Paris. He is best known for his work on the calculus of variations and analytical mechanics.

(g) Use part (f) and establish the **Triangle Inequality**

$$\left\{\sum_{j=1}^{n} (a_j + b_j)^2\right\}^{1/2} \le \left\{\sum_{j=1}^{n} a_j^2\right\}^{1/2} + \left\{\sum_{j=1}^{n} b_j^2\right\}^{1/2}.$$

8.β. In this project, let $\{a_1, a_2, \ldots, a_n\}$, and so forth, be sets of n positive real numbers.

(a) It can be proved (for example, by using the Mean Value Theorem) that if a and b are positive and $0 < \alpha < 1$, then

$$a^\alpha b^{1-\alpha} \le \alpha a + (1-\alpha)b$$

and that the equality holds if and only if $a = b$. Assuming this, let $r > 1$ and let s satisfy

$$\frac{1}{r} + \frac{1}{s} = 1,$$

(so that $s > 1$ and $r + s = rs$). Show that if A and B are positive, then

$$AB \le \frac{A^r}{r} + \frac{B^s}{s},$$

and that the equality holds if and only if $A^r = B^s$.

(b) Let $\{a_1, \ldots, a_n\}$ and $\{b_1, \ldots, b_n\}$ be positive real numbers. If $r, s > 1$ and $(1/r) + (1/s) = 1$, establish **Hölder's Inequality**†

$$\sum_{j=1}^{n} a_j b_j \le \left\{\sum_{j=1}^{n} a_j^r\right\}^{1/r} \left\{\sum_{j=1}^{n} b_j^s\right\}^{1/s}.$$

(Hint: Let $A = \{\sum a_j^r\}^{1/r}$ and $B = \{\sum b_j^s\}^{1/s}$ and apply part (a) to a_j/A and b_j/B.)

(c) Using Hölder's Inequality, establish the **Minkowski Inequality**‡

$$\left\{\sum_{j=1}^{n} (a_j + b_j)^r\right\}^{1/r} \le \left\{\sum_{j=1}^{n} a_j^r\right\}^{1/r} + \left\{\sum_{j=1}^{n} b_j^r\right\}^{1/r}.$$

(Hint: $(a+b)^r = (a+b)(a+b)^{r/s} = a(a+b)^{r/s} + b(a+b)^{r/s}$.)

(d) Using Hölder's Inequality, prove that

$$(1/n) \sum_{j=1}^{n} a_j \le \left\{(1/n) \sum_{j=1}^{n} a_j^r\right\}^{1/r}.$$

(e) If $a_1 \le a_2$ and $b_1 \le b_2$, then $(a_1 - a_2)(b_1 - b_2) \ge 0$ and hence

$$a_1 b_1 + a_2 b_2 \ge a_1 b_2 + a_2 b_1.$$

Show that if $a_1 \le a_2 \le \cdots \le a_n$ and $b_1 \le b_2 \le \cdots \le b_n$, then

$$n \sum_{j=1}^{n} a_j b_j \ge \left\{\sum_{j=1}^{n} a_j\right\}\left\{\sum_{j=1}^{n} b_j\right\}.$$

† Otto Hölder (1859–1937) studied at Göttingen and taught at Leipzig. He worked in both algebra and analysis.

‡ Hermann Minkowski (1864–1909) was professor at Königsberg and Göttingen. He is best known for his work on convex sets and the "geometry of numbers."

(f) Suppose that $0 \le a_1 \le a_2 \le \cdots \le a_n$ and $0 \le b_1 \le b_2 \le \cdots \le b_n$ and $r \ge 1$. Establish the **Chebyshev Inequality**†

$$\left\{(1/n) \sum_{j=1}^{n} a_j^r\right\}^{1/r}\left\{(1/n) \sum_{j=1}^{n} b_j^r\right\}^{1/r} \le \left\{(1/n) \sum_{j=1}^{n} (a_j b_j)^r\right\}^{1/r}.$$

Show that this inequality must be reversed if $\{a_j\}$ is increasing and $\{b_j\}$ is decreasing.

Section 9   Open and Closed Sets

Many of the deepest properties of real analysis depend on certain topological notions. In the next few sections we shall introduce the basic concepts and derive some of the most crucial topological properties of the space $\boldsymbol{R}^p$. These results will be frequently used in the following chapters.

9.1 DEFINITION. A set G in $\boldsymbol{R}^p$ is said to be **open in $\boldsymbol{R}^p$** (or merely **open**) if, for each point x in G, there is a real number $r > 0$ such that every point y in $\boldsymbol{R}^p$ satisfying $\|x - y\| < r$ also belongs to the set G. (See Figure 9.1.)

By using Definition 8.9, we can rephrase this definition by saying that a set G is open if every point in G is the center of some open ball entirely contained in G.

9.2 EXAMPLES. (a) The entire set $\boldsymbol{R}^p$ is open, since we can take $r = 1$ for any x.

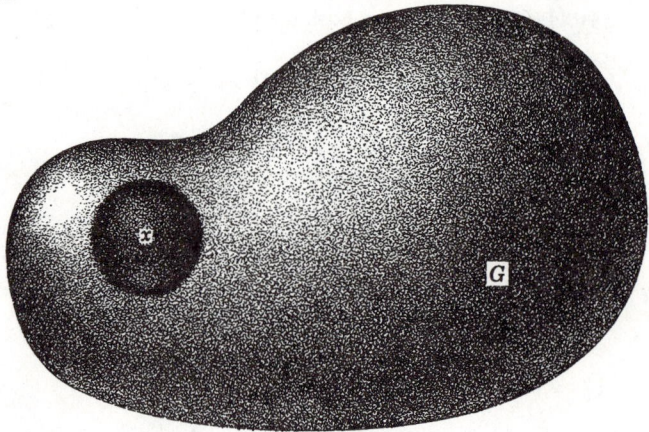

Figure 9.1.   An open set.

† PAFNUTI L. CHEBYSHEV (1821–1894) was a professor at St. Petersburg. He made many contributions to mathematics, but his most important work was in number theory, probability, and approximation theory.

(b) The set $G = \{x \in \mathbf{R} : 0 < x < 1\}$ is open in $\mathbf{R} = \mathbf{R}^1$. The set $F = \{x \in \mathbf{R} : 0 \le x \le 1\}$ is not open in $\mathbf{R}$. (Why?)

(c) The sets $G = \{(x, y) \in \mathbf{R}^2 : x^2 + y^2 < 1\}$ and $H = \{(x, y) : 0 < x^2 + y^2 < 1\}$ are open, but the set $F = \{(x, y) : x^2 + y^2 \le 1\}$ is not open in $\mathbf{R}^2$. (Why?)

(d) The set $G = \{(x, y) \in \mathbf{R}^2 : 0 < x < 1, \ y = 0\}$ is not open in $\mathbf{R}^2$. [Compare this with (b).] The set $H = \{(x, y) \in \mathbf{R}^2 : 0 < y < 1\}$ is open, but the set $K = \{(x, y) \in \mathbf{R}^2 : 0 \le y < 1\}$ is not open in $\mathbf{R}^2$.

(e) The set $G = \{(x, y, z) \in \mathbf{R}^3 : z > 0\}$ is open in $\mathbf{R}^3$ as is the set $H = \{(x, y, z) \in \mathbf{R}^3 : x > 0, \ y > 0, \ z > 0\}$. On the other hand, the set $F = \{(x, y, z) \in \mathbf{R}^3 : x = y = z\}$ is not open.

(f) The empty set $\emptyset$ is open in $\mathbf{R}^p$, since it contains no points at all, and hence the requirement in Definition 9.1 is trivially satisfied.

(g) If B is the open ball with center z and radius $a > 0$ and if $x \in B$, then the ball with center x and radius $a - \|z - x\|$ is contained in B. Thus B is open in $\mathbf{R}^p$.

We now state the basic properties of open sets in $\mathbf{R}^p$. In courses on topology this next result is summarized by saying that the open sets, as defined in Definition 9.1, form a **topology** for $\mathbf{R}^p$.

9.3 OPEN SET PROPERTIES. (a) *The empty set $\emptyset$ and the entire space $\mathbf{R}^p$ are open in $\mathbf{R}^p$.*

(b) *The intersection of any two open sets is open in $\mathbf{R}^p$.*

(c) *The union of any collection of open sets is open in $\mathbf{R}^p$.*

PROOF. We have already commented on the open character of the sets $\emptyset$ and $\mathbf{R}^p$.

To prove (b), let G_1, G_2 be open and let $G_3 = G_1 \cap G_2$. To show that G_3 is open, let $x \in G_3$. Since x belongs to the open set G_1, there exists $r_1 > 0$ such that if $\|x - z\| < r_1$, then $z \in G_1$. Similarly, there exists $r_2 > 0$ such that if $\|x - w\| < r_2$, then $w \in G_2$. Choosing r_3 to be the minimum of r_1 and r_2, we conclude that if $y \in \mathbf{R}^p$ is such that $\|x - y\| < r_3$, then y belongs to both G_1 and G_2. Hence such elements y belong to $G_3 = G_1 \cap G_2$, showing that G_3 is open in $\mathbf{R}^p$.

To prove (c), let $\{G_\alpha, G_\beta, \ldots\}$ be a collection of sets which are open and let G be their union. To show that G is open, let $x \in G$. By definition of the union, it follows that for some set, say for G_λ, we have $x \in G_\lambda$. Since G_λ is open, there exists a ball with center x which is entirely contained in G_λ. Since $G_\lambda \subseteq G$, this ball is entirely contained in G, showing that G is open in $\mathbf{R}^p$.   Q.E.D.

By induction, it follows from property (b) above that the intersection of any *finite* collection of sets which are open is also open in $\mathbf{R}^p$. That the intersection of an infinite collection of open sets may not be open can be seen from the example

$$(9.1) \qquad G_n = \left\{ x \in \boldsymbol{R} : -\frac{1}{n} < x < 1 + \frac{1}{n} \right\}, \qquad n \in \boldsymbol{N}.$$

The intersection of the sets G_n is the set $F = \{x \in \boldsymbol{R} : 0 \le x \le 1\}$, which is not open.

Closed Sets

We now introduce the important notion of a closed set in $\boldsymbol{R}^p$.

9.4 DEFINITION. A set F in $\boldsymbol{R}^p$ is said to be **closed in** $\boldsymbol{R}^p$ (or merely **closed**) in case its complement $\mathscr{C}(F) = \boldsymbol{R}^p \setminus F$ is open in $\boldsymbol{R}^p$.

9.5 EXAMPLES. (a) The entire set $\boldsymbol{R}^p$ is closed in $\boldsymbol{R}^p$, since its complement is the empty set, which was seen in 9.2(f) to be open in $\boldsymbol{R}^p$.

(b) The empty set $\emptyset$ is closed in $\boldsymbol{R}^p$, since its complement in $\boldsymbol{R}^p$ is all of $\boldsymbol{R}^p$ which was seen in 9.2(a) to be open in $\boldsymbol{R}^p$.

(c) The set $F = \{x \in \boldsymbol{R} : 0 \le x \le 1\}$ is closed in $\boldsymbol{R}$. One way of seeing this is by noting that the complement of F in $\boldsymbol{R}$ is the union of the two sets $\{x \in \boldsymbol{R} : x < 0\}$, $\{x \in \boldsymbol{R} : x > 1\}$, each of which is open. Similarly, the set $\{x \in \boldsymbol{R} : 0 \le x\}$ is closed.

(d) The set $F = \{(x, y) \in \boldsymbol{R}^2 : x^2 + y^2 \le 1\}$ is closed, since its complement in $\boldsymbol{R}^2$ is the set

$$\{(x, y) \in \boldsymbol{R}^2 : x^2 + y^2 > 1\}.$$

which is seen to be open.

(e) The set $H = \{(x, y, z) \in \boldsymbol{R}^3 : x \ge 0\}$ is closed in $\boldsymbol{R}^3$, as is the set $F = \{(x, y, z) \in \boldsymbol{R}^3 : x = y = z\}$.

(f) The closed ball B with center x in $\boldsymbol{R}^p$ and radius $r > 0$ is a closed set of $\boldsymbol{R}^p$. For, if $z \notin B$, then the open ball with center z and radius $\|z - x\| - r$ is contained in $\mathscr{C}(B)$. Therefore, $\mathscr{C}(B)$ is open and B is closed in $\boldsymbol{R}^p$.

In ordinary parlance, when applied to doors, windows, and minds, the words "open" and "closed" are antonyms. However, when applied to subsets of $\boldsymbol{R}^p$, these words are not antonyms. For example, we noted above that the sets $\emptyset$, $\boldsymbol{R}^p$ are *both* open and closed in $\boldsymbol{R}^p$. (The reader will probably be relieved to learn that there are no other subsets of $\boldsymbol{R}^p$ which have both properties.) In addition, there are many subsets of $\boldsymbol{R}^p$ which are *neither* open nor closed; in fact, most subsets of $\boldsymbol{R}^p$ have this neutral character. As a simple example, we cite the set

$$(9.2) \qquad A = \{x \in \boldsymbol{R} : 0 \le x < 1\}.$$

This set A fails to be open in $\boldsymbol{R}$, since it contains the point 0. Similarly, it fails to be closed in $\boldsymbol{R}$, because its complement in $\boldsymbol{R}$ is the set $\{x \in \boldsymbol{R} : x < 0 \text{ or } x \ge 1\}$, which is not open since it contains the point 1. The reader should construct other examples of sets which are neither open nor closed in $\boldsymbol{R}^p$.

We now state the fundamental properties of closed sets. The proof of this result follows directly from Theorem 9.3 by using DeMorgan's laws (Theorem 1.8 and Exercise 1.L).

9.6 CLOSED SET PROPERTIES. (a) *The empty set $\emptyset$ and the entire space $\mathbf{R}^p$ are closed in $\mathbf{R}^p$.*

(b) *The union of any two closed sets is closed in $\mathbf{R}^p$.*

(c) *The intersection of any collection of closed sets is closed in $\mathbf{R}^p$.*

Neighborhoods

We now introduce some additional topological notions that will be useful and permit us to characterize open and closed sets in other terms.

9.7 DEFINITION. (a) If $x \in \mathbf{R}^p$, then any set which contains an open set containing x is called a **neighborhood** of x.

(b) A point $x \in \mathbf{R}^p$ is called an **interior point** of a set $A \subseteq \mathbf{R}^p$ in case there is a neighborhood of x which is entirely contained in A.

(c) A point $x \in \mathbf{R}^p$ is called a **boundary point** of a set $A \subseteq \mathbf{R}^p$ in case every neighborhood of x contains a point in A and a point in $\mathscr{C}(A)$.

(d) A point $x \in \mathbf{R}^p$ is called an **exterior point** of a set $A \subseteq \mathbf{R}^p$ in case there exists a neighborhood of x which is entirely contained in $\mathscr{C}(A)$.

It should be noted that given $x \in \mathbf{R}^p$ and $A \subseteq \mathbf{R}^p$, there are three mutually exclusive possibilities: (i) x is an interior point of A, (ii) x is a boundary point of A, or (iii) x is an exterior point of A.

9.8 EXAMPLES. (a) A set U is a neighborhood of a point x if and only if there exists a ball with center x entirely contained in U.

(b) A point x is an interior point of A if and only if there exists a ball with center x entirely contained in A.

(c) A point x is a boundary point of A if and only if for each natural number n there exist points $a_n \in A$ and $b_n \in \mathscr{C}(A)$ such that $\|x - a_n\| < 1/n$ and $\|x - b_n\| < 1/n$.

(d) Every point of the interval $(0, 1) \subseteq \mathbf{R}$ is an interior point. The points 0, 1 are the boundary points of $(0, 1)$.

(e) Let $A = [0, 1] \subseteq \mathbf{R}$. Then the interior points of A are the points in the open interval $(0, 1)$. The points 0, 1 are the boundary points of A.

(f) The boundary points of the open and the closed balls with center $x \in \mathbf{R}^p$ and radius $r > 0$, are the points in the sphere with center x and radius r. (See Definition 8.9.)

We now characterize open sets in terms of neighborhoods and interior points.

9.9   THEOREM.   *If $B \subseteq \mathbf{R}^p$, then the following statements are equivalent:*
(a) *B is open;*
(b) *every point of B is an interior point of B;*
(c) *B is a neighborhood of each of its points.*

PROOF.   If (a) holds and $x \in B$, then the open set B is a neighborhood of x and x is therefore an interior point of B.

It is trivial that (b) implies (c).

If (c) holds, then for each $x \in B$, there is an open set $G_x \subseteq B$ with $x \in G_x$. Hence $B = \bigcup \{G_x : x \in B\}$, so that it follows from Theorem 9.3(c) that B is open in $\mathbf{R}^p$.   Q.E.D.

It follows from what we have shown that an open set contains *none* of its boundary points.   Closed sets are the other extreme in this respect.

9.10   THEOREM.   *A set $F \subseteq \mathbf{R}^p$ is closed if and only if it contains all of its boundary points.*

PROOF.   Suppose that F is closed and that x is a boundary point of F. If $x \notin F$, then the open set $\mathscr{C}(F)$ contains x and no points of F, contrary to the hypothesis that x is a boundary point of F.   Hence we must have $x \in F$.

Conversely, suppose that F contains all of its boundary points.   If $y \notin F$, then y is neither a point of F or a boundary point of F; hence it is an exterior point.   Therefore there exists a neighborhood M of y entirely contained in $\mathscr{C}(F)$.   Since this is true for all $y \notin F$, we infer that $\mathscr{C}(F)$ is open, whence F is closed in $\mathbf{R}^p$.   Q.E.D.

Open Sets in **R**

We close this section by characterizing the form of an arbitrary open subset of **R**.

9.11   THEOREM.   *A subset of **R** is open if and only if it is the union of a countable collection of open intervals.*

PROOF.   Since an open interval is open (why?), it follows from 9.3(c) that the union of any countable union of open intervals is open.

Conversely, let $G \neq \emptyset$ be an open set in **R** and let $\{r_n : n \in \mathbf{N}\}$ be an enumeration of all of the rational points in G.   For each $n \in \mathbf{N}$ let m_n be the smallest natural number such that the interval $J_n = (r_n - 1/m_n, r_n + 1/m_n)$ is entirely contained in G.   It follows that

$$\bigcup_{n \in \mathbf{N}} J_n \subseteq G.$$

Now let x be an arbitrary point in G and let $m \in \mathbf{N}$ be such that $(x - 2/m, x + 2/m) \subseteq G$.   It follows from Theorem 6.10 that there exists a rational number y in $(x - 1/m, x + 1/m)$; hence $y \in G$ and so $y = r_n$ for some natural number n. If x does not belong to $J_n = (r_n - 1/m_n, r_n + 1/m_n)$, then we must have $1/m_n < 1/m$; but since it is readily seen that

$$\left(r_n - \frac{1}{m}, r_n + \frac{1}{m} \right) \subseteq \left(x - \frac{2}{m}, x + \frac{2}{m} \right) \subseteq G,$$

this contradicts the choice of the m_n. Therefore we have $x \in J_n$ for this value of n. Since $x \in G$ is arbitrary, we infer that

$$G \subseteq \bigcup_{n \in N} J_n.$$

Therefore G is equal to this union. Q.E.D.

It does *not* follow from Theorem 9.11 that a subset of R is closed if and only if it is the intersection of a countable collection of closed intervals. (Why?) It does *not* follow that the countable union of closed intervals must be closed, nor does every closed set have this property.

A generalization of this result is given in Exercise 9.G.

Exercises

9.A. Justify the assertion made about the set G, F made in Example 9.2(b).

9.B. Justify the assertions made in Example 9.2(c).

9.C. Prove that the intersection of any *finite* collection of open sets is open in R^p. (Hint: use 9.3(b) and induction.)

9.D. What are the interior, boundary, and exterior points in R of the set $[0, 1)$. Conclude that it is neither open nor closed.

9.E. Give an example in R^2 which is neither open nor closed. Prove your assertion.

9.F. Write out the details of the proof of Theorem 9.6.

9.G. Show that a subset of R^p is open if and only if it is the union of a countable collection of open balls. (Hint: the set of all points in R^p all of whose coordinates are rational numbers is countable.)

9.H. Every open subset of R^p is the union of a countable collection of closed sets.

9.I. Every closed subset of R^p is the intersection of a countable collection of open sets.

9.J. If A is any subset of R^p, let A° denote the union of all open sets which are contained in A; the set A° is called the **interior** of A. Note that A° is an open set; prove that it is the largest open set contained in A. Prove that

$$A^\circ \subseteq A, \qquad (A^\circ)^\circ = A^\circ$$

$$(A \cap B)^\circ = A^\circ \cap B^\circ, \qquad (R^p)^\circ = R^p.$$

Give an example to show that $(A \cup B)^\circ = A^\circ \cup B^\circ$ may not hold.

9.K. Prove that a point belongs to A° if and only if it is an interior point of A.

9.L. If A is any subset of R^p, let A^- denote the intersection of all closed sets containing A; the set A^- is called the **closure** of A. Note that A^- is a closed set; prove that it is the smallest closed set containing A. Prove that

$$A \subseteq A^-, \qquad (A^-)^- = A^-$$
$$(A \cup B)^- = A^- \cup B^-, \qquad \emptyset^- = \emptyset.$$

Give an example to show that $(A \cap B)^- = A^- \cap B^-$ may not hold.

9.M. Prove that a point belongs to A^- if and only if it is either an interior or a boundary point of A.

9.N. Give an example of a set A in R^p such that $A^\circ = \emptyset$ and $A^- = R^p$. Can such a set A be countable?

9.O. Let A and B be non-void subsets of R. The Cartesian product $A \times B$ is open in R^2 if and only if A and B are open in R.

9.P. Let A and B be non-void subsets of R. The Cartesian product $A \times B$ is closed in R^2 if and only if A and B are closed in R.

9.Q. Interpret the concepts introduced in this section for the Cantor set F of Definition 7.4. In particular:

(a) Show that F is closed in R.

(b) There are no interior points in F.

(c) There are no non-empty open sets contained in F.

(d) Every point of F is a boundary point.

(e) The set F cannot be expressed as the union of a countable collection of closed intervals.

(f) The complement of F can be expressed as the union of a countable collection of open intervals.

Section 10 The Nested Cells and Bolzano-Weierstrass Theorems

In this section we shall present two very important results that will be often used in later chapters. In a sense they can be regarded as the Completeness Property for R^p, when $p > 1$.

We recall from Section 7 that if $a \leq b$, then the open cell in R, denoted by (a, b), is the set defined by

$$(a, b) = \{x \in R : a < x < b\}.$$

It is readily seen that such a set is open in R. Similarly, the closed cell $[a, b]$ in R is the set

$$[a, b] = \{x \in R : a \leq x \leq b\},$$

which is closed in R. The Cartesian product of two intervals is usually called a **rectangle** and the Cartesian product of three intervals is often called a

parallelepiped. For simplicity, we shall employ the term "cell" regardless of the dimension of the space.

10.1 DEFINITION.   An **open cell** J in $\mathbf{R}^p$ is the Cartesian product of p open cells of real numbers. Hence J has the form

$$J = \{x = (x_1, \ldots, x_p) \in \mathbf{R}^p : a_i < x_i < b_i, \quad \text{for} \quad i = 1, 2, \ldots, p\}.$$

Similarly, a **closed cell** I in $\mathbf{R}^p$ is the Cartesian product of p closed cells of real numbers. Hence I has the form

$$I = \{x = (x_1, \ldots, x_p) \in \mathbf{R}^p : a_i \leq x_i \leq b_i, \quad \text{for} \quad i = 1, 2, \ldots, p\}.$$

A subset of $\mathbf{R}^p$ is **bounded** if it is contained in some cell.

As an exercise, show that an open cell in $\mathbf{R}^p$ is an open set, and a closed cell is a closed set. Also, a subset of $\mathbf{R}^p$ is bounded if and only if it is contained in some ball. It will be observed that this terminology for bounded sets is consistent with that introduced in Section 6 for the case $p = 1$.

The reader will recall from Section 7 that the Supremum Property of the real number system implies that every nested sequence of non-empty closed cells in $\mathbf{R}$ has a common point. We shall now prove that this property carries over to the space $\mathbf{R}^p$.

10.2   NESTED CELLS THEOREM.   *Let (I_k) be a sequence of non-empty closed cells in $\mathbf{R}^p$ which is nested in the sense that $I_1 \supseteq I_2 \supseteq \cdots \supseteq I_k \supseteq \cdots$. Then there exists a point in $\mathbf{R}^p$ which belongs to all of the cells.*

PROOF.   Suppose that I_k is the cell

$$I_k = \{(x_1, \ldots, x_p) : a_{k1} \leq x_1 \leq b_{k1}, \ldots, a_{kp} \leq x_p \leq b_{kp}\}.$$

It is easy to see that the cells $[a_{k1}, b_{k1}]$, $k \in \mathbf{N}$, form a nested sequence of non-empty closed cells of real numbers and hence by the completeness of the real number system $\mathbf{R}$, there is a real number y_1 which belongs to all of these cells. Applying this argument to each coordinate, we obtain a point $y = (y_1, \ldots, y_p)$ of $\mathbf{R}^p$ such that if j satisfies $j = 1, 2, \ldots, p$, then y_j belongs to all the cells $\{[a_{kj}, b_{kj}] : k \in \mathbf{N}\}$. Hence the point y belongs to all of the cells (I_k).   Q.E.D.

Cluster Points and Bolzano-Weierstrass

10.3   DEFINITION.   A point $x \in \mathbf{R}^p$ is a **cluster point** (or a **point of accumulation**) of a subset $A \subseteq \mathbf{R}^p$ in case every neighborhood of x contains at least one point of A distinct from x.

We shall consider some examples.

10.4   EXAMPLES.   (a) A point $x \in R^p$ is a cluster point of A if and only if for every natural number n there exists an element $a_n \in A$ such that $0 < \|x - a_n\| < 1/n$.

(b) If a boundary point of a set does not belong to the set, then it is a cluster point of the set.

(c) Every point of the unit interval I of R is a cluster point of I.

(d) Let $A = (0, 1)$, then every point of A is both an interior and a cluster point of A.   The points 0, 1 are cluster points (but not interior points) of A.

(e) Let $B = I \cap Q$ be the set of all rational numbers in the unit interval. Every point of I is a cluster point of B in R, but there are no interior points of B.

(f) A finite subset of R^p has no cluster points.   (Why?)

(g) The infinite set of integers $Z \subseteq R$ has no cluster points.   (Why?)

10.5   THEOREM   *A set $F \subseteq R^p$ is closed if and only if it contains all of its cluster points.*

PROOF   Suppose that F is closed and that x is a cluster point of F.   If $x \notin F$, then the open set $\mathscr{C}(F)$ is a neighborhood of x and so must contain at least one point of F.   But this is impossible, so we conclude that $x \in F$.

Conversely, if F contains all of its cluster points, we shall show that $\mathscr{C}(F)$ is open.   For, if $y \in \mathscr{C}(F)$, then y is not a cluster point of F.   Therefore, there exists a neighborhood V_y of y such that $F \cap V_y = \emptyset$.   Therefore $V_y \subseteq \mathscr{C}(F)$. Since this is true for every $y \in \mathscr{C}(F)$, we infer that $\mathscr{C}(F)$ is open in R^p.   Q.E.D.

The next result is one of the most important results in this book.   It is of basic importance and will be frequently used.   It should be noted that the conclusion may fail if either hypothesis is removed [see Examples 10.4(f, g)].

10.6.   BOLZANO-WEIERSTRASS† THEOREM.   *Every bounded infinite subset of R^p has a cluster point.*

PROOF.   If B is a bounded set with an infinite number of elements, let I_1 be a closed cell containing B.   We divide I_1 into 2^p closed cells by bisecting each of its sides.   Since I_1 contains infinitely many points of B, at least one part obtained in this subdivision will also contain infinitely many

---

† BERNARD BOLZANO (1781–1848) was professor of the philosophy of religion at Prague, but he had deep thoughts about mathematics.   Like Cauchy, he was a pioneer in introducing a higher standard of rigor in mathematical analysis.   His treatise on the paradoxes of the infinite appeared after his death.

KARL WEIERSTRASS (1815–1897) was for many years a professor at Berlin and exercised a profound influence on the development of analysis.   Always insisting on rigorous proof he developed, but did not publish, an introduction to the real number system.   He also made important contributions to real and complex analysis, differential equations, and the calculus of variations.

points of B.   (For if each of the 2^p parts contained only a finite number of points of the set B, then B must be a finite set, contrary to hypothesis.)   Let I_2 be one of these parts in the subdivision of I_1 which contains infinitely many elements of B.   Now divide I_2 into 2^p closed cells by bisecting each of its sides.   Again, one of these subcells of I_2 must contain an infinite number of points of B, for otherwise I_2 could contain only a finite number, contrary to its construction.   Let I_3 be a subcell of I_2 containing infinitely many points of B.   Continuing this process, we obtain a nested sequence (I_k) of non-empty closed cells of $\mathbf{R}^p$.   According to the Nested Cells Theorem, there is a point y which belongs to all of the cells I_k, $k = 1, 2, \ldots$.   We shall now show that y is a cluster point of B and this will complete the proof of the assertion.

First, we note that if $I_1 = [a_1, b_1] \times \cdots \times [a_p, b_p]$ with $a_k < b_k$, and if $l(I_1) = \sup \{b_1 - a_1, \ldots, b_p - a_p\}$, then $l(I_1) > 0$ is the length of the largest side of I_1. According to the above construction of the sequence (I_k), we have

$$0 < l(I_k) = \frac{1}{2^{k-1}} l(I_1)$$

for $k \in \mathbf{N}$.   Suppose that V is any neighborhood of the common point y and suppose that all points z in $\mathbf{R}^p$ with $\|y - z\| < r$ belong to V.   We now choose k so large that $I_k \subseteq V$; such a choice is possible since if w is any other point of I_k, then it follows from Theorem 8.10 that

$$\|y - w\| \le \sqrt{p}\, l(I_k) = \frac{\sqrt{p}}{2^{k-1}} l(I_1).$$

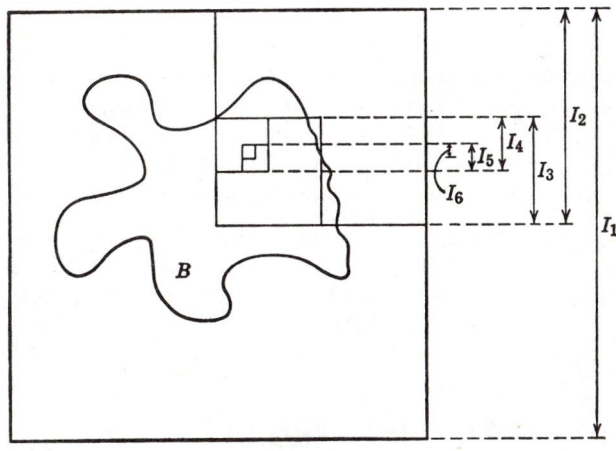

Figure 10.1

According to Exercise 6.L, it follows that if k is sufficiently large, then

$$\frac{\sqrt{p}}{2^{k-1}} l(I_1) < r.$$

For such a value of k we have $I_k \subseteq V$. Since I_k contains infinitely many elements of B, it follows that V contains at least one element of B different from y. Therefore, y is a cluster point of B.   Q.E.D.

Exercises

10.A. Let $I_n \subseteq \mathbf{R}^p$ be the open cells given by $I_n = (0, 1/n) \times \cdots \times (0, 1/n)$. Show that these cells are nested but that they do not contain any common point.

10.B. Let $J_n \subseteq \mathbf{R}^p$ be the closed intervals given by $J_n = [n, +\infty) \times \cdots \times [n, +\infty)$. Show that these intervals are nested, but that they do not contain any common point.

10.C. A point x is a cluster point of a set $A \subseteq \mathbf{R}^p$ if and only if every neighborhood of x contains infinitely many points of A.

10.D. Let $A = \{1/n : n \in \mathbf{N}\}$. Show that every point of A is a boundary point in $\mathbf{R}$, but that 0 is the only cluster point of A in R.

10.E. Let A, B be subsets of $\mathbf{R}^p$ and let x be a cluster point of $A \cap B$ in $\mathbf{R}^p$. Prove that x is a cluster point of both A and B.

10.F. Let A, B be subsets of $\mathbf{R}^p$ and let x be a cluster point of $A \cup B$ in $\mathbf{R}^p$. Prove that x is either a cluster point of A or of B.

10.G. Show that every point in the Cantor set F is a cluster point of both F and $\mathscr{C}(F)$.

10.H. If A is any subset of $\mathbf{R}^p$, then there exists a countable subset C of A such that if $x \in A$ and $\varepsilon > 0$, then there is an element $z \in C$ such that $\|x - z\| < \varepsilon$. Hence every element of A is either in C or is a cluster point of C.

Projects

10.α. Let M be a set and d be a metric on M as defined in Exercise 8.S. Reexamine the definitions and theorems of Sections 9 and 10, in order to determine which carry over for sets that have a metric. It will be seen, for example, that the notions of open, closed, and bounded set carry over. The Bolzano-Weierstrass fails for suitable M and d, however. Whenever possible, either show that the theorem extends or give a counterexample to show that it may fail.

10.β. Let $\mathscr{T}$ be a family of subsets of a set X which (i) contains $\emptyset$ and X, (ii) contains the intersection of any finite family of sets in $\mathscr{T}$, and (iii) contains the union of any family of sets in $\mathscr{T}$. We call $\mathscr{T}$ a **topology** for X, and refer to the sets in $\mathscr{T}$ as the **open sets.** Reexamine the definitions and theorems of Sections 9 and 10, trying to determine which carry over for sets X which have a topology $\mathscr{T}$.

Section 11   The Heine-Borel Theorem

The Nested Cells Theorem 10.2 and the Bolzano-Weierstrass Theorem 10.6 are intimately related to the very important notion of compactness, which we shall discuss in the present section. Although it is possible to obtain most of the results of the later sections without knowing the Heine-Borel Theorem, we cannot go much farther in analysis without requiring this theorem, so it is false economy to avoid exposure to this deep result.

11.1 DEFINITION. A set K is said to be **compact** if, whenever it is contained in the union of a collection $\mathcal{G} = \{G_\alpha\}$ of open sets, then it is also contained in the union of some *finite* number of the sets in $\mathcal{G}$.

A collection $\mathcal{G}$ of open sets whose union contains K is often called a **covering** of K. Thus the requirement that K be compact is that every covering $\mathcal{G}$ of K can be replaced by a *finite* covering of K, using only sets in $\mathcal{G}$. We note that in order to apply this definition to prove that a set K is compact, we need to examine an arbitrary collection of open sets whose union contains K and show that K is contained in the union of some finite subcollection of each such collection. On the other hand, to show that a set H is not compact, it is sufficient to exhibit only *one* covering which cannot be replaced by a finite subcollection which still covers H.

11.2 EXAMPLES. (a) Let $K = \{x_1, x_2, \ldots, x_m\}$ be a finite subset of $\mathbf{R}^p$. It is clear that if $\mathcal{G} = \{G_\alpha\}$ is a collection of open sets in $\mathbf{R}^p$, and if every point of K belongs to some subset of $\mathcal{G}$. then at most m carefully selected subsets of $\mathcal{G}$ will also have the property that their union contains K. Hence K is a compact subset of $\mathbf{R}^p$.

(b) In $\mathbf{R}$ we consider the subset $H = \{x \in \mathbf{R} : x \geq 0\}$. Let $G_n = (-1, n)$, $n \in \mathbf{N}$, so that $\mathcal{G} = \{G_n : n \in \mathbf{N}\}$ is a collection of open subsets of $\mathbf{R}$ whose union contains H. If $\{G_{n_1}, G_{n_2}, \ldots, G_{n_k}\}$ is a finite subcollection of $\mathcal{G}$, let $M = \sup\{n_1, n_2, \ldots, n_k\}$ so that $G_{n_j} \subseteq G_M$, for $j = 1, 2, \ldots, k$. It follows that G_M is the union of $\{G_{n_1}, G_{n_2}, \ldots, G_{n_k}\}$. However, the real number M does not belong to G_M and hence does not belong to

$$\bigcup_{j=1}^{k} G_{n_j}.$$

Therefore, no finite union of the sets $\mathcal{G}$ can contain H, and H is not compact.

(c) Let $H = (0, 1)$ in $\mathbf{R}$. If $G_n = (1/n, 1 - 1/n)$ for $n > 2$, then the collection $\mathcal{G} = \{G_n : n > 2\}$ of open sets is a covering of H. If $\{G_{n_1}, \ldots, G_{n_k}\}$ is a finite subcollection of $\mathcal{G}$, let $M = \sup\{n_1, \ldots, n_k\}$ so that $G_{n_j} \subseteq G_M$ for $j = 1, 2, \ldots, k$. It follows that G_M is the union of the sets $\{G_{n_1}, \ldots, G_{n_k}\}$. However, the real number $1/M$ belongs to H but does not belong to G_M. Therefore, no finite subcollection of $\mathcal{G}$ can form a covering of H, so that H is not compact.

(d) Consider the set $I = [0, 1]$; we shall show that I is compact. Let $\mathcal{G} = \{G_\alpha\}$ be a collection of open subsets of $\mathbf{R}$ whose union contains I. The real number $x = 0$ belongs to some open set in the collection $\mathcal{G}$ and so do numbers x satisfying $0 \le x < \varepsilon$, for some $\varepsilon > 0$. Let x^* be the supremum of those points x in I such that the cell $[0, x]$ is contained in the union of a finite number of sets in $\mathscr{G}$. Since x^* belongs to I, it follows that x^* is an element of some open set in $\mathscr{G}$. Hence for some $\varepsilon > 0$, the cell $[x^* - \varepsilon, x^* + \varepsilon]$ is contained in a set G_0 in the collection $\mathscr{G}$. But (by the definition of x^*) the cell $[0, x^* - \varepsilon]$ is contained in the union of a finite number of sets in $\mathscr{G}$. Hence by adding the single set G_0 to the finite number already needed to cover $[0, x^* - \varepsilon]$, we infer that the set $[0, x^* + \varepsilon]$ is contained in the union of a finite number of sets in $\mathscr{G}$. Thus $x^* = 1$ and I is compact.

It is usually not an easy matter to prove that a set is compact, using the definition only. We now present a remarkable and important theorem which completely characterizes compact subsets of $\mathbf{R}^p$. In fact, part of the importance of the Heine-Borel Theorem† is due to the simplicity of the conditions for compactness in $\mathbf{R}^p$.

11.3 HEINE-BOREL THEOREM.   *A subset of $\mathbf{R}^p$ is compact if and only if it is closed and bounded.*

PROOF.   First we show that if K is compact in $\mathbf{R}^p$, then K is closed. Let x belong to $\mathscr{C}(K)$ and for each natural number m, let G_m be the set defined by

$$G_m = \{y \in \mathbf{R}^p : \|y - x\| > 1/m\}.$$

It is readily seen that each set G_m, $m \in \mathbf{N}$, is open in $\mathbf{R}^p$. Also, the union of all the sets G_m, $m \in \mathbf{N}$, consists of all points of $\mathbf{R}^p$ except x. Since $x \notin K$, each point of K belongs to some set G_m. In view of the compactness of K, it follows that there exists a natural number M such that K is contained in the union of the sets

$$G_1, G_2, \ldots, G_M.$$

Since the sets G_m increase with m, then K is contained in G_M. Hence the neighborhood $\{z \in \mathbf{R}^p : \|z - x\| < 1/M\}$ does not intersect K, showing that $\mathscr{C}(K)$ is open. Therefore, K is closed in $\mathbf{R}^p$. (See Figure 11.1 where the closed balls complementary to the G_m are depicted.)

Next we show that if K is compact in $\mathbf{R}^p$, then K is bounded (that is, K is contained in some set $\{x \in \mathbf{R}^p : \|x\| < r\}$ for sufficiently large r). In fact, for

† EDUARD HEINE (1821–1881) studied at Berlin under Weierstrass and later taught at Bonn and Halle. In 1872 he proved that a continuous function on a closed interval is uniformly continuous.

(F. E. J.) ÉMILE BOREL (1871–1956), a student of Hermite, was professor at Paris and one of the most influential mathematicians of his day. He made numerous and deep contributions to analysis and probability. In 1895 he proved that if a countable collection of open intervals cover a closed interval, then they have a finite subcovering.

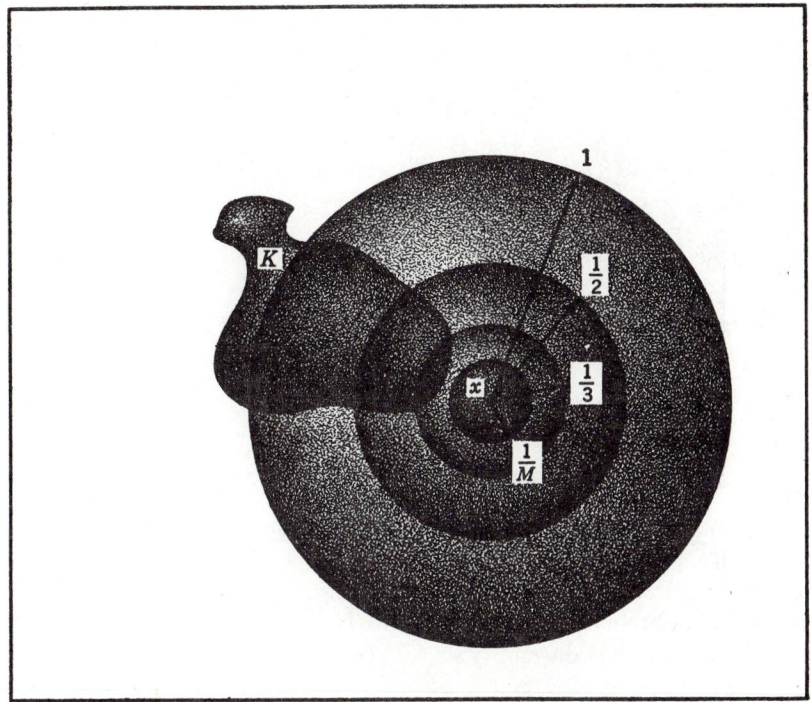

Figure 11.1. A compact set is closed.

each natural number m, let H_m be the open set defined by

$$H_m = \{x \in \mathbf{R}^p : \|x\| < m\}.$$

The entire space $\mathbf{R}^p$, and hence K, is contained in the union of the increasing sets H_m, $m \in \mathbf{N}$. Since K is compact, there exists a natural number M such that $K \subseteq H_M$. This proves that K is bounded.

To complete the proof of this theorem we need to show that if K is a closed and bounded set which is contained in the union of a collection $\mathcal{G} = \{G_\alpha\}$ of open sets in $\mathbf{R}^p$, then it is contained in the union of some finite number of sets in $\mathcal{G}$. Since the set K is bounded, we may enclose it in a closed cell I_1 in $\mathbf{R}^p$. For example, we may take $I_1 = \{(x_1, \ldots, x_p) : |x_k| \le r, \ k = 1, \ldots, p\}$ for suitably large $r > 0$. For the purpose of obtaining a contradiction, we shall assume that K is not contained in the union of any finite number of the sets in $\mathcal{G}$. Therefore, at least one of the 2^p closed cells obtained by bisecting the sides of I_1 contains points of K and is such that the part of K in it is not contained in the union of any finite number of the sets in $\mathcal{G}$. (For, if each of the 2^p parts of K were contained in the union of a finite number of sets in $\mathcal{G}$, then K would be contained in the union of a finite number of sets in $\mathcal{G}$,

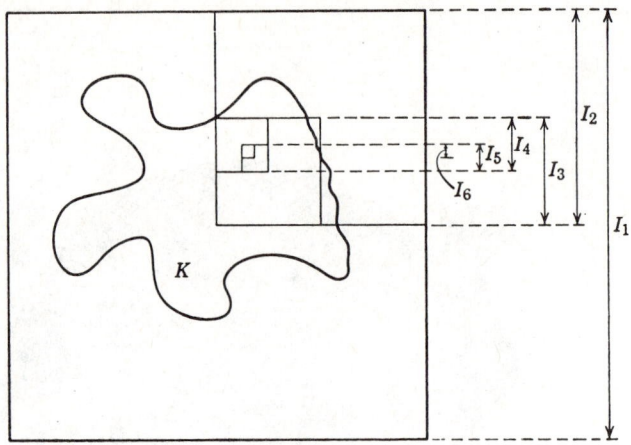

Figure 11.2

contrary to hypothesis.) Let I_2 be any one of the subcells in this subdivision of I_1 which is such that the non-empty set $K \cap I_2$ is not contained in the union of any finite number of sets in $\mathscr{G}$. We continue this process by bisecting the sides of I_2 to obtain 2^p closed subcells of I_2 and we let I_3 be one of these subcells such that the non-empty set $K \cap I_3$ is not contained in the union of a finite number of sets in $\mathscr{G}$, and so on.

In this way we obtain a nested sequence (I_n) of non-empty cells (see Figure 11.2); according to the Nested Cells Theorem there is a point y common to the I_n. Since each I_n contains points in K, the common element y is a cluster point of K. Since K is closed, then y belongs to K and is contained in some open set G_λ in $\mathscr{G}$. Therefore, there exists a number $\varepsilon > 0$ such that all points w with $\|y - w\| < \varepsilon$ belong to G_λ. On the other hand, the cells I_k, $k \geq 2$, are obtained by successive bisection of the sides of the cell $I_1 = \{(x_1 \ldots, x_p) : |x_i| \leq r\}$ so the length of the side of I_k is $r/2^{k-2}$. It follows from Theorem 8.10 that if $w \in I_k$, then $\|y - w\| \leq r\sqrt{p}/2^{k-3}$. Hence, if k is chosen so large that $r\sqrt{p}/2^{k-3} < \varepsilon$, then all points in I_k are contained in the single set G_λ. But this contradicts the construction of I_k as a set such that $K \cap I_k$ is not contained in the union of a finite number of sets in $\mathscr{G}$. This contradiction shows that the assumption that the closed bounded set K requires an infinite number of sets in $\mathscr{G}$ to enclose it is untenable.   Q.E.D.

Some Applications

As a consequence of the Heine-Borel Theorem, we obtain the next result, which is due to Cantor. It is a strengthening of the Nested Cells Theorem, since general closed sets are considered here and not just closed cells.

11.4 CANTOR INTERSECTION THEOREM. *Let F_1 be a non-empty closed, bounded subset of $\mathbf{R}^p$ and let*

$$F_1 \supseteq F_2 \supseteq \cdots \supseteq F_n \supseteq \cdots$$

be a sequence of non-empty closed sets. Then there exists a point belonging to all of the sets $\{F_k : k \in \mathbf{N}\}$.

PROOF. Since F_1 is closed and bounded, it follows from the Heine-Borel Theorem that it is compact. For each $k \in \mathbf{N}$, let G_k be the complement of F_k in $\mathbf{R}^p$. Since F_k is assumed to be closed, G_k is open in $\mathbf{R}^p$. If, contrary to the theorem, there is no point belonging to all of the sets F_k, $k \in \mathbf{N}$, then the union of the sets G_k, $k \in \mathbf{N}$, contains the compact set F_1. Therefore, the set F_1 is contained in the union of a finite number of the sets G_k; say, in G_1, $G_2, \ldots, G_K$. Since the G_k increase, we have $G_1 \cup \cdots \cup G_K = G_K$. Since $F_1 \subseteq G_K$, it follows that $F_1 \cap F_K = \emptyset$. By hypothesis $F_1 \supseteq F_K$, so $F_1 \cap F_K = F_K$. Our assumption leads to the conclusion that $F_K = \emptyset$, which contradicts the hypothesis and establishes the theorem. Q.E.D.

11.5 LEBESGUE COVERING THEOREM. *Suppose that $\mathcal{G} = \{G_\alpha\}$ is a covering of a compact subset K of $\mathbf{R}^p$. There exists a strictly positive number λ such that if x, y belong to K and $\|x - y\| < \lambda$, then there is a set in $\mathcal{G}$ containing both x and y.*

PROOF. For each point u in K, there is an open set $G_{\alpha(u)}$ in $\mathcal{G}$ containing u. Let $\delta(u) > 0$ be such that if $\|v - u\| < 2\delta(u)$, then v belongs to $G_{\alpha(u)}$. Consider the open set $S(u) = \{v \in \mathbf{R}^p : \|v - u\| < \delta(u)\}$ and the collection $\mathcal{S} = \{S(u) : u \in K\}$ of open sets. Since $\mathcal{S}$ is a covering of the compact set K, then K is contained in the union of a finite number of sets in $\mathcal{S}$, say in $S(u_1), \ldots, S(u_n)$. We now define λ to be the strictly positive real number

$$\lambda = \inf \{\delta(u_1), \ldots, \delta(u_n)\}.$$

If x, y belong to K and $\|x - y\| < \lambda$, then x belongs to $S(u_j)$ for some j with $1 \leq j \leq n$, so $\|x - u_j\| < \delta(u_j)$. Since $\|x - y\| < \lambda$, we have $\|y - u_j\| \leq \|y - x\| + \|x - u_j\| < 2\delta(u_j)$. According to the definition of $\delta(u_j)$, we infer that both x and y belong to the set $G_{\alpha(u_j)}$. Q.E.D.

We remark that a positive number λ having the property stated in the theorem is sometimes called a **Lebesgue† number** of the covering $\mathcal{G}$.

Although we shall make use of arguments based on compactness in later sections, it seems appropriate to insert here two results which appear intuitively clear, but whose proof seems to require use of some type of compactness argument.

† HENRI LEBESGUE (1875–1941) is best known for his pioneering work on the modern theory of the integral which is named for him and which is basic to present-day analysis.

11.6 NEAREST POINT THEOREM. *Let F be a non-void closed subset of* R^p *and let x be a point outside of F. Then there exists at least one point y belonging to F such that* $\|z - x\| \geq \|y - x\|$ *for all* $z \in F$.

PROOF. Since F is closed and $x \notin F$, then (cf. Exercise 11.H) the distance from x to F, which is defined to be $d = \inf\{\|x - z\| : z \in F\}$ satisfies $d > 0$. Let $F_k = \{z \in F : \|x - z\| \leq d + 1/k\}$ for $k \in N$. According to Example 9.5(f) these sets are closed in R^p and it is clear that F_1 is bounded and that $F_1 \supseteq F_2 \supseteq \cdots \supseteq F_k \supseteq \cdots$.

Furthermore, by the definition of d and F_k, it is seen that F_k is non-empty. It follows from the Cantor Intersection Theorem 11.4 that there is a point y belonging to all F_k, $k \in N$. It is readily seen that $\|x - y\| = d$, so that y satisfies the conclusion. (See Figure 11.3.)   Q.E.D.

A variant of the next theorem is of considerable importance in the theory of analytic functions. We shall state the result only for $p = 2$ and use intuitive ideas as to what it means for a set to be surrounded by a closed curve (that is, a curve which has no end points).

11.7 CIRCUMSCRIBING CONTOUR THEOREM. *Let F be a closed and bounded set in* R^2 *and let G be an open set which contains F. Then there exists a closed curve C, lying entirely in G and made up of arcs of a finite number of circles, such that F is surrounded by C.*

PARTIAL PROOF. If x belongs to $F \subseteq G$, there exists a number $\delta(x) > 0$ such that if $\|y - x\| < \delta(x)$, then y also belongs to G. Now let $G(x) = \{y \in R^2 : \|y - x\| < \frac{1}{2}\delta(x)\}$ for each x in F. Since the collection $\mathscr{G} = \{G(x) : x \in F\}$ constitutes a covering of the compact set F, the union of a

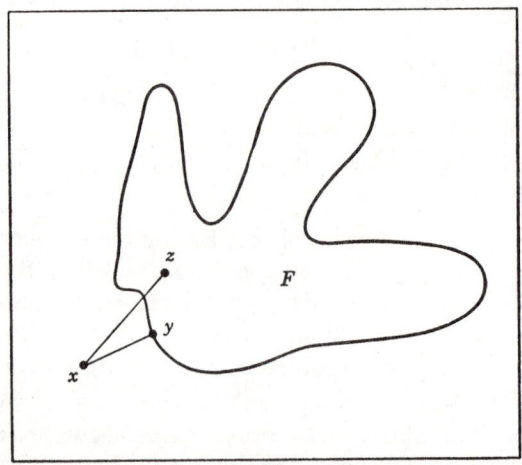

Figure 11.3

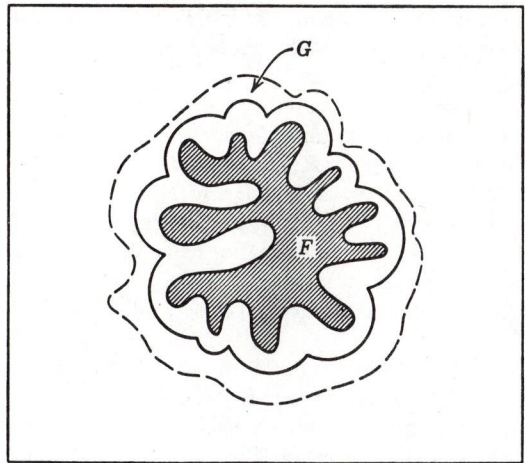

Figure 11.4

finite number of the sets in $\mathscr{G}$, say $G(x_1), \ldots, G(x_k)$, contains the compact set F. By using arcs from the circles with centers x_j and radii $\frac{1}{2}\delta(x_j)$, we obtain the desired curve C. (See Figure 11.4). The detailed construction of the curve will not be given here.   Q.E.D.

Exercises

11.A. Show directly from the definition (i.e., without using the Heine-Borel Theorem) that the open ball given by $\{(x, y): x^2 + y^2 < 1\}$ is not compact in $\mathbf{R}^2$.

11.B. Show directly that the entire space $\mathbf{R}^2$ is not compact.

11.C. Prove directly that if K is compact in $\mathbf{R}^p$ and $F \subseteq K$ is a closed set, then F is compact in $\mathbf{R}^p$.

11.D. Prove that if K is a compact subset of $\mathbf{R}$, then K is compact when regarded as a subset of $\mathbf{R}^2$.

11.E. By modifying the argument in Example 11.2(d), prove that the interval $J = \{(x, y): 0 \le x \le 1, 0 \le y \le 1\}$ is compact in $\mathbf{R}^2$.

11.F. Locate the places where the hypotheses that the set K is bounded and that it is closed were used in the proof of the Heine-Borel Theorem.

11.G. Prove the Cantor Intersection Theorem by selecting a point x_n from F_n and then applying the Bolzano-Weierstrass Theorem 10.6 to the set $\{x_n : n \in \mathbf{N}\}$.

11.H. If $F \ne \emptyset$ is closed in $\mathbf{R}^p$ and if

$$d(x, F) = \inf \{\|x - z\| : z \in F\} = 0,$$

then x belongs to F.

11.I. Does the Nearest Point Theorem in $\mathbf{R}$ imply that there is a strictly positive real number nearest zero?

11.J. If F is a non-empty closed set in $\mathbf{R}^p$ and if $x \notin F$, is there a *unique* point of F that is nearest to x?

11.K. If K is a compact subset of $\mathbf{R}^p$ and x is a point of $\mathbf{R}^p$, then the set $K_x = \{x + y : y \in K\}$ is also compact. (This set K_x is sometimes called the **translation** of the set K by x.)

11.L. The intersection of two open sets is compact if and only if it is empty. Can the intersection of an infinite collection of open sets be a non-empty compact set?

11.M. If F is a compact subset of $\mathbf{R}^2$ and G is an open set which contains F, then there exists a closed polygonal curve C lying entirely in G which surrounds F.

11.N. Let $\{H_n : n \in \mathbf{N}\}$ be a family of closed subsets of $\mathbf{R}^p$ with the property that no set H_n contains a non-void open set. (For example, H_n is a point or a line in $\mathbf{R}^2$.) Let $G \neq \emptyset$ be an open set.

(a) If $x_1 \in G \setminus H_1$, show that there exists a closed ball B_1 with center x_1 such that $B_1 \subseteq G$ and $H_1 \cap B_1 = \emptyset$.

(b) If $x_2 \notin H_2$ belongs to the interior of B_1, show that there exists a closed ball B_2 with center x_2 such that B_2 is contained in the interior of B_1 and $H_2 \cap B_2 = \emptyset$.

(c) Continue this process to obtain a nested family of closed balls such that $H_n \cap B_n = \emptyset$. By the Cantor Intersection Theorem 11.4 there is a point x_0 common to all of the B_n. Conclude that $x_0 \in G \setminus \bigcap H_n$, so that G cannot be contained in $\bigcup H_n$. This result is a form of what is often called "the Baire† Category Theorem."

11.O. A **line** in $\mathbf{R}^2$ is a set of points (x, y) which satisfy an equation of the form $ax + by + c = 0$ where $(a, b) \neq (0, 0)$. Use the preceding exercise to show that $\mathbf{R}^2$ is not the union of a countable collection of lines.

11.P. The set $\mathscr{C}(\mathbf{Q})$ of irrational numbers in $\mathbf{R}$ is not the union of a countable family of closed sets, none of which contains a non-empty open set.

11.Q. The set $\mathbf{Q}$ of rational numbers is not the intersection of a countable collection of open sets in $\mathbf{R}$.

Section 12   Connected Sets

We shall now introduce the notion of a connected set which will be used occasionally in the following.

12.1. DEFINITION. A subset $D \subseteq \mathbf{R}^p$ is said to be **disconnected** if there exist two open sets A, B such that $A \cap D$ and $B \cap D$ are disjoint, non-empty, and have union D. In this case the pair A, B is said to form a **disconnection** of D. A subset which is not disconnected is said to be **connected**. (See Figure 12.1.)

12.2 EXAMPLES. (a) The set $\mathbf{N} \subseteq \mathbf{R}$ is disconnected, since we can take $A = \{x \in \mathbf{R} : x < 3/2\}$ and $B = \{x \in \mathbf{R} : x > 3/2\}$.

(b) The set $H = \{1/n : n \in \mathbf{N}\}$ is disconnected.

(c) The set S consisting of all positive rational numbers is disconnected in $\mathbf{R}$ since we can take $A = \{x \in \mathbf{R} : x < \sqrt{2}\}$ and $B = \{x \in \mathbf{R} : x > \sqrt{2}\}$.

---

† RENÉ LOUIS BAIRE (1874–1932) was a professor at Dijon. He worked in set-theory and real analysis.

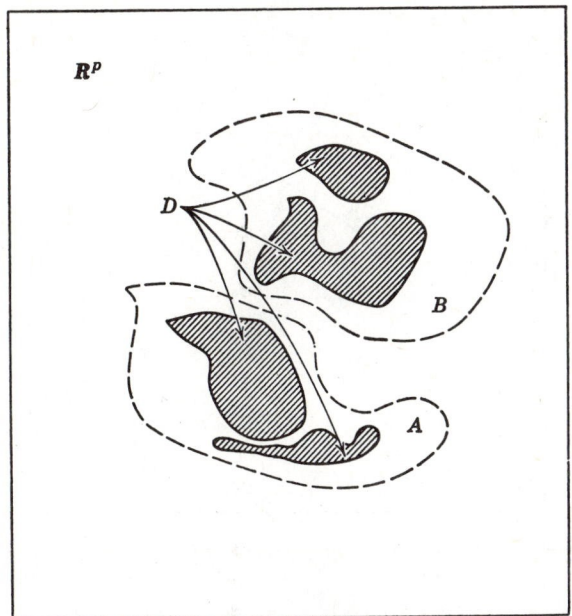

Figure 12.1.   A disconnected set.

(d) If $0 < c < 1$, then the sets $A = \{x \in \mathbf{R}, x \leq c\}$, $B = \{x \in \mathbf{R} : x > c\}$ split the unit interval $\mathbf{I} = \{x \in \mathbf{R} : 0 \leq x \leq 1\}$ into disjoint, non-empty sets with union $\mathbf{I}$. However since A is not open, this example does *not* show that $\mathbf{I}$ is disconnected. In fact, we shall show below that the set $\mathbf{I}$ is connected.

12.3   THEOREM.   *The closed unit interval* $\mathbf{I} = [0, 1]$ *is a connected subset of* $\mathbf{R}$.

PROOF.   We proceed by contradiction and suppose that A, B are open sets forming a disconnection of $\mathbf{I}$. Thus $A \cap \mathbf{I}$ and $B \cap \mathbf{I}$ are non-empty bounded disjoint sets whose union is $\mathbf{I}$. Since A and B are open, the sets $A \cap \mathbf{I}$ and $B \cap \mathbf{I}$ cannot consist of only one point. (Why?) For the sake of definiteness, we suppose that there exist points $a \in A$, $b \in B$ such that $0 < a < b < 1$. Applying the Supremum Property 6.4, we let $c = \sup \{x \in A : x < b\}$ so that $0 < c < 1$; hence $c \in A \cup B$. If $c \in A$, then $c \neq b$ and since A is open there is a point $a_1 \in A$, $c < a_1$, such that the interval $[c, a_1]$ is contained in $\{x \in A : x < b\}$, contrary to the definition of c. Similarly, if $c \in B$, then since B is open there is a point $b_1 \in B$, $b_1 < c$, such that the interval $[b_1, c]$ is contained in $B \cap \mathbf{I}$, contrary to the definition of c. Hence the hypothesis that $\mathbf{I}$ is disconnected leads to a contradiction.

Q.E.D.

The reader should note that the same proof can be used to show that the open interval $(0, 1)$ is connected in $\mathbf{R}$.

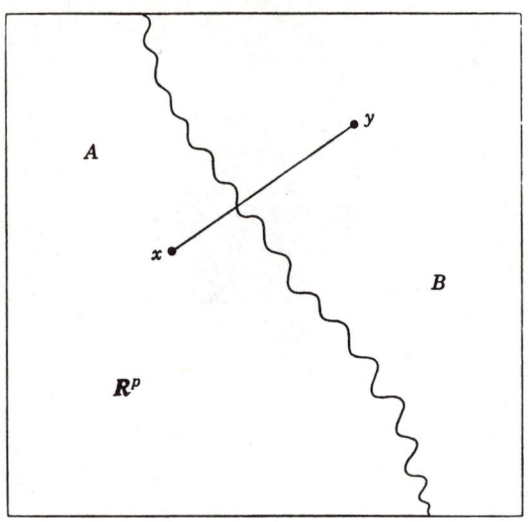

Figure 12.2

12.4 THEOREM. *The entire space $\mathbf{R}^p$ is connected.*

PROOF.   If not, then there exist two disjoint non-empty open sets A, B whose union is $\mathbf{R}^p$. (See Figure 12.2.) Let $x \in A$ and $y \in B$ and consider the line segment S joining x and y; namely,

$$S = \{x + t(y-x) : t \in \mathbf{I}\}.$$

Let $A_1 = \{t \in \mathbf{R} : x + t(y-x) \in A\}$ and let $B_1 = \{t \in \mathbf{R} : x + t(y-x) \in B\}$.   It is easily seen that A_1 and B_1 are disjoint non-empty open subsets of $\mathbf{R}$ and provide a disconnection for $\mathbf{I}$, contradicting Theorem 12.3.   Q.E.D.

12.5 COROLLARY. *The only subsets of $\mathbf{R}^p$ which are both open and closed are $\emptyset$ and $\mathbf{R}^p$.*

PROOF.   For if A is both open and closed in $\mathbf{R}^p$, then $B = \mathbf{R}^p \setminus A$ is also. If A is not empty and not all of $\mathbf{R}^p$, then the pair A, B forms a disconnection for $\mathbf{R}^p$, contradicting the theorem.   Q.E.D.

Connected Open Sets

In certain areas of analysis, connected open sets play an especially important role.   By using the definition it is easy to establish the next result.

12.6 LEMMA. *An open subset of $\mathbf{R}^p$ is connected if and only if it cannot be expressed as the union of two disjoint non-empty open sets.*

It is sometimes useful to have another characterization of open connected sets. In order to give such a characterization, we shall introduce some terminology. If x and y are two points in $\mathbf{R}^p$, then a **polygonal curve** joining x and y is a set P obtained as the union of a finite number of ordered line segments $(L_1, L_2, \ldots, L_n)$ in $\mathbf{R}^p$ such that the line segment L_1 has end points x, z_1; the line segment L_2 has end points $z_1, z_2; \ldots$; and the line segment L_n has end points z_{n-1}, y. (See Figure 12.3.)

12.7 THEOREM. *Let G be an open set in $\mathbf{R}^p$. Then G is connected if and only if any pair of points x, y in G can be joined by a polygonal curve lying entirely in G.*

PROOF. Assume that G is not connected and that A, B is a disconnection for G. Let $x \in A \cap G$ and $y \in B \cap G$ and let $P = (L_1, L_2, \ldots, L_n)$ be a polygonal curve lying entirely in G and joining x and y. Let k be the smallest natural number such that the end point z_{k-1} of L_k belongs to $A \cap G$ and the end point z_k belongs to $B \cap G$ (see Figure 12.4). If we define A_1 and B_1 by

$$A_1 = \{t \in \mathbf{R} : z_{k-1} + t(z_k - z_{k-1}) \in A \cap G\},$$
$$B_1 = \{t \in \mathbf{R} : z_{k-1} + t(z_k - z_{k-1}) \in B \cap G\},$$

then it is easily seen that A_1 and B_1 are disjoint non-empty open subsets of $\mathbf{R}$. Hence the pair A_1, B_1 form a disconnection for the unit interval $\mathbf{I}$, contradicting Theorem 12.3. Therefore, if G is not connected, there exist two points in G which cannot be joined by a polygonal curve in G.

Next, suppose that G is a connected open set in $\mathbf{R}^p$ and that x belongs to G. Let G_1 be the subset of G consisting of all points in G which can be

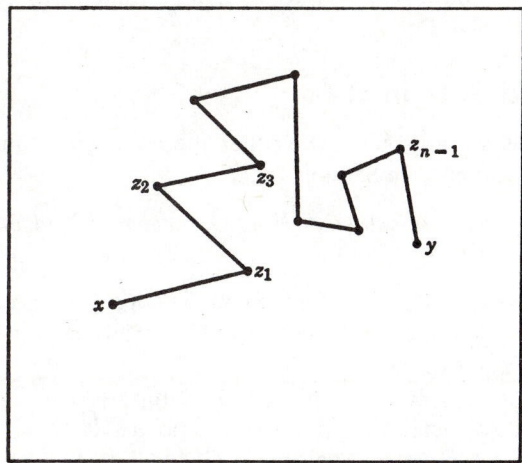

Figure 12.3. A polygonal curve.

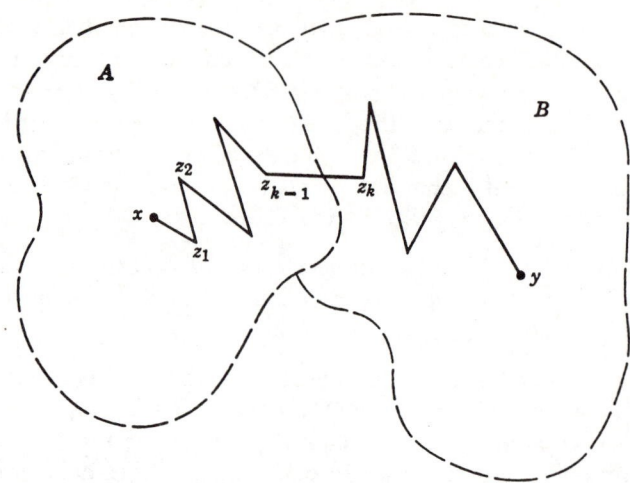

Figure 12.4

joined to x by a polygonal curve which lies entirely in G; let G_2 consist of all the points in G which cannot be joined to x by a polygonal curve lying in G. It is clear that $G_1 \cap G_2 = \emptyset$. The set G_1 is not empty since it contains the point x. We shall now show that G_1 is open in $\mathbf{R}^p$. If y belongs to G_1, it follows from the fact that G is open that for some real number $r > 0$, then $\|w - y\| < r$ implies that $w \in G$. By definition of G_1, the point y can be joined to x by a polygonal curve and by adding a segment from y to w, we infer that w belongs to G_1. Hence G_1 is an open subset of $\mathbf{R}^p$. Similarly, the subset G_2 is open in $\mathbf{R}^p$. If G_2 is not empty, then the sets G_1, G_2 form a disconnection of G, contrary to the hypothesis that G is connected. Therefore, $G_2 = \emptyset$ and every point of G can be joined to x by a polygonal curve lying entirely in G.   Q.E.D.

Connected Sets in $\mathbf{R}$

We close this section by showing that the connected subsets of $\mathbf{R}$ are precisely the intervals (see Section 7).

12.8   THEOREM.   _A subset of $\mathbf{R}$ is connected if and only if it is an interval.

PARTIAL PROOF.   The proof given in Theorem 12.3 can be readily modified to establish the connectedness of an arbitrary non-void interval. We leave the details to the reader.

Conversely, let $C \subseteq \mathbf{R}$ be connected and suppose that $C \neq \emptyset$. We note that C has the property that if a, $b \in C$ and $a < b$, then any number c satisfying $a < c < b$ must also belong to C; for if $c \notin C$, then the sets $A = \{x \in \mathbf{R} : x < c\}$ and $B = \{x \in \mathbf{R} : x > c\}$ form a disconnection of C.

(i) Now suppose that C is bounded above and below, and let $a = \inf C$ and $b = \sup C$. We shall show that C must have one of the four forms

$$[a, b], \qquad [a, b), \qquad (a, b], \qquad (a, b).$$

Indeed, if $a \in C$ and $b \in C$, then we have seen in the preceding paragraph that $[a, b] \subseteq C$ and the fact that $C \subseteq [a, b]$ follows from the fact that a and b are lower and upper bounds, respectively, of C.

If $a \in C$ but $b \notin C$, let b' be any number with $a \le b' < b$. Since $b = \sup C$, there must be an element $b'' \in C$ such that $a \le b' < b''$. Therefore the number b' must belong to C and, since b' is any number satisfying $a \le b' < b$, we infer that $C = [a, b)$.

Similarly, if $a \notin C$ but $b \in C$, we infer that $C = (a, b]$, while if $a \notin C$ and $b \notin C$, then we deduce that $C = (a, b)$.

(ii) Now suppose that C is bounded below but not bounded above, and let $a = \inf C$, so that $C \subseteq [a, +\infty)$. If $a \in C$ and if x is any real number with $a \le x$, then since C is not bounded above there exist $c \in C$ such that $x \le c$ whence it follows from the above property that $x \in C$. Since x is an arbitrary number satisfying $a \le x$, we conclude that $C = [a, +\infty)$.

Similarly, if $a \notin C$ we conclude that $C = (a, +\infty)$.

(iii) If C is not bounded below but is bounded above and if $b = \sup C$, then there are the two cases $C = (-\infty, b]$ or $C = (-\infty, b)$ according as $b \in C$ or $b \notin C$.

(iv) Finally, if C is neither bounded below nor bounded above, then we have the case $C = (-\infty, +\infty)$. Q.E.D.

Exercises

12.A. If A and B are connected subsets of $\mathbf{R}^p$, give examples to show that $A \cup B$, $A \cap B$, $A \setminus B$ can be either connected or disconnected.

12.B. If $C \subseteq \mathbf{R}^p$ is connected and x is a cluster point of C, then $C \cup \{x\}$ is connected.

12.C. If $C \subseteq \mathbf{R}^p$ is connected, show that its closure C^- (see Exercise 9.L) is also connected.

12.D A set $D \subseteq \mathbf{R}^p$ is disconnected if and only if $D = E \cup F$, where E, F are non-empty and $E \cap F^- = \emptyset$, $E^- \cap F = \emptyset$.

12.E. If $K \subseteq \mathbf{R}^p$ is convex (see Exercise 8.Q), then K is connected.

12.F. The Cantor set $\mathbf{F}$ is wildly disconnected. Show that if $x, y \in \mathbf{F}$, $x \ne y$, then there is a disconnection A, B of $\mathbf{F}$ such that $x \in A$, $y \in B$.

12.G. If C_1 and C_2 are connected subsets of $\mathbf{R}$, then the product $C_1 \times C_2$ is a connected subset of $\mathbf{R}^2$.

12.H. Show that the set

$$A = \{(x, y) \in \mathbf{R}^2 : 0 < y \le x^2, x \ne 0\} \cup \{(0, 0)\}$$

is connected in $\mathbf{R}^2$. However there does not exist a polygonal curve lying entirely in A joining $(0, 0)$ to other points in the set.

12.I. Show that the set

$$S = \left\{(x, y) \in \mathbf{R}^2 : y = \sin\frac{1}{x}, x \neq 0 \right\} \cup \left\{(0, y) : -1 \leq y \leq 1\right\}$$

is connected in $\mathbf{R}^2$. However, it is not always possible to join two points in S by a polygonal curve (or any "continuous" curve) lying entirely in S.

Section 13   The Complex Number System

Once the real number system is at hand, it is a simple matter to create the complex number system. We shall indicate in this section how the complex field can be constructed.†

As seen before, the real number system is a field which satisfies certain additional properties. In Section 8, we constructed the Cartesian space $\mathbf{R}^p$ and introduced some algebraic operations in the p-fold Cartesian product of $\mathbf{R}$. However, we did *not* make $\mathbf{R}^p$ into a field. It may come as a surprise that it is not possible to define a multiplication which makes $\mathbf{R}^p$, $p \geq 3$, into a field. Nevertheless, it is possible to define a multiplication operation in $\mathbf{R} \times \mathbf{R}$ which makes this set into a field. We now introduce the desired operations.

13.1   DEFINITION.   The complex number system C consists of all ordered pairs (x, y) of real numbers with the operation of **addition** defined by

$$(x, y) + (x', y') = (x + x', y + y'),$$

and the operation of **multiplication** defined by

$$(x, y) \cdot (x', y') = (xx' - yy', xy' + x'y).$$

Thus the complex number system C has the same elements as the two-dimensional space $\mathbf{R}^2$. It has the same addition operation, but it possesses a multiplication as $\mathbf{R}^2$ does not. Therefore, considered merely as sets, C and $\mathbf{R}^2$ are equal since they have the same elements; however, from the standpoint of algebra, they are not the same since they possess different operations.

An element of C is called a **complex number** and is often denoted by a single letter such as z. If $z = (x, y)$, then we refer to the real number x as the **real part** of z and to y as the **imaginary part** of z, in symbols,

$$x = \text{Re } z, \qquad y = \text{Im } z.$$

The complex number $\bar{z} = (x, -y)$ is called the **conjugate** of $z = (x, y)$.

† This section may be omitted on a first reading.

It is an important fact that the definition of addition and multiplication given above for elements of C makes it a "field" in the sense of abstract algebra. That is, it satisfies the algebraic properties listed in 4.1 provided the number 0 in (A3) is replaced by the pair $(0, 0)$, the element corresponding to $-a$ in (A4) is the pair $(-x, -y)$, the number 1 in (M3) is replaced by the pair $(1, 0)$, and the number corresponding to $1/a$ is the pair

$$\left(\frac{x}{x^2+y^2}, \frac{-y}{x^2+y^2}\right)$$

when $(x, y) \neq (0, 0)$.

Sometimes it is convenient to adopt part of the notation of Section 8 and write

$$az = a(x, y) = (ax, ay),$$

when a is a real number and $z = (x, y)$ is in C. With this notation, it is clear that each element in C has a unique representation in the form of a sum of a product of a real number with $(1, 0)$ and of the product of a real number with $(0, 1)$. Thus we can write

$$z = (x, y) = x(1, 0) + y(0, 1).$$

Since the element $(1, 0)$ is the identity element of C, it is natural to denote it by 1 (or to suppress it entirely when it is a factor). For the sake of brevity it is convenient to introduce a symbol for $(0, 1)$ and i is the conventional choice. With this notation, we write

$$z = (x, y) = x + iy.$$

In addition, we have $\bar{z} = (x, -y) = x - iy$ and

$$x = \operatorname{Re} z = \frac{z + \bar{z}}{2}, \qquad y = \operatorname{Im} z = \frac{z - \bar{z}}{2i}$$

By Definition 13.1, we have $(0, 1)(0, 1) = (-1, 0)$ which can be written as $i^2 = -1$. Thus in C the quadratic equation

$$z^2 + 1 = 0,$$

has a solution. The historical reason for the development of the complex number system was to obtain a system of "numbers" in which every quadratic equation has a solution. It was realized that not every equation with real coefficients has a real solution, and so complex numbers were invented to remedy this defect. It is a well-known fact that not only do the complex numbers suffice to produce solutions for every quadratic equation with real coefficients, but they also suffice to guarantee solutions for any polynomial equation of arbitrary degree and with coefficients which may be complex numbers. This result is called the Fundamental Theorem of Algebra and was proved first by the great Gauss† in 1799.

Although C cannot be given the order properties discussed in Section 5, it is easy to endow it with the metric and topological structure of Sections 8 and 9. For, if $z = (x, y)$ belongs to C, we define the *absolute value* of z to be

$$|z| = (x^2 + y^2)^{1/2}.$$

It is readily seen that the absolute value just defined has the properties:

    (i) $|z| \geq 0$;

    (ii) $|z| = 0$ if and only if $z = 0$;

    (iii) $|wz| = |w| \, |z|$;

    (iv) $||w| - |z|| \leq |w \pm z| \leq |w| + |z|$.

It will be observed that the absolute value of the complex number $z = (x, y)$ is precisely the same as the norm of the element (x, y) in $\mathbf{R}^2$. Therefore, all of the topological properties of the Cartesian spaces that were introduced and studied in Sections 9 to 12 are meaningful and valid for C. In particular, the notions of open and closed sets in C are exactly as for the Cartesian space $\mathbf{R}^2$. Furthermore, the Bolzano-Weierstrass Theorem 10.6, and the Heine-Borel Theorem 11.3 and its consequences, also hold in C, as does Theorem 12.7.

The reader should keep these remarks in mind throughout the remaining section of this book. It will be seen that *all of the succeeding material which applies to Cartesian spaces of dimension exceeding one, applies equally well to the complex number system.* Thus most of the results to be obtained pertaining to sequences, continuous functions, derivatives, integrals, and infinite series are also valid for C without change either in statement or in proof. The only exceptions to this statement are those properties which are based on the order properties of $\mathbf{R}$.

In this sense complex analysis is a special case of real analysis; however, there are a number of deep and important new features to the study of analytic functions that have no general counterpart in the realm of real analysis. Hence only the fairly superficial aspects of complex analysis are subsumed in what we shall do.

Exercises

13.A. Show that the complex number iz is obtained from z by a counter-clockwise rotation of $\pi/2$ radians ($= 90°$) around the origin.

13.B. If $c = (\cos \theta, \sin \theta) = \cos \theta + i \sin \theta$, then the number cz is obtained from z by a counter-clockwise rotation of θ radians around the origin.

† CARL FRIEDRICH GAUSS (1777–1855), the prodigious son of a day laborer, was one of the greatest of all mathematicians, but is also remembered for his work in astronomy, physics, and geodesy. He became professor and director of the Observatory at Göttingen.

13.C. Describe the geometrical relation between the complex numbers z and $az + b$, where $a \neq 0$. Show that the mapping defined for $z \in C$, by $f(z) = az + b$, sends circles into circles and lines into lines.

13.D. Describe the geometrical relations among the complex numbers z, $\bar{z}$ and $1/z$ for $z \neq 0$. Show that the mapping defined by $g(z) = \bar{z}$ sends circles into circles and lines into lines. Which circles and lines are left fixed under g?

13.E. Show that the **inversion mapping**, defined by $h(z) = 1/z$, sends circles and lines into circles and lines. Which circles are sent into lines? Which lines are sent into circles? Examine the images under h of the vertical lines given by the equation Re z = constant, the horizontal lines Im z = constant, the circles $|z|$ = constant.

13.F. Investigate the geometrical character of the mapping defined by $g(z) = z^2$. Determine if the mapping g is one-one and if it maps C onto all of C. Examine the inverse images under g of the lines

$$\text{Re } z = \text{constant}, \qquad \text{Im } z = \text{constant},$$

and the circles $|z|$ = constant.

III
CONVERGENCE

The material in the preceding two chapters should provide an adequate understanding of the real number system and the Cartesian spaces. Now that these algebraic and topological foundations have been laid, we are prepared to pursue questions of a more analytic nature. We shall begin with a study of the convergence of sequences. Some of the results in this chapter may be familiar to the reader from other courses in analysis, but the presentation given here is intended to be rigorous and to give certain more profound results than are usually discussed in earlier courses.

We shall first introduce the meaning of the convergence of a sequence of elements in R^p and establish some elementary (but useful) results about convergent sequences. We then present some important criteria for convergence. Next we study the convergence and uniform convergence of sequences of functions. After a brief section on the limit superior we append a final section which, though interesting, can be omitted without loss of continuity since the results will not be applied later.

Because of the linear limitations inherent in a book, we have decided to follow this chapter with a study of continuity, differentiation, and integration. This has the unfortunate aspect of deferring a full presentation of series until much later. The instructor is encouraged to give at least a brief introduction to series along with this chapter, or he can go directly to the first part of Chapter VI after Section 16, if he prefers to do so.

Section 14   Introduction to Sequences

Although the theory of convergence can be presented on a very abstract level, we prefer to discuss the convergence of sequences in a Cartesian spaces R^p, paying special attention to the case of the real line. The reader should interpret the ideas by drawing diagrams in R and R^2.

14.1 DEFINITION. If S is any set, a **sequence** in S is a function on the set $N = \{1, 2, \ldots\}$ of natural numbers and whose range is in S. In particular, a sequence in R^p is a function whose domain is N and whose range is contained in R^p.

In other words, a sequence in R^p assigns to each natural number $n = 1, 2, \ldots$, a uniquely determined element of R^p. Traditionally, the element of R^p which is assigned to a natural number n is denoted by a symbol such as x_n and, although this notation is at variance with that employed for most functions, we shall adhere to the conventional symbolism. [To be consistent with earlier notation, if $X : N \to R^p$ is a sequence, the value of X at $n \in N$ should be symbolized by $X(n)$, rather than by x_n.]

While we accept the traditional notation, we also wish to distinguish between the function X and its values $X(n) = x_n$. Hence when the elements of the sequence (that is, the values of the function) are denoted by x_n, we shall denote the function by the notation $X = (x_n)$ or by $X = (x_n : n \in N)$. We use parentheses to indicate that the ordering induced by that in N is a matter of importance. Thus we are distinguishing notationally between the sequence $X = (x_n : n \in N)$ and the set $\{x_n : n \in N\}$ of values of this sequence.

In defining sequences we often list in order the elements of the sequence, stopping when the rule of formation seems evident. Thus we may write

$$(2, 4, 6, 8, \ldots)$$

for the sequence of even integers. A more satisfactory method is to specify a formula for the general term of the sequence, such as

$$(2n : n \in N).$$

In practice it is often more convenient to specify the value x_1 and a method of obtaining x_{n+1}, $n \geq 1$, when x_n is known. Still more generally, we may specify x_1 and a rule for obtaining x_{n+1} from $x_1, x_2, \ldots, x_n$. We shall refer to either of these methods as *inductive* definitions of the sequence. In this way we might define the sequence of even natural numbers by the definition

$$x_1 = 2, \qquad x_{n+1} = x_n + 2, \qquad n \geq 1.$$

or by the (apparently more complicated) definition

$$x_1 = 2, \qquad x_{n+1} = x_n + x_1, \qquad n \geq 1.$$

Clearly, many other methods of defining this sequence are possible.

We now introduce some methods of constructing new sequences from given ones.

14.2 DEFINITION. If $X = (x_n)$ and $Y = (y_n)$ are sequences in R^p, then we define their **sum** to be the sequence $X + Y = (x_n + y_n)$ in R^p, their **difference** to be the sequence $X - Y = (x_n - y_n)$, and their **inner product** to be the sequence $X \cdot Y = (x_n \cdot y_n)$ in R which is obtained by taking the inner product of corresponding terms. Similarly, if $X = (x_n)$ is a sequence in R and if $Y = (y_n)$ is a sequence in R^p, we define the **product** of X and Y to be the sequence in R^p denoted by $XY = (x_n y_n)$; or, if $c \in R$ and $X = (x_n)$, we define $cX = (cx_n)$. Finally, if $Y = (y_n)$ is a sequence in R with $y_n \neq 0$, we can define the **quotient** of a sequence $X = (x_n)$ in R^p by Y to be the sequence $X/Y = (x_n/y_n)$.

For example, if X, Y are the sequences in R given by

$$X = (2, 4, 6, \ldots, 2n, \ldots), \qquad Y = \left(1, \frac{1}{2}, \frac{1}{3}, \ldots, \frac{1}{n}, \ldots\right),$$

then we have

$$X + Y = \left(3, \frac{9}{2}, \frac{19}{3}, \ldots, \frac{2n^2 + 1}{n}, \ldots\right),$$

$$X - Y = \left(1, \frac{7}{2}, \frac{17}{3}, \ldots, \frac{2n^2 - 1}{n}, \ldots\right),$$

$$XY = (2, 2, 2, \ldots, 2, \ldots),$$

$$3X = (6, 12, 18, \ldots, 6n, \ldots),$$

$$\frac{X}{Y} = (2, 8, 18, \ldots, 2n^2, \ldots).$$

Similarly, if Z denotes the sequence in R given by

$$Z = \left(1, 0, 1, \ldots, \frac{1 - (-1)^n}{2}, \ldots\right),$$

then we have defined $X + Z$, $X - Z$ and XZ; but X/Z is not defined, since some of the elements in Z are zero.

We now come to the notion of the limit of a sequence.

14.3 DEFINITION. Let $X = (x_n)$ be a sequence in R^p. An element x of R^p is said to be a **limit of** X if, for each neighborhood V of x there is a natural number K_V such that for all $n \geq K_V$, then x_n belongs to V. *If x is a limit of X, we also say that X **converges** to x. If a sequence has a limit, we say that the sequence is **convergent**. If a sequence has no limit then we say that it is **divergent**.*

The notation K_V is used to suggest that the choice of K will depend on V. It is clear that a small neighborhood V will usually require a large value of K_V in order to guarantee that $x_n \in V$ for all $n \geq K_V$.

We have defined the limit of a sequence $X = (x_n)$ in terms of neighborhoods. It is often convenient to use the norm in R^p to give an equivalent definition, which we now state as a theorem.

14.4 THEOREM. *Let $X = (x_n)$ be a sequence in R^p. An element x of R^p is a limit of X if and only if for each $\varepsilon > 0$ there is a natural number $K(\varepsilon)$ such that for all $n \geq K(\varepsilon)$, then $\|x_n - x\| < \varepsilon$.*

PROOF. Suppose that x is a limit of the sequence X according to Definition 14.3. Now let $\varepsilon > 0$ and consider the open ball $V(\varepsilon) = \{y \in \mathbf{R}^p : \|y - x\| < \varepsilon\}$, which is a neighborhood of x. By Definition 14.3 there is a natural number $K_{V(\varepsilon)}$ such that if $n \geq K_{V(\varepsilon)}$, then $x_n \in V(\varepsilon)$. Hence if $n \geq K_{V(\varepsilon)}$, then $\|x_n - x\| < \varepsilon$. This shows that the stated property holds when x is a limit of X.

Conversely, suppose that the property in the theorem holds for all $\varepsilon > 0$; we must show that Definition 14.3 is satisfied. To do this, let V be any neighborhood of x; then there is a number $\varepsilon > 0$ such that the open ball $V(\varepsilon)$ with center x and radius ε is contained in V. According to the property in the theorem, there is a natural number $K(\varepsilon)$ such that if $n \geq K(\varepsilon)$, then $\|x_n - x\| < \varepsilon$. Stated differently, if $n \geq K(\varepsilon)$, then $x_n \in V(\varepsilon)$; hence $x_n \in V$ and the requirement in Definition 14.3 is satisfied. Q.E.D.

14.5 UNIQUENESS OF LIMITS. *A sequence in $\mathbf{R}^p$ can have at most one limit.*

PROOF. Suppose, on the contrary that x', x'' are limits of $X = (x_n)$ and that $x' \neq x''$. Let V', V'' be disjoint neighborhoods of x', x'', respectively, and let K', K'' be natural numbers such that if $n \geq K'$ then $x_n \in V'$ and if $n \geq K''$ then $x_n \in V''$. Let $K = \sup \{K', K''\}$ so that both $x_K \in V'$ and $x_K \in V''$. We infer that x_K belongs to $V' \cap V''$, contrary to the supposition that V' and V'' are disjoint. Q.E.D.

When a sequence $X = (x_n)$ in $\mathbf{R}^p$ has a limit x, we often write

$$x = \lim X, \qquad \text{or} \qquad x = \lim_n (x_n),$$

or sometimes use the symbolism $x_n \to x$.

We say that a sequence $X = (x_n)$ in $\mathbf{R}^p$ is **bounded** if there exists $M > 0$ such that $\|x_n\| < M$ for all $n \in \mathbf{N}$.

14.6 LEMMA. *A convergent sequence in $\mathbf{R}^p$ is bounded.*

PROOF. Let $x = \lim (x_n)$ and let $\varepsilon = 1$. By Theorem 14.4 there exists a natural number $K = K(1)$ such that if $n \geq K$, then $\|x_n - x\| \leq 1$. By using the Triangle Inequality, we infer that if $n \geq K$, then $\|x_n\| \leq \|x\| + 1$. If we set $M = \sup \{\|x_1\|, \|x_2\|, \ldots, \|x_{K-1}\|, \|x\| + 1\}$, then $\|x_n\| \leq M$ for all $n \in \mathbf{N}$. Q.E.D.

It might be suspected that the theory of convergence of sequences in $\mathbf{R}^p$ is more complicated than in $\mathbf{R}$, but this is not the case (except for notational matters). In fact, the next result is important in that it shows that questions of convergence in $\mathbf{R}^p$ can be reduced to the idential questions in $\mathbf{R}$ for each of the coordinate sequences.

Before stating this result, we recall that a typical element x in $\mathbf{R}^p$ is represented in coordinate fashion by "p-tuple"

$$x = (x_1, x_2, \ldots, x_p).$$

Hence each element in a sequence (x_n) in $\mathbf{R}^p$ has a similar representation; thus $x_n = (x_{1n}, x_{2n}, \ldots, x_{pn})$. In this way, the sequence (x_n) generates p sequences of real numbers; namely, $(x_{1n}), (x_{2n}), \ldots, (x_{pn})$. We shall now show that the convergence of the sequence (x_n) is faithfully reflected by the convergence of these p sequences of coordinates.

14.7 THEOREM. *A sequence (x_n) in $\mathbf{R}^p$ with*

$$x_n = (x_{1n}, x_{2n}, \ldots, x_{pn}), \qquad n \in \mathbf{N},$$

converges to an element $y = (y_1, y_2, \ldots, y_p)$ if and only if the corresponding p sequences of real numbers

$$(14.1) \qquad\qquad (x_{1n}), (x_{2n}), \ldots, (x_{pn}),$$

converge to $y_1, y_2, \ldots, y_p$ respectively.

PROOF. If $x_n \to y$, then $\|x_n - y\| < \varepsilon$ for $n \ge K(\varepsilon)$. In view of Theorem 8.10, for each $j = 1, 2, \ldots, p$, we have

$$|x_{jn} - y_j| \le \|x_n - y\| < \varepsilon, \qquad \text{for} \quad n \ge K(\varepsilon).$$

Hence each of the p coordinate sequences must converge to the corresponding real number.

Conversely, suppose that the sequences in (14.1) converge to y_j for $j = 1, 2, \ldots, p$. Given $\varepsilon > 0$, there is a natural number $M(\varepsilon)$ such that if $n \ge M(\varepsilon)$, then

$$|x_{jn} - y_j| < \varepsilon/\sqrt{p} \qquad \text{for} \quad j = 1, 2, \ldots, p.$$

From this it follows that, when $n \ge M(\varepsilon)$, then

$$\|x_n - y\|^2 = \sum_{j=1}^{p} |x_{jn} - y_j|^2 \le \varepsilon^2,$$

so that the sequence (x_n) converges to y. Q.E.D.

Some Examples

We shall now present some examples, establishing the convergence of a sequence using only the methods we presently have available. It will be noticed that in order to proceed, we must have "guessed" the value of the limit by previous examination of the sequence. All of the examples to be presented next involve some manipulative skill and "trickery," but the results we obtain will be very useful to us in establishing (by less tricky procedures) the convergence of other sequences. So we are as interested in the results as in the methods.

14.8 EXAMPLES. (a) Let (x_n) be the sequence in $\mathbf{R}$ where $x_n = 1/n$. We shall show that $\lim(1/n) = 0$. To do this let $\varepsilon > 0$; according to Corollary 6.7(b) (of the Archimedean Property) there exists a natural number $K(\varepsilon)$ such that $1/K(\varepsilon) < \varepsilon$. Then, if $n \geq K(\varepsilon)$ we have

$$0 < x_n = \frac{1}{n} \leq \frac{1}{K(\varepsilon)} < \varepsilon,$$

whence it follows that $|x_n - 0| < \varepsilon$ for $n \geq K(\varepsilon)$. Since $\varepsilon > 0$ is arbitrary this proves that $\lim(1/n) = 0$.

(b) Let $a > 0$ and consider the sequence $X = (1/(1 + na))$ in $\mathbf{R}$. We shall show that $\lim X = 0$. First we note that

$$0 < \frac{1}{1 + na} < \frac{1}{na}.$$

We want the dominant term to be smaller than a given $\varepsilon > 0$ when n is sufficiently large. By Corollary 6.7(b) again, there exists a natural number $K(\varepsilon)$ such that $1/K(\varepsilon) < a\varepsilon$. Then if $n \geq K(\varepsilon)$ we have

$$0 < \frac{1}{1 + na} < \frac{1}{na} \leq \frac{1}{K(\varepsilon)a} < \varepsilon,$$

whence it follows that $|1/(1 + na) - 0| < \varepsilon$ for $n \geq K(\varepsilon)$. Since $\varepsilon > 0$ is arbitrary this shows that $\lim X = 0$.

(c) Let $b \in \mathbf{R}$ satisfy $0 < b < 1$ and consider the sequence (b^n). We shall show that $\lim(b^n) = 0$. To do this, it is convenient to write b in the form

$$b = \frac{1}{1 + a}$$

where $a > 0$, and to use Bernoulli's Inequality $(1 + a)^n \geq 1 + na$ for $n \in \mathbf{N}$. (See Exercise 5.C.) Hence

$$0 < b^n = \frac{1}{(1 + a)^n} \leq \frac{1}{1 + na} < \frac{1}{na}.$$

As in the preceding example, if $\varepsilon > 0$ is given, then there is a natural number $K(\varepsilon)$ such that $|b^n - 0| < \varepsilon$ when $n \geq K(\varepsilon)$. Therefore we have $\lim(b^n) = 0$.

(d) Let $c > 0$ and consider the sequence $(c^{1/n})$. We shall show that $\lim(c^{1/n}) = 1$.

First suppose that $c > 1$. Then $c^{1/n} = 1 + d_n$ with $d_n > 0$ and hence by Bernoulli's Inequality

$$c = (1 + d_n)^n \geq 1 + nd_n.$$

It follows that $c - 1 \geq nd_n$. Since $c > 1$, we have $c - 1 > 0$. Hence given

$\varepsilon > 0$, then there is a natural number $K(\varepsilon)$ such that if $n \geq K(\varepsilon)$, then

$$0 < c^{1/n} - 1 = d_n \leq \frac{c-1}{n} < \varepsilon.$$

Therefore $|c^{1/n} - 1| < \varepsilon$ when $n \geq K(\varepsilon)$, as desired.

Now suppose that $0 < c < 1$ (for the case $c = 1$ is obvious). Then $c^{1/n} = 1/(1 + h_n)$ with $h_n > 0$ and hence by Bernoulli's Inequality

$$c = \frac{1}{(1+h_n)^n} \leq \frac{1}{1+nh_n} < \frac{1}{nh_n}.$$

It follows that $0 < h_n < 1/nc$. But since $c > 0$, given $\varepsilon > 0$ there is a natural number $K(\varepsilon)$ such that if $n \geq K(\varepsilon)$ then

$$0 < 1 - c^{1/n} = \frac{h_n}{1+h_n} < h_n < \frac{1}{nc} < \varepsilon.$$

Therefore $|c^{1/n} - 1| < \varepsilon$ when $n \geq K(\varepsilon)$, as desired.

(e) Consider the sequence $X = (n^{1/n})$; we shall show that $\lim X = 1$, a rather non-obvious fact. Write $n^{1/n} = 1 + k_n$ with $k_n > 0$ for $n > 1$; hence $n = (1 + k_n)^n$. By the Binomial Theorem, when $n > 1$ we have

$$n = 1 + nk_n + \frac{n(n-1)}{2} k_n^2 + \cdots > \frac{n(n-1)}{2} k_n^2.$$

It follows that $k_n^2 < 2/(n-1)$, so that

$$k_n < \sqrt{\frac{2}{n-1}}.$$

Now let $\varepsilon > 0$ be given. Then there exists $K(\varepsilon)$ such that if $n \geq K(\varepsilon)$, then $1/(n-1) < \varepsilon^2/2$; whence it follows that $0 < k_n < \varepsilon$ and so

$$0 < n^{1/n} - 1 = k_n < \varepsilon$$

for $n \geq K(\varepsilon)$. Since $\varepsilon > 0$ is arbitrary, this proves that $\lim (n^{1/n}) = 1$.

These examples show that a body of results which will make the ingenuity employed here unnecessary would prove highly useful. We shall obtain such results in the following two sections; but we close this section with a result which is very often useful.

14.9 **THEOREM.** *Let $X = (x_n)$ be a sequence in $\mathbf{R}^p$ and let $x \in \mathbf{R}^p$. Let $A = (a_n)$ be a sequence in $\mathbf{R}$ which is such that*
  (i) $\lim (a_n) = 0$,
  (ii) $\|x_n - x\| \leq C |a_n|$ *for some $C > 0$ and all $n \in \mathbf{N}$.*
Then $\lim (x_n) = x$.

PROOF. Let $\varepsilon > 0$ be given. Since $\lim (a_n) = 0$, there exists a natural number $K(\varepsilon)$ such that if $n \ge K(\varepsilon)$ then

$$C\,|a_n| = C\,|a_n - 0| \le \varepsilon.$$

It follows that

$$\|x_n - x\| \le C\,|a_n| \le \varepsilon$$

for all $n \ge K(\varepsilon)$. Since $\varepsilon > 0$ is arbitrary, we infer that $\lim (x_n) = x$.   Q.E.D.

Exercises

14.A. Let $b \in \mathbf{R}$; show that $\lim (b/n) = 0$.

14.B. Show that $\lim (1/n - 1/(n+1)) = 0$.

14.C. Let $X = (x_n)$ be a sequence in $\mathbf{R}^p$ which is convergent to x, and let $c \in \mathbf{R}$. Show that $\lim (cx_n) = cx$.

14.D. Let $X = (x_n)$ be a sequence in $\mathbf{R}^p$ which is convergent to x. Show that $\lim (\|x_n\|) = \|x\|$. (Hint: use the Triangle Inequality.)

14.E. Let $X = (x_n)$ be a sequence in $\mathbf{R}^p$ and let $\lim (\|x_n\|) = 0$. Show that $\lim (x_n) = 0$. However, give an example in $\mathbf{R}$ to show that the convergence of $(|x_n|)$ may not imply the convergence of (x_n).

14.F. Show that $\lim (1/\sqrt{n}) = 0$. In fact, if (x_n) is a sequence of positive numbers and $\lim (x_n) = 0$, then $\lim (\sqrt{x_n}) = 0$.

14.G. Let $d \in \mathbf{R}$ satisfy $d > 1$. Use Bernoulli's Inequality to show that the sequence (d^n) is not bounded in $\mathbf{R}$. Hence it is not convergent.

14.H. Let $b \in R$ satisfy $0 < b < 1$; show that $\lim (nb^n) = 0$. (Hint: use the Binomial Theorem as in Example 14.8(e).)

14.I. Let $X = (x_n)$ be a sequence of strictly positive real numbers such that $\lim (x_{n+1}/x_n) < 1$. Show that for some r with $0 < r < 1$ and some $C > 0$, then we have $0 < x_n < Cr^n$ for all sufficiently large $n \in \mathbf{N}$. Use this to show that $\lim (x_n) = 0$.

14.J. Let $X = (x_n)$ be a sequence of strictly positive real numbers such that $\lim (x_{n+1}/x_n) > 1$. Show that X is not a bounded sequence and hence is not convergent.

14.K. Give an example of a convergent sequence (x_n) of strictly positive real numbers such that $\lim (x_{n+1}/x_n) = 1$. Give an example of a divergent sequence with this property.

14.L. Apply the results of Exercises 14.I and 14.J to the following sequences. (Here $0 < a < 1$, $1 < b$, $c > 0$.)

(a) (a^n),
(b) (na^n),
(c) (b^n),
(d) (b^n/n),
(e) $(c^n/n!)$,
(f) $(2^{3n}/3^{2n})$.

14.M. Let $X = (x_n)$ be a sequence of strictly positive real numbers such that $\lim (x_n^{1/n}) < 1$. Show that for some r with $0 < r < 1$, then $0 < x_n < r^n$ for all sufficiently large $n \in \mathbf{N}$. Use this to deduce that $\lim (x_n) = 0$.

14.N. Let $X = (x_n)$ be a sequence of strictly positive real numbers such that $\lim (x_n^{1/n}) > 1$. Show that X is not a bounded sequence and hence is not convergent.

14.O. Give an example of a convergent sequence (x_n) of strictly positive real numbers such that $\lim (x_n^{1/n}) = 1$. Give an example of a divergent sequence with this property.

14.P. Reexamine the convergence of the sequences in Exercise 14.L in the light of Exercises 14.M and 14.N.

14.Q. Examine the convergence of the following sequences in R.

(a) $\left(\dfrac{(-1)^n}{n}\right)$,

(b) $\left(\dfrac{1}{n^2}\right)$,

(c) $\left(\dfrac{n^2}{n+1}\right)$.

(d) $((-1)^n)$.

Section 15 Subsequences and Combinations

This section gives some information about the convergence of sequences obtained in various ways from sequences which are known to be convergent. It will help to enable us to expand our collection of convergent sequences rather extensively.

15.1 DEFINITION. If $X = (x_n)$ is a sequence in R^p and if $r_1 < r_2 < \cdots < r_n < \cdots$ is a strictly increasing sequence of natural numbers, then the sequence X' in R^p given by

$$(x_{r_1}, x_{r_2}, \ldots, x_{r_n}, \ldots),$$

is called a **subsequence** of X.

It may be helpful to connect the notion of a subsequence with that of the composition of two functions. Let g be a function with domain N and range in N and let g be **strictly increasing** in the sense that if $n < m$, then $g(n) < g(m)$. Then g defines a subsequence of $X = (x_n)$ by the formula

$$X \circ g = (x_{g(n)} : n \in N).$$

Conversely, every subsequence of X has the form $X \circ g$ for some strictly increasing function g with $D(g) = N$ and $R(g) \subseteq N$.

It is clear that a given sequence has many different subsequences. Although the next result is very elementary, it is of sufficient importance that it must be made explicit.

15.2 LEMMA. *If a sequence X in R^p converges to an element x, then any subsequence of X also converges to x.*

PROOF. Let V be a neighborhood of the limit element x; by definition, there exists a natural number K_V such that for all $n \geq K_V$, then x_n belongs to V. Now let X' be a subsequence of X; say

$$X' = (x_{r_1}, x_{r_2}, \ldots, x_{r_n}, \ldots).$$

Since $r_n \geq n$, then $r_n \geq K_V$ and hence x_{r_n} belongs to V. This proves that X' also converges to x.

Q.E.D.

15.3 COROLLARY. *If $X = (x_n)$ is a sequence which converges to an element x of $\mathbf{R}^p$ and if m is any natural number, then the sequence $X' = (x_{m+1}, x_{m+2}, \ldots)$ also converges to x.*

PROOF. Since X' is a subsequence of X, the result follows directly from the preceding lemma.

Q.E.D.

The preceding results have been mostly directed towards proving that a sequence converges to a given point. It is also important to know precisely what it means to say that a sequence X does *not* converge to x. The next result is elementary but not trivial and its verification is an important part of everyone's education. Therefore, we leave its detailed proof to the reader.

15.4 THEOREM. *If $X = (x_n)$ is a sequence in $\mathbf{R}^p$, then the following statements are equivalent:*

(a) *X does not converge to x.*

(b) *There exists a neighborhood V of x such that if n is any natural number, then there is a natural number $m = m(n) \geq n$ such that x_m does not belong to V.*

(c) *There exists a neighborhood V of x and a subsequence X' of X such that none of the elements of X' belongs to V.*

15.5 EXAMPLES. (a) Let X be the sequence in $\mathbf{R}$ consisting of the natural numbers

$$X = (1, 2, \ldots, n, \ldots).$$

Let x be any real number and consider the neighborhood V of x consisting of the open interval $(x - 1, x + 1)$. According to the Archimedean Property 6.6 there exists a natural number k_0 such that $x + 1 < k_0$; hence, if $n \geq k_0$, it follows that $x_n = n$ does not belong to V. Therefore the subsequence $X' = (k_0, k_0 + 1, \ldots)$ of X has no points in V, showing that X does not converge to x.

(b) Let $Y = (y_n)$ be the sequence in $\mathbf{R}$ consisting of $Y = (-1, 1, \ldots, (-1)^n, \ldots)$. We leave it to the reader to show that no point y, except possibly $y = \pm 1$, can be a limit of Y. We shall show that the point $y = -1$ is not a limit of Y; the consideration for $y = +1$ is entirely similar. Let V be the neighborhood of $y = -1$ consisting of the open interval $(-2, 0)$. Then, if n is even, the element $y_n = (-1)^n = +1$ does not belong to V. Therefore, the subsequence Y' of Y corresponding to $r_n = 2n$, $n \in \mathbf{N}$, avoids the neighborhood V, showing that $y = -1$ is not a limit of Y.

(c) Let $Z = (z_n)$ be a sequence in $\mathbf{R}$ with $z_n \geq 0$, for $n \geq 1$. We conclude that no number $z < 0$ can be a limit for Z. In fact, the open set $V = \{x \in \mathbf{R} : x < 0\}$ is a neighborhood of z containing none of the elements of Z. This shows (why?) that z cannot be the limit of Z. Hence if Z has a limit, this limit must be positive.

Combinations of Sequences

The next theorem enables one to use the algebraic operations of Definitions 14.2 to form new sequences whose convergence can be predicted from the convergence of the given sequences.

15.6 THEOREM. (a) *Let X and Y be sequences in $\mathbf{R}^p$ which converge to x and y, respectively. Then the sequences $X + Y$, $X - Y$, and $X \cdot Y$ converge to $x + y$, $x - y$, and $x \cdot y$, respectively.*

(b) *Let $X = (x_n)$ be a sequence in $\mathbf{R}^p$ which converges to x and let $A = (a_n)$ be a sequence in $\mathbf{R}$ which converges to a. Then the sequence $(a_n x_n)$ in $\mathbf{R}^p$ converges to ax.*

(c) *Let $X = (x_n)$ be a sequence in $\mathbf{R}^p$ which converges to x and let $B = (b_n)$ be a sequence of non-zero real numbers which converges to a non-zero number b. Then the sequence $(b_n^{-1} x_n)$ in $\mathbf{R}^p$ converges to $b^{-1}x$.*

PROOF. (a) To show that $(x_n + y_n) \to x + y$, we need to appraise the magnitude of $\|(x_n + y_n) - (x + y)\|$. To do this, we use the Triangle Inequality to obtain

$$(15.1) \qquad \|(x_n + y_n) - (x + y)\| = \|(x_n - x) + (y_n - y)\|$$
$$\leq \|x_n - x\| + \|y_n - y\|.$$

By hypothesis, if $\varepsilon > 0$ we can choose K_1 such that if $n \geq K_1$, then $\|x_n - x\| < \varepsilon/2$ and we choose K_2 such that if $n \geq K_2$, then $\|y_n - y\| < \varepsilon/2$. Hence if $K_0 = \sup\{K_1, K_2\}$ and $n \geq K_0$, then we conclude from (15.1) that

$$\|(x_n + y_n) - (x + y)\| < \varepsilon/2 + \varepsilon/2 = \varepsilon.$$

Since this can be done for arbitrary $\varepsilon > 0$, we infer that $X + Y$ converges to $x + y$. Precisely the same argument can be used to show that $X - Y$ converges to $x - y$.

To prove that $X \cdot Y$ converges to $x \cdot y$, we make the estimate

$$|x_n \cdot y_n - x \cdot y| = |(x_n \cdot y_n - x_n \cdot y) + (x_n \cdot y - x \cdot y)|$$
$$\leq |x_n \cdot (y_n - y)| + |(x_n - x) \cdot y|.$$

Using the Schwarz Inequality, we obtain

$$(15.2) \qquad |x_n \cdot y_n - x \cdot y| \leq \|x_n\| \, \|y_n - y\| + \|x_n - x\| \, \|y\|.$$

According to Lemma 14.6, there exists a number $M > 0$ which is an upper bound for $\{\|x_n\|, \|y\|\}$. In addition, from the convergence of X, Y, we conclude that if $\varepsilon > 0$ is given, then there exist natural numbers K_1, K_2 such that if $n \geq K_1$, then $\|y_n - y\| < \varepsilon/2M$ and if $n \geq K_2$ then $\|x_n - x\| < \varepsilon/2M$. Now choose $K = \sup\{K_1, K_2\}$; then, if $n \geq K$, we infer from (15.2) that

$$|x_n \cdot y_n - x \cdot y| \leq M \|y_n - y\| + M \|x_n - x\|$$

$$< M\left(\frac{\varepsilon}{2M} + \frac{\varepsilon}{2M}\right) = \varepsilon.$$

This proves that $X \cdot Y$ converges to $x \cdot y$.

Part (b) is proved in the same way.

To prove (c), we estimate as follows:

$$\left\|\frac{1}{b_n} x_n - \frac{1}{b} x\right\| = \left\|\left(\frac{1}{b_n} x_n - \frac{1}{b} x_n\right) + \left(\frac{1}{b} x_n - \frac{1}{b} x\right)\right\|$$

$$\leq \left|\frac{1}{b_n} - \frac{1}{b}\right| \|x_n\| + \frac{1}{|b|} \|x_n - x\|$$

$$= \frac{|b - b_n|}{|b_n b|} \|x_n\| + \frac{1}{|b|} \|x_n - x\|.$$

Now let $M > 0$ such that

$$\frac{1}{M} < |b| \qquad \text{and} \qquad \|x\| < M.$$

It follows that there exists a natural number K_0 such that if $n \geq K_0$, then

$$\frac{1}{M} < |b_n| \qquad \text{and} \qquad \|x_n\| < M.$$

Hence if $n \geq K_0$, the above estimate yields

$$\left\|\frac{1}{b_n} x_n - \frac{1}{b} x\right\| \leq M^3 |b_n - b| + M \|x_n - x\|.$$

Therefore, if $\varepsilon > 0$ is a preassigned real number, then there are natural numbers K_1, K_2 such that if $n \geq K_1$, then $|b_n - b| < \varepsilon/2M^3$ and if $n \geq K_2$, then $\|x_n - x\| < \varepsilon/2M$. Letting $K = \sup\{K_0, K_1, K_2\}$ we conclude that if $n \geq K$, then

$$\left\|\frac{1}{b_n} x_n - \frac{1}{b} x\right\| < M^3 \frac{\varepsilon}{2M^3} + M \frac{\varepsilon}{2M} = \varepsilon,$$

which proves that (x_n/b_n) converges to x/b.

$$\text{Q.E.D.}$$

15.7 APPLICATIONS. Again we restrict attention to sequences in $\mathbf{R}$. (a) Let $X = (x_n)$ be the sequence in $\mathbf{R}$ defined by

$$x_n = \frac{2n + 1}{n + 5}, \qquad n \in \mathbf{N}.$$

We note that we can write x_n in the form

$$x_n = \frac{2+1/n}{1+5/n};$$

thus X can be regarded as the quotient of $Y = (2+1/n)$ and $Z = (1+5/n)$. Since the latter sequence consists of non-zero terms and has limit 1 (why?), the preceding theorem applies to allow us to conclude that

$$\lim X = \frac{\lim Y}{\lim Z} = \frac{2}{1} = 2.$$

(b) If $X = (x_n)$ is a sequence in $\mathbf{R}$ which converges to x and if p is a polynomial, then the sequence defined by $(p(x_n) : n \in \mathbf{N})$ converges to $p(x)$. (Hint: use Theorem 15.6 and induction.)

(c) Let $X = (x_n)$ be a sequence in $\mathbf{R}$ which converges to x and let r be a rational function; that is, $r(y) = p(y)/q(y)$, where p and q are polynomials. Suppose that $q(x_n)$ and $q(x)$ are non-zero, then the sequence $(r(x_n) : n \in \mathbf{N})$ converges to $r(x)$. (Hint: use part (b) and Theorem 15.6.)

We conclude this section with a result which is often useful. It is sometimes described by saying that one "passes to the limit in an inequality."

15.8 LEMMA. *Suppose that $X = (x_n)$ is a convergent sequence in $\mathbf{R}^p$ with limit x. If there exists an element c in $\mathbf{R}^p$ and a number $r > 0$ such that $\|x_n - c\| \le r$ for n sufficiently large, then $\|x - c\| \le r$.*

PROOF. The set $V = \{y \in \mathbf{R}^p : \|y - c\| > r\}$ is an open subset of $\mathbf{R}^p$. If $x \in V$, then V is a neighborhood of x and so $x_n \in V$ for sufficiently large values of n, contrary to the hypothesis. Therefore $x \notin V$ and hence we have $\|x - c\| \le r$.   Q.E.D.

It is important to note that we have assumed the existence of the limit in this result, for the remaining hypotheses are not sufficient to enable us to prove its existence.

Exercises

15.A. If (x_n) and (y_n) are convergent sequences of real numbers and if $x_n \le y_n$ for all $n \in \mathbf{N}$, then $\lim (x_n) \le \lim (y_n)$.

15.B. If $X = (x_n)$ and $Y = (y_n)$ are sequences of real numbers which both converge to c and if $Z = (z_n)$ is a sequence such that $x_n \le z_n \le y_n$ for $n \in \mathbf{N}$, then Z also converges to c.

15.C. For x_n given by the following formulas, either establish the convergence or the divergence of the sequence $X = (x_n)$:

(a) $x_n = \dfrac{n}{n+1}$,

(b) $x_n = \dfrac{(-1)^n n}{n+1}$,

(c) $x_n = \dfrac{2n}{3n^2+1}$,

(d) $x_n = \dfrac{2n^2+3}{3n^2+1}$,

(e) $x_n = n^2 - n$,

(f) $x_n = \sin n$.

15.D. If X and Y are sequences in $\mathbf{R}^p$ and if $X + Y$ converges, do X and Y converge and have $\lim (X + Y) = \lim X + \lim Y$?

15.E. If X and Y are sequences in $\mathbf{R}^p$ and if $X \cdot Y$ converges, do X and Y converge and have $\lim X \cdot Y = (\lim X) \cdot (\lim Y)$?

15.F. If $X = (x_n)$ is a positive sequence which converges to x, then $(\sqrt{x_n})$ converges to $\sqrt{x}$. (Hint: $\sqrt{x_n} - \sqrt{x} = (x_n - x)/(\sqrt{x_n} + \sqrt{x})$ when $x \neq 0$.)

15.G. If $X = (x_n)$ is a sequence of real numbers such that $Y = (x_n^2)$ converges to 0, then does X converge to 0?

15.H. If $x_n = \sqrt{n+1} - \sqrt{n}$, do the sequences $X = (x_n)$ and $Y = (\sqrt{n}\, x_n)$ converge?

15.I. Let (x_n) be a sequence in $\mathbf{R}^p$ such that the subsequences (x_{2n}) and (x_{2n+1}) converge to $x \in \mathbf{R}^p$. Prove that (x_n) converges to x.

15.J. Let (x_n) and (y_n) be sequences in $\mathbf{R}$ such that $\lim (x_n) \neq 0$ and $\lim (x_n y_n)$ exists. Prove that $\lim (y_n)$ exists.

15.K. Does Exercise 15.J remain true in $\mathbf{R}^2$?

15.L. If $0 < a \leq b$ and if $x_n = (a^n + b^n)^{1/n}$, then $\lim (x_n) = b$.

15.M. Every irrational number in $\mathbf{R}$ is the limit of a sequence of rational numbers. Every rational number in $\mathbf{R}$ is the limit of a sequence of irrational numbers.

15.N. Let $A \subseteq \mathbf{R}^p$ and $x \in \mathbf{R}^p$. Then x is a boundary point of A if and only if there is a sequence (a_n) of elements in A and a sequence (b_n) of elements in $\mathscr{C}(A)$ such that

$$\lim (a_n) = x = \lim (b_n).$$

15.O. Let $A \subseteq \mathbf{R}^p$ and $x \in \mathbf{R}^p$. Then x is a cluster point of A if and only if there is a sequence (a_n) of distinct elements in A such that $x = \lim (a_n)$.

15.P. If $x = \lim (x_n)$ and if $\|x_n - c\| < r$ for all $n \in \mathbf{N}$, does it follow that $\|x - c\| < r$?

Projects

15.α. Let d be a metric on a set M in the sense of Exercise 8.S. If $X = (x_n)$ is a sequence in M, then an element $x \in M$ is said to be a **limit** of X if, for each $\varepsilon > 0$ there exists a number $K(\varepsilon)$ in $\mathbf{N}$ such that for all $n \geq K(\varepsilon)$, then $d(x_n, x) < \varepsilon$. Use this definition and show that Theorems 14.5, 14.6, 15.2, 15.3, and 15.4 can be extended to metric spaces. Show that the metrics d_1, d_2, d_∞ in $\mathbf{R}^p$ give rise to the same convergent sequences in $\mathbf{R}^p$. Show that if d is the discrete metric on a set, then the only sequences which converge relative to d are those which are "constant after some natural number."

15.β. Let m denote the collection of all bounded sequences in $\mathbf{R}$; let c denote the collection of all convergent sequences in $\mathbf{R}$; and let c_0 denote the collection of all sequences in $\mathbf{R}$ which converge to zero.

(a) With the sum $X + Y$ and product cX as given in Definition 14.2, show that each of the above collections is a vector space in which the zero element is the sequence $0 = (0, 0, \ldots)$.

(b) In each of the collections m, c, c_0, define the norm of $X = (x_n)$ by $\|x\| = \sup\{|x_n| : n \in N\}$. Show that this definition actually yields a norm.

(c) If X and Y belong to either m, c, or c_0, then the product XY also belongs to it and $\|XY\| \le \|X\| \|Y\|$. Give an example to show that equality may hold in this last relation, and one to show that equality may fail.

(d) Show that the metric induced by the norm in part (b) in these spaces is given by $d(X, Y) = \sup\{|x_n - y_n| : n \in N\}$.

(e) Show that if a sequence (X_k) converges to Y relative to the metric in (d), then each "coordinate sequence" converges to the corresponding coordinate of Y. (Warning: X_k is a sequence in $\mathbf{R}$, while (X_k) is a sequence in m, c, or c_0; that is, a "sequence of sequences" in $\mathbf{R}$.)

(f) Give an example of a sequence (X_k) in c_0 where each coordinate sequence converges to 0, but where $d(X_k, 0)$ does not converge to 0.

Section 16   Two Criteria for Convergence

Until now the main method available for showing that a sequence is convergent is to identify it as a subsequence or an algebraic combination of convergent sequences. When this can be done, we are able to calculate the limit using the results of the preceding section. However, when this cannot be done, we have to fall back on Definition 14.3 or Theorem 14.4 in order to establish the existence of the limit. The use of these latter tools has the noteworthy disadvantage that we must already know (or at least suspect) the correct value of the limit and we then verify that our suspicion is correct.

There are many cases, however, where there is no obvious candidate for the limit of a given sequence, even though a preliminary analysis has led to the belief that convergence does take place. In this section we give some results which are deeper than those in the preceding sections and which can be used to establish the convergence of a sequence when no particular element presents itself as the value of the limit. The first result in this direction is very important. Although it can be generalized to $\mathbf{R}^p$, it is convenient to restrict its statement to the case of sequences in $\mathbf{R}$.

16.1   Monotone Convergence Theorem.   Let $X = (x_n)$ be a sequence of real numbers which is monotone increasing in the sense that

$$x_1 \le x_2 \le \cdots \le x_n \le x_{n+1} \le \cdots.$$

Then the sequence X converges if and only if it is bounded, in which case

$$\lim (x_n) = \sup \{x_n\}.$$

PROOF.   It was seen in Lemma 14.6 that a convergent sequence is bounded. If $x = \lim (x_n)$ and $\varepsilon > 0$, then there exists a natural number

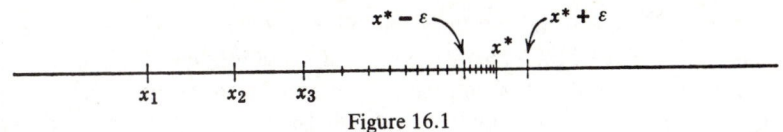

Figure 16.1

$K(\varepsilon)$ such that if $n \geq K(\varepsilon)$, then

$$x - \varepsilon \leq x_n \leq x + \varepsilon.$$

Since X is monotone, this relation yields

$$x - \varepsilon \leq \sup\{x_n\} \leq x + \varepsilon,$$

whence it follows that $|x - \sup\{x_n\}| \leq \varepsilon$. Since this holds for all $\varepsilon > 0$, we infer that $\lim(x_n) = x = \sup\{x_n\}$.

Conversely, suppose that $X = (x_n)$ is a bounded monotone increasing sequence of real numbers. According to the Supremum Principle, the supremum $x^* = \sup\{x_n\}$ exists; we shall show that it is the limit of X. Since x^* is an upper bound of the elements in X, then $x_n \leq x^*$ for $n \in \mathbf{N}$. Since x^* is the supremum of X, if $\varepsilon > 0$ the number $x^* - \varepsilon$ is not an upper bound of X and exists a natural number $K(\varepsilon)$ such that

$$x^* - \varepsilon < x_{K(\varepsilon)}.$$

In view of the monotone character of X, for all $n \geq K(\varepsilon)$, then

$$x^* - \varepsilon < x_n \leq x^*,$$

whence it follows that $|x_n - x^*| < \varepsilon$. Recapitulating, the number $x^* = \sup\{x_n\}$ has the property that, given $\varepsilon > 0$ there is a natural number $K(\varepsilon)$ (depending on ε) such that $|x_n - x^*| < \varepsilon$ whenever $n \geq K(\varepsilon)$. This shows that $x^* = \lim X$.   Q.E.D.

16.2 COROLLARY. *Let $X = (x_n)$ be a sequence of real numbers which is monotone decreasing in the sense that*

$$x_1 \geq x_2 \geq \cdots \geq x_n \geq x_{n+1} \geq \cdots.$$

Then the sequence X converges if and only if it is bounded, in which case

$$\lim(x_n) = \inf\{x_n\}.$$

PROOF. Let $y_n = -x_n$ for $n \in \mathbf{N}$. Then the sequence $Y = (y_n)$ is readily seen to be a monotone increasing sequence. Moreover, Y is bounded if and only if X is bounded. Therefore, the conclusion follows from the theorem.   Q.E.D.

16.3 EXAMPLES. (a) We return to the sequence $X = (1/n)$ discussed in Example 14.8(a). It is clear that

$$\frac{1}{1} > \frac{1}{2} > \cdots > \frac{1}{n} > \cdots > 0;$$

it therefore follows from Corollary 16.2 that $X = (1/n)$ converges. We can establish the value of lim $(1/n)$ if we can calculate inf $\{1/n\}$. Alternatively, once the convergence of X is assured we can often evaluate its limit by using Lemma 15.2 and Theorem 15.6. In the case at hand, if $X' = (1/2, 1/4, \ldots, 1/2n, \ldots)$, then it follows that

$$\lim X = \lim X' = \tfrac{1}{2} \lim X.$$

We conclude, therefore, that lim $X = 0$.

(b) Let $Y = (y_n)$ be the sequence in $\boldsymbol{R}$ defined inductively by

$$y_1 = 1, \qquad y_{n+1} = (2y_n + 3)/4 \qquad \text{for} \quad n \in N.$$

Direct calculation shows that $y_1 < y_2 < 2$. If $y_{n-1} < y_n < 2$, then

$$2y_{n-1} + 3 < 2y_n + 3 < 2 \cdot 2 + 3,$$

from which it follows that $y_n < y_{n+1} < 2$. By induction, the sequence Y is monotone increasing and bounded above by the number 2. It follows from the Monotone Convergence Theorem that the sequence Y converges to a limit which is no greater than 2. In this case it might not be so easy to evaluate $y = \lim Y$ by calculating sup $\{y_n\}$. However, once we know that the limit exists, there is another way to calculate its value. According to Lemma 15.2, we have $y = \lim (y_n) = \lim (y_{n+1})$. Using Theorem 15.6, the limit y must satisfy the relation

$$y = (2y + 3)/4.$$

Therefore, we conclude that $y = \tfrac{3}{2}$.

(c) Let $Z = (z_n)$ be the sequence in $\boldsymbol{R}$ defined by

$$z_1 = 1, \qquad z_{n+1} = \sqrt{2z_n} \qquad \text{for} \quad n \in N.$$

It is clear that $z_1 < z_2 < 2$. If $z_n < z_{n+1} < 2$, then $2z_n < 2z_{n+1} < 4$ so that $z_{n+1} = \sqrt{2z_n} < z_{n+2} = \sqrt{2z_{n+1}} < 2 = \sqrt{4}$. This shows that Z is a monotone increasing sequence which is bounded above by 2; hence Z converges to a number z. It may be shown directly that $2 = \sup \{z_n\}$ so that the limit $z = 2$. Alternatively, we can use the method of the preceding example. Knowing that the sequence has a limit z, we conclude from the relation $z_{n+1} = \sqrt{2z_n}$ that z must satisfy $z = \sqrt{2z}$. To find the roots of this last equation, we square to obtain $z^2 = 2z$, which has roots 0, 2. Evidently 0 cannot be the limit (why?); hence this limit must equal 2.

(d) Let $U = (u_n)$ be the sequence of real numbers defined by $u_n = (1 + 1/n)^n$ for $n \in N$. Applying the Binomial Theorem, we can write

$$u_n = 1 + \frac{n}{1}\frac{1}{n} + \frac{n(n-1)}{2!}\frac{1}{n^2} + \frac{n(n-1)(n-2)}{3!}\frac{1}{n^3}$$

$$+ \cdots + \frac{n(n-1)\cdots 2 \cdot 1}{n!}\frac{1}{n^n}.$$

Dividing the powers of n into the numerators of the binomial coefficients, we have

$$u_n = 1+1+\frac{1}{2!}\left(1-\frac{1}{n}\right)+\frac{1}{3!}\left(1-\frac{1}{n}\right)\left(1-\frac{2}{n}\right)$$
$$+\cdots+\frac{1}{n!}\left(1-\frac{1}{n}\right)\left(1-\frac{2}{n}\right)\cdots\left(1-\frac{n-1}{n}\right).$$

Expressing u_{n+1} in the same way, we have

$$u_{n+1} = 1+1+\frac{1}{2!}\left(1-\frac{1}{n+1}\right)+\frac{1}{3!}\left(1-\frac{1}{n+1}\right)\left(1-\frac{2}{n+1}\right)$$
$$+\cdots+\frac{1}{n!}\left(1-\frac{1}{n+1}\right)\left(1-\frac{2}{n+1}\right)\cdots\left(1-\frac{n-1}{n+1}\right)$$
$$+\frac{1}{(n+1)!}\left(1-\frac{1}{n+1}\right)\left(1-\frac{2}{n+1}\right)\cdots\left(1-\frac{n}{n+1}\right).$$

Note that the expression for u_n contains $n+1$ terms and that for u_{n+1} contains $n+2$ terms. An elementary examination shows that each term in u_n is no greater than the corresponding term in u_{n+1} and the latter has one more positive term. Therefore, we have

$$u_1 < u_2 < \cdots < u_n < u_{n+1} < \cdots.$$

To show that the sequence is bounded, we observe that if $p = 1, 2, \ldots, n$, then $(1-p/n) < 1$. Moreover, $2^{p-1} \le p!$ (why?) so that $1/p! \le 1/2^{p-1}$. From the above expression for u_n, these estimates yield

$$2 < u_n < 1+1+\frac{1}{2}+\frac{1}{2^2}+\cdots+\frac{1}{2^{n-1}} < 3, \qquad n > 2.$$

It follows that the monotone sequence U is bounded above by 3. The Monotone Convergence Theorem implies that the sequence U converges to a real number which is at most 3. As is probably well-known to the reader, the limit of U is the fundamental number e. By refining our estimates we can find closer rational approximations to the value of e, but we cannot evaluate it exactly in this way since it is irrational—although it is possible to calculate as many decimal places as desired. (This illustrates that a result such as the Monotone Convergence Theorem, which only establishes the existence of the limit of a sequence, can be of great use even when the exact value cannot be easily obtained.)

The Bolzano-Weierstrass Theorem

The Monotone Convergence Theorem is extraordinarily useful and important, but it has the drawback that it applies only to sequences which are monotone. It behooves us, therefore, to find a condition which will imply convergence in $\boldsymbol{R}$ or $\boldsymbol{R}^p$ without using the monotone property. This desired condition is the Cauchy Criterion, which will be introduced below. However, we shall first give a form of the Bolzano-Weierstrass Theorem 10.6 that is particularly applicable for sequences.

16.4 Bolzano-Weierstrass Theorem. *A bounded sequence in* $\mathbf{R}^p$ *has a convergent subsequence.*

PROOF. Let $X = (x_n)$ be a bounded sequence in $\mathbf{R}^p$. If there are only a finite number of distinct values in the sequence X, then at least one of these values must occur infinitely often. If we define a subsequence of X by selecting this element each time it appears, we obtain a convergent subsequence of X.

On the other hand, if the sequence X contains an infinite number of distinct values in $\mathbf{R}^p$, then since these points are bounded, the Bolzano-Weierstrass Theorem 10.6 for sets implies that there is at least one cluster point, say x^*. Let x_{n_1} be an element of X such that

$$\|x_{n_1} - x^*\| < 1.$$

Consider the neighborhood $V_2 = \{y : \|y - x^*\| < \frac{1}{2}\}$. Since the point x^* is a cluster point of the set $S_1 = \{x_m : m \geq 1\}$, it is also a cluster point of the set $S_2 = \{x_m : m > n_1\}$ obtained by deleting a finite number of elements of S_1. (Why?) Therefore, there is an element x_{n_2} of S_2 (whence $n_2 > n_1$) belonging to V_2. Now let V_3 be the neighborhood $V_3 = \{y : \|y - x^*\| < \frac{1}{3}\}$ and let $S_3 = \{x_m : m > n_2\}$. Since x^* is a cluster point of S_3 there must be an element x_{n_3} of S_3 (whence $n_3 > n_2$) belonging to V_3. By continuing in this way we obtain a subsequence $X' = (x_{n_1}, x_{n_2}, \ldots)$ of X with

$$\|x_{n_r} - x^*\| < 1/r,$$

so that $\lim X' = x^*$.   Q.E.D.

16.5 Corollary. *If* $X = (x_n)$ *is a sequence in* $\mathbf{R}^p$ *and* x^* *is a cluster point of the set* $\{x_n : n \in \mathbf{N}\}$, *then there is a subsequence* X' *of* X *which converges to* x^*.

In fact, this is what the second part of the proof of 16.4 established.

Cauchy Sequences

We now introduce the important notion of a Cauchy sequence in $\mathbf{R}^p$. It will turn out that a sequence in $\mathbf{R}^p$ is convergent if and only if it is a Cauchy sequence.

16.6 Definition. A sequence $X = (x_n)$ in $\mathbf{R}^p$ is said to be a **Cauchy sequence** in case for every $\varepsilon > 0$ there is a natural number $M(\varepsilon)$ such that for all $m, n \geq M(\varepsilon)$, then $\|x_m - x_n\| < \varepsilon$.

In order to help motivate the notion of a Cauchy sequence, we shall show that every convergent sequence in $\mathbf{R}^p$ is a Cauchy sequence.

16.7 Lemma. *If* $X = (x_n)$ *is a convergent sequence in* $\mathbf{R}^p$, *then* X *is a Cauchy sequence.*

PROOF. If $x = \lim X$; then given $\varepsilon > 0$ there is a natural number $K(\varepsilon/2)$ such that if $n \geq K(\varepsilon/2)$, then $\|x_n - x\| < \varepsilon/2$. Thus if $M(\varepsilon) = K(\varepsilon/2)$ and if $m, n \geq M(\varepsilon)$, then

$$\|x_m - x_n\| \leq \|x_m - x\| + \|x - x_n\| < \varepsilon/2 + \varepsilon/2 = \varepsilon.$$

Hence the convergent sequence X is a Cauchy sequence. Q.E.D.

In order to apply the Bolzano-Weierstrass Theorem, we shall require the following result.

16.8 LEMMA. *A Cauchy sequence in $\mathbf{R}^p$ is bounded.*

PROOF. Let $X = (x_n)$ be a Cauchy sequence and let $\varepsilon = 1$. If $m = M(1)$ and $n \geq M(1)$, then $\|x_m - x_n\| < 1$. From the Triangle Inequality this implies that $\|x_n\| < \|x_m\| + 1$ for $n \geq M(1)$. Therefore, if

$$B = \sup \{\|x_1\|, \ldots, \|x_{m-1}\|, \|x_m\| + 1\},$$

then we have $\|x_n\| \leq B$ for all $n \in \mathbf{N}$. Thus the Cauchy sequence X is bounded. Q.E.D.

16.9 LEMMA. *If a subsequence X' of a Cauchy sequence X in $\mathbf{R}^p$ converges to an element x, then the entire sequence X converges to x.*

PROOF. Since $X = (x_n)$ is a Cauchy sequence, given $\varepsilon > 0$ there is a natural number $M(\varepsilon/2)$ such that if $m, n \geq M(\varepsilon/2)$, then

$$(*) \qquad \|x_m - x_n\| < \varepsilon/2.$$

If the sequence $X' = (x_{n_j})$ converges to x, there is a natural number $K \geq M(\varepsilon/2)$, belonging to the set $\{n_1, n_2, \ldots\}$ and such that

$$\|x - x_K\| < \varepsilon/2.$$

Now let n be any natural number such that $n \geq M(\varepsilon/2)$. It follows that $(*)$ holds for this value of n and for $m = K$. Thus

$$\|x - x_n\| \leq \|x - x_K\| + \|x_K - x_n\| < \varepsilon,$$

when $n \geq M(\varepsilon/2)$. Therefore, the sequence X converges to the element x, which is the limit of the subsequence X'. Q.E.D.

We are now prepared to obtain the important Cauchy Criterion. Our proof is deceptively short, but the reader will note that the work has already been done and we are merely putting the pieces together.

16.10 CAUCHY CONVERGENCE CRITERION. *A sequence in $\mathbf{R}^p$ is convergent if and only if it is a Cauchy sequence.*

PROOF. It was seen in Lemma 16.7 that a convergent sequence must be a Cauchy sequence.

Conversely, suppose that X is a Cauchy sequence in R^p. It follows from Lemma 16.8 that the sequence X is bounded in R^p. According to the Bolzano-Weierstrass Theorem 16.4, the bounded sequence X has a convergent subsequence X'. By Lemma 16.9 the entire sequence X converges to the limit of X'.

Q.E.D.

16.11 EXAMPLES. (a) Let $X = (x_n)$ be the sequence in R defined by

$$x_1 = 1, \ x_2 = 2, \ldots, \ x_n = \tfrac{1}{2}(x_{n-2} + x_{n-1}) \qquad \text{for} \quad n > 2.$$

It can be shown by induction that

$$1 \leq x_n \leq 2 \qquad \text{for} \quad n \in N,$$

but the sequence X is neither monotone decreasing nor increasing. (Actually the terms with odd subscript form an increasing sequence and those with even subscript form a decreasing sequence.) Since the terms in the sequence are formed by averaging, it is readily seen that

$$|x_n - x_{n+1}| = \frac{1}{2^{n-1}} \qquad \text{for} \quad n \in N.$$

Thus if $m > n$, we employ the Triangle Inequality to obtain

$$|x_n - x_m| \leq |x_n - x_{n+1}| + \cdots + |x_{m-1} - x_m|$$

$$= \frac{1}{2^{n-1}} + \cdots + \frac{1}{2^{m-2}} = \frac{1}{2^{n-1}}\left(1 + \frac{1}{2} + \cdots + \frac{1}{2^{m-n-1}}\right) < \frac{1}{2^{n-2}}.$$

Given $\varepsilon > 0$, if n is chosen so large that $1/2^n < \varepsilon/4$ and if $m \geq n$, it follows that

$$|x_n - x_m| < \varepsilon.$$

Therefore, X is a Cauchy sequence in R and, by the Cauchy Criterion, the sequence X converges to a number x. To evaluate the limit we note that taking the limit in the rule of definition yields the valid, but uninformative, result

$$x = \tfrac{1}{2}(x + x).$$

However, since the sequence X converges, so does the subsequence with odd indices. By induction we can establish that

$$x_1 = 1, \qquad x_3 = 1 + \frac{1}{2}, \qquad x_5 = 1 + \frac{1}{2} + \frac{1}{2^3}, \ldots,$$

$$x_{2n+1} = 1 + \frac{1}{2} + \frac{1}{2^3} + \cdots + \frac{1}{2^{2n-1}}, \ldots$$

It follows that

$$x_{2n+1} = 1 + \frac{1}{2}\left(1 + \frac{1}{4} + \cdots + \frac{1}{4^{n-1}}\right)$$

$$= 1 + \frac{1}{2} \cdot \frac{1 - 1/4^n}{1 - 1/4} = 1 + \frac{2}{3}\left(1 - \frac{1}{4^n}\right).$$

Therefore, the subsequence with odd indices converges to $\frac{5}{3}$; hence the entire sequence has the same limit.

(b) Let $X = (x_n)$ be the real sequence given by

$$x_1 = \frac{1}{1!}, \quad x_2 = \frac{1}{1!} - \frac{1}{2!}, \ldots, \quad x_n = \frac{1}{1!} - \frac{1}{2!} + \cdots + \frac{(-1)^{n+1}}{n!}, \ldots$$

Since this sequence is not monotone, a direct application of the Monotone Convergence Theorem is not possible. Observe that if $m > n$, then

$$x_m - x_n = \frac{(-1)^{n+2}}{(n+1)!} + \frac{(-1)^{n+3}}{(n+2)!} + \cdots + \frac{(-1)^{m+1}}{m!}.$$

Recalling that $2^{r-1} \le r!$, we find that

$$|x_m - x_n| \le \frac{1}{(n+1)!} + \frac{1}{(n+2)!} + \cdots + \frac{1}{m!}$$

$$\le \frac{1}{2^n} + \frac{1}{2^{n+1}} + \cdots + \frac{1}{2^{m-1}} < \frac{1}{2^{n-1}}.$$

Therefore the sequence is a Cauchy sequence in $\mathbf{R}$.

(c) If $X = (x_n)$ is the sequence in $\mathbf{R}$ defined by

$$x_n = \frac{1}{1} + \frac{1}{2} + \cdots + \frac{1}{n} \qquad \text{for} \quad n \in \mathbf{N},$$

and if $m > n$, then

$$x_m - x_n = \frac{1}{n+1} + \frac{1}{n+2} + \cdots + \frac{1}{m}.$$

Since each of these $m - n$ terms exceeds $1/m$, this difference exceeds $(m - n)/m = 1 - n/m$. In particular, if $m = 2n$, we have

$$x_{2n} - x_n > \tfrac{1}{2}$$

This shows that X is not a Cauchy sequence, whence we conclude that X is divergent. (We have just proved that the "harmonic series" is divergent.)

Exercises

16.A. Let $x_1 \in \mathbf{R}$ satisfy $x_1 > 1$ and let $x_{n+1} = 2 - 1/x_n$ for $n \in \mathbf{N}$. Show that the sequence (x_n) is monotone and bounded. What is its limit?

16.B. Let $y_1 = 1$ and $y_{n+1} = (2 + y_n)^{1/2}$ for $n \in N$. Show that (y_n) is monotone and bounded. What is its limit?

16.C. Let $a > 0$ and let $z_1 > 0$. Define $z_{n+1} = (a + z_n)^{1/2}$ for $n \in N$. Show that (z_n) converges.

16.D. If a satisfies $0 < a < 1$, show that the sequence $X = (a^n)$ is convergent. Since $Y = (a^{2n})$ is a subsequence, we have $\lim X = \lim Y = (\lim X)^2$, and that $\lim X = 0$.

16.E. Show that every sequence in R either has a monotone increasing subsequence or a monotone decreasing subsequence.

16.F. Use Exercise 16.E to prove the Bolzano-Weierstrass Theorem for sequences in R.

16.G. Determine the convergence or divergence of the sequence (x_n), where

$$x_n = \frac{1}{n+1} + \frac{1}{n+2} + \cdots + \frac{1}{2n} \qquad \text{for} \qquad n \in N.$$

16.H. Let $X = (x_n)$ and $Y = (y_n)$ be sequences in R^p and let $Z = (z_n)$ be the "shuffled" sequence defined by $z_1 = x_1$, $z_2 = y_1$, ..., $z_{2n} = x_n$, $z_{2n+1} = y_n$, ... Is it true that Z is convergent if and only if X and Y are convergent and $\lim X = \lim Y$?

16.I. Show directly that the following are Cauchy sequences:

(a) $\left(\dfrac{1}{n}\right)$,   (b) $\left(\dfrac{n+1}{n}\right)$,   (c) $\left(1 + \dfrac{1}{1!} + \cdots + \dfrac{1}{n!}\right)$.

16.J. Show directly that the following are not Cauchy sequences:

(a) $((-1)^n)$,   (b) $(n + (-1)^n/n)$,   (c) (n^2).

16.K. Let $X = (x_n)$ be a sequence of strictly positive real numbers, let $\lim (x_{n+1}/x_n) = L$, and let $0 < \varepsilon < L$. Show that there exist $A > 0$, $B > 0$, and $K \in N$ such that $A(L - \varepsilon)^n \leq x_n \leq B(L + \varepsilon)^n$ for $n \geq K$. Then show that $\lim (x_n^{1/n}) = L$.

16.L. Apply Example 16.3(d) and the preceding exercise to the sequence $(n^n/n!)$ to show that $\lim (n/(n!)^{1/n}) = e$.

16.M. Establish the convergence and the limits of the following sequences:

(a) $((1 + 1/n)^{n+1})$,   (b) $((1 + 1/2n)^n)$,

(c) $((1 + 2/n)^n)$,   (d) $((1 + 1/(n+1))^{3n})$.

16.N. Let $0 < a_1 < b_1$ and define, for $n \in N$,

$$a_{n+1} = (a_n b_n)^{1/2}, \qquad b_{n+1} = \tfrac{1}{2}(a_n + b_n).$$

By induction show that $a_n < b_n$. Show that (a_n) and (b_n) converge to the same limit.

16.O. Give a proof of the Cantor Intersection Theorem 11.4 by taking a point $x_n \in F_n$ and applying the Bolzano-Weierstrass Theorem 16.4.

16.P. Give a proof of the Nearest Point Theorem 11.6 by using the Bolzano-Weierstrass Theorem 16.4.

16.Q. Prove that if K_1 and K_2 are compact subsets of R^p, then there exist points $x_1 \in K_1$, $x_2 \in K_2$ such that if $z_1 \in K_1$, $z_2 \in K_2$, then $\|z_1 - z_2\| \geq \|x_1 - x_2\|$.

Project

16.α. In this project, let m, c, and c_0 designate the collections of real sequences that were introduced in Project 15.β and let d denote the metric defined in part (d) of that project.

(a) If $r \in I$ and $r = 0 . r_1 r_2 \cdots r_n \cdots$ is its decimal expansion, consider the element $X_r = (r_n)$ in m. Conclude that there is an uncountable subset A of m such that if X_r and X_s are distinct elements of A, then $d(X_r, X_s) \geq 1$.

(b) Suppose that B is a subset of c with the property that if X and Y are distinct elements of B, then $d(X, Y) \geq 1$. Prove that B is a countable set.

(c) If $j \in N$, let $Z_j = (z_{nj} : n \in N)$ be the sequence whose first j elements are 1 and whose remaining elements are 0. Observe that Z_j belongs to each of the metric spaces m, c, and c_0 and that $d(Z_j, Z_k) = 1$ for $j \neq k$. Show that the sequence $(Z_j : j \in N)$ is monotone in the sense that each coordinate sequence $(z_{nj} : j \in N)$ is monotone. Show that the sequence (Z_j) does not converge with respect to the metric d in any of the three spaces.

(d) Show that there is a sequence (X_j) in m, c, and c_0 which is bounded (in the sense that there exists a constant K such that $d(X_j, 0) \leq K$ for all $j \in N$) but which possesses no convergent subsequence.

(e) (If d is a metric on a set M, we say that a sequence (X_j) in M is a **Cauchy sequence** if for every $\varepsilon > 0$ there exists $K(\varepsilon) \in N$ such that $d(X_j, X_k) < \varepsilon$ whenever $j, k \geq K(\varepsilon)$. We say that M is **complete** with respect to d in case every Cauchy sequence in M converges to an element of M.) Prove that the sets m, c, and c_0 are complete with respect to the metric d we have been considering.

(f) Let f be the collection of all real sequences which have only a finite number of non-zero elements and define d as before. Show that d is a metric on f, but that f is not complete with respect to d.

Section 17 Sequences of Functions

In the preceding three sections we have considered the convergence of sequences of elements of R^p; in the present section we shall consider sequences of *functions*. After some simple preliminaries, we shall introduce the rather subtle, but basic, notion of uniform convergence of a sequence of functions.

Let $D \subseteq R^p$ be given and suppose that for each natural number $n \in N$ there is a function f_n with domain D and range in R^q; we shall say (f_n) is a *sequence of functions* on $D \subseteq R^p$ to R^q. It should be understood that, for any point x in D such a sequence of functions gives a sequence of elements in R^q; namely, the sequence

$$(17.1) \qquad (f_n(x))$$

which is obtained by evaluating each of the functions at x. For certain points x in D the sequence (17.1) may converge and for other points x in $D this sequence may diverge. For each of those points x for which the sequence (17.1) converges there is, by Theorem 14.5, a uniquely determined point of R^q. In general, the value of this limit, when it exists, will depend on the choice of the point x. In this way, there arises a function whose domain consists of all points x in $D \subseteq R^p$ for which the sequence (17.1) converges in R^q.

We shall now collect these introductory words in a formal definition of convergence of a sequence of functions.

17.1 DEFINITION. Let (f_n) be a sequence of functions on $D \subseteq R^p$ to R^q, let D_0 be a subset of D, and let f be a function with domain containing D_0 and range in R^q. We say that the **sequence** (f_n) **converges on** D_0 **to** f if, for each x in D_0 the sequence $(f_n(x))$ converges in R^q to $f(x)$. In this case we call the function f **the limit on** D_0 **of the sequence** (f_n). When such a function f exists we say that the sequence (f_n) **converges to** f **on** D_0, or simply that the sequence is **convergent on** D_0.

It follows from Theorem 14.5 that, except for possible change in the domain D_0, the limit function is uniquely determined. Ordinarily, we choose D_0 to be the largest set possible; that is, the set of all x in D for which (17.1) converges. In order to symbolize that the sequence (f_n) converges on D_0 to f we sometimes write

$$f = \lim (f_n) \quad \text{on } D_0, \qquad \text{or} \qquad f_n \to f \quad \text{on } D_0.$$

We shall now consider some examples of this idea. For simplicity, we shall treat the special case $p = q = 1$.

17.2 EXAMPLES. (a) For each natural number n, let f_n be defined for x in $D = R$ by $f_n(x) = x/n$. Let f be defined for all x in $D = R$ by $f(x) = 0$. (See Figure 17.1). The statement that the sequence (f_n) converges on R to f is equivalent to the statement that for each real number x the numerical sequence (x/n) converges to 0. To see that this is the case, we apply Example 14.8(a) and Theorem 15.6(b).

(b) Let $D = \{x \in R : 0 \le x \le 1\}$ and for each natural number n let f_n be defined by $f_n(x) = x^n$ for all x in D and let f be defined by

$$f(x) = 0, \qquad 0 \le x < 1,$$
$$= 1, \qquad x = 1.$$

(See Figure 17.2.) It is clear that when $x = 1$, then $f_n(x) = f_n(1) = 1^n = 1$ so that $f_n(1) \to f(1)$. We have shown in Example 14.8(c), that if $0 \le x < 1$, then $f_n(x) = x^n \to 0$. Therefore, we conclude that (f_n) converges on D to f. (It is not difficult to prove that if $x > 1$ then $(f_n(x))$ does not converge at all.)

(c) Let $D = R$ and for each natural number n, let f_n be the function

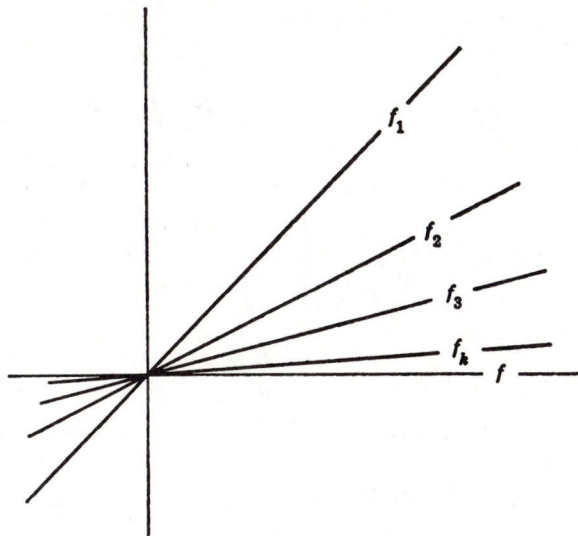

Figure 17.1

defined for x in D by

$$f_n(x) = \frac{x^2 + nx}{n},$$

and let $f(x) = x$. (See Figure 17.3.) Since $f_n(x) = (x^2/n) + x$, it follows from Example 14.8(a) and Theorem 15.6(b) that $(f_n(x))$ converges to $f(x)$ for all $x \in \mathbf{R}$.

(d) Let $D = \mathbf{R}$ and, for each natural number n, let f_n be defined to be $f_n(x) = (1/n) \sin(nx + n)$. (See Figure 17.4.) (A rigorous definition of the

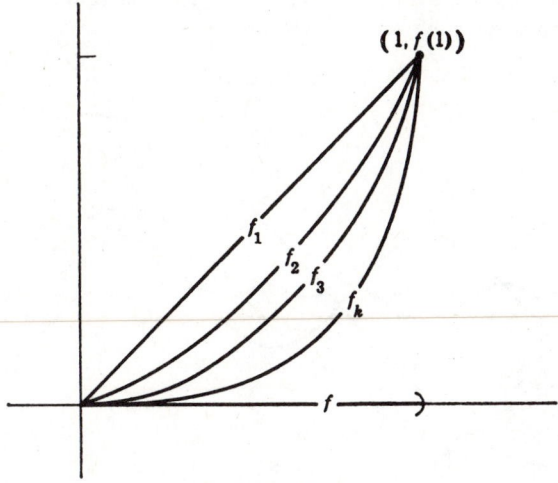

Figure 17.2

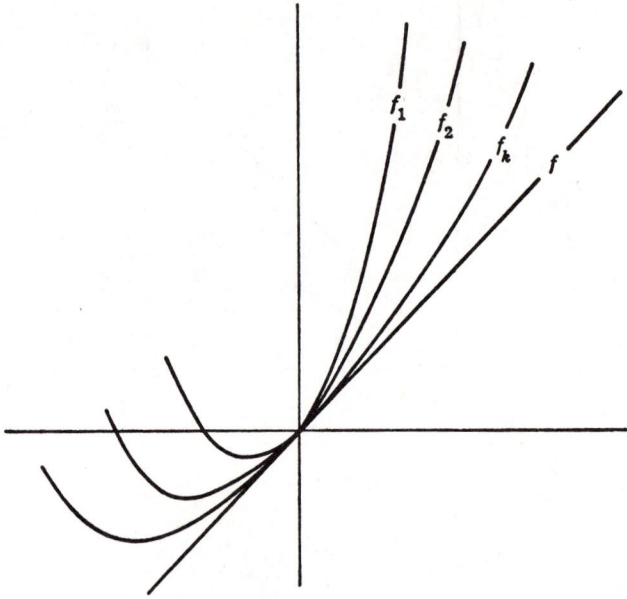

Figure 17.3

sine function is not needed here; in fact, all we require is that $|\sin y| \le 1$ for any real number y.) If f is defined to be the zero function $f(x) = 0$, $x \in \mathbf{R}$, then $f = \lim (f_n)$ on $\mathbf{R}$. Indeed, for any real number x, we have

$$|f_n(x) - f(x)| = \frac{1}{n} |\sin (nx + n)| \le \frac{1}{n}.$$

If $\varepsilon > 0$, there exists a natural number $K(\varepsilon)$ such that if $n \ge K(\varepsilon)$, then $1/n < \varepsilon$. Hence for such n we conclude that

$$|f_n(x) - f(x)| < \varepsilon$$

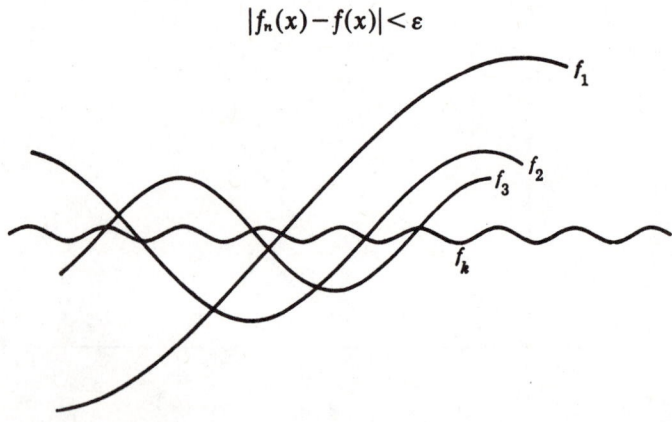

Figure 17.4

no matter what the value of x. Therefore, we infer that the sequence (f_n) converges to f. (Note that by choosing n sufficiently large, we can make the differences $|f_n(x) - f(x)|$ less than ε for all values of x simultaneously!)

Partly to reinforce Definition 17.1 and partly to prepare the way for the important notion of uniform convergence, we formulate the following restatement of Definition 17.1.

17.3 LEMMA. *A sequence (f_n) of functions on $D \subseteq \mathbf{R}^p$ to $\mathbf{R}^q$ converges to a function f on $D_0 \subseteq D$ if and only if for each $\varepsilon > 0$ and each x in D_0 there is a natural number $K(\varepsilon, x)$ such that for all $n \geq K(\varepsilon, x)$, then*

$$(17.2) \qquad \|f_n(x) - f(x)\| < \varepsilon.$$

Since this is just a reformulation of Definition 17.1, we shall not go through the details of the proof, but leave them to the reader as an exercise. We wish only to point out that the value of n required in inequality (17.2) will depend, in general, on *both* $\varepsilon > 0$ and $x \in D_0$. An alert reader will have already noted that, in Examples 17.2(a–c) the value of n required to obtain (17.2) does depend on both $\varepsilon > 0$ and $x \in D_0$. However, in Example 17.2(d) the inequality (17.2) can be satisfied for all x in D_0 provided n is chosen sufficiently large but dependent on ε alone.

It is precisely this rather subtle difference which distinguishes between the notions of "ordinary" convergence of a sequence of functions (in the sense of Definition 17.1) and "uniform" convergence, which we now define.

17.4 DEFINITION. A sequence (f_n) of functions on $D \subseteq \mathbf{R}^p$ to $\mathbf{R}^q$ **converges uniformly** on a subset D_0 of D to a function f in case for each $\varepsilon > 0$ there is a natural number $K(\varepsilon)$ (depending on ε but **not** on $x \in D_0$) such that for all $n \geq K(\varepsilon)$ and $x \in D_0$, then

$$(17.3) \qquad \|f_n(x) - f(x)\| < \varepsilon.$$

In this case we say that the sequence is **uniformly convergent on** D_0. (See Figure 17.5.)

It follows immediately that if the sequence (f_n) is uniformly convergent on D_0 to f, then this sequence of functions also converges to f in the sense of Definition 17.1. That the converse is not true is seen by a careful examination of Examples 17.2(a–c); other examples will be given below. Before we proceed, it is useful to state a necessary and sufficient condition for the sequence (f_n) to *fail* to converge uniformly on D_0 to f.

17.5 LEMMA. *A sequence (f_n) does not converge uniformly on D_0 to f if and only if for some $\varepsilon_0 > 0$ there is a subsequence (f_{n_k}) of (f_n) and a sequence (x_k) in D_0 such that*

$$(17.4) \qquad \|f_{n_k}(x_k) - f(x_k)\| \geq \varepsilon_0 \qquad for \quad k \in \mathbf{N}.$$

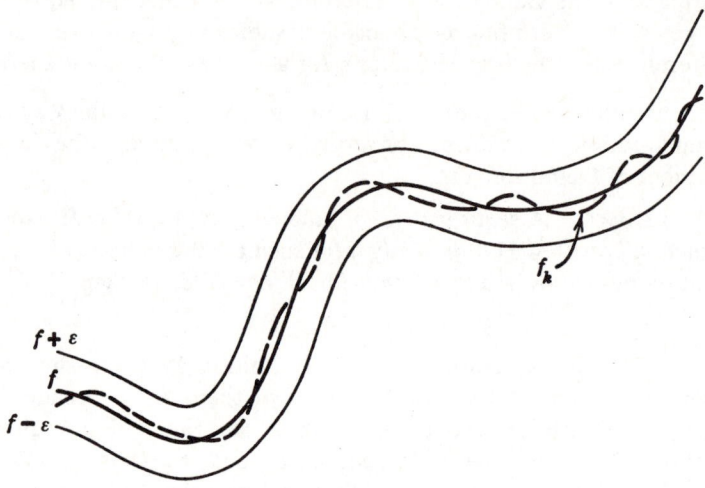

Figure 17.5

The proof of this result merely requires that the reader negate Definition 17.4. It will be left to the reader as an *essential exercise*. The preceding lemma is useful to show that Examples 17.2(a–c) do not converge uniformly on the given sets D_0.

17.6 EXAMPLES. (a) We consider Example 17.2(a). If $n_k = k$ and $x_k = k$, then $f_k(x_k) = 1$ so that

$$|f_k(x_k) - f(x_k)| = |1 - 0| = 1.$$

This shows that the sequence (f_n) does not converge uniformly on R to f.

(b) We consider Example 17.2(b). If $n_k = k$ and $x_k = (\tfrac{1}{2})^{1/k}$, then

$$|f_k(x_k) - f(x_k)| = |f_k(x_k)| = \tfrac{1}{2}.$$

Therefore, we infer that the sequence (f_n) does not converge uniformly on $[0, 1]$ to f.

(c) We consider Example 17.2(c). If $n_k = k$ and $x_k = k$, then

$$|f_k(x_k) - f(x_k)| = k,$$

showing that (f_k) does not converge uniformly on R to f.

(d) We consider Example 17.2(d). Then since

$$|f_n(x) - f(x)| \le 1/n$$

for all x in R, the sequence (f_n) converges uniformly on R to f.

The Uniform Norm

In discussing uniform convergence it is often useful to employ a certain norm on a vector space of functions.

If $D \subseteq \mathbf{R}^p$ and $f : D \to \mathbf{R}^q$, we say that f is **bounded** in case there exists $M > 0$ such that $\|f(x)\| \le M$ for all $x \in D$. If $f : D \to \mathbf{R}^q$ is bounded, then it follows that the number $\|f\|_D$ defined by

$$(17.5) \qquad \|f\|_D = \sup \{\|f(x)\| : x \in D\}$$

exists in $\mathbf{R}$. (We note that the norm on the right side of this equation is the norm in the space $\mathbf{R}^q$.).

17.7 DEFINITION. If $D \subseteq \mathbf{R}^p$, then the collection of all bounded functions on D to $\mathbf{R}^q$ is denoted by $B_{pq}(D)$ or (when p and q are understood) by $B(D)$.

In the space $B_{pq}(D)$ we define the **vector sum** of two functions f, g and the **scalar multiple** of $c \in R$ and f by

$$(17.6) \qquad (f + g)(x) = f(x) + g(x), \qquad (cf)(x) = cf(x)$$

for all $x \in D$. We define the **zero function** to be the function $0 : D \to \mathbf{R}^q$ defined for all $x \in D$ by $0(x) = 0$. We now connect this terminology with the notions presented in Section 8.

17.8 LEMMA. (a) *The set $B_{pq}(D)$ is a vector space under the vector operations defined in equation* (17.6).

(b) *The function $f \mapsto \|f\|_D$ defined on $B_{pq}(D)$ in equation* (17.5) *is a norm on $B_{pq}(D)$.*

PROOF. The proof of (a) requires only routine calculations.

To prove (b), we need to establish the four properties of a norm given in Definition 8.5. (i) It is clear from (17.5) that $\|f\|_D \ge 0$. (ii) Clearly $\|0\|_D = \sup \{\|0(x)\| : x \in D\} = 0$. Conversely, if $\|f\|_D = 0$, then since $0 \le \|f(x)\| \le \|f\|_D = 0$, we infer that $\|f(x)\| = 0$ and hence $f(x) = 0$ for all $x \in D$ so that $f = 0$. (iii) The fact that $\|cf\|_D = |c| \|f\|_D$ is readily seen. (iv) Since

$$\|(f + g)(x)\| = \|f(x) + g(x)\| \le \|f(x)\| + \|g(x)\|$$
$$\le \|f\|_D + \|g\|_D$$

for all $x \in D$, it follows that $\|f\|_D + \|g\|_D$ is an upper bound for the set $\{\|(f + g)(x)\| : x \in D\}$. Therefore we have

$$\|f + g\|_D = \sup \{\|(f + g)(x)\| : x \in D\}$$
$$\le \|f\|_D + \|g\|_D. \qquad \text{Q.E.D.}$$

Sometimes the norm $f \mapsto \|f\|_D$ is called the **uniform norm** (or the **supremum norm**) **on** $B_{pq}(D)$. We shall now show that the uniform convergence of functions in $B_{pq}(D)$ is equivalent to convergence in the uniform norm.

17.9 THEOREM. *A sequence (f_n) in $B_{pq}(D)$ converges uniformly on D to*

$f \in B_{pq}(D)$ *if and only if*

$$\|f_n - f\|_D \to 0.$$

PROOF. If the sequence (f_n) converges uniformly to f on D, then for any $\varepsilon > 0$ there is a natural number $K(\varepsilon)$ such that if $n \geq K(\varepsilon)$ and $x \in D$ then $\|f_n(x) - f(x)\| < \varepsilon$. This implies that

$$\|f_n - f\|_D = \sup \{\|(f_n - f)(x)\| : x \in D\} \leq \varepsilon.$$

Since $\varepsilon > 0$ is arbitrary, this implies that $\|f_n - f\|_D \to 0$.

Conversely, if $\|f_n - f\|_D \to 0$, then given $\varepsilon > 0$ there exists $K(\varepsilon)$ such that if $n \geq K(\varepsilon)$ then $\|f_n - f\|_D \leq \varepsilon$. This implies that if $x \in D$, then

$$\|f_n(x) - f(x)\| = \|(f_n - f)(x)\| \leq \|f_n - f\|_D \leq \varepsilon.$$

Therefore the sequence (f_n) converges uniformly on D to f.   Q.E.D.

We now illustrate the use of this lemma as a tool in examining a sequence of functions for uniform convergence. We observe first that the norm has been defined only for bounded functions; hence we can employ it (directly, at least) only when the sequence consists of bounded functions.

17.10 EXAMPLES. (a) We cannot apply Lemma 17.9 to the example considered in 17.2(a) and 17.6(a) for the reason that the functions f_n, defined to be $f_n(x) = x/n$, are not bounded on $\mathbf{R}$, which was given as the domain. For the purpose of illustration, we change the domain to obtain a bounded sequence on the new domain. For convenience, let us take $E = [0, 1]$. Although the sequence (x/n) did not converge uniformly to the zero function on the domain $\mathbf{R}$ [as was seen in Example 17.6(a)], the convergence is uniform on $E = [0, 1]$. To see this, we calculate

$$\|f_n - f\|_E = \sup \left\{ \left| \frac{x}{n} - 0 \right| : 0 \leq x \leq 1 \right\} = \frac{1}{n};$$

hence $\|f_n - f\|_E = 1/n \to 0$.

(b) We now consider the sequence discussed in Examples 17.2(b) and 17.6(b). Here $D = [0, 1]$, $f_n(x) = x^n$, and the limit function f is equal to 0 for $0 \leq x < 1$ and equal to 1 for $x = 1$. Calculating the norm of the difference $f_n - f$, we have

$$\|f_n - f\|_D = \sup \left\{ \begin{matrix} x^n, & 0 \leq x < 1 \\ 0, & x = 1 \end{matrix} \right\} = 1 \qquad \text{for} \quad n \in \mathbf{N}.$$

Since this norm does not converge to zero, we infer that the sequence (f_n) does not converge uniformly on $D = [0, 1]$ to f. This bears out our earlier considerations.

(c) We consider Example 17.2(c). Once again we cannot apply Lemma 17.9, since the functions are not bounded. Again, we choose a smaller domain, taking $E = [0, a]$ with $a > 0$. Since

$$|f_n(x) - f(x)| = \left|\frac{x^2 + nx}{n} - x\right| = \frac{x^2}{n},$$

we have

$$\|f_n - f\|_E = \sup \{|f_n(x) - f(x)| : 0 \le x \le a\} = \frac{a^2}{n}.$$

Hence the sequence converges uniformly to f on the interval $[0, a]$. (Why does this not contradict the result obtained in Example 17.6(c)?)

(d) Referring to Example 17.2(d), we consider the function $f_n(x) = (1/n) \sin (nx + n)$ on $D = \mathbf{R}$. Here the limit function $f(x) = 0$ for all $x \in D$. In order to establish the uniform convergence of this sequence, note that

$$\|f_n - f\|_D = \sup \{(1/n) |\sin (nx + n)| : x \in \mathbf{R}\}$$

But since $|\sin y| \le 1$, we conclude that $\|f_n - f\|_D = 1/n$. Hence (f_n) converges uniformly on $\mathbf{R}$, as was established in Example 17.6(d).

One of the more useful aspects of the norm is that it facilitates the formulation of a Cauchy Criterion for the uniform convergence of a sequence of bounded functions.

17.11 CAUCHY CRITERION FOR UNIFORM CONVERGENCE. *Let (f_n) be a sequence of functions in $B_{pq}(D)$. Then there is a function $f \in B_{pq}(D)$ to which (f_n) is uniformly convergent on D if and only if for each $\varepsilon > 0$ there is a natural number $M(\varepsilon)$ such that for all $m, n \ge M(\varepsilon)$, then $\|f_m - f_n\|_D < \varepsilon$.*

PROOF. Suppose that the sequence (f_n) converges uniformly on D to a function $f \in B_{pq}(D)$. Then, for $\varepsilon > 0$ there is a natural number $K(\varepsilon)$ such that if $n \ge K(\varepsilon)$, then $\|f_n - f\|_D < \varepsilon/2$. Hence if both $m, n \ge K(\varepsilon)$, we conclude that

$$\|f_m - f_n\|_D \le \|f_m - f\|_D + \|f - f_n\|_D < \varepsilon.$$

Conversely, suppose the Cauchy Criterion is satisfied and that for $\varepsilon > 0$ there is a natural number $M(\varepsilon)$ such that $\|f_m - f_n\|_D < \varepsilon$ when $m, n \ge M(\varepsilon)$. Now for each $x \in D$ we have

$$(17.6) \qquad \|f_m(x) - f_n(x)\| \le \|f_m - f_n\|_D < \varepsilon \qquad \text{for} \quad m, n \ge M(\varepsilon).$$

Hence the sequence $(f_n(x))$ is a Cauchy sequence in $\mathbf{R}^q$ and so converges to some element of $\mathbf{R}^q$. We define f for x in D by

$$f(x) = \lim (f_n(x)).$$

From (17.6) we conclude that if m is a fixed natural number satisfying $m \ge M(\varepsilon)$ and if n is any natural number with $n \ge M(\varepsilon)$, then for all x in D

we have

$$\|f_m(x) - f_n(x)\| < \varepsilon.$$

If we apply Lemma 15.8 it follows that if $m \geq M(\varepsilon)$ and $x \in D$, then $\|f_m(x) - f(x)\| \leq \varepsilon$. Since f_m is a bounded function, it follows readily from this (how?) that f is bounded and hence belongs to $B_{pq}(D)$. Moreover we conclude that (f_n) converges uniformly to f on D.   Q.E.D.

Exercises

In these exercises you may make use of the elementary properties of the trigonometric and exponential functions from earlier courses.

17.A. For each $n \in N$, let f_n be defined for $x > 0$ by $f_n(x) = 1/(nx)$. For what values of x does $\lim (f_n(x))$ exist?

17.B. For each $n \in N$, let g_n be defined for $x \geq 0$ by the formula

$$g_n(x) = nx, \qquad 0 \leq x \leq 1/n,$$

$$= \frac{1}{nx}, \qquad 1/n < x,$$

Show that $\lim (g_n(x)) = 0$ for all $x > 0$.

17.C. Show that $\lim ((\cos \pi x)^{2n})$ exists for all values of x. What is its limit?

17.D. Show that, if we define f_n on R by

$$f_n(x) = \frac{nx}{1 + n^2 x^2},$$

then (f_n) converges on R.

17.E. Let h_n be defined on the interval $I = [0, 1]$ by the formula

$$h_n(x) = 1 - nx, \qquad 0 \leq x \leq 1/n,$$

$$= 0, \qquad 1/n < x \leq 1.$$

Show that $\lim (h_n)$ exists on I.

17.F. Let g_n be defined on I by

$$g_n(x) = nx, \qquad\qquad 0 \leq x \leq 1/n,$$

$$= \frac{n}{n-1}(1-x), \qquad 1/n < x \leq 1.$$

Show that $\lim (g_n)$ exists on I.

17.G. Show that if f_n is defined on R by

$$f_n(x) = \frac{2}{\pi} \text{ Arc tan } (nx),$$

then $f = \lim (f_n)$ exists on R. In fact the limit is given by

$$f(x) = 1, \qquad x > 0,$$

$$= 0, \qquad x = 0,$$

$$= -1, \qquad x < 0.$$

17.H. Show that $\lim (e^{-nx})$ exists for $x \geq 0$. Also consider the existence of $\lim (xe^{-nx})$.

17.I. Suppose that (x_n) is a convergent sequence of points which lies, together with its limit x, in a set $D \subseteq \mathbf{R}^p$. Suppose that (f_n) converges on D to the function f. Is it true that $f(x) = \lim (f_n(x_n))$?

17.J. Consider the preceding exercise with the additional hypothesis that the convergence of the (f_n) is uniform on D.

17.K. Prove that the convergence in Exercise 17.A is not uniform on the entire set of convergence, but that it is uniform for $x \geq 1$.

17.L. Show that the convergence in Exercise 17.B is not uniform on the domain $x \geq 0$, but that it is uniform on a set $x \geq c$, where $c > 0$.

17.M. Is the convergence in Exercise 17.D uniform on $\mathbf{R}$?

17.N. Is the convergence in Exercise 17.E uniform on $\mathbf{I}$?

17.O. Is the convergence in Exercise 17.F uniform on $\mathbf{I}$? Is it uniform on $[c, 1]$ for $c > 0$?

17.P. Does the sequence (xe^{-nx}) converge uniformly for $x \geq 0$?

17.Q. Does the sequence $(x^2 e^{-nx})$ converge uniformly for $x \geq 0$?

17.R. Let (f_n) be a sequence of functions which converges on D to a function f. If A and B are subsets of D and it is known that the convergence is uniform on A and also on B, show that the convergence is uniform on $A \cup B$.

17.S. Give an example of a sequence (f_n) in $B_{pq}(\mathbf{I})$ such that $\|f_n\|_{\mathbf{I}} \leq 1$ for all $n \in \mathbf{N}$ which does not have a uniformly convergent subsequence. (Hence the Bolzano-Weierstrass Theorem does not hold in $B_{pq}(\mathbf{I})$.)

Section 18 The Limit Superior

In Section 6 we introduced the notion of the supremum of a non-empty bounded set of real numbers, and we have made important use of this notion many times. However, in dealing with a bounded infinite set $S \subseteq \mathbf{R}$ it is sometimes also of interest to consider the largest cluster point s^* of S. This point s^* is the infimum of all real numbers which are exceeded by at most a finite number of elements of S. We shall adapt this notion to bounded sequences in $\mathbf{R}$ to obtain the frequently useful concept of the "limit superior."

18.1 DEFINITION. Let $X = (x_n)$ be a bounded sequence in $\mathbf{R}$.

(a) The **limit superior** of X, which we denote by

$$\limsup X, \quad \limsup (x_n), \quad \text{or} \quad \overline{\lim} (x_n),$$

is the infimum of the set V of $v \in \mathbf{R}$ such that there are at most a finite number of $n \in \mathbf{N}$ such that $v < x_n$.

(b) The **limit inferior** of X, which we denote by

$$\liminf X, \quad \liminf (x_n), \quad \text{or} \quad \underline{\lim} (x_n),$$

Figure 18.1

is the supremum of the set W of $w \in R$ such that there are at most a finite number of $m \in N$ such that $x_m < w$.

While a bounded sequence need not have a limit, it always has a unique limit superior (and a unique limit inferior). That is clear from the fact that the number $v = \sup\{x_n : n \in N\}$ belongs to the set V, while the number $\inf\{x_n : n \in N\} - 1$ is a lower bound for V.

There are several equivalent ways, which are often useful, that one can define the limit superior of a bounded sequence. (The reader is strongly urged to attempt to prove this result before reading the proof.)

18.2 THEOREM. *If* $X = (x_n)$ *is a bounded sequence in* R, *then the following statements are equivalent for a real number* x^*.

(a) $x^* = \lim \sup (x_n)$.

(b) *If* $\varepsilon > 0$, *there are at most a finite number of* $n \in N$ *such that* $x^* + \varepsilon < x_n$, *but there are an infinite number such that* $x^* - \varepsilon < x_n$.

(c) *If* $v_m = \sup\{x_n : n \geq m\}$, *then* $x^* = \inf\{v_m : n \in N\}$.

(d) *If* $v_m = \sup\{x_n : n \geq m\}$, *then* $x^* = \lim (v_m)$.

(e) *If* L *is the set of* $v \in R$ *such that there exists a subsequence of* X *which converges to* v, *then* $x^* = \sup L$.

PROOF. Let $x^* = \lim \sup (x_n)$ and let $\varepsilon > 0$. By Definition 18.1, there exists a $v \in V$ with $x^* \leq v \leq x^* + \varepsilon$. Therefore $x^* + \varepsilon$ also belongs to V, so there can be at most a finite number of $n \in N$ such that $x^* + \varepsilon < x_n$. On the other hand $x^* - \varepsilon$ is not in V, so there are an infinite number of $n \in N$ such that $x^* - \varepsilon < x_n$. Hence (a) implies (b).

If (b) holds, given $\varepsilon > 0$, then for all sufficiently large m we have $v_m \leq x^* + \varepsilon$; therefore $\inf\{v_m : m \in N\} \leq x^* + \varepsilon$. But since there is an infinite number of $n \in N$ such that $x^* - \varepsilon < x_n$, then $x^* - \varepsilon < v_m$ for all $m \in N$ and hence $x^* - \varepsilon \leq \inf\{v_m : m \in N\}$. Since $\varepsilon > 0$ is arbitrary, we deduce that $x^* = \inf\{v_m : m \in N\}$ and (c) holds.

If the sequence (v_m) is defined as in (c), then it is monotone decreasing and hence $\inf (v_m) = \lim (v_n)$, so that (c) implies (d).

Now suppose that x^* satisfies (d) and $X' = (x_{n_k})$ is a convergent subsequence of X; since $n_k \geq k$ we have $x_{n_k} \leq v_k$ and hence $\lim X' \leq \lim (v_k) = x^*$. Conversely, note that there exists $n_1 \in N$ such that $v_1 - 1 < x_{n_1} \leq v_1$. Inductively choose $n_{k+1} > n_k$ such that

$$v_k - \frac{1}{k+1} < x_{n_{k+1}} \leq v_k.$$

Since $\lim (v_k) = x^*$, it follows that $x^* = \lim (x_{n_k})$. Therefore (d) implies (e).

Finally, let $w = \sup L$. If $\varepsilon > 0$ is given, then there can be at most a finite number of $n \in N$ with $w + \varepsilon < x_n$ (by the Bolzano Weierstrass Theorem 16.4). Therefore $w + \varepsilon \in V$ and $\lim \sup X \le w + \varepsilon$. On the other hand there exists a subsequence X' converging to some number exceeding $w - \varepsilon$; hence $w - \varepsilon$ is not in V and so $w - \varepsilon \le \lim \sup X$. Since $\varepsilon > 0$ is arbitrary, we infer that $w = \lim \sup X$. Therefore (e) implies (a).   Q.E.D.

Both of the characterizations (d) and (e) can be regarded as justifying the term "limit superior." There are corresponding characterizations for the limit inferior of a bounded sequence which the reader should write out and prove.

We now establish the basic algebraic properties of the limit superior and the limit inferior of bounded sequences.

18.3   THEOREM.   *Let* $X = (x_n)$ *and* $Y = (y_n)$ *be bounded sequences of real numbers. Then the following relations hold:*

(a) $\lim \inf (x_n) \le \lim \sup (x_n)$.

(b) *If* $c \ge 0$, *then* $\lim \inf (cx_n) = c \lim \inf (x_n)$ *and* $\lim \sup (cx_n) = c \lim \sup (x_n)$.

(b') *If* $c \le 0$, *then* $\lim \inf (cx_n) = c \lim \sup (x_n)$ *and* $\lim \sup (cx_n) = c \lim \inf (x_n)$.

(c) $\lim \inf (x_n) + \lim \inf (y_n) \le \lim \inf (x_n + y_n)$.

(d) $\lim \sup (x_n + y_n) \le \lim \sup (x_n) + \lim \sup (y_n)$.

(e) *If* $x_n \le y_n$ *for all* n, *then* $\lim \inf (x_n) \le \lim \inf (y_n)$ *and also* $\lim \sup (x_n) \le \lim \sup (y_n)$.

PROOF.   (a) If $w < \lim \inf (x_n)$ and $v > \lim \sup (x_n)$, then there are infinitely many $n \in N$ such that $w \le x_n$, while there are only a finite number such that $v < x_n$. Therefore, we must have $w \le v$, which implies (a).

(b) If $c \ge 0$, then multiplication by c preserves all inequalities of the form $w \le x_n$, etc.

(b') If $c \le 0$, then multiplication by c reverses inequalities and converts the limit superior into the limit inferior, and conversely.

Statement (c) is dual to (d) and can be derived directly from (d) or proved by using the same type of argument. To prove (d), let $v > \lim \sup (x_n)$ and $u > \lim \sup (y_n)$; by definition there are only a finite number of $n \in N$ such that $v < x_n$ and a finite number such that $u < y_n$. Therefore there can be only a finite number of n such that $v + u < x_n + y_n$, showing that $\lim \sup (x_n + y_n) \le v + u$. This proves statement (d).

We now prove the second assertion in (e). If $u > \lim \sup (y_n)$, then there can be only a finite number of natural numbers n such that $u < y_n$. Since $x_n \le y_n$, then $\lim \sup (x_n) \le u$, and so $\lim \sup (x_n) \le \lim \sup (y_n)$.   Q.E.D.

Each of the equivalent conditions given in Theorem 18.2 can be used to prove the parts of Theorem 18.3. It is suggested that some of these alternative proofs be written out as an exercise.

It might be asked whether the inequalities in Theorem 18.3 can be replaced by equalities. In general, the answer is no. For, if $X = ((-1)^n)$, then $\liminf X = -1$ and $\limsup X = +1$. If $Y = ((-1)^{n+1})$, then $X + Y = (0)$ so that

$$\liminf X + \liminf Y = -2 < 0 = \liminf (X + Y),$$

$$\limsup (X + Y) = 0 < 2 = \limsup X + \limsup Y.$$

We have seen that the limit inferior and the limit superior exist for any bounded sequence, regardless of whether the sequence is convergent. We now show that the existence of $\lim X$ is equivalent to the equality of $\liminf X$ and $\limsup X$.

18.4 LEMMA. *Let X be a bounded sequence of real numbers. Then X is convergent if and only if $\liminf X = \limsup X$ in which case $\lim X$ is the common value.*

PROOF. If $x = \lim X$, then for each $\varepsilon > 0$ there is a natural number $N(\varepsilon)$ such that

$$x - \varepsilon < x_n < x + \varepsilon, \qquad n \geq N(\varepsilon).$$

The second inequality shows that $\limsup X \leq x + \varepsilon$ and the first inequality shows that $x - \varepsilon \leq \liminf X$. Hence $0 \leq \limsup X - \liminf X \leq 2\varepsilon$, and since $\varepsilon > 0$ is arbitrary, we have the stated equality.

Conversely, suppose that $x = \liminf X = \limsup X$. If $\varepsilon > 0$, it follows from Theorem 18.2(b) that there exists a natural number $N_1(\varepsilon)$ such that if $n \geq N_1(\varepsilon)$, then $x_n < x + \varepsilon$. Similarly, there exists a natural number $N_2(\varepsilon)$ such that if $n \geq N_2(\varepsilon)$, then $x - \varepsilon < x_n$. Let $N(\varepsilon) = \sup \{N_1(\varepsilon), N_2(\varepsilon)\}$; if $n \geq N(\varepsilon)$, then $|x_n - x| < \varepsilon$, showing that $x = \lim X$.   Q.E.D.

Unbounded Sequences

Sometimes it is convenient to have the limit superior and the limit inferior defined for arbitrary (that is, not necessarily bounded) sequences in $\mathbf{R}$. To do this we need to introduce the symbols $+\infty$ and $-\infty$, but it is to be emphasized that we do not consider them to be real numbers; they are merely convenient symbols.

If S is a *non-empty* set in $\mathbf{R}$ which is not bounded above, we define $\sup S = +\infty$; if T is a *non-empty* set in $\mathbf{R}$ which is not bounded below, we define $\inf T = -\infty$. As remarked after Definition 6.1, every real number is an upper bound of the empty set $\emptyset$, so we define $\sup \emptyset = -\infty$. Similarly, every real number is a lower bound of $\emptyset$, so we define $\inf \emptyset = +\infty$.

Now let $X = (x_n)$ be a sequence in $\mathbf{R}$ which is not bounded above; then the set V of numbers $v \in R$ such that there are at most a finite number of $n \in N$ such that $v < x_n$ is empty. Hence the inf $V = +\infty$. Thus if $X = (x_n)$ is a sequence in R which is not bounded above, then we have

$$\limsup (x_n) = +\infty.$$

Similarly, if $Y = (y_n)$ is a sequence in R which is not bounded below, then we have

$$\liminf (y_n) = -\infty.$$

We note that if $X = (x_n)$ is a sequence in R that is not bounded above, then the sets $\{x_n : n \ge m\}$ are not bounded above and so

$$v_m = \sup \{x_n : n \ge m\} = +\infty$$

for all $m \in N$.

Infinite Limits

If $X = (x_n)$ is a sequence in R, we say that $X = (x_n)$ **diverges to** $+\infty$, and write $\lim (x_n) = +\infty$, if for every $\alpha \in R$ there is a $K(\alpha) \in N$ such that if $n \ge K(\alpha)$ then $x_n > \alpha$.

Similarly we say that $X = (x_n)$ **diverges to** $-\infty$, and write $\lim (x_n) = -\infty$, in case for every $\alpha \in R$ there is a $K(\alpha) \in N$ such that if $n \ge K(\alpha)$ then $x_n < \alpha$.

It is an exercise to show that $X = (x_n)$ diverges to $+\infty$ if and only if

$$\liminf (X_n) = \limsup (x_n) = +\infty,$$

and that $X = (x_n)$ diverges to $-\infty$ if and only if

$$\liminf (x_n) = \limsup (x_n) = -\infty.$$

Exercises

18.A. Determine the limit superior and the limit inferior of the following bounded sequences in R.

(a) $((-1)^n)$, (b) $((-1)^n/n))$,

(c) $((-1)^n + 1/n)$, (d) $(\sin n)$.

18.B. If $X = (x_n)$ is a bounded sequence in R, show that there is a subsequence of X which converges to $\liminf X$.

18.C. Formulate and prove directly the theorem corresponding to Theorem 18.2 for the limit inferior.

18.D. Give a direct proof of Theorem 18.3(c).

18.E. Prove Theorem 18.3(d) by using 18.2(b) as the definition of the limit superior. Do the same using 18.2(d) and 18.2(e).

18.F. If $X = (x_n)$ is a bounded sequence of strictly positive elements in R show that $\limsup (x_n^{1/n}) \le \limsup (x_{n+1}/x_n)$.

18.G. Determine the limit superior and the limit inferior of the following sequences in R.

(a) $((-1)^n n)$,   (b) $(n \sin n)$,

(c) $(n(\sin n)^2)$,   (d) $(n \tan n)$.

18.H. Show that the sequence $X = (x_n)$ in R diverges to $+\infty$ if and only if $\liminf X = +\infty$.

18.I. Show that $\limsup X = +\infty$ if and only if there is a subsequence X' of X such that $\lim X' = +\infty$.

18.J. Interpret Theorem 18.3 for unbounded sequences.

Section 19 Some Extensions

It is frequently important in analysis to estimate the "order of magnitude" of a sequence or to compare two sequences relative to their magnitude. In doing so we discard terms which make no "essential contribution." For example, if $x_n = 2n + 17$, then when $n \in N$ is large, the dominant contribution comes from the term $2n$. If $y_n = n^2 - 5n$, then when $n \in N$ is large, the dominant term is n^2. And, although the first few terms of (y_n) are smaller than those of (x_n), the terms of this sequence ultimately grow more rapidly than those of (x_n).

We shall now introduce some terminology to make this idea more precise and some notation, due to Landau,† that is often useful.

19.1 DEFINITION. Let $X = (x_n)$ and $Y = (y_n)$ be sequences in R and suppose that $y_n \neq 0$ for all sufficiently large $n \in N$. We say that X and Y are **equivalent** and write

$$X \sim Y \qquad \text{or} \qquad (x_n) \sim (y_n)$$

in case $\lim (x_n/y_n) = 1$. We say that X is of a **lower order of magnitude** than Y and write

$$X = o(Y) \qquad \text{or} \qquad x_n = o(y_n)$$

in case $\lim (x_n/y_n) = 0$. We say that X is **dominated** by Y and write

$$X = O(Y) \qquad \text{or} \qquad x_n = O(y_n)$$

in case the sequence (x_n/y_n) is bounded.

It is clear that either $X \sim Y$ or $X = o(Y)$ implies that $X = O(Y)$. Various properties of these notations will be given in the exercises.

Cesàro Summation

We have already defined what is meant by the convergence of a sequence $X = (x_n)$ in R^p to an element x. However, it may be possible to attach

---

† EDMUND (G. H.) LANDAU (1877–1938) was a professor at Göttingen and is known for his research and books on number theory and analysis. These books are noted for their rigor and brevity of style (and their elementary German).

x to the sequence X as a sort of "generalized limit," even though the sequence X does not converge to x in the sense of Definition 14.3. There are many ways in which one can generalize the idea of the limit of a sequence and to give very much of an account of some of them would take us far beyond the scope of this book. However, there is a method which is both elementary in nature and useful in applications to oscillatory sequences. Since it is of some importance and the proof of the main result is typical of many analytical arguments, we inject here a brief introduction to the theory of Cesàro† summability.

19.2 Definition. If $X = (x_n)$ is a sequence of elements in $\mathbf{R}^p$, then the sequence $S = (\sigma_n)$ defined by

$$\sigma_1 = x_1, \quad \sigma_2 = \frac{x_1 + x_2}{2}, \dots, \quad \sigma_n = \frac{x_1 + x_2 + \cdots + x_n}{n}, \dots,$$

is called the **sequence of arithmetic means** of X.

In other words, the elements of S are found by averaging the terms in X. Since this average tends to smooth out occasional fluctuations in X, it is reasonable to expect that the sequence S has more chance of converging than the original sequence X. In case the sequence S of arithmetic means converges to an element y, we say that the sequence X is **Cesàro summable** to y, or that y is the $(C, 1)$-**limit** of the sequence X.

For example, let X be the non-convergent real sequence $X = (1, 0, 1, 0, \dots)$; it is readily seen that if n is an even natural number, then $\sigma_n = \frac{1}{2}$ and if n is odd then $\sigma_n = (n+1)/2n$. Since $\frac{1}{2} = \lim (\sigma_n)$, the sequence X is Cesàro summable to $\frac{1}{2}$, which is not the limit of X but seems like the most natural "generalized limit" we might try to attach to X.

It seems reasonable, in generalizing the notion of the limit of a sequence, to require that the generalized limit give the usual value of the limit whenever the sequence is convergent. We now show that the Cesàro method has this property.

19.3 Theorem. *If the sequence $X = (x_n)$ converges to x, then the sequence $S = (\sigma_n)$ of arithmetic means also converges to x.*

proof. We need to estimate the magnitude of

(19.1)
$$\sigma_n - x = \frac{1}{n}(x_1 + x_2 + \cdots + x_n) - x$$
$$= \frac{1}{n}\{(x_1 - x) + (x_2 - x) + \cdots + (x_n - x)\}.$$

Since $x = \lim (x_n)$, then given $\varepsilon > 0$ there is a natural number $N(\varepsilon)$ such that if $m \geq N(\varepsilon)$, then $\|x_m - x\| < \varepsilon$. Also, since the sequence $X = (x_n)$ is

† Ernesto Cesàro (1859–1906) studied in Rome and taught at Naples. He did work in geometry and algebra as well as analysis.

convergent, there is a real number A such that $\|x_k - x\| < A$ for all k. If $n \geq N = N(\varepsilon)$, we break the sum on the right side of (19.1) into a sum from $k = 1$ to $k = N$ plus a sum from $k = N + 1$ to $k = n$. We apply the estimate $\|x_k - x\| < \varepsilon$ to the latter $n - N$ terms to obtain

$$\|\sigma_n - x\| \leq \frac{NA}{n} + \frac{n - N}{n} \varepsilon \qquad \text{for} \quad n \geq N(\varepsilon).$$

If n is sufficiently large, then $NA/n < \varepsilon$ and since $(n - N)/n < 1$, we find that $\|\sigma_n - x\| < 2\varepsilon$ for n sufficiently large. Hence $x = \lim (\sigma_n)$.   Q.E.D.

We shall not pursue the theory of summability any further, but refer the reader to books on divergent series and summability. For example, see the book by Knopp listed in the References. One of the most interesting elementary applications of Cesàro summability is the celebrated theorem of Fejér which asserts that a continuous function can be recovered from its Fourier series by the process of Cesàro summability, even though it cannot always be recovered from this series by ordinary convergence. (See Theorem 38.12.)

Double and Iterated Sequences

We recall that a sequence in $\mathbf{R}^p$ is a function defined on the set $\mathbf{N}$ of natural numbers and with range in $\mathbf{R}^p$. A double sequence in $\mathbf{R}^p$ is a function X with domain $\mathbf{N} \times \mathbf{N}$ consisting of all ordered pairs of natural numbers and range in $\mathbf{R}^p$. In other words, at each ordered pair (m, n) of natural numbers the value of the double sequence X is an element of $\mathbf{R}^p$ which we shall typically denote by x_{mn}. Generally we shall use a symbolism such as $X = (x_{mn})$ to represent X, but sometimes it is convenient to list the elements in an array such as

$$(19.2) \qquad X = \begin{bmatrix} x_{11} & x_{12} & \cdots & x_{1n} & \cdots \\ x_{21} & x_{22} & \cdots & x_{2n} & \cdots \\ \cdot & \cdot & \cdot & \cdot & \cdot \\ x_{m1} & x_{m2} & \cdots & x_{mn} & \cdots \\ \cdot & \cdot & \cdot & \cdot & \cdot \end{bmatrix}.$$

Observe that, in this array, the first index refers to the row in which the element x_{mn} appears and the second index refers to the column.

19.4   DEFINITION.   If $X = (x_{mn})$ is a double sequence in $\mathbf{R}^p$, then an element x is said to be a **limit** (or a **double limit**) of X if for each positive number ε there is a natural number $N(\varepsilon)$ such that for all $m, n \geq N(\varepsilon)$ then $\|x_{mn} - x\| < \varepsilon$. In this case we say that the double sequence **converges** to x and write

$$x = \lim_{mn} (x_{mn}) \qquad \text{or} \qquad x = \lim X.$$

Much of the elementary theory of limits of sequences carries over with little change to double sequences. In particular, the fact that the double limit is uniquely determined (when it exists) is proved in exactly the same manner as in Theorem 14.5. Similarly, one can define algebraic operations for double sequences and obtain results exactly parallel to those discussed in Theorem 15.6. There is also a Cauchy Criterion for the convergence of a double sequence which we will state, but whose proof we leave to the reader.

19.5 CAUCHY CRITERION. *If $X = (x_{mn})$ is a double sequence in $\mathbf{R}^p$, then X is convergent if and only if for each $\varepsilon > 0$ there is a natural number $M(\varepsilon)$ such that for all $m, n, r, s \geq M(\varepsilon)$, then*

$$\|x_{mn} - x_{rs}\| < \varepsilon.$$

We shall not pursue in any more detail that part of the theory of double sequences which is parallel to the theory of (single) sequences. Rather, we propose to look briefly at the relation between the limit as defined in 19.4 and the "iterated" limits.

To begin with, we note that a double sequence can be regarded (in at least two ways) as giving a *sequence of sequences*! On one hand, we can regard each row in the array given in (19.2) as a sequence in $\mathbf{R}^p$. Thus the first row in the array (19.2) yields the sequence $Y_1 = (x_{1n} : n \in N) = (x_{11}, x_{12}, \ldots, x_{1n}, \ldots)$; the second row in (19.2) yields the sequence $Y_2 = (x_{2n} : n \in N)$; etc. It makes perfectly good sense to consider the limits of the row sequences $Y_1, Y_2, \ldots, Y_m, \ldots$ (when these limits exist). Supposing that these limits exist and denoting them by $y_1, y_2, \ldots, y_m, \ldots$, we obtain a sequence of elements in $\mathbf{R}^p$ which might well be examined for convergence. Thus we are considering the existence of $y = \lim (y_m)$. Since the elements y_m are given by $y_m = \lim Y_m$ where $Y_m = (x_{mn} : n \in N)$, we are led to denote the limit $y = \lim (y_m)$ (when it exists) by the expression

$$y = \lim_m \lim_n (x_{mn}).$$

We shall refer to y as an **iterated limit** of the double sequence (or more precisely as the **row iterated limit** of this double sequence).

What has been done for rows can equally well be done for columns. Thus we form the sequences

$$Z_1 = (x_{m1} : m \in N), \qquad Z_2 = (x_{m2} : m \in N),$$

and so forth. Supposing that the limits $z_1 = \lim Z_1, z_2 = \lim Z_2, \ldots$, exist, we can then consider $z = \lim (z_n)$. When this latter limit exists, we denote it by

$$z = \lim_n \lim_m (x_{mn}),$$

and refer to z as an **iterated limit,** or the **column iterated limit** of the double sequence $X = (x_{mn})$.

The first question we might ask is: if the double limit of the sequence $X = (x_{mn})$ exists, then do the iterated limits exist? The answer to this question may come as a surprise to the reader; it is negative. To see this, let X be the double sequence in $\mathbf{R}$ which is given by $x_{mn} = (-1)^{m+n}(1/m + 1/n)$, then it is readily seen that the double limit of this sequence exists and is 0. However, it is also readily verified that none of the sequences

$$Y_1 = (x_{1n} : n \in \mathbf{N}), \ldots, Y_m = (x_{mn} : n \in \mathbf{N}), \ldots$$

has a limit. Hence neither iterated limit can possibly exist, since none of the "inner" limits exists.

The next question is: if the double limit exists and if one of the iterated limits exists, then does this iterated limit equal the double limit? This time the answer is affirmative. In fact, we shall now establish a somewhat stronger result.

19.6 DOUBLE LIMIT THEOREM. *If the double limit* $x = \lim_{mn}(x_{mn})$ *exists, and if for each natural number m the limit* $y_m = \lim_n(x_{mn})$ *exists, then the iterated limit* $\lim_m \lim_n (x_{mn})$ *exists and equals x.*

PROOF. By hypothesis, given $\varepsilon > 0$ there is a natural number $N(\varepsilon)$ such that if $m, n \geq N(\varepsilon)$, then $\|x_{mn} - x\| < \varepsilon$. Again by hypothesis, the limits $y_m = \lim_n (x_{mn})$ exist, and from the above inequality and Lemma 15.8 it follows that $\|y_m - x\| \leq \varepsilon$ for all $m \geq N(\varepsilon)$. Therefore, we conclude that $x = \lim (y_m)$.   Q.E.D.

The preceding result shows that if the double limit exists, then the only thing that can prevent the iterated limits from existing and being equal to the double limit is that the "inner" limits may not exist. More precisely, we have the following result.

19.7 COROLLARY. *Suppose the double limit exists and that the limits*

$$y_m = \lim_n (x_{mn}), \qquad z_n = \lim_m (x_{mn})$$

exist for all natural numbers m, n. Then the iterated limits

$$\lim_m \lim_n (x_{mn}), \qquad \lim_n \lim_m (x_{mn})$$

exist and equal the double limit.

We next inquire as to whether the existence and equality of the two iterated limits implies the existence of the double limit. The answer is no. This is seen by examining the double sequence $X = (x_{mn})$ in $\mathbf{R}$ defined by $x_{mn} = 1$ when $m \neq n$ and $x_{mn} = 0$ when $m = n$. Here both iterated limits exist and are equal, but the double limit does not exist. However, under some additional conditions, we can establish the existence of the double limit from the existence of one of the iterated limits.

19.8 DEFINITION. For each natural number m, let $Y_m = (x_{mn})$ be a sequence in $\mathbf{R}^p$ which converges to y_m. We say that the sequences $\{Y_m : m \in \mathbf{N}\}$ are **uniformly convergent** if, for each $\varepsilon > 0$ there is a natural number $N(\varepsilon)$ such that if $n \ge N(\varepsilon)$, then $\|x_{mn} - y_m\| < \varepsilon$ for all natural numbers m.

The reader will do well to compare this definition with Definition 17.4 and observe that they are of the same character. Partly in order to motivate Theorem 19.10 to follow, we show that if each of the sequences Y_m is convergent, then the existence of the double limit implies that the sequences $\{Y_m : m \in \mathbf{N}\}$ are uniformly convergent.

19.9 LEMMA. *If the double limit of the double sequence $X = (x_{mn})$ exists and if, for each natural number m, the sequence $Y_m = (x_{mn} : n \in \mathbf{N})$ is convergent, then this collection is uniformly convergent.*

PROOF. Since the double limit exists, given $\varepsilon > 0$ there is a natural number $N(\varepsilon)$ such that if $m, n \ge N(\varepsilon)$, then $\|x_{mn} - x\| < \varepsilon$. By hypothesis, the sequence $Y_m = (x_{mn} : n \in \mathbf{N})$ converges to an element y_m and, applying Lemma 15.8, we find that if $m \ge N(\varepsilon)$, then $\|y_m - x\| \le \varepsilon$. Thus if $m, n \ge N(\varepsilon)$, we infer that

$$\|x_{mn} - y_m\| \le \|x_{mn} - x\| + \|x - y_m\| < 2\varepsilon.$$

In addition, for $m = 1, 2, \ldots, N(\varepsilon) - 1$ the sequence Y_m converges to y_m; hence there is a natural number $K(\varepsilon)$ such that if $n \ge K(\varepsilon)$, then

$$\|x_{mn} - y_m\| < \varepsilon, \qquad m = 1, 2, \ldots, N(\varepsilon) - 1.$$

Letting $M(\varepsilon) = \sup\{N(\varepsilon), K(\varepsilon)\}$, we conclude that if $n \ge M(\varepsilon)$, then for any value of m we have

$$\|x_{mn} - y_m\| < 2\varepsilon.$$

This establishes the uniformity of the convergence of the sequences $\{Y_m : m \in \mathbf{N}\}$. Q.E.D.

The preceding lemma shows that, under the hypothesis that the sequences Y_m converge, then the uniform convergence of this collection of sequences is a necessary condition for the existence of the double limit. We now establish a result in the converse direction.

19.10 ITERATED LIMIT THEOREM. *Suppose that the single limits*

$$y_m = \lim_n (x_{mn}), \qquad z_n = \lim_m (x_{mn}),$$

exist for all $m, n \in N$, *and that the convergence of one of these collections is uniform. Then both iterated limits and the double limit exist and all three are equal.*

PROOF.   Suppose that the convergence of the collection $\{Y_m : m \in N\}$ is uniform. Hence given $\varepsilon > 0$, there is a natural number $N(\varepsilon)$ such that if $n \geq N(\varepsilon)$, then

(19.3) $$\|x_{mn} - y_m\| < \varepsilon$$

for all natural numbers m. To show that $\lim (y_m)$ exists, take a fixed number $q \geq N(\varepsilon)$. Since $z_q = \lim (x_{rq} : r \in N)$ exists, we know that if $r, s \geq R(\varepsilon, q)$, then

$$\|y_r - y_s\| \leq \|y_r - x_{rq}\| + \|x_{rq} - x_{sq}\| + \|x_{sq} - y_s\| < 3\varepsilon.$$

Therefore, (y_r) is a Cauchy sequence and converges to an element y in R^p. This establishes the existence of the iterated limit

$$y = \lim_m (y_m) = \lim_m \lim_n (x_{mn}).$$

We now show that the double limit exists. Since $y = \lim (y_m)$, given $\varepsilon > 0$ there is an $M(\varepsilon)$ such that if $m \geq M(\varepsilon)$, then $\|y_m - y\| < \varepsilon$. Letting $K(\varepsilon) = \sup \{N(\varepsilon), M(\varepsilon)\}$, we again use (19.3) to conclude that if $m, n \geq K(\varepsilon)$, then

$$\|x_{mn} - y\| \leq \|x_{mn} - y_m\| + \|y_m - y\| < 2\varepsilon.$$

This proves that the double limit exists and equals y.

Finally, to show that the other iterated limit exists and equals y, we make use of Theorem 19.6 or its corollary.   Q.E.D.

It might be conjectured that, although the proof just given makes use of the existence of both collections of single limits and the uniformity of one of them, the conclusion may follow with the existence (and uniformity) of just one collection of single limits. We leave it to the reader to investigate the truth or falsity of this conjecture.

Exercises

19.A. Establish the following relations:
(a) $(n^2 + 2) \sim (n^2 - 3)$,   (b) $(n^2 + 2) = o(n^3)$,
(c) $((-1)^n n^2) = O(n^2)$,   (d) $((-1)^n n^2) = o(n^3)$,
(e) $(\sqrt{n+1} - \sqrt{n}) \sim (1/2\sqrt{n})$,   (f) $(\sin n) = O(1)$.

19.B. Let X, Y, and Z be sequences with non-zero elements. Show that:
(a) $X \sim X$.
(b) If $X \sim Y$, then $Y \sim X$.
(c) If $X \sim Y$ and $Y \sim Z$, then $X \sim Z$.

19.C. If $X_1 = O(Y)$ and $X_2 = O(Y)$, we conclude that $X_1 \pm X_2 = O(Y)$ and summarize this in the "equation"

(a) $O(Y) \pm O(Y) = O(Y)$. Give similar interpretations for and prove that

(b) $o(Y) \pm o(Y) = o(Y)$.

(c) If $c \neq 0$, then $o(cY) = o(Y)$ and $O(cY) = O(Y)$.

(d) $O(o(Y)) = o(Y)$, $o(O(Y)) = o(Y)$.

(e) $O(X)O(Y) = O(XY)$, $O(X)o(Y) = o(XY)$, $o(X)o(Y) = o(XY)$.

19.D. Show that $X = o(Y)$ and $Y = o(X)$ cannot hold simultaneously. Give an example of sequences such that $X = O(Y)$ but $Y \neq O(X)$.

19.E. If X is a monotone sequence in $\mathbf{R}$, show that the sequence of arithmetic means is monotone.

19.F. If $X = (x_n)$ is a sequence in $\mathbf{R}$ and (σ_n) is the sequence of arithmetic means, then $\lim \sup (\sigma_n) \le \lim \sup (x_n)$. Give an example where inequality holds.

19.G. If $X = (x_n)$ is a sequence of positive real numbers, then is (σ_n) monotone increasing?

19.H. If a sequence $X = (x_n)$ in $\mathbf{R}^p$ is Cesàro summable, then $X = o(n)$. (Hint: $x_n = n\sigma_n - (n-1)\sigma_{n-1}$.)

19.I. Let X be a monotone sequence in $\mathbf{R}$. Is it true that X is Cesàro summable if and only if it is convergent?

19.J. Give a proof of Theorem 19.5.

19.K. Consider the existence of the double and the iterated limits of the double sequences (x_{mn}), where x_{mn} is given by

(a) $(-1)^{m+n}$,

(b) $\dfrac{1}{m+n}$,

(c) $\dfrac{1}{m} + \dfrac{1}{n}$,

(d) $\dfrac{m}{m+n}$,

(e) $(-1)^m \left(\dfrac{1}{m} + \dfrac{1}{n} \right)$,

(f) $\dfrac{mn}{m^2 + n^2}$.

19.L. Is a convergent double sequence bounded?

19.M. If $X = (x_{mn})$ is a convergent double sequence of real numbers, and if for each $m \in \mathbf{N}$, the limit $y_m = \lim \sup_n (x_{mn})$ exists, then we have $\lim_{mn} (x_{mn}) = \lim_m (y_m)$.

19.N. Which of the double sequences in Exercise 19.K are such that the collection $\{Y_m = \lim (x_{mn}) : m \in \mathbf{N}\}$ is uniformly convergent?

19.O. Let $X = (x_{mn})$ be a bounded double sequence in $\mathbf{R}$ with the property that for each $m \in \mathbf{N}$ the sequence $Y_m = (x_{mn} : n \in \mathbf{N})$ is monotone increasing and for each $n \in \mathbf{N}$ the sequence $Z_n = (x_{mn} : m \in \mathbf{N})$ is monotone increasing. Is it true that the iterated limits exist and are equal? Does the double limit need to exist?

19.P. Discuss the problem posed in the final paragraph of this section.

IV
CONTINUOUS FUNCTIONS

We now begin our study of the most important class of functions in analysis: the continuous functions. In this chapter, we shall blend the results of Chapters II and III and reap a rich harvest of theorems which have considerable depth and utility.

Section 20 introduces and examines the notion of continuity. In Section 21 we introduce the important class of linear functions. The fundamental Section 22 studies the properties of continuous functions on compact and connected sets, and Section 23 discusses the notion of uniform continuity. The results of these four sections will be used repeatedly throughout the remainder of the book. Sequences of continuous functions are studied in Section 24, and upper and lower limits are studied in 25. The final section presents some interesting and important results, but these results will not be applied in later sections.

It is not assumed that the reader has any previous familiarity with a rigorous treatment of continuous functions. However, in some of the examples and exercises, we use the exponential, the logarithm, and the trigonometric functions in order to give some non-trivial examples.

Section 20 Local Properties of Continuous Functions

We shall suppose that f is a function with domain $D(f)$ contained in $\mathbf{R}^p$ and with range $R(f)$ contained in $\mathbf{R}^q$. In general we shall not require that $D(f) = \mathbf{R}^p$ or that $p = q$. We shall first define continuity in terms of neighborhoods and then mention a few equivalent conditions.

20.1 DEFINITION. If $a \in D(f)$, then we say that f is **continuous at** a

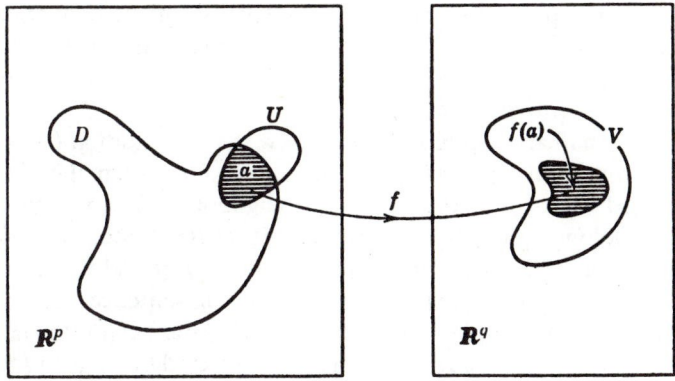

Figure 20.1

if for every neighborhood V of $f(a)$ there exists a neighborhood U (depending on V) of a such that if x is any element of $U \cap D(f)$, then $f(x)$ is an element of V. (See Figure 20.1.) If $A \subseteq D(f)$, then we say that f is **continuous on** A in case it is continuous at every point of A.

Sometimes it is said that a continuous function is one which "sends neighboring points into neighboring points." This intuitive phrase is to be avoided if it leads one to believe that the image of a neighborhood of a need be a neighborhood of $f(a)$. (Consider $x \mapsto |x|$ at $x = 0$.)

We now give two equivalent statements which could have been used as the definition.

20.2 THEOREM. *Let a be a point in the domain $D(f)$ of the function f. The following statements are equivalent:*

(a) *f is continuous at a.*

(b) *If $\epsilon > 0$, there exists a number $\delta(\epsilon) > 0$ such that if $x \in D(f)$ is any element such that $\|x - a\| < \delta(\epsilon)$, then $\|f(x) - f(a)\| < \epsilon$.*

(c) *If (x_n) is any sequence of elements of $D(f)$ which converges to a, then the sequence $(f(x_n))$ converges to $f(a)$.*

PROOF. Suppose that (a) holds and that $\epsilon > 0$, then the ball $V_\epsilon = \{y \in \mathbf{R}^q : \|y - f(a)\| < \epsilon\}$ is a neighborhood of the point $f(a)$. By Definition 20.1 there is a neighborhood U of a, such that if $x \in U \cap D(f)$, then $f(x) \in V_\epsilon$. Since U is a neighborhood of a, there is a positive real number $\delta(\epsilon)$ such that the open ball with radius $\delta(\epsilon)$ and center a is contained in U. Therefore, condition (a) implies (b).

Suppose that (b) holds and let (x_n) be a sequence of elements in $D(f)$ which converges to a. Let $\epsilon > 0$ and invoke condition (b) to obtain a

$\delta(\varepsilon) > 0$ with the property stated in (b). Because of the convergence of (x_n) to a, there exists a natural number $N(\delta(\varepsilon))$ such that if $n \geq N(\delta(\varepsilon))$, then $\|x_n - a\| < \delta(\varepsilon)$. Since each $x_n \in D(f)$ it follows from (b) that $\|f(x_n) - f(a)\| < \varepsilon$, proving that (c) holds.

Finally, we shall argue indirectly and show that if condition (a) does not hold, then condition (c) does not hold. If (a) fails, then there exists a neighborhood V_0 of $f(a)$ such that for any neighborhood U of a, there is an element x_U belonging to $D(f) \cap U$ but such that $f(x_U)$ does not belong to V_0. For each natural number n consider the neighborhood U_n of a defined by $U_n = \{x \in \mathbf{R}^p : \|x - a\| < 1/n\}$; from the preceding sentence, for each n in $\mathbf{N}$ there is an element x_n belonging to $D(f) \cap U_n$ but such that $f(x_n)$ does not belong to V_0. The sequence (x_n) just constructed belongs to $D(f)$ and converges to a, yet none of the elements of the sequence $(f(x_n))$ belongs to the neighborhood V_0 of $f(a)$. Hence we have constructed a sequence for which the condition (c) does not hold. This shows that part (c) implies (a).

Q.E.D.

The following useful discontinuity criterion is a consequence of what we have just done.

20.3 DISCONTINUITY CRITERION. *The function f is not continuous at a point a in $D(f)$ if and only if there is a sequence (x_n) of elements in $D(f)$ which converges to a but such that the sequence $(f(x_n))$ of images does not converge to $f(a)$.*

The next result is a simple reformulation of the definition. We recall from Definition 2.12 that the **inverse image** $f^{-1}(H)$ of a subset H of $\mathbf{R}^q$ under f is defined by

$$f^{-1}(H) = \{x \in D(f) : f(x) \in H\}.$$

20.4 THEOREM. *The function f is continuous at a point a in $D(f)$ if and only if for every neighborhood V of $f(a)$ there is a neighborhood V_1 of a such that*

$$(20.1) \qquad\qquad V_1 \cap D(f) = f^{-1}(V).$$

PROOF If V_1 is a neighborhood of a satisfying this equation, then we can take $U = V_1$. Conversely, if Definition 20.1 is satisfied, then we take $V_1 = U \cup f^{-1}(V)$ to obtain equation (20.1). Q.E.D.

Before we push the theory any further, we shall pause to give some examples. For simplicity, most of the examples are for the case where $\mathbf{R}^p = \mathbf{R}^q = \mathbf{R}$.

20.5 EXAMPLES. (a) Let $D(f) = \mathbf{R}$ and let f be the "constant" function defined to be equal to the real number c for all real numbers x. Then f is continuous at every point of $\mathbf{R}$; in fact, we can take the neighborhood U

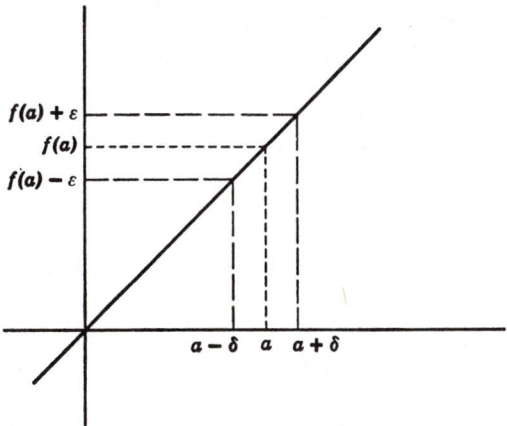

Figure 20.2

of Definition 20.1 to be equal to R for any point a in $D(f)$. Similarly, the function g defined by

$$g(x) = 1, \qquad 0 \le x \le 1,$$
$$= 2, \qquad 2 \le x \le 3,$$

is continuous at each point in its domain.

(b) Let $D(f) = R$ and let f be the "identity" function defined by $f(x) = x$, $x \in R$. (See Figure 20.2.) If a is a given real number, let $\varepsilon > 0$ and let $\delta(\varepsilon) = \varepsilon$. Then, if $|x - a| < \delta(\varepsilon)$, we have $|f(x) - f(a)| = |x - a| < \varepsilon$.

(c) Let $D(f) = R$ and let f be the "squaring" function defined by $f(x) = x^2$, $x \in R$. Let a belong to R and let $\varepsilon > 0$; then $|f(x) - f(a)| = |x^2 - a^2| = |x - a| |x + a|$. We wish to make the above expression less than ε by making $|x - a|$ sufficiently small. If $a = 0$, then we choose $\delta(\varepsilon) = \sqrt{\varepsilon}$. If $a \ne 0$, then we want to obtain a bound for $|x + a|$ on a neighborhood of a. For example, if $|x - a| < |a|$, then $0 < |x| < 2|a|$ and $|x + a| \le |x| + |a| < 3|a|$. Hence

$$(20.2) \qquad |f(x) - f(a)| \le 3|a| |x - a|,$$

provided that $|x - a| < |a|$. Thus if we define $\delta(\varepsilon) = \inf\{|a|, \varepsilon/3|a|\}$, then when $|x - a| < \delta(\varepsilon)$, the inequality (20.2) holds and we have $|f(x) - f(a)| < \varepsilon$.

(d) We consider the same function as in (c) but use a slightly different technique. Instead of factoring $x^2 - a^2$, we write it as a polynomial in $x - a$. Thus

$$x^2 - a^2 = (x^2 - 2ax + a^2) + (2ax - 2a^2) = (x - a)^2 + 2a(x - a).$$

Using the Triangle Inequality, we obtain

$$|f(x)-f(a)| \le |x-a|^2+2\,|a|\,|x-a|.$$

If $\delta \le 1$ and $|x-a|<\delta$, then $|x-a|^2<\delta^2 \le \delta$ and the term on the right side is dominated by $\delta+2\,|a|\,\delta = \delta(1+2\,|a|)$. Hence we are led to choose

$$\delta(\varepsilon) = \inf\left\{1, \frac{\varepsilon}{1+2\,|a|}\right\}.$$

(e) Consider $D(f)=\{x \in \mathbf{R} : x \ne 0\}$ and let f be defined by $f(x)=1/x$, $x \in D(f)$. If $a \in D(f)$, then

$$|f(x)-f(a)| = |1/x-1/a| = \frac{|x-a|}{|ax|}.$$

Again we wish to find a bound for the coefficient of $|x-a|$ which is valid in a neighborhood of $a \ne 0$. We note that if $|x-a|<\frac{1}{2}|a|$, then $\frac{1}{2}|a|<|x|$, and we have

$$|f(x)-f(a)| \le \frac{2}{|a|^2}|x-a|.$$

Thus we are led to take $\delta(\varepsilon) = \inf\{\frac{1}{2}|a|, \frac{1}{2}\varepsilon\,|a|^2\}$.

(f) Let f be defined for $D(f) = \mathbf{R}$ by

$$f(x) = 0, \qquad x \le 0,$$
$$= 1, \qquad x > 0.$$

It may be seen that f is continuous at all points $a \ne 0$. We shall show that f is not continuous at 0 by using the Discontinuity Criterion 20.3. In fact, if $x_n = 1/n$, then the sequence $(f(1/n))=(1)$ does not converge to $f(0)$. (See Figure 20.3.)

(g) Let $D(f) = \mathbf{R}$ and let f be Dirichlet's† discontinuous function defined by

$$f(x) = 1, \qquad \text{if } x \text{ is rational,}$$
$$= 0, \qquad \text{if } x \text{ is irrational.}$$

If a is a rational number, let $X = (x_n)$ be a sequence of irrational numbers converging to a. (Theorem 6.10 assures us of the existence of such a sequence.) Since $f(x_n) = 0$ for all $n \in \mathbf{N}$, the sequence $(f(x_n))$ does not converge to $f(a) = 1$ and f is not continuous at the rational number a. On the other hand, if b is an irrational number, then there exists a sequence $Y = (y_n)$ of rational numbers converging to b. The sequence $(f(y_n))$ does not converge to $f(b)$, so f is not continuous at b. Therefore, Dirichlet's function is *not continuous at any point*.

† PETER GUSTAV LEJEUNE DIRICHLET (1805–1859) was born in the Rhineland and taught at Berlin for almost thirty years before going to Göttingen as Gauss' successor. He made fundamental contributions to number theory and analysis.

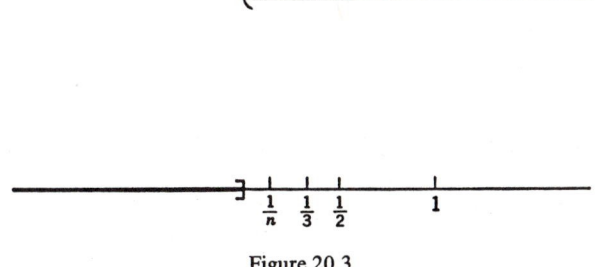

Figure 20.3

(h) Let $D(f) = \{x \in \mathbf{R} : x > 0\}$. For any irrational number $x > 0$, we define $f(x) = 0$; for a rational number of the form m/n, with the natural numbers m, n having no common factor except 1, we define $f(m/n) = 1/n$. We shall show that f is continuous at every irrational number in $D(f)$ and discontinuous at every rational number in $D(f)$. The latter statement follows by taking a sequence of irrational numbers converging to the given rational number and using the Discontinuity Criterion. Let a be an irrational number and $\varepsilon > 0$; then there is a natural number n such that $1/n < \varepsilon$. If δ is chosen so small that the interval $(a - \delta, a + \delta)$ contains no rational number with denominator less than n, then it follows that for x in this interval we have $|f(x) - f(a)| = |f(x)| \le 1/n < \varepsilon$. Thus f is continuous at the irrational number a. Therefore, this function is *continuous precisely at the irrational points in its domain*.

(i) This time, let $D(f) = \mathbf{R}^2$ and let f be the function on $\mathbf{R}^2$ with values in $\mathbf{R}^2$ defined by

$$f(x, y) = (2x + y, x - 3y).$$

Let (a, b) be a fixed point in $\mathbf{R}^2$; we shall show that f is continuous at this point. To do this, we need to show that we can make the expression

$$\|f(x, y) - f(a, b)\| = \{(2x + y - 2a - b)^2 + (x - 3y - a + 3b)^2\}^{1/2}$$

arbitrarily small by choosing (x, y) sufficiently close to (a, b). Since $\{p^2 + q^2\}^{1/2} \le \sqrt{2} \sup \{|p|, |q|\}$, it is evidently enough to show that we can make the terms

$$|2x + y - 2a - b|, \qquad |x - 3y - a + 3b|,$$

arbitrarily small by choosing (x, y) sufficiently close to (a, b) in $\mathbf{R}^2$. In fact, by the Triangle Inequality,

$$|2x + y - 2a - b| = |2(x - a) + (y - b)| \le 2|x - a| + |y - b|.$$

Now $|x - a| \le \{(x - a)^2 + (y - b)^2\}^{1/2} = \|(x, y) - (a, b)\|$, and similarly for $|y - b|$; hence we have

$$|2x + y - 2a - b| \le 3\|(x, y) - (a, b)\|.$$

Similarly,

$$|x - 3y - a + 3b| \le |x - a| + 3|y - b| \le 4\|(x, y) - (a, b)\|.$$

Therefore, if $\varepsilon > 0$, we can take $\delta(\varepsilon) = \varepsilon/(4\sqrt{2})$ and be certain that if $\|(x, y) - (a, b)\| < \delta(\varepsilon)$, then $\|f(x, y) - f(a, b)\| < \varepsilon$, although a larger value of δ can be attained by a more refined analysis (for example, by using the Schwarz Inequality 8.7).

(j) Again let $D(f) = \mathbf{R}^2$ and let f be defined by

$$f(x, y) = (x^2 + y^2, 2xy)$$

If (a, b) is a fixed point in $\mathbf{R}^2$, then

$$\|f(x, y) - f(a, b)\| = \{(x^2 + y^2 - a^2 - b^2)^2 + (2xy - 2ab)^2\}^{1/2}.$$

As in (i), we examine the two terms on the right side separately. It will be seen that we need to obtain elementary estimates of magnitude. From the Triangle Inequality, we have

$$|x^2 + y^2 - a^2 - b^2| \le |x^2 - a^2| + |y^2 - b^2|.$$

If the point (x, y) is within a distance of 1 of (a, b), then $|x| \le |a| + 1$ whence $|x + a| \le 2|a| + 1$ and $|y| \le |b| + 1$ so that $|y + b| \le 2|b| + 1$. Thus we have

$$|x^2 + y^2 - a^2 - b^2| \le |x - a|(2|a| + 1) + |y - b|(2|b| + 1)$$
$$\le 2(|a| + |b| + 1)\|(x, y) - (a, b)\|.$$

In a similar fashion, we have

$$|2xy - 2ab| = 2|xy - xb + xb - ab| \le 2|x||y - b| + 2|b||x - a|$$
$$\le 2(|a| + |b| + 1)\|(x, y) - (a, b)\|.$$

Therefore, we set

$$\delta(\varepsilon) = \inf\left\{1, \frac{\varepsilon}{2\sqrt{2}(|a| + |b| + 1)}\right\};$$

if $\|(x, y) - (a, b)\| < \delta(\varepsilon)$, then we have $\|f(x, y) - f(a, b)\| < \varepsilon$, proving that f is continuous at the point (a, b).

Combinations of Functions

The next result is a direct consequence of Theorems 15.6 and 20.2(c), so we shall not write out the details. Alternatively, it could be proved directly by using arguments quite parallel to those employed in the proof of Theorem 15.6. We recall that if f and g are functions with domains $D(f)$ and $D(g)$ in $\mathbf{R}^p$ and ranges in $\mathbf{R}^q$, then we define their **sum** $f + g$, their **difference** $f - g$ and their **inner product** $f \cdot g$ for each x in $D(f) \cap D(g)$ by the formulas

$$f(x) + g(x), \qquad f(x) - g(x), \qquad f(x) \cdot g(x).$$

Similarly, if c is a real number and if φ is a function with domain $D(\varphi)$ in $\mathbf{R}^p$ and range in $\mathbf{R}$, we define the products cf for x in $D(f)$ and φf for x in

$D(\varphi) \cap D(f)$ by the formulas

$$cf(x), \qquad \varphi(x)f(x).$$

In particular, if $\varphi(x) \neq 0$ for $x \in D_0$, then we can define the **quotient** f/φ for x in $D(f) \cap D_0$ by

$$f(x)/\varphi(x).$$

With these definitions, we now state the result.

20.6 THEOREM. *If the functions f, g, φ are continuous at a point, then the algebraic combinations*

$$f+g, \qquad f-g, \qquad f \cdot g, \qquad cf, \qquad \varphi f \qquad and \qquad f/\varphi$$

are also continuous at this point.

There is another algebraic combination that is often useful. If f is defined on $D(f)$ in $\mathbf{R}^p$ to $\mathbf{R}$, we define the **absolute value** $|f|$ of f to be the function with range in the real numbers $\mathbf{R}$ whose value at x in $D(f)$ is given by $|f(x)|$.

20.7 THEOREM. *If f is continuous at a point, then $|f|$ is also continuous there.*

PROOF. From the Triangle Inequality, we have,

$$\left| |f(x)| - |f(a)| \right| \leq |f(x) - f(a)|,$$

from which the result is immediate. Q.E.D.

We recall the notion of the composition of two functions. Let f have domain $D(f)$ in $\mathbf{R}^p$ and range in $\mathbf{R}^q$ and let g have domain $D(g)$ in $\mathbf{R}^q$ and range in $\mathbf{R}^r$. In Definition 2.2, we defined the composition $h = g \circ f$ to have domain $D(h) = \{x \in D(f) : f(x) \in D(g)\}$ and for x in $D(h)$ we set $h(x) = g[f(x)]$. Thus $h = g \circ f$ is a function mapping $D(h)$, which is a subset of $D(f) \subseteq \mathbf{R}^p$, into $\mathbf{R}^r$. We now establish the continuity of this function.

20.8 THEOREM. *If f is continuous at a and g is continuous at $b = f(a)$, then the composition $g \circ f$ is continuous at a.*

PROOF. Let W be a neighborhood of the point $c = g(b)$. Since g is continuous at b, there is a neighborhood V of b such that if y belongs to $V \cap D(g)$, then $g(y) \in W$. Since f is continuous at a, there exists a neighborhood U of a such that if x belongs to $U \cap D(f)$, then $f(x)$ is in V. Therefore, if x belongs to $U \cap D(g \circ f)$, then $f(x)$ is in $V \cap D(g)$ and $g[f(x)]$ belongs to W. (See Figure 20.4.) This shows that $h = g \circ f$ is continuous at a. .Q.E.D.

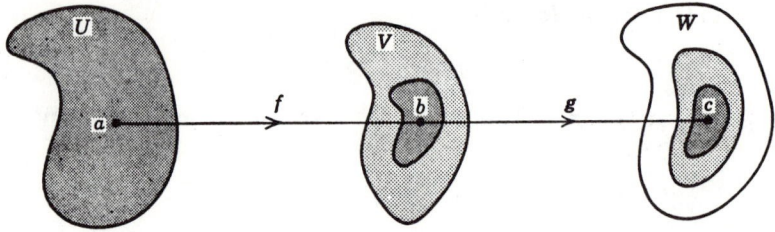

Figure 20.4

Exercises

20.A. Prove that if f is defined for $x \geq 0$ by $f(x) = \sqrt{x}$, then f is continuous at every point of its domain.

20.B. Show that a "polynomial function"; that is, a function f with the form

$$f(x) = a_n x^n + a_{n-1} x^{n-1} + \cdots + a_1 x + a_0 \qquad \text{for } x \in R,$$

is continuous at every point of R.

20.C. Show that a "rational function" (that is, the quotient of two polynomial functions) is continuous at every point where it is defined.

20.D. Use the Schwarz Inequality to show that one can take $\delta(\varepsilon) = \varepsilon/\sqrt{15}$ in Example 20.5(i).

20.E. Let f be the function on R to R defined by

$$f(x) = x, \qquad x \text{ irrational,}$$
$$= 1 - x, \qquad x \text{ rational.}$$

Show that f is continuous at $x = \frac{1}{2}$ and discontinuous elsewhere.

20.F. Let f be continuous on R to R. Show that if $f(x) = 0$ for rational x then $f(x) = 0$ for all x in R.

20.G. Let f and g be continuous on R to R. Is it true that $f(x) = g(x)$ for $x \in R$ if and only if $f(y) = g(y)$ for all rational numbers y in R?

20.H. Use the inequality $|\sin x| \leq |x|$ for $x \in R$ to show that the sine function is continuous at $x = 0$. Use this fact, together with the identity

$$\sin x - \sin u = 2 \sin \tfrac{1}{2}(x - u) \cos \tfrac{1}{2}(x + u),$$

to prove that the sine function is continuous at any point of R.

20.I. Using the results of the preceding exercise, show that the function g, defined on R to R by

$$g(x) = x \sin (1/x), \qquad x \neq 0,$$
$$= 0, \qquad x = 0,$$

is continuous at every point. Sketch a graph of this function.

20.J. Let h be defined for $x \neq 0$, $x \in R$, by

$$h(x) = \sin (1/x), \qquad x \neq 0.$$

Show that no matter how h is defined at $x = 0$, it will be discontinuous at $x = 0$.

20.K. Let $F: \mathbf{R}^2 \to \mathbf{R}$ be defined by

$$F(x, y) = x^2 + y^2 \quad \text{if both } x, y \in \mathbf{Q},$$
$$= 0 \qquad \text{otherwise.}$$

Determine the points where F is continuous.

20.L. We say that a function f on $\mathbf{R}$ to $\mathbf{R}$ is **additive** if it satisfies

$$f(x + y) = f(x) + f(y)$$

for all $x, y \in \mathbf{R}$. Show that an additive function which is continuous at $x = 0$ is continuous at any point of $\mathbf{R}$. Show that a monotone additive function is continuous at every point.

20.M. Suppose that f is a continuous additive function on $\mathbf{R}$. If $c = f(1)$, show that $f(x) = cx$ for all x in $\mathbf{R}$. (Hint: first show that if r is a rational number, then $f(r) = cr$.)

20.N. Let $g: \mathbf{R} \to \mathbf{R}$ satisfy the relation

$$g(x + y) = g(x)g(y) \qquad \text{for } x, y \in \mathbf{R}.$$

Show that, if g is continuous at $x = 0$, then g is continuous at every point. Also, if $g(a) = 0$ for some $a \in \mathbf{R}$, then $g(x) = 0$ for all $x \in \mathbf{R}$.

20.O. If $|f|$ is continuous at a point, then is it true that f is also continuous at this point?

20.P. Let $f, g: \mathbf{R}^p \to \mathbf{R}$ be continuous at a point $a \in \mathbf{R}^p$ and let h, k be defined on $\mathbf{R}^p$ to $\mathbf{R}$ by

$$h(x) = \sup \{f(x), g(x)\}, \qquad k(x) = \inf \{f(x), g(x)\}.$$

Show that h and k are continuous at a. (Hint: note that $\sup \{b, c\} = \frac{1}{2}(b + c + |b - c|)$ and $\inf \{b, c\} = \frac{1}{2}(b + c - |b - c|)$.)

20.Q. If $x \in \mathbf{R}$, we often define $[x]$ to be the greatest integer $n \in \mathbf{Z}$ such that $n \le x$. The map $x \mapsto [x]$ is called the **greatest integer** function. Sketch the graphs and determine the points of continuity of the functions defined for $x \in \mathbf{R}$ by

(a) $f(x) = [x]$,
(b) $g(x) = x - [x]$,
(c) $h(x) = [2 \sin x]$,
(d) $k(x) = \sin \frac{1}{2} \pi [x]$.

20.R. A function f defined on an interval $I \subseteq \mathbf{R}$ to $\mathbf{R}$ is said to be **increasing** on I if $x \le x'$, $x, x' \in I$ imply that $f(x) \le f(x')$. It is said to be **strictly increasing** on I if $x < x'$, $x, x' \in I$ imply that $f(x) < f(x')$. Similar definitions can be given for **decreasing** and **strictly decreasing** functions. A function which is either increasing or decreasing on an interval is said to be **monotone** on this interval.

(a) If f is increasing on I, then f is continuous at an interior point $c \in I$ if and only if for every $\varepsilon > 0$ there are points $x_1, x_2 \in I$, $x_1 < c < x_2$, such that $f(x_2) - f(x_1) < \varepsilon$.

(b) If f is increasing on I, then f is continuous at an interior point $c \in I$ if and only if

$$\sup \{f(x): x < c\} = f(c) = \inf \{f(x): x > c\}.$$

20.S. Suppose that f is increasing on $I = [a, b]$ in the sense of the preceding exercise. Let

$$j_c = \inf\{f(x): x > c\} - \sup\{f(x): x < c\}.$$

If $j_c > 0$, we say that f has a **jump of** j_c at the point c.

(a) If $n \in \mathbf{N}$, show that there can only be a finite set of points in I at which f has a jump exceeding $1/n$.

(b) Show that an increasing function can have at most a countable set of points of discontinuity.

Projects

20.α. Let g be a function on $\mathbf{R}$ to $\mathbf{R}$ which is not identically zero and which satisfies the functional equation

(*) $$g(x+y) = g(x)g(y) \qquad \text{for } x, y \in \mathbf{R}.$$

The purpose of this project is to show that g must be an "exponential function."

(a) Show that g is continuous at every point of $\mathbf{R}$ if and only if it is continuous at the point $x = 0$.

(b) Show that $g(x) > 0$ for all $x \in \mathbf{R}$.

(c) Prove that $g(0) = 1$. If $a = g(1)$, then $a > 0$ and $g(r) = a^r$ for all $r \in \mathbf{Q}$.

(d) The function g is strictly increasing, is constant, or is strictly decreasing according as $g(1) > 1$, $g(1) = 1$, or $0 < g(1) < 1$, when g is continuous.

(e) If $g(x) > 1$ for x in some interval $(0, \delta)$, $\delta > 0$, then g is strictly increasing and continuous on $\mathbf{R}$.

(f) If $a > 0$, then there exists at most one continuous function g satisfying (*) such that $g(1) = a$.

(g) Suppose that $a > 1$. Referring to Project 6.β, show that there exists a unique continuous function satisfying (*) such that $g(1) = a$.

20.β. Let $P = \{x \in \mathbf{R}: x > 0\}$ and let $h: P \to \mathbf{R}$ be a function not identically zero which satisfies the functional equation

(†) $$h(xy) = h(x) + h(y) \qquad \text{for } x, y \in P.$$

The purpose of this project is to show that h must be a "logarithmic function."

(a) Show that h is continuous at every point of P if and only if it is continuous at the point $x = 1$.

(b) Show that h cannot be defined at $x = 0$ to satisfy (†) for $\{x \in \mathbf{R}: x \geq 0\}$.

(c) Prove that $h(1) = 0$. If $x > 0$ and $r \in \mathbf{Q}$, then $h(x^r) = rh(x)$.

(d) Show that if $h(x) > 0$ on some interval $(1, \delta)$ with $\delta > 1$, then h is strictly increasing and continuous on P.

(e) If h is continuous, show that $h(x) \neq 0$ for $x \neq 1$. Also, either $h(x) > 0$ for $x > 1$, or $h(x) < 0$ for $x > 1$.

(f) If $b > 1$, show that there exists at most one continuous function on P which satisfies (†) and is such that $h(b) = 1$.

(g) Suppose that $b > 1$. Referring to Project 6.γ, show that there exists a unique continuous function satisfying (†) such that $h(b) = 1$.

Section 21 Linear Functions

The preceding discussion pertained to arbitrary functions defined on a part of R^p to R^q. Before we continue that discussion we want to introduce a relatively simple but extremely important class of functions, namely the "linear functions," which arise in very many applications.

21.1 DEFINITION. A function f with domain R^p and range in R^q is said to be **linear** in case

$$(21.1) \qquad f(ax + by) = af(x) + bf(y)$$

for all a, b in R and x, y in R^p.

It follows from (21.1) by induction that if $a, b \ldots, c$ are $n \in N$ real numbers and $x, y, \ldots, z$ are n elements of R^p, then

$$f(ax + by + \cdots + cz) = af(x) + bf(y) + \cdots + cf(z).$$

It is readily seen that the functions in Examples 20.6(b) and 20.6(i) are linear functions for the case $p = q = 1$ and $p = q = 2$, respectively. In fact it is not difficult to characterize the most general linear function from R^p to R^q.

21.2 THEOREM. *If f is a linear function with domain R^p and range in R^q, then there are pq real numbers (c_{ij}), $1 \le i \le q$, $1 \le j \le p$, such that if $x = (x_1, x_2, \ldots, x_p)$ is any point in R^p, and if $y = (y_1, y_2, \ldots, y_q) = f(x)$ is its image under f, then*

$$
\begin{aligned}
y_1 &= c_{11}x_1 + c_{12}x_2 + \cdots + c_{1p}x_p, \\
y_2 &= c_{21}x_1 + c_{22}x_2 + \cdots + c_{2p}x_p, \\
& \cdots \cdots \cdots \cdots \cdots \\
y_q &= c_{q1}x_1 + c_{q2}x_2 + \cdots + c_{qp}x_p.
\end{aligned}
\tag{21.2}
$$

Conversely, if (c_{ij}) is a collection of pq real numbers, then the function which assigns to x in R^p the element y in R^q according to the equations (21.2) is a linear function with domain R^p and range in R^q.

PROOF. Let $e_1, e_2, \ldots, e_p$ be the elements of R^p given by $e_1 = (1, 0, \ldots, 0)$, $e_2 = (0, 1, \ldots, 0), \ldots,$ $e_p = (0, 0, \ldots, 1)$. We examine the images of these vectors under the linear function f. Suppose that

$$
\begin{aligned}
f(e_1) &= (c_{11}, c_{21}, \ldots, c_{q1}), \\
f(e_2) &= (c_{12}, c_{22}, \ldots, c_{q2}), \\
& \cdots \cdots \cdots \cdots \cdots \\
f(e_p) &= (c_{1p}, c_{2p}, \ldots, c_{qp}).
\end{aligned}
\tag{21.3}
$$

Thus the real number c_{ij} is the ith coordinate of the point $f(e_j)$.

An arbitrary element $x = (x_1, x_2, \ldots, x_p)$ of $\mathbf{R}^p$ can be expressed simply in terms of the vectors $e_1, e_2, \ldots, e_p$; in fact,

$$x = x_1 e_1 + x_2 e_2 + \cdots + x_p e_p.$$

Since f is linear, it follows that

$$f(x) = x_1 f(e_1) + x_2 f(e_2) + \cdots + x_p f(e_p).$$

If we use the equations (21.3), we have

$$
\begin{aligned}
f(x) &= x_1(c_{11}, c_{21}, \ldots, c_{q1}) + x_2(c_{12}, c_{22}, \ldots, c_{q2}) \\
&\quad + \cdots + x_p(c_{1p}, c_{2p}, \ldots, c_{qp}) \\
&= (c_{11}x_1, c_{21}x_1, \ldots, c_{q1}x_1) + (c_{12}x_2, c_{22}x_2, \ldots, c_{q2}x_2) \\
&\quad + \cdots + (c_{1p}x_p, c_{2p}x_p, \ldots, c_{qp}x_p) \\
&= (c_{11}x_1 + c_{12}x_2 + \cdots + c_{1p}x_p, c_{21}x_1 + c_{22}x_2 + \cdots + c_{2p}x_p, \\
&\quad \ldots, c_{q1}x_1 + c_{q2}x_2 + \cdots + c_{qp}x_p).
\end{aligned}
$$

This shows that the coordinates of $f(x)$ are given by the relations (21.2), as asserted.

Conversely, it is easily verified by direct calculation that if the relations (21.2) are used to obtain the coordinates y_i of y from the coordinates x_i of x, then the resulting function satisfies the relation (21.1) and so is linear. We shall omit this calculation, since it is straightforward.   Q.E.D.

It should be mentioned that the rectangular array of numbers

$$
(21.4) \qquad
\begin{bmatrix}
c_{11} & c_{12} & \cdots & c_{1p} \\
c_{21} & c_{22} & \cdots & c_{2p} \\
\cdot & \cdot & \cdots & \cdot \\
c_{q1} & c_{q2} & \cdots & c_{qp}
\end{bmatrix},
$$

consisting of q rows and p columns, is often called the **matrix** corresponding to the linear function f. There is a one-one correspondence between linear functions of $\mathbf{R}^p$ into $\mathbf{R}^q$ and $q \times p$ matrices of real numbers. As we have seen, the action of f is completely described in terms of its matrix. We shall not find it necessary to develop any of the extensive theory of matrices, however, but will regard the matrix (21.4) as being shorthand for a more elaborate description of the linear function f.

We shall now prove that a linear function from $\mathbf{R}^p$ to $\mathbf{R}^q$ is automatically continuous. To do this, we first restate the Schwarz Inequality in the form

$$|a_1 b_1 + a_2 b_2 + \cdots + a_p b_p|^2 \le \{a_1^2 + a_2^2 + \cdots + a_p^2\}\{b_1^2 + b_2^2 + \cdots + b_p^2\}.$$

We apply this inequality to each expression in equation (21.2) to obtain, for $1 \le i \le q$, the estimate

$$|y_i|^2 \le (|c_{i1}|^2 + |c_{i2}|^2 + \cdots + |c_{ip}|^2) \, \|x\|^2 = \sum_{j=1}^{p} |c_{ij}|^2 \, \|x\|^2.$$

Adding these inequalities, we have

$$\|y\|^2 \le \left\{ \sum_{i=1}^{q} \sum_{j=1}^{p} |c_{ij}|^2 \right\} \|x\|^2,$$

from which we conclude that

(21.5) $$\|y\| = \|f(x)\| \le \left\{ \sum_{i=1}^{q} \sum_{j=1}^{p} |c_{ij}|^2 \right\}^{1/2} \|x\|.$$

21.3 **THEOREM.** *If f is a linear function with domain $\mathbf{R}^p$ and range in $\mathbf{R}^q$, then there exists a positive constant A such that if u, v are any two vectors in $\mathbf{R}^p$, then*

(21.6) $$\|f(u) - f(v)\| \le A \, \|u - v\|.$$

Therefore, a linear function on $\mathbf{R}^p$ to $\mathbf{R}^q$ is continuous at every point.

PROOF. We have seen, in deriving formula (21.5) that there exists a constant A such that if x is any element of $\mathbf{R}^p$ then $\|f(x)\| \le A \, \|x\|$. Now let $x = u - v$ and use the linearity of f to obtain $f(x) = f(u - v) = f(u) - f(v)$. Therefore, the formula (21.6) results. It is clear that this relation implies the continuity of f, for we can make $\|f(u) - f(v)\| < \varepsilon$ by taking $\|u - v\| < \varepsilon/A$ if $A > 0$. Q.E.D.

It is an exercise to show that if f and g are linear functions on $\mathbf{R}^p$ to $\mathbf{R}^q$, then $f + g$ is a linear function on $\mathbf{R}^p$ to $\mathbf{R}^q$. Similarly, if $c \in \mathbf{R}$, then cf is a linear function. We leave it to the reader to show that the collection $\mathscr{L}(\mathbf{R}^p, \mathbf{R}^q)$ of all linear functions on $\mathbf{R}^p$ to $\mathbf{R}^q$ is a vector space under these vector operations. In the exercises we will show how to define a norm on this vector space.

Exercises

21.A. Show that $f: \mathbf{R}^p \to \mathbf{R}^q$ is a linear function if and only if $f(ax) = af(x)$ and $f(x + y) = f(x) + f(y)$ for all $a \in \mathbf{R}$ and all $x, y \in \mathbf{R}^p$.

21.B. If f is a linear function of $\mathbf{R}^p$ into $\mathbf{R}^q$, show that the columns of the matrix representation (21.4) of f indicate the elements in $\mathbf{R}^q$ into which the elements $e_1 = (1, 0, \ldots, 0)$, $e_2 = (0, 1, \ldots, 0)$, $\ldots$, $e_p = (0, 0; \ldots, 1)$ of $\mathbf{R}^p$ are mapped by f.

21.C. Let f be a linear function of $\mathbf{R}^2$ into $\mathbf{R}^3$ which sends the elements $e_1 = (1, 0)$, $e_2 = (0, 1)$ of $\mathbf{R}^2$ into the vectors $f(e_1) = (2, 1, 0)$, $f(e_2) = (1, 0, -1)$ of $\mathbf{R}^3$. Give the matrix representation of f. What vectors in $\mathbf{R}^3$ are the images under f of the elements $(2, 0)$, $(1, 1)$, and $(1, 3)$?

21.D. If f denotes the linear function of Exercise 21.C, show that not every vector in $\mathbf{R}^3$ is the image under f of a vector in $\mathbf{R}^2$.

21.E. Let g be any linear function on $\boldsymbol{R}^2$ to $\boldsymbol{R}^3$. Show that not every element of $\boldsymbol{R}^3$ is the image under g of a vector in $\boldsymbol{R}^2$.

21.F. Let h be any linear function on $\boldsymbol{R}^3$ to $\boldsymbol{R}^2$. Show that there exist non-zero vectors in $\boldsymbol{R}^3$ which are mapped into the zero vector of $\boldsymbol{R}^2$ by h.

21.G. Let f be a linear function on $\boldsymbol{R}^2$ to $\boldsymbol{R}^2$ and let the matrix representation of f be given by

$$\begin{bmatrix} a & b \\ c & d \end{bmatrix}$$

Show that $f(x) \neq 0$ when $x \neq 0$ if and only if $\Delta = ad - bc \neq 0$.

21.H. Let f be as in Exercise 21.G. Show that f maps $\boldsymbol{R}^2$ onto $\boldsymbol{R}^2$ if and only if $\Delta = ad - bc \neq 0$. Show that if $\Delta \neq 0$, then the inverse function f^{-1} is linear and has the matrix representation

$$\begin{bmatrix} d/\Delta & -b/\Delta \\ -c/\Delta & a/\Delta \end{bmatrix}$$

21.I. Let g be a linear function from $\boldsymbol{R}^p$ to $\boldsymbol{R}^q$. Show that g is one-one if and only if $g(x) = 0$ implies that $x = 0$.

21.J. If h is a one-one linear function from $\boldsymbol{R}^p$ onto $\boldsymbol{R}^p$, show that the inverse h^{-1} is a linear function from $\boldsymbol{R}^p$ onto $\boldsymbol{R}^p$.

21.K. Show that the sum and the composition of two linear functions are linear functions.

21.L. If f is a linear map on $\boldsymbol{R}^p$ to $\boldsymbol{R}^q$, define

$$\|f\|_{pq} = \sup \{\|f(x)\| : x \in \boldsymbol{R}^p, \|x\| \leq 1\}.$$

Show that the mapping $f \mapsto \|f\|_{pq}$ defines a norm on the vector space $\mathscr{L}(\boldsymbol{R}^p, \boldsymbol{R}^q)$ of all linear functions on $\boldsymbol{R}^p$ to $\boldsymbol{R}^q$. Show that $\|f(x)\| \leq \|f\|_{pq} \|x\|$ for all $x \in \boldsymbol{R}^p$.

21.M. If f is a linear map on $\boldsymbol{R}^p$ to $\boldsymbol{R}^q$, define

$$M(f) = \inf \{M > 0 : \|f(x)\| \leq M \|x\|, x \in \boldsymbol{R}^p\}.$$

Show that $M(f) = \|f\|_{pq}$.

21.N. If f and g are in $\mathscr{L}(\boldsymbol{R}^p, \boldsymbol{R}^p)$ show that $f \circ g$ is also in $\mathscr{L}(\boldsymbol{R}^p, \boldsymbol{R}^p)$ and that $\|f \circ g\|_{pp} \leq \|f\|_{pp} \|g\|_{pp}$. Show that the inequality can be strict for certain f and g.

21.O. Give an example of a linear map f in $\mathscr{L}(\boldsymbol{R}^p, \boldsymbol{R}^q)$ with matrix representation $[c_{ij}]$ where we have

$$\|f\|_{pq} < \left\{ \sum_{i=1}^{q} \sum_{j=1}^{p} c_{ij}^2 \right\}^{1/2}.$$

21.P. If (21.4) gives the matrix for f, show that $|c_{ij}| \leq \|f\|_{pq}$ for all i, j.

Section 22 Global Properties of Continuous Functions

In Section 20 we considered "local" continuity; that is, we were concerned with continuity at a point. In this section we shall be concerned with establishing some deeper properties of continuous functions. Here we shall be concerned with "global" continuity in the sense that we will assume that the functions are continuous at every point of their domain.

Unless there is a special mention to the contrary, f will denote a function with domain $D(f)$ contained in $\mathbf{R}^p$ and with range in $\mathbf{R}^q$. We recall that if B is a subset of the range space $\mathbf{R}^q$, the *inverse image* of B under f is the set

$$f^{-1}(B) = \{x \in D(f) : f(x) \in B\}.$$

Observe that $f^{-1}(B)$ is automatically a subset of $D(f)$ even though B is not necessarily a subset of the range of f.

In topology courses, where one is more concerned with global than local continuity, the next result is often taken as the definition of (global) continuity. Its importance will soon be evident.

22.1 GLOBAL CONTINUITY THEOREM. *The following statements are equivalent:*

(a) *f is continuous on its domain $D(f)$.*

(b) *If G is any open set in $\mathbf{R}^q$, then there exists an open set G_1 in $\mathbf{R}^p$ such that $G_1 \cap D(f) = f^{-1}(G)$.*

(c) *If H is any closed set in $\mathbf{R}^q$, then there exists a closed set H_1 in $\mathbf{R}^p$ such that $H_1 \cap D(f) = f^{-1}(H)$.*

PROOF. First, we shall suppose that (a) holds and let G be an open subset of $\mathbf{R}^q$. If a belongs to $f^{-1}(G)$, then since G is a neighborhood of $f(a)$, it follows from the continuity of f at a that there is an open set $U(a)$ such that if $x \in D(f) \cap U(a)$, then $f(x) \in G$. Select $U(a)$ for each a in $f^{-1}(G)$ and let G_1 be the union of the sets $U(a)$. By Theorem 9.3(c), the set G_1 is open and it is plain that $G_1 \cap D(f) = f^{-1}(G)$. Hence (a) implies (b).

We shall now show that (b) implies (a). If a is an arbitrary point of $D(f)$ and G is an open neighborhood of $f(a)$, then condition (b) implies that there exists an open set G_1 in $\mathbf{R}^p$ such that $G_1 \cap D(f) = f^{-1}(G)$. Since $f(a) \in G$, it follows that $a \in G_1$, so G_1 is a neighborhood of a. If $x \in G_1 \cap D(f)$, then $f(x) \in G$ whence f is continuous at a. This proves that condition (b) implies (a).

We now prove the equivalence of conditions (b) and (c). First we observe that if B is any subset of $\mathbf{R}^q$ and if $C = \mathbf{R}^q \setminus B$, then we have $f^{-1}(B) \cap f^{-1}(C) = \emptyset$ and

(22.1) $$D(f) = f^{-1}(B) \cup f^{-1}(C).$$

If B_1 is a subset of $\mathbf{R}^p$ such that $B_1 \cap D(f) = f^{-1}(B)$ and $C_1 = \mathbf{R}^p \setminus B_1$, then $C_1 \cap f^{-1}(B) = \emptyset$ and

(22.2) $$D(f) = (B_1 \cap D(f)) \cup (C_1 \cap D(f)) = f^{-1}(B) \cup (C_1 \cap D(f)).$$

The formulas (22.1) and (22.2) are two representations of $D(f)$ as the union of $f^{-1}(B)$ with another set with which it has no common points. Therefore, we have $C_1 \cap D(f) = f^{-1}(C)$.

Suppose that (b) holds and that H is closed in $\mathbf{R}^q$. Apply the argument just completed in the case where $B = \mathbf{R}^q \setminus H$ and $C = H$. Then B and B_1 are open sets in $\mathbf{R}^q$ and $\mathbf{R}^p$, respectively, so $C_1 = \mathbf{R}^p \setminus B_1$ is closed in $\mathbf{R}^p$. This shows that (b) implies (c).

To see that (c) implies (b), use the above argument with $B = \mathbf{R}^q \setminus G$, where G is an open set in $\mathbf{R}^q$.   Q.E.D.

In the case where $D(f) = \mathbf{R}^p$, the preceding result simplifies to some extent.

22.2 COROLLARY. *Let f be defined on all of $\mathbf{R}^p$ and with range in $\mathbf{R}^q$. Then the following statements are equivalent:*
  (a) *f is continuous on $\mathbf{R}^p$;*
  (b) *if G is open in $\mathbf{R}^q$, then $f^{-1}(G)$ is open in $\mathbf{R}^p$;*
  (c) *if H is closed in $\mathbf{R}^q$, then $f^{-1}(H)$ is closed in $\mathbf{R}^p$.*

It should be emphasized that the Global Continuity Theorem 22.1 does *not* say that if f is continuous and if G is an open set in $\mathbf{R}^p$, then the direct image $f(G) = \{f(x) : x \in G\}$ is open in $\mathbf{R}^q$. In general, a continuous function need not send open sets to open sets or closed sets to closed sets. For example, the function f on $\mathbf{R}$ to $\mathbf{R}$, defined by

$$f(x) = \frac{1}{1+x^2}$$

is continuous on $\mathbf{R}$. [In fact, it was seen in Examples 20.5(a) and (c) that the functions $f_1(x) = 1$, and $f_2(x) = x^2$, for $x \in \mathbf{R}$, are continuous at every point. From Theorem 15.6, it follows that

$$f_3(x) = 1 + x^2, \qquad x \in \mathbf{R},$$

is continuous at every point and, since f_3 never vanishes, this same theorem implies that the function f given above is continuous on $\mathbf{R}$.] If G is the open set $G = (-1, 1)$, then $f(G) = (\frac{1}{2}, 1]$, which is not open in $\mathbf{R}$. Similarly, if H is the closed set $H = \{x \in \mathbf{R} : x \geq 1\}$, then $f(H) = (0, \frac{1}{2}]$, which is not closed in $\mathbf{R}$. Similarly, the function f maps the set $\mathbf{R}$, which is both open and closed in $\mathbf{R}$, into the set $f(\mathbf{R}) = (0, 1]$, which is neither open nor closed in $\mathbf{R}$.

The moral of the preceding remarks is that the property of a set being open or closed is not necessarily preserved under mapping by a continuous function. However, there are important properties of a set which are preserved under continuous mapping. For example, we shall now show that the properties of connectedness and compactness of sets have this character.

Preservation of Connectedness

We recall from Definition 12.1 that a set H in $\mathbf{R}^p$ is disconnected if there exist open sets A, B in $\mathbf{R}^p$ such that $A \cap H$ and $B \cap H$ are disjoint non-empty sets whose union is H. A set is connected if it is not disconnected.

22.3 PRESERVATION OF CONNECTEDNESS. *If $H \subseteq D(f)$ is connected in $\mathbf{R}^p$ and f is continuous on H, then $f(H)$ is connected in $\mathbf{R}^q$.*

PROOF. Let h be the restriction of f to the set H so that $D(h) = H$ and $h(x) = f(x)$ for all $x \in H$. We note that $f(H) = h(H)$ and that h is continuous on H.

If $f(H) = h(H)$ is disconnected in $\mathbf{R}^q$, then there exist open sets A, B in $\mathbf{R}^q$ such that $A \cap h(H)$ and $B \cap h(H)$ are disjoint non-empty sets whose union is $h(H)$. By the Global Continuity Theorem 22.1, there exist open sets A_1, B_1 in $\mathbf{R}^p$ such that

$$A_1 \cap H = h^{-1}(A), \qquad B_1 \cap H = h^{-1}(B).$$

These intersections are non-empty and their disjointness follows from the disjointness of the sets $A \cap h(H)$ and $B \cap h(H)$. The assumption that the union of $A \cap h(H)$ and $B \cap h(H)$ is $h(H)$ implies that the union of $A_1 \cap H$ and $B_1 \cap H$ is H. Therefore, the disconnectedness of $f(H) = h(H)$ implies the disconnectedness of H.   Q.E.D.

The very word "continuous" suggests that there are no sudden "breaks" in the graph of the function; hence the next result is by no means unexpected. However, the reader is invited to attempt to provide a different proof of this theorem and he will come to appreciate its depth.

22.4 BOLZANO'S INTERMEDIATE VALUE THEOREM. *Let $H \subseteq D(f)$ be a connected subset of $\mathbf{R}^p$ and let f be continuous on H and with values in $\mathbf{R}$. If k is any real number satisfying*

$$\inf \{f(x) : x \in H\} < k < \sup \{f(x) : x \in H\},$$

then there is at least one point of H where f takes the value k.

PROOF. If $k \notin f(H)$, then the sets $A = \{t \in \mathbf{R} : t < k\}$, $B = \{t \in \mathbf{R} : t > k\}$ form a disconnection of $f(H)$, contrary to the previous theorem.   Q.E.D.

Preservation of Compactness

We now demonstrate that the important property of compactness is preserved under continuous mapping. We recall that it is a consequence of the important Heine-Borel Theorem 11.3 that a subset K of $\mathbf{R}^p$ is compact if and only if it is both closed and bounded in $\mathbf{R}^p$. Thus the next result could be rephrased by saying that if K is closed and bounded in $\mathbf{R}^p$ and if f is continuous on K and with range in $\mathbf{R}^q$, then $f(K)$ is closed and bounded in $\mathbf{R}^q$.

22.5 PRESERVATION OF COMPACTNESS. *If* $K \subseteq D(f)$ *is compact and* f *is continuous on* K, *then* $f(K)$ *is compact.*

FIRST PROOF. We assume that K is closed and bounded in $\mathbf{R}^p$ and shall show that $f(K)$ is closed and bounded in $\mathbf{R}^q$. If $f(K)$ is not bounded, for each $n \in \mathbf{N}$ there exists a point x_n in K with $\|f(x_n)\| \geq n$. Since K is bounded, the sequence $X = (x_n)$ is bounded; hence it follows from the Bolzano-Weierstrass Theorem 16.4 that there is a subsequence of X which converges to an element x. Since $x_n \in K$ for $n \in \mathbf{N}$, the point x belongs to the closed set K. Hence f is continuous at x, so f is bounded by $\|f(x)\| + 1$ on a neighborhood of x. Since this contradicts the assumption that $\|f(x_n)\| \geq n$, the set $f(K)$ is bounded.

We shall prove that $f(K)$ is closed by showing that any cluster point y of $f(K)$ must be contained in this set. In fact, if n is a natural number, there is a point z_n in K such that $\|f(z_n) - y\| < 1/n$. By the Bolzano-Weierstrass Theorem 16.4, the sequence $Z = (z_n)$ has a subsequence $Z' = (z_{n(k)})$ which converges to an element z. Since K is closed, then $z \in K$ and f is continuous at z. Therefore,

$$f(z) = \lim_k (f(z_{n(k)})) = y,$$

which proves that y belongs to $f(K)$. Hence $f(K)$ is closed.

SECOND PROOF. By restricting f to K we may assume that $D(f) = K$. Now assume that $\mathcal{G} = \{G_\alpha\}$ is a family of open sets in $\mathbf{R}^q$ whose union contains $f(K)$. By the Global Continuity Theorem 22.1, for each set G_α in $\mathcal{G}$ there is an open subset C_α of $\mathbf{R}^p$ such that $C_\alpha \cap D = f^{-1}(G_\alpha)$. The family $\mathcal{C} = \{C_\alpha\}$ consists of open subsets of $\mathbf{R}^p$; we claim that the union of these sets contains K. For, if $x \in K$, then $f(x)$ is contained in $f(K)$; hence $f(x)$ belongs to some set G_α and by construction x belongs to the corresponding set C_α. Since K is compact, it is contained in the union of a finite number of sets in $\mathcal{C}$ and its image $f(K)$ is contained in the union of the corresponding finite number of sets in $\mathcal{G}$. Since this holds for arbitrary family $\mathcal{G}$ of open sets covering $f(K)$, the set $f(K)$ is compact in $\mathbf{R}^q$.   Q.E.D.

When the range of the function is $\mathbf{R}$, the next theorem is sometimes reformulated by saying that *a continuous real-valued function on a compact set attains its maximum and minimum values.*

22.6 MAXIMUM AND MINIMUM VALUE THEOREM. *Let* $K \subseteq D(f)$ *be compact in* $\mathbf{R}^p$ *and let* f *be a continuous real valued function. Then there are* *points x^* and x_* in K such that*

$$f(x^*) = \sup \{f(x) : x \in K\}, \qquad f(x_*) = \inf \{f(x) : x \in K\}.$$

FIRST PROOF. Since K is compact in $\mathbf{R}^p$, it follows from the preceding theorem that $f(K)$ is bounded in $\mathbf{R}$. Let $M = \sup f(K)$ and let (x_n) be a sequence in K such that

$$f(x_n) \geq M - 1/n, \qquad n \in \mathbf{N}.$$

By the Bolzano-Weierstrass Theorem 16.4, some subsequence $(x_{n(k)})$ converges to a limit $x^* \in K$. Since f is continuous at x^*, we must have $f(x^*) = \lim(f(x_{n(k)})) = M$. The proof of the existence of x_* is quite similar.

SECOND PROOF. By restricting f to K, we may assume that $D(f) = K$. We set $M = \sup f(K)$. Then for each $n \in \mathbf{N}$, let $G_n = \{u \in \mathbf{R} : u < M - 1/n\}$. Since G_n is open, it follows from the Global Continuity Theorem 22.1 that there exists an open set C_n in $\mathbf{R}^p$ such that

$$C_n \cap K = \{x \in K : f(x) < M - 1/n\}.$$

Now if the value M is not attained, then the union of the family $\mathscr{C} = \{C_n\}$ of open sets contains all of K. Since K is compact and the family $\{C_n \cap K\}$ is increasing, there is an $r \in \mathbf{N}$ such that $K \subseteq C_r$. But then we have $f(x) < M - 1/r$ for all $x \in K$, contrary to the fact that $M = \sup f(K)$.   Q.E.D.

If f has range in $\mathbf{R}^q$ with $q > 1$, the following corollary is sometimes useful.

22.7 COROLLARY. *Let f be a function on $D(f) \subseteq \mathbf{R}^p$ to $\mathbf{R}^q$ and let $K \subseteq D(f)$ be compact. If f is continuous on K, then there are points x^* and x_* in K such that*

$$\|f(x^*)\| = \sup \{\|f(x)\| : x \in K\}, \qquad \|f(x_*)\| = \inf \{\|f(x)\| : x \in K\}.$$

It follows from Theorem 21.2 that if $f : \mathbf{R}^p \to \mathbf{R}^q$ is linear, then there exists a constant $M > 0$ such that $\|f(x)\| \leq M \|x\|$ for all $x \in \mathbf{R}^p$. However it is not always true that there exists a constant $m > 0$ such that $\|f(x)\| \geq m \|x\|$ for all $x \in \mathbf{R}^p$. We now show that this is the case if and only if f is an injective linear function.

22.8 COROLLARY. *Let $f : \mathbf{R}^p \to \mathbf{R}^q$ be a linear function. Then f is injective if and only if there exists $m > 0$ such that $\|f(x)\| \geq m \|x\|$ for all $x \in \mathbf{R}^p$.*

PROOF. Suppose that f is injective, and let $S = \{x \in \mathbf{R}^p : \|x\| = 1\}$ be the compact unit sphere in $\mathbf{R}^p$.

By Corollary 22.7 there exists $x_* \in S$ such that $\|f(x_*)\| = m = \inf \{\|f(x)\| : x \in S\}$. Since f is injective, $m = \|f(x_*)\| > 0$. Hence $\|f(x)\| \geq m > 0$ for all $x \in S$. Now, if $u \in \mathbf{R}^p$, $u \neq 0$, then $u/\|u\|$ belongs to S and by the linearity of f we have

$$\frac{1}{\|u\|}\|f(u)\| = \left\|f\left(\frac{u}{\|u\|}\right)\right\| \geq m,$$

whence it follows that $\|f(u)\| \geq m\|u\|$ for all $u \in \mathbf{R}^p$ (since the result is trivial for $u = 0$).

Conversely, suppose $\|f(x)\| \geq m\|x\|$ for all $x \in \mathbf{R}^p$. If $f(x_1) = f(x_2)$, then we have

$$0 = \|f(x_1) - f(x_2)\| = \|f(x_1 - x_2)\| \geq m\|x_1 - x_2\|,$$

which implies that $x_1 = x_2$. Therefore f is injective.   Q.E.D.

One of the most striking consequences of Theorem 22.5 is that if f is continuous and injective on a compact domain, then the inverse function f^{-1} is automatically continuous.

22.9 CONTINUITY OF THE INVERSE FUNCTION. *Let K be a compact subset of $\mathbf{R}^p$ and let f be a continuous injective function with domain K and range $f(K)$ in $\mathbf{R}^q$. Then the inverse function is continuous with domain $f(K)$ and range K.*

PROOF. We observe that since K is compact, then Theorem 22.5 implies that $f(K)$ is compact and hence closed. Since f is injective by hypothesis, the inverse function $g = f^{-1}$ is defined. Let H be any closed set in $\mathbf{R}^p$ and consider $H \cap K$; since this set is bounded and closed (by Theorem 9.6(c)), the Heine-Borel Theorem assures that $H \cap K$ is a compact subset of $\mathbf{R}^p$. By Theorem 22.5, we conclude that $H_1 = f(H \cap K)$ is compact and hence closed in $\mathbf{R}^q$. Now if $g = f^{-1}$, then

$$H_1 = f(H \cap K) = g^{-1}(H).$$

Since H_1 is a subset of $f(K) = D(g)$, we can write this last equation as

$$H_1 \cap D(g) = g^{-1}(H).$$

From the Global Continuity Theorem 22.1(c), we infer that $g = f^{-1}$ is continuous.   Q.E.D.

We shall close this section with the introduction of some notation that will be convenient.

22.10 DEFINITION. If $D \subseteq \mathbf{R}^p$, then the collection of all **continuous functions** on D to $\mathbf{R}^q$ is denoted by $C_{pq}(D)$. The collection of all **bounded continuous functions** on D to $\mathbf{R}^q$ is denoted by $BC_{pq}(D)$. When p and q are understood, we will denote these collections merely by $C(D)$ and $BC(D)$.

The first part of the following result is a consequence of Theorem 20.6, and the second part is proved in the same way that Lemma 17.8 was proved.

22.11 THEOREM. (a) *The spaces* $C_{pq}(D)$ *and* $BC_{pq}(D)$ *are vector spaces under the vector operations*

$$(f+g)(x) = f(x) + g(x), \qquad (cf)(x) = cf(x) \qquad for\ x \in D.$$

(b) *The space* $BC_{pq}(D)$ *is a normed space under the norm*

$$\|f\|_D = \sup\{\|f(x)\| : x \in D\}.$$

Of course, in the special case where D is a compact subset of $\mathbf{R}^p$, then $C_{pq}(D) = BC_{pq}(D)$.

Exercises

22.A. Interpret the Global Continuity Theorem 22.1 for the real-valued functions $f(x) = x^2$ and $g(x) = 1/x$, $x \neq 0$. Take various open and closed sets and determine their inverse images under f and g.

22.B. Let $H: \mathbf{R} \to \mathbf{R}$ be defined by

$$h(x) = 1, \qquad 0 \le x \le 1,$$
$$= 0, \qquad \text{otherwise.}$$

Exhibit an open set G such that $h^{-1}(G)$ is not open in $\mathbf{R}$, and a closed set F such that $h^{-1}(F)$ is not closed in $\mathbf{R}$.

22.C. If f is bounded and continuous on $\mathbf{R}^p$ to $\mathbf{R}$ and if $f(x_0) > 0$, show that f is strictly positive on some neighborhood of x_0. Does the same conclusion hold if f is merely continuous at x_0?

22.D. If $p: \mathbf{R}^2 \to \mathbf{R}$ is a polynomial and $c \in \mathbf{R}$, show that the set $\{(x, y) : p(x, y) < c\}$ is open in $\mathbf{R}^2$.

22.E. If $f: \mathbf{R}^p \to \mathbf{R}$ is continuous on $\mathbf{R}^p$ and $\alpha < \beta$, show that the set $\{x \in \mathbf{R}^p : \alpha \le f(x) \le \beta\}$ is closed in $\mathbf{R}^p$.

22.F. A subset $D \subseteq \mathbf{R}^p$ is disconnected if and only if there exists a continuous function $f: D \to \mathbf{R}$ such that $f(D) = \{0, 1\}$.

22.G. Let f be continuous on $\mathbf{R}^2$ to $\mathbf{R}^q$. Define the functions g_1, g_2 on $\mathbf{R}$ to $\mathbf{R}^q$ by

$$g_1(t) = f(t, 0), \qquad g_2(t) = f(0, t).$$

Show that g_1 and g_2 are continuous.

22.H. Let f, g_1, g_2 be related by the formulas in the preceding exercise. Show that from the continuity of g_1 and g_2 at $t = 0$ one cannot prove the continuity of f at $(0, 0)$.

22.I. Give an example of a function on $I = [0, 1]$ to $\mathbf{R}$ which is not bounded.

22.J. Give an example of a bounded function f on I to $\mathbf{R}$ which does not take on either of the numbers $\sup\{f(x) : x \in I\}$ or $\inf\{f(x) : x \in I\}$.

22.K. Give an example of a bounded and continuous function g on $\mathbf{R}$ to $\mathbf{R}$ which does not take on either of the numbers $\sup\{g(x) : x \in \mathbf{R}\}$ or $\inf\{g(x) : x \in \mathbf{R}\}$.

22.L. Show that every polynomial of odd degree and real coefficients has a real root. Show that the polynomial $p(x) = x^4 + 7x^3 - 9$ has at least two real roots.

22.M. If $c > 0$ and n is a natural number, there exists a unique positive number b such that $b^n = c$.

22.N. Let f be continuous on I to R with $f(0)<0$ and $f(1)>0$. If $N = \{x \in I : f(x) < 0\}$ and if $c = \sup N$, show that $f(c) = 0$.

22.O. Let f be a continuous function on R to R which is strictly increasing (in the sense that if $x' < x''$ then $f(x') < f(x'')$). Prove that f is injective and that its inverse function f^{-1} is continuous and strictly increasing.

22.P. Let f be a continuous function on R to R which does not take on any of its values twice. Is it true that f must either be strictly increasing or strictly decreasing?

22.Q. Let g be a function on I to R. Prove that if g takes on each of its values exactly twice, then g cannot be continuous at every point of I.

22.R. Let f be continuous on the interval $[0, 2\pi]$ to R and such that $f(0) = f(2\pi)$. Prove that there exists a point c in this interval such that $f(c) = f(c + \pi)$. (Hint: consider $g(x) = f(x) - f(x + \pi)$.) Conclude that there are, at any time, antipodal points on the equator of the earth which have the same temperature.

22.S. Let $\varphi : [0, 2\pi) \to R^2$ be defined by $\varphi(t) = (\cos t, \sin t)$ for $t \in [0, 2\pi)$. Then φ is an injective continuous map of $[0, 2\pi)$ onto the unit circle $S = \{(x, y) \in R^2 : x^2 + y^2 = 1\}$. Show that $\varphi^{-1} : S \to [0, 2\pi)$ cannot be continuous. (We conclude that Theorem 22.9 may fail if the domain is not compact.)

Project

22.α. The purpose of this project is to show that many of the results of Section 22 hold for continuous functions whose domains and ranges are contained in metric spaces. (In establishing these results we may either observe that earlier definitions apply to metric spaces or can be reformulated to do so.)

(a) Show that Theorem 20.2 can be reformulated for a function from one metric space to another one.

(b) Show that the Global Continuity Theorem 22.1 holds without change.

(c) Prove that the Preservation of Connectedness Theorem 22.3 holds.

(d) Prove that the Preservation of Compactness Theorem 22.5 holds.

Section 23   Uniform Continuity and Fixed Points

Let f be defined on a subset $D(f)$ of R^p to R^q. Then it is readily seen that the following statements are equivalent:

(i) f is continuous at every point in $D(f)$.

(ii) Given $\varepsilon > 0$ and $u \in D(f)$, there is a $\delta(\varepsilon, u) > 0$ such that if x belongs to $D(f)$ and $\|x - u\| \le \delta$, then $\|f(x) - f(u)\| \le \varepsilon$.

The thing that is to be noted here is that the δ depends, in general, on both ε and u. That δ depends on u is a reflection of the fact that the function f may change its values rapidly near certain points and slowly near others.

Now it can happen that a function is such that the number δ can be chosen to be independent of the point u in $D(f)$ and depending only on ε. For example, if $f(x) = 2x$, then

$$|f(x) - f(u)| = 2|x - u|$$

and so we can choose $\delta(\varepsilon, u) = \varepsilon/2$ for all values of u.

On the other hand, if $g(x) = 1/x$ for $x > 0$, then

$$g(x) - g(u) = \frac{u - x}{ux}.$$

If $0 < \delta < u$ and $|x - u| \le \delta$, then we leave it to the reader to show that

$$|g(x) - g(u)| \le \frac{\delta}{u(u - \delta)}$$

and that this inequality cannot be improved, since equality actually holds for $x = u - \delta$. If we want to make $|g(x) - g(u)| \le \varepsilon$, then the largest value of δ we can select is

$$\delta(\varepsilon, u) = \frac{\varepsilon u^2}{1 + \varepsilon u}.$$

Thus if $u > 0$, then g is continuous at u because we can select $\delta(\varepsilon, u) = \varepsilon u^2/(1 + \varepsilon u)$, and this is the largest value we can choose. Since

$$\inf \left\{ \frac{\varepsilon u^2}{1 + \varepsilon u} : u > 0 \right\} = 0,$$

we cannot obtain a $\delta(\varepsilon, u) > 0$ which is independent of the choice of u for *all points* $u > 0$.

We shall now restrict g to a smaller domain. In fact, let $a > 0$ and define $h(x) = 1/x$ for $x \ge a$. Then the analysis just made shows that we can use the same value of $\delta(\varepsilon, u)$. However, this time the domain is smaller and

$$\inf \left\{ \frac{\varepsilon u^2}{1 + \varepsilon u} : u \ge a \right\} = \frac{\varepsilon a^2}{1 + \varepsilon a} > 0.$$

Hence if we define $\delta(\varepsilon) = \varepsilon a^2/(1 + \varepsilon a)$, then we can use this number for all points $u \ge a$.

In order to help fix these ideas, the reader should look over Examples 20.5 and determine in which examples the δ was chosen to depend on the point in question and in which ones it was chosen independently of the point.

With these preliminaries we now introduce the formal definition.

23.1 DEFINITION. Let f have domain $D(f)$ in $\mathbf{R}^p$ and range in $\mathbf{R}^q$. We say that f is **uniformly continuous** on a set $A \subseteq D(f)$ if for each $\varepsilon > 0$ there is a $\delta(\varepsilon) > 0$ such that if x and u belong to A and $\|x - u\| \le \delta(\varepsilon)$, then $\|f(x) - f(u)\| \le \varepsilon$.

It is clear that if f is uniformly continuous on A, then it is continuous at every point of A. In general, however, the converse does not hold. It is useful to have in mind what is meant by saying that a function is *not* uniformly continuous, so we state such a criterion, leaving its proof to the reader.

23.2 LEMMA. *A necessary and sufficient condition that the function f is not uniformly continuous on $A \subseteq D(f)$ is that there exist $\varepsilon_0 > 0$, and two*

160   CONTINUOUS FUNCTIONS

sequences $X = (x_n)$, $Y = (y_n)$ *in A such that if* $n \in N$, *then* $\|x_n - y_n\| \le 1/n$ *and* $\|f(x_n) - f(y_n)\| > \varepsilon_0$.

As an exercise the reader should apply this criterion to show that $g(x) = 1/x$ is not uniformly continuous on $D(g) = \{x : x > 0\}$.

We now present a very useful result which assures that a continuous function is automatically uniformly continuous on any compact set in its domain.

23.3   UNIFORM CONTINUITY THEOREM.   *Let f be a continuous function with domain $D(f)$ in $\mathbf{R}^p$ and range in $\mathbf{R}^q$. If $K \subseteq D(f)$ is compact, then f is uniformly continuous on K.*

FIRST PROOF.   Suppose that f is not uniformly continuous on K. By Lemma 23.2 there exists $\varepsilon_0 > 0$ and two sequences (x_n) and (y_n) in K such that if $n \in N$, then

(23.1) $\|x_n - y_n\| \le 1/n$, $\|f(x_n) - f(y_n)\| > \varepsilon_0$.

Since K is compact in $\mathbf{R}^p$, the sequence X is bounded; by the Bolzano-Weierstrass Theorem 16.4, there is a subsequence $(x_{n(k)})$ of (x_n) which converges to an element z. Since K is closed, the limit z belongs to K and f is continuous at z. It is clear that the corresponding subsequence $(y_{n(k)})$ of Y also converges to z.

It follows from Theorem 20.2(c) that both sequences $(f(x_{n(k)}))$ and $(f(y_{n(k)}))$ converge to $f(z)$. Therefore, when k is sufficiently great, we have $\|f(x_{n(k)}) - f(y_{n(k)})\| < \varepsilon_0$. But this contradicts the second relation in (23.1).

SECOND PROOF.   (A shorter proof could be based on the Lebesgue Covering Theorem 11.5, but we prefer to use the definition of compactness.) Suppose that f is continuous at every point of the compact set K. According to Theorem 20.2(b), given $\varepsilon > 0$ and u in K there is a number $\delta(\frac{1}{2}\varepsilon, u) > 0$ such that if $x \in K$ and $\|x - u\| < \delta(\frac{1}{2}\varepsilon, u)$, then $\|f(x) - f(u)\| < \frac{1}{2}\varepsilon$. For each u in K, form the open ball $G(u) = \{x \in \mathbf{R}^p : \|x - u\| < \frac{1}{2}\delta(\frac{1}{2}\varepsilon, u)\}$; then the set K is certainly contained in the union of the family $\mathcal{G} = \{G(u) : u \in K\}$, since to each point u in K there is an open ball $G(u)$ which contains it. Since K is compact, it is contained in the union of a finite number of sets in the family $\mathcal{G}$, say, $G(u_1), \ldots, G(u_N)$. We now define

$$\delta(\varepsilon) = \frac{1}{2} \inf \{\delta(\tfrac{1}{2}\varepsilon, u_1), \ldots, \delta(\tfrac{1}{2}\varepsilon, u_N)\},$$

and we shall show that $\delta(\varepsilon)$ has the desired property. For, suppose that x, u belong to K and that $\|x - u\| \le \delta(\varepsilon)$. Then there exists a natural number k with $1 \le k \le N$ such that x belongs to the set $G(u_k)$; that is, $\|x - u_k\| < \frac{1}{2}\delta(\frac{1}{2}\varepsilon, u_k)$. Since $\delta(\varepsilon) \le \frac{1}{2}\delta(\frac{1}{2}\varepsilon, u_k)$, it follows that

$$\|u - u_k\| \le \|u - x\| + \|x - u_k\| < \delta(\tfrac{1}{2}\varepsilon, u_k).$$

Therefore, we have the relations

$$\|f(x)-f(u_k)\|<\tfrac{1}{2}\varepsilon, \qquad \|f(u)-f(u_k)\|<\tfrac{1}{2}\varepsilon,$$

whence $\|f(x)-f(u)\|<\varepsilon$. We have shown that if x, u are any two points of K for which $\|x-u\|\le\delta(\varepsilon)$, then $\|f(x)-f(u)\|<\varepsilon$. Q.E.D.

In later sections we shall make use of the idea of uniform continuity on many occasions, so we shall not give any applications here. However, we shall introduce here another property which is often available and is sufficient to guarantee uniform continuity.

23.4 Definition. If f has domain $D(f)$ contained in $\mathbf{R}^p$ and range in $\mathbf{R}^q$, we say that f satisfies a **Lipschitz**† **condition** if there exists a constant $A>0$ such that

$$(23.2) \qquad \|f(x)-f(u)\|\le A\,\|x-u\|$$

for all points x, u in $D(f)$. In case the inequality (23.2) holds with a constant $A<1$, the function is called a **contraction.**

It is clear that if relation (23.2) holds, then on setting $\delta(\varepsilon)=\varepsilon/A$ one can establish the uniform continuity of f on $D(f)$. Therefore, if f satisfies a Lipschitz condition, then f is uniformly continuous. The converse, however, is not true, as may be seen by considering the function defined for $D(f)=I$ by $f(x)=\sqrt{x}$. If (23.2) holds, then setting $u=0$ one must have $|f(x)|\le A\,|x|$ for some constant A, but it is readily seen that the latter inequality cannot hold.

By recalling Theorem 21.3, we see that a linear function with domain $\mathbf{R}^p$ and range in $\mathbf{R}^q$ satisfies a Lipschitz condition. Moveover, it will be seen in Section 27 that any real function with a bounded derivative also satisfies a Lipschitz condition.

Fixed Point Theorems

If f is a function with domain $D(f)$ and range in the same space $\mathbf{R}^p$, then a point u in $D(f)$ is said to be a **fixed point** of f in case $f(u)=u$. A number of important results can be proved on the basis of the existence of fixed points of functions so it is of importance to have some affirmative criteria in this direction. The first theorem we give is elementary in character, yet it is often useful and has the important advantage that it provides a construction of the fixed point. For simplicity, we shall first state the result when the domain of the function is the entire space.

† Rudolph Lipschitz (1832–1903) was a professor at Bonn. He made contributions to algebra, number theory, differential geometry, and analysis.

23.5 FIXED POINT THEOREM FOR CONTRACTIONS. *Let f be a contraction with domain $\mathbf{R}^p$ and range contained in $\mathbf{R}^p$. Then f has a unique fixed point.*

PROOF. We are supposing that there exists a constant C with $0 < C < 1$ such that $\|f(x) - f(y)\| \leq C \|x - y\|$ for all x, y in $\mathbf{R}^p$. Let x_1 be an arbitrary point in $\mathbf{R}^p$ and set $x_2 = f(x_1)$; inductively, set

$$(23.3) \qquad\qquad x_{n+1} = f(x_n), \qquad n \in \mathbf{N}.$$

We shall show that the sequence (x_n) converges to a unique fixed point u of f and estimate the rapidity of the convergence.

To do this, we observe that

$$\|x_3 - x_2\| = \|f(x_2) - f(x_1)\| \leq C \|x_2 - x_1\|,$$

and, inductively, that

$$(23.4) \quad \|x_{n+1} - x_n\| = \|f(x_n) - f(x_{n-1})\| \leq C \|x_n - x_{n-1}\| \leq C^{n-1} \|x_2 - x_1\|.$$

If $m \geq n$, then repeated use of (23.4) yields

$$\|x_m - x_n\| \leq \|x_m - x_{m-1}\| + \|x_{m-1} - x_{m-2}\| + \cdots + \|x_{n+1} - x_n\|$$
$$\leq \{C^{m-2} + C^{m-3} + \cdots + C^{n-1}\} \|x_2 - x_1\|.$$

Hence it follows that, for $m \geq n$, then

$$(23.5) \qquad\qquad \|x_m - x_n\| \leq \frac{C^{n-1}}{1-C} \|x_2 - x_1\|.$$

Since $0 < C < 1$, the sequence (C^{n-1}) converges to zero. Therefore, (x_n) is a Cauchy sequence. If $u = \lim (x_n)$, then it is clear from (23.3) that u is a fixed point of f. From (23.5) and Lemma 15.8, we obtain the estimate

$$(23.6) \qquad\qquad \|u - x_n\| \leq \frac{C^{n-1}}{1-C} \|x_2 - x_1\|$$

for the rapidity of the convergence.

Finally, we show that there is only one fixed point for f. In fact, if u, v are two distinct fixed points of f, then

$$\|u - v\| = \|f(u) - f(v)\| \leq C \|u - v\|.$$

Since $u \neq v$, then $\|u - v\| \neq 0$, so this relation implies that $1 \leq C$, contrary to the hypothesis that $C < 1$.   Q.E.D.

It will be observed that we have actually established the following result.

23.6 COROLLARY. *If f is a contraction with constant $C < 1$, if x_1 is an arbitrary point in $\mathbf{R}^p$, and if the sequence $X = (x_n)$ is defined by equation (23.3), then X converges to the unique fixed point u of f with the rapidity estimated by (23.6).*

In case the function f is not defined on all of $\mathbf{R}^p$, then somewhat more care needs to be exercised to assure that the iterative definition (23.3) of the sequence can be carried out and that the points remain in the domain of f. Although some other formulations are possible, we shall content ourselves with the following one.

23.7 THEOREM. *Suppose that f is a contraction with constant C which is defined for $D(f) = \{x \in \mathbf{R}^p : \|x\| \le B\}$ and that $\|f(0)\| \le B(1-C)$. Then the sequence*

$$x_1 = 0, \, x_2 = f(x_1), \ldots, x_{n+1} = f(x_n), \ldots$$

converges to the unique fixed point of f which lies in the set $D(f)$.

PROOF. Indeed, if $x \in D = D(f)$, then $\|f(x) - f(0)\| \le C \|x - 0\| \le CB$, whence it follows that

$$\|f(x)\| \le \|f(0)\| + CB \le (1 - C)B + CB = B.$$

Therefore $f(D) \subseteq D$. Thus the sequence (x_n) can be defined and remains in D, so the previous proof applies. Q.E.D.

The Contraction Theorem established above has certain advantages: it is constructive, the error of approximation can be estimated, and it guarantees a unique fixed point. However, it has the disadvantage that the requirement that f be a contraction is a very severe restriction. It is a deep and important fact, first proved in 1910 by L. E. J. Brouwer,[†] that *any* continuous function with domain $D = \{x \in \mathbf{R}^p : \|x\| \le B\}$ and range contained in D must have at least one fixed point.

23.8 BROUWER FIXED POINT THEOREM. *Let $B > 0$ and let $D = \{x \in \mathbf{R}^p : \|x\| \le B\}$. Then any continuous function with domain D and range contained in D has at least one fixed point.*

The proof of this result when $p = 1$ will be given as an exercise. For the case $p > 1$, however, the proof would take us too far afield. For a proof based on elementary notions only, see Dunford-Schwartz, pages 467–470. For a more systematic account of fixed point and related theorems, consult the book of Lefschetz.

Exercises

23.A. Examine each of the functions in Example 20.5 and either show that the function is uniformly continuous on its domain or that it is not.

23.B. Give a proof of the Uniform Continuity Theorem 23.3 by using the Lebesgue Covering Theorem 11.5.

---

† L. E. J. BROUWER (1881–1966) was professor at Amsterdam and dean of the Dutch school of mathematics. In addition to his early contributions to topology, he is noted for his work on the foundations of mathematics.

23.C. If B is bounded in $\mathbf{R}^p$ and $f : B \to \mathbf{R}^q$ is uniformly continuous, show that f is bounded on B. Show that this conclusion fails if B is not bounded in $\mathbf{R}^p$.

23.D. Show that the functions, defined for $x \in \mathbf{R}$ by

$$f(x) = \frac{1}{1 + x^2}, \qquad g(x) = \sin x,$$

are uniformly continuous on $\mathbf{R}$.

23.E. Show that the functions, defined for $D = \{x \in \mathbf{R} : x \geq 0\}$, by

$$h(x) = x, \qquad k(x) = e^{-x},$$

are uniformly continuous on D.

23.F. Show that the following functions are not uniformly continuous on their domains.

(a) $f(x) = 1/x^2$, $D(f) = \{x \in \mathbf{R} : x > 0\}$,
(b) $g(x) = \tan x$, $D(g) = \{x \in \mathbf{R} : 0 \leq x < \pi/2\}$,
(c) $h(x) = e^x$, $D(h) = \mathbf{R}$.
(d) $k(x) = \sin(1/x)$, $D(k) = \{x \in \mathbf{R} : x > 0\}$.

23.G. A function $g : \mathbf{R} \to \mathbf{R}^q$ is **periodic** if there exists a number $p > 0$ such that $g(x + p) = g(x)$ for all $x \in \mathbf{R}$. Show that a continuous periodic function is bounded and uniformly continuous on $\mathbf{R}$.

23.H. Let f be defined on $D \subseteq \mathbf{R}^p$ to $\mathbf{R}^q$, and suppose that f is uniformly continuous on D. If (x_n) is a Cauchy sequence in D, show that $(f(x_n))$ is a Cauchy sequence in $\mathbf{R}^q$.

23.I. Suppose that $f : (0, 1) \to \mathbf{R}$ is uniformly continuous on $(0, 1)$. Show that f can be defined at $x = 0$ and $x = 1$ in such a way that it becomes continuous on $[0, 1]$.

23.J. Let $D = \{x \in \mathbf{R}^p : \|x\| < 1\}$. Show that $f : D \to \mathbf{R}^q$ can be extended to a continuous function on $D_1 = \{x \in \mathbf{R}^p : \|x\| \leq 1\}$ to $\mathbf{R}^q$ if and only if f is uniformly continuous on D.

23.K. If f and g are uniformly continuous on $\mathbf{R}$ to $\mathbf{R}$, show that $f + g$ is uniformly continuous on $\mathbf{R}$, but that fg need not be uniformly continuous on $\mathbf{R}$ even when one of f and g are bounded.

23.L. If $f : I \to I$ is continuous, show that f has a fixed point in I. (Hint: Consider $g(x) = f(x) - x$.)

23.M. Give an example of a function $f : \mathbf{R}^p \to \mathbf{R}^p$ such that $\|f(x) - f(u)\| \leq \|x - u\|$ for all $x, u \in \mathbf{R}^p$ which does not have a fixed point. (Why does this not contradict Theorem 23.5?)

23.N. Let f and g be continuous functions on $[a, b]$ such that the range $R(f) \subseteq R(g) = [0, 1]$. Prove that there exists a point $c \in [a, b]$ such that $f(c) = g(c)$.

Project

23.α. This project introduces the notion of the "oscillation" of a function on a set and at a point. Let $I = [a, b] \subseteq \mathbf{R}$ and let $f : I \to \mathbf{R}$ be bounded. If $\emptyset \neq A \subseteq I$ we define the **oscillation of f on A** to be the number

$$\Omega_f(A) = \sup\{f(x) - f(y) : x, y \in A\}.$$

(a) Show that $0 \leq \Omega_f(A) \leq 2 \sup\{|f(x)| : x \in A\}$. If $\emptyset \neq A \subseteq B \subseteq I$, then $\Omega_f(A) \leq \Omega_f(B)$.

(b) If $c \in I$ we define the **oscillation of** f **at** c to be the number

$$\omega_f(c) = \inf_\delta \Omega_f(N_\delta)$$

where $N_\delta = \{x \in I : |x - x_0| < \delta\}$. Show that (cf. Section 25)

$$\omega_f(c) = \lim_{\delta \to 0} \Omega_f(N_\delta).$$

Also, if $\omega_f(c) < \alpha$, then there exists $\delta > 0$ such that $\Omega_f(N_\delta) < \alpha$.

(c) Show that f is continuous at $c \in I$ if and only if $\omega_f(c) = 0$.

(d) If $\alpha > 0$ and if $\omega_f(x) < \alpha$ for all $x \in I$, then there exists $\delta > 0$ such that if $A \subseteq I$ is such that its diameter $d(A) = \sup \{|x - y| : x, y \in A\}$ is less than δ, then $\Omega_f(A) < \alpha$.

(e) If $\alpha > 0$, then the set $D_\alpha = \{x \in I : \omega_f(x) \ge \alpha\}$ is a closed set in $\mathbf{R}$. Show that

$$D = \bigcup_{\alpha > 0} D_\alpha = \bigcup_{n \in N} D_{1/n}$$

is the set of points at which f is discontinuous. Hence the set of points of discontinuity of a function is the union of a countable family of closed sets. (Such a set is called an F_σ-**set**.)

(f) Extend these definitions and results to a function defined on a closed cell in $\mathbf{R}^p$.

Section 24   Sequences of Continuous Functions

There are many times when one needs to consider a *sequence of continuous functions*. In this section we shall present several interesting and important theorems about such sequences. Theorem 24.1 will be used very often in the following and is a key result. The remaining theorems will not be used frequently, but the reader should be familiar with their statements, at least.

In this section the importance of uniform convergence should become clearer. We recall that a sequence (f_n) of functions on a subset D of $\mathbf{R}^p$ to $\mathbf{R}^q$ is said to converge uniformly on D to f if for every $\varepsilon > 0$ there is an $N(\varepsilon)$ such that if $n \ge N(\varepsilon)$ and $x \in D$, then $\|f_n(x) - f(x)\| < \varepsilon$. We recall from Theorem 17.9 that this is true if and only if $\|f_n - f\|_D \to 0$, when (f_n) is a bounded sequence.

Interchange of Limit and Continuity

We observe that the limit of a sequence of continuous functions may not be continuous. It is very easy to see this; for $n \in N$, and $x \in I$, let $f_n(x) = x^n$. We have seen, in Example 17.2(b), that the sequence (f_n) converges on I to the function f defined by

$$f(x) = 0, \quad 0 \le x < 1,$$
$$= 1, \quad x = 1.$$

Thus, despite the simple character of the continuous functions f_n, the limit function is not continuous at the point $x = 1$.

Although the extent of the discontinuity of the limit function in the example just given is not very great, it is evident that more complicated examples can be constructed which will produce more extensive discontinuity. It would be interesting to investigate exactly how discontinuous the limit of a sequence of continuous functions can be, but this investigation would take us too far afield. Furthermore, for most applications it is more important to find additional conditions which will guarantee that the limit function is continuous.

We shall now establish the important fact that uniform convergence of a sequence of continuous functions is sufficient to guarantee the continuity of the limit function.

24.1 THEOREM. *Let $F = (f_n)$ be a sequence of continuous functions with domain D in $\mathbf{R}^p$ and range in $\mathbf{R}^q$ and let this sequence converge uniformly on D to a function f. Then f is continuous on D.*

PROOF. Since (f_n) converges uniformly on D to f, given $\varepsilon > 0$ there is a natural number $N = N(\varepsilon/3)$ such that $\|f_N(x) - f(x)\| < \varepsilon/3$ for all x in D. To show that f is continuous at a point a in D, we note that

$$(24.1) \quad \|f(x) - f(a)\| \le \|f(x) - f_N(x)\| + \|f_N(x) - f_N(a)\| + \|f_N(a) - f(a)\|$$
$$\le \varepsilon/3 + \|f_N(x) - f_N(a)\| + \varepsilon/3.$$

Since f_N is continuous, there exists a number $\delta = \delta(\varepsilon/3, a, f_N) > 0$ such that if $\|x - a\| < \delta$ and $x \in D$, then $\|f_N(x) - f_N(a)\| < \varepsilon/3$. (See Figure 24.1.) Therefore, for such x we have $\|f(x) - f(a)\| < \varepsilon$. This establishes the continuity of the limit function f at the arbitrary point a in D. Q.E.D.

We remark that, although the uniform convergence of the sequence of continuous functions is sufficient for the continuity of the limit function, it is

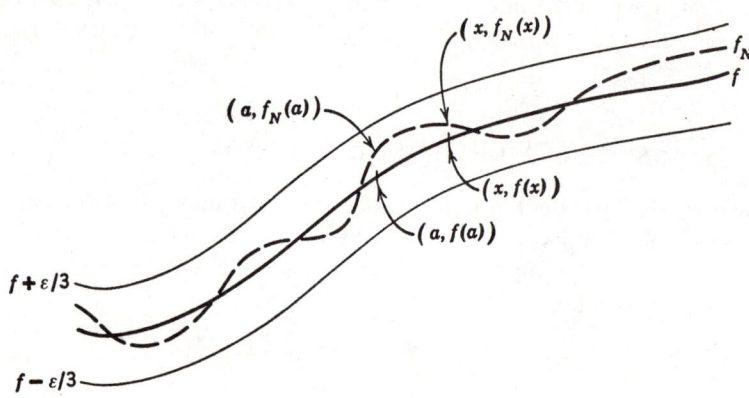

Figure 24.1

not necessary. Thus if (f_n) is a sequence of continuous functions which converges to a continuous function f, then it does *not* follow that the convergence is uniform (see Exercise 24.A).

As we have seen in Theorem 17.9, uniform convergence on a set D of a sequence of functions is implied by convergence in the uniform norm on D. Hence Theorem 24.1 has the following formulation.

24.2 THEOREM. *If (f_n) is a sequence of functions in $BC_{pq}(D)$ such that $\|f_n - f\|_D \to 0$, then $f \in BC_{pq}(D)$.*

Approximation Theorems

For many applications it is convenient to "approximate" continuous functions by functions of an elementary nature. Although there are several reasonable definitions that one can use to make the word "approximate" more precise, one of the most natural as well as one of the most important is to require that at every point of the given domain the approximating function shall not differ from the given function by more than the preassigned error. This sense is sometimes referred to as "uniform approximation" and it is intimately connected with uniform convergence. We suppose that f is a given function with domain $D = D(f)$ contained in R^p and range in R^q. We say that a function g **approximates f uniformly on D to within** $\varepsilon > 0$, if

$$\|g(x) - f(x)\| \le \varepsilon \qquad \text{for all } x \in D;$$

or, what amounts to the same thing, if

$$\|g - f\|_D = \sup\{\|g(x) - f(x)\| : x \in D\} \le \varepsilon.$$

Here we have used the norm which was introduced in equation (17.5). We say that the function f can be **uniformly approximated on D** by functions in a class $\mathcal{G}$ if, for each number $\varepsilon > 0$ there is a function g_ε in $\mathcal{G}$ such that $\|g_\varepsilon - f\|_D < \varepsilon$; or, equivalently, if there exists a sequence of functions in $\mathcal{G}$ which converges uniformly on D to f.

24.3 DEFINITION. A function g with domain R^p and range in R^q is called a **step function** if it assumes only a finite number of distinct values in R^q, each non-zero value being taken on an interval in R^p.

For example, if $p = q = 1$, then the function g defined explicitly by

$$
\begin{aligned}
g(x) &= 0, & x &\le -2, \\
&= 1, & -2 &< x \le 0, \\
&= 3, & 0 &< x < 1, \\
&= -5, & 1 &\le x \le 3, \\
&= 0, & x &> 3.
\end{aligned}
$$

is a step function.

We now show that a continuous function whose domain is a compact cell can be uniformly approximated by step functions.

24.4   THEOREM.   *Let f be a continuous function whose domain D is a compact cell in $\mathbf{R}^p$ and whose values belong to $\mathbf{R}^q$. Then f can be uniformly approximated on D by step functions.*

PROOF.   Let $\varepsilon > 0$ be given; since f is uniformly continuous (Theorem 23.3), there is a number $\delta(\varepsilon) > 0$ such that if x, y belong to D and $\|x - y\| < \delta(\varepsilon)$, then $\|f(x) - f(y)\| < \varepsilon$.   Divide the domain D of f into disjoint cells $I_1, \ldots, I_n$ such that if x, y belong to I_k, then $\|x - y\| < \delta(\varepsilon)$.   (How?) Let x_k be any point belonging to the cell I_k, $k = 1, \ldots, n$ and define $g_\varepsilon(x) = f(x_k)$ for $x \in I_k$ and $g_\varepsilon(x) = 0$ for $x \notin D$.   Then it is clear that $\|g_\varepsilon(x) - f(x)\| < \varepsilon$ for $x \in D$ so that g_ε approximates f uniformly on D to within ε.   (See Figure 24.2.)   Q.E.D.

It is natural to expect that a continuous function can be uniformly approximated by simple functions which are also continuous (as the step functions are not).   For simplicity, we shall establish the next result only in the case where $p = q = 1$ although there evidently is a generalization for higher dimensions.

We say that a function g defined on a compact cell $J = [a, b]$ of $\mathbf{R}$ with values in $\mathbf{R}$ is **piecewise linear** if there are a finite number of points c_k with $a = c_0 < c_1 < c_2 < \cdots < c_n = b$ and corresponding real numbers $A_k, B_k, k = 0, 1, \ldots, n$, such that when x satisfies the relation $c_{k-1} < x < c_k$, the function g has the form

$$g(x) = A_k x + B_k, \qquad k = 0, 1, \ldots, n.$$

If g is continuous on J, then the constants A_k, B_k must satisfy certain relations, of course.

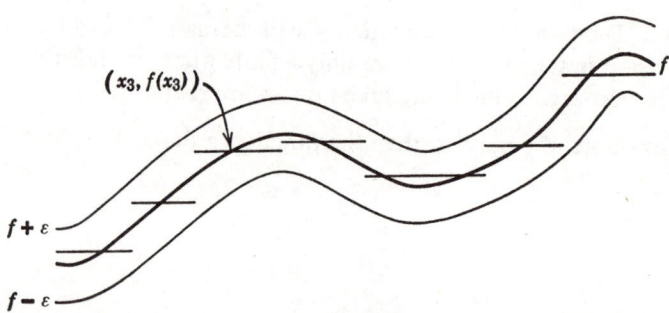

Figure 24.2.   Approximation by a step function.

24.5 THEOREM. *Let f be a continuous function whose domain is a compact cell J in $\mathbf{R}$. Then f can be uniformly approximated on J by continuous piecewise linear functions.*

PROOF. As before, f is uniformly continuous on the compact set J. Therefore, given $\varepsilon > 0$, we divide $J = [a, b]$ into cells by adding intermediate points c_k, $k = 0, 1, \ldots, n$, with $a = c_0 < c_1 < c_2 < \cdots < c_n = b$ so that $c_k - c_{k-1} < \delta(\varepsilon)$. Connect the points $(c_k, f(c_k))$ by line segments, and define the resulting continuous piecewise linear function g_ε. It is clear that g_ε approximates f uniformly on J within ε.   Q.E.D.

Approximation by Polynomials

We shall now prove a deeper, more useful, and more interesting result concerning the approximation by polynomials. First, we prove the Weierstrass Approximation Theorem for $p = q = 1$, by using the polynomials of S. Bernsteın.†

24.6 DEFINITION. Let f be a function with domain $I = [0, 1]$ and range in $\mathbf{R}$. The nth **Bernstein polynomial** for f is defined to be

$$(24.2) \qquad B_n(x) = B_n(x; f) = \sum_{k=0}^{n} f\left(\frac{k}{n}\right)\binom{n}{k}x^k(1-x)^{n-k}.$$

These Bernstein polynomials are not as terrifying as they look at first glance. A reader with some experience with probability should see the Binomial Distribution lurking in the background. Even without such experience, the reader should note that the value $B_n(x; f)$ of the polynomial at the point x is calculated from the values $f(0), f(1/n), f(2/n), \ldots, f(1)$, with certain non-negative weight factors $\varphi_k(x) = \binom{n}{k}x^k(1-x)^{n-k}$ which may be seen to be very small for those values of k for which k/n is far from x. In fact, the function φ_k is non-negative on I and takes its maximum value at the point k/n. Moreover, as we shall see below, the sum of all the $\varphi_k(x)$, $k = 0, 1, \ldots, n$, is 1 for each x in I.

We recall that the Binomial Theorem asserts that

$$(24.3) \qquad (s+t)^n = \sum_{k=0}^{n} \binom{n}{k}s^k t^{n-k},$$

where $\binom{n}{k}$ denotes the binomial coefficient

$$\binom{n}{k} = \frac{n!}{k!(n-k)!}.$$

† SERGE N. BERNSTEIN (1880–1968) made profound contributions to analysis, approximation theory, and probability. He was born in Odessa and was a professor in Leningrad and Moscow.

By direct inspection we observe that

$$(24.4) \qquad \binom{n-1}{k-1} = \frac{(n-1)!}{(k-1)!(n-k)!} = \frac{k}{n}\binom{n}{k},$$

$$(24.5) \qquad \binom{n-2}{k-2} = \frac{(n-2)!}{(k-2)!(n-k)!} = \frac{k(k-1)}{n(n-1)}\binom{n}{k}.$$

Now let $s = x$ and $t = 1-x$ in (24.3), to obtain

$$(24.6) \qquad 1 = \sum_{k=0}^{n}\binom{n}{k}x^k(1-x)^{n-k}.$$

Writing (24.6) with n replaced by $n-1$ and k by j, we have

$$1 = \sum_{j=0}^{n-1}\binom{n-1}{j}x^j(1-x)^{n-1-j}.$$

Multiply this last relation by x and apply the identity (24.4) to obtain

$$x = \sum_{j=0}^{n-1}\frac{j+1}{n}\binom{n}{j+1}x^{j+1}(1-x)^{n-(j+1)}.$$

Now let $k = j+1$, whence

$$x = \sum_{k=1}^{n}\frac{k}{n}\binom{n}{k}x^k(1-x)^{n-k}.$$

We also note that the term corresponding to $k=0$ can be included, since it vanishes. Hence we have

$$(24.7) \qquad x = \sum_{k=0}^{n}\frac{k}{n}\binom{n}{k}x^k(1-x)^{n-k}.$$

A similar calculation, based on (24.6) with n replaced by $n-2$ and identity (24.5), shows that

$$(n^2-n)x^2 = \sum_{k=0}^{n}(k^2-k)\binom{n}{k}x^k(1-x)^{n-k}.$$

Therefore we conclude that

$$(24.8) \qquad \left(1-\frac{1}{n}\right)x^2+\frac{1}{n}x = \sum_{k=0}^{n}\left(\frac{k}{n}\right)^2\binom{n}{k}x^k(1-x)^{n-k}.$$

Multiplying (24.6) by x^2, (24.7) by $-2x$, and adding them to (24.8), we obtain

$$(24.9) \qquad (1/n)x(1-x) = \sum_{k=0}^{n}(x-k/n)^2\binom{n}{k}x^k(1-x)^{n-k},$$

which is an estimate that will be needed below.

Examining Definition 24.6, formula (24.6) says that the nth Bernstein polynomial for the constant function $f_0(x) = 1$ coincides with f_0. Formula

(24.7) says the same thing for the function $f_1(x) = x$. Formula (24.8) asserts that the nth Bernsteĭn polynomial for the function $f_2(x) = x^2$ is

$$B_n(x; f_2) = (1 - 1/n)x^2 + (1/n)x,$$

which converges uniformly on I to f_2. We shall now prove that if f is any continuous function on I to R, then the sequence of Bernsteĭn polynomials has the property that it converges uniformly on I to f. This will give us a constructive proof of the Weierstrass Approximation Theorem. In the process of proving this theorem we shall need formula (24.9).

24.7 BERNSTEĬN APPROXIMATION THEOREM. *Let f be continuous on I with values in R. Then the sequence of Bernsteĭn polynomials for f, defined in equation (24.2), converges uniformly on I to f.*

PROOF. On multiplying formula (24.6) by $f(x)$, we get

$$f(x) = \sum_{k=0}^{n} f(x)\binom{n}{k}x^k(1-x)^{n-k}.$$

Therefore, we obtain the relation

$$f(x) - B_n(x) = \sum_{k=0}^{n} \{f(x) - f(k/n)\}\binom{n}{k}x^k(1-x)^{n-k}$$

from which it follows that

(24.10) $\qquad |f(x) - B_n(x)| \le \sum_{k=0}^{n} |f(x) - f(k/n)| \binom{n}{k}x^k(1-x)^{n-k}.$

Now f is bounded, say by M, and also uniformly continuous. Note that if k is such that k/n is near x, then the corresponding term in the sum (24.10) is small because of the continuity of f at x; on the other hand, if k/n is far from x, the factor involving f can only be said to be less than $2M$ and any smallness must arise from the other factors. We are led, therefore, to break (24.10) into two parts: those values of k where $x - k/n$ is small and those for which $x - k/n$ is large.

Let $\varepsilon > 0$ and let $\delta(\varepsilon)$ be as in the definition of uniform continuity for f. It turns out to be convenient to choose n so large that

(24.11) $\qquad n \ge \sup\{(\delta(\varepsilon))^{-4}, M^2/\varepsilon^2\},$

and break (24.10) into two sums. The sum taken over those k for which $|x - k/n| < n^{-1/4} \le \delta(\varepsilon)$ yields the estimate

$$\sum_{k} \varepsilon\binom{n}{k}x^k(1-x)^{n-k} \le \varepsilon\sum_{k=1}^{n} \binom{n}{k}x^k(1-x)^{n-k} = \varepsilon.$$

The sum taken over those k for which $|x - k/n| \ge n^{-1/4}$, that is, $(x - k/n)^2 \ge n^{-1/2}$, can be estimated by using formula (24.9). For this part of the sum in

(24.10) we obtain the upper bound

$$\sum_k 2M\binom{n}{k}x^k(1-x)^{n-k} = 2M\sum_k \frac{(x-k/n)^2}{(x-k/n)^2}\binom{n}{k}x^k(1-x)^{n-k}$$

$$\leq 2M\sqrt{n}\sum_{k=1}^{n}(x-k/n)^2\binom{n}{k}x^k(1-x)^{n-k}$$

$$\leq 2M\sqrt{n}\left\{\frac{1}{n}x(1-x)\right\} \leq \frac{M}{2\sqrt{n}},$$

since $x(1-x) \leq \frac{1}{4}$ on the interval I. Recalling the determination (24.11) for n, we conclude that each of these two parts of (24.10) is bounded above by ε. Hence, for n chosen in (24.11) we have

$$|f(x) - B_n(x)| < 2\varepsilon,$$

independently of the value of x. This shows that the sequence (B_n) converges uniformly on I to f. Q.E.D.

As a direct corollary of the theorem of Bernstein, we have the following important result.

24.8 WEIERSTRASS APPROXIMATION THEOREM. *Let f be a continuous function on a compact interval of* R *and with values in* R. *Then f can be uniformly approximated by polynomials.*

PROOF. If f is defined on $[a, b]$, then the function g defined on $I = [0, 1]$ by

$$g(t) = f((b-a)t + a), \qquad t \in I,$$

is continuous. Hence g can be uniformly approximated by Bernstein polynomials and a simple change of variable yields a polynomial approximation to f. Q.E.D.

We have chosen to go through the details of the Bernstein Theorem 24.7 because it gives a constructive method of finding a sequence of polynomials which converges uniformly on I to the given continuous function. In addition, the method of proof of Theorem 24.6 is characteristic of many analytic arguments and it is important to develop an understanding of such arguments. Finally, although we shall establish more general approximation results in Section 26, in order to do so we shall need to know that the absolute value function can be uniformly approximated on a compact interval by polynomials. Although it would be possible to show this special case directly, the argument is not so simple. For a more complete discussion of approximation the reader is referred to the book of E. Cheney listed in the References.

Exercises

24.A. Give an example of a sequence of continuous functions which converges to a continuous function but where the convergence is not uniform.

24.B. Give an example of a sequence of everywhere discontinuous functions which converges uniformly to a continuous function.

24.C. Give an example of a sequence of continuous functions which converges on a compact set to a function that has an infinite number of discontinuities.

24.D. Let (f_n) be a sequence of continuous functions on $D \subseteq R^p$ to R^q such that (f_n) converges uniformly to f on D, and let (x_n) be a sequence of elements in D which converges to $x \in D$. Does it follow that $(f_n(x_n))$ converges to $f(x)$?

24.E. Consider the sequences (f_n) defined on $D = \{x \in R : x \geq 0\}$ to R by the following formulas:

(a) $\dfrac{x^n}{n}$, (b) $\dfrac{x^n}{1+x^n}$, (c) $\dfrac{x^n}{n+x^n}$,

(d) $\dfrac{x^{2n}}{1+x^n}$, (e) $\dfrac{x^n}{1+x^{2n}}$, (f) $\dfrac{x}{n}e^{-x/n}$.

Discuss the convergence and the uniform convergence of these sequences and the continuity of the limit functions. In case of non-uniform convergence on D, consider appropriate intervals in D.

24.F. Let (f_n) be a sequence on $D \subseteq R^p$ to R^q which converges on D to f. Suppose that each f_n is continuous at c and that the sequence converges uniformly on some neighborhood of c. Prove that f is continuous at c.

24.G. Let (f_n) be a sequence of continuous functions on $D \subseteq R^p$ to R which is monotone decreasing in the sense that if $x \in D$, then

$$f_1(x) \geq f_2(x) \geq \cdots \geq f_n(x) \geq f_{n+1}(x) \geq \cdots$$

If $\lim (f_n(c)) = 0$ for some $c \in D$ and $\varepsilon > 0$, show that there exists $m \in N$ and a neighborhood U of c such that if $n \geq m$ and $x \in U \cap D$, then $f_n(x) < \varepsilon$.

24.H. Use the preceding exercise to prove the following result of U. Dini.† If (f_n) is a monotone sequence of continuous functions which converges at each point of a compact set K in R^p to a function f which is continuous on K, then the convergence is uniform on K.

24.I. Show, by examples, that Dini's Theorem fails if we drop either of the hypotheses that K is compact or that f is continuous.

24.J. Prove the following theorem of G. Pólya.‡ If for each $n \in N$ the function f_n on I to R is monotone increasing and if $f(x) = \lim (f_n(x))$ is continuous on I, then the convergence is uniform on I. (Observe that it is not assumed that f_n is continuous.)

---

† ULISSE DINI (1845–1918) studied and taught at Pisa. He worked on geometry and analysis, particularly Fourier series.

‡ GEORGE PÓLYA (1887–   ) was born in Budapest and taught at Zürich and Stanford. He is widely known for his work in complex analysis, probability, number theory, and the theory of inference.

24.K. Let (f_n) be a sequence of continuous functions on $D \subseteq \mathbf{R}^p$ to $\mathbf{R}^q$ and let $f(x) = \lim (f_n(x))$ for $x \in D$. Show that f is continuous at a point $c \in D$ if and only if for each $\varepsilon > 0$ there exists $m \in \mathbf{N}$ and a neighborhood U of c such that if $x \in D \cap U$, then $\|f_m(x) - f(x)\| < \varepsilon$.

24.L. Suppose that $f : \mathbf{R} \to \mathbf{R}$ is uniformly continuous on $\mathbf{R}$ and for $n \in \mathbf{N}$, let $f_n(x) = f(x + 1/n)$ for $x \in \mathbf{R}$. Show that (f_n) converges uniformly on $\mathbf{R}$ to f.

24.M. If $f_2(x) = x^2$ for $x \in \mathbf{I}$, how large must n be so that the nth Bernstein polynomial B_n for f_2 satisfies $|f_2(x) - B_n(x)| \le 1/1000$ for all $x \in \mathbf{I}$?

24.N. If $f_3(x) = x^3$ for $x \in \mathbf{I}$, calculate the nth Bernstein polynomial for f_3. Show directly that this sequence of polynomials converges uniformly to f_3 on $\mathbf{I}$.

24.O. Differentiate equation (24.3) once with respect to s and substitute $s = x$, $t = 1 - x$ to give another derivation of equation (24.7).

24.P. Differentiate equation (24.3) twice with respect to s to give another derivation of equation (24.8).

24.Q. (a) Let J be a compact interval in $\mathbf{R}$, and let $a \in R$, $c \in J$. Draw a graph of the function $\varphi : J \to \mathbf{R}$ defined by $\varphi(x) = a + m(|x - c| + x + c)$.

(b) Show that every continuous piecewise linear function can be written as the sum of a finite number of functions $\varphi_1, \ldots, \varphi_n$ having the form given in part (a).

(c) Assuming that, on any compact interval, the absolute value function $A(x) = |x|$ is the uniform limit of a sequence of polynomials in x, use the observation in part (b) to give another proof of the Weierstrass Approximation Theorem. (This method of proof is due to Lebesgue.)

24.R. Prove that the function $x \mapsto e^x$ on $\mathbf{R}$ is *not* the uniform limit on $\mathbf{R}$ of a sequence of polynomials. Hence the Weierstrass Approximation Theorem may fail for infinite intervals.

24.S. Show that the Weierstrass Approximation Theorem fails for bounded open intervals.

Section 25 Limits of Functions

Although it is not possible to give a precise definition, the field of "mathematical analysis" is generally understood to be that body of mathematics in which systematic use is made of various limiting concepts. If this is a reasonably accurate statement, it may seem odd to the reader that we have waited this long before inserting a section dealing with limits. There are several reasons for this delay, the main one being that elementary analysis deals with several different types of limit operations. We have already discussed the convergence of sequences and the limiting implicit in the study of continuity. In the next chapters, we shall bring in the limiting operations connected with the derivative and the integral. Although all of these limit notions are special cases of a more general one, the general notion is of a rather abstract character. For that reason, we prefer to introduce and discuss the notions separately, rather than to develop the general limiting idea first and then specialize. Once the special cases are well understood it is not difficult to comprehend the abstract notion. For an excellent exposition of this abstract limit, see the expository article of E. J. McShane cited in the References.

In this section we shall be concerned with the limit of a function at a point and some slight extensions of this idea. Often this idea is studied *before* continuity; in fact, the very definition of a continuous function is sometimes expressed in terms of this limit instead of using the definition we have given in Section 20. One of the reasons why we have chosen to study continuity separately from the limit is that we shall introduce two slightly different definitions of the limit of a function at a point. Since both definitions are widely used, we shall present them both and attempt to relate them to each other.

Unless there is specific mention to the contrary, we shall let f be a function with domain D contained in $\boldsymbol{R}^p$ and values in $\boldsymbol{R}^q$ and we shall consider the limiting character of f at a cluster point c of D. Therefore, every neighborhood of c contains infinitely many points of D.

25.1 DEFINITION. (i) An element b of $\boldsymbol{R}^q$ is said to be the **deleted limit of** f **at** c if for every neighborhood V of b there is a neighborhood U of c such that if x belongs to $U \cap D$ and $x \neq c$, then $f(x)$ belongs to V. In this case we write

$$(25.1) \qquad b = \lim_c f \qquad \text{or} \qquad b = \lim_{x \to c} f(x).$$

(ii) An element b of $\boldsymbol{R}^q$ is said to be the **non-deleted limit of** f **at** c if for every neighborhood V of b there is a neighborhood U of c such that if x belongs to $U \cap D$, then $f(x)$ belongs to V. In this case we write

$$(25.2) \qquad b = \operatorname{Lim}_c f \qquad \text{or} \qquad b = \operatorname{Lim}_{x \to c} f(x).$$

It is important to observe that the difference between these two notions centers on whether the value $f(c)$, when it exists, is considered or not. Note also the rather subtle notational distinction we have introduced in equations (25.1) and (25.2). It should be realized that most authors introduce only one of these notions, in which case they refer to it merely as "the limit" and generally employ the notation in (25.1). Since the deleted limit is the more popular, we have chosen to preserve the conventional symbolism in referring to it.

The uniqueness of either limit, when it exists, is readily established. We content ourself with the following statement.

25.2 LEMMA. (a) *If either of the limits* $\lim_c f$ *and* $\operatorname{Lim}_c f$ *exists, then it is uniquely determined.*

(b) *If the non-deleted limit exists, then the deleted limit exists and*

$$\lim_c f = \operatorname{Lim}_c f.$$

(c) *If c does not belong to the domain D of f, then the deleted limit exists if and only if the non-deleted limit exists.*

Part (b) of the lemma just stated shows that the notion of the non-deleted limit is somewhat more restrictive than that of the deleted limit. Part (c) shows that they can be different only in the case where c belongs to D. To give an example where these notions differ, consider the function f on $\mathbf{R}$ to $\mathbf{R}$ defined by

$$(25.3) \qquad \begin{aligned} f(x) &= 0, & x &\neq 0, \\ &= 1, & x &= 0. \end{aligned}$$

If $c = 0$, then the deleted limit of f at $c = 0$ exists and equals 0, while the non-deleted limit does not exist.

We now state some necessary and sufficient conditions for the existence of the limits, leaving their proof to the reader. It should be realized that in part (c) of both results the limit refers to the limit of a *sequence*, which was discussed in Section 14.

25.3 THEOREM. *The following statements, pertaining to the deleted limit, are equivalent.*
(a) *The deleted limit $b = \lim_c f$ exists.*
(b) *If $\varepsilon > 0$, there is a $\delta > 0$ such that if $x \in D$ and $0 < \|x - c\| < \delta$, then $\|f(x) - b\| < \varepsilon$.*
(c) *If (x_n) is any sequence in D such that $x_n \neq c$ and $c = \lim (x_n)$, then $b = \lim (f(x_n))$.*

25.4 THEOREM. *The following statements, pertaining to the non-deleted limit, are equivalent.*
(a) *The non-deleted limit $b = \text{Lim}_c f$ exists.*
(b) *If $\varepsilon > 0$, there is a $\delta > 0$ such that if $x \in D$ and $\|x - c\| < \delta$, then $\|f(x) - b\| < \varepsilon$.*
(c) *If (x_n) is any sequence in D such that $c = \lim (x_n)$, then we have $b = \lim (f(x_n))$.*

The next result yields an instructive connection between these two limits and continuity of f at c.

25.5 THEOREM. *If c is a cluster point belonging to the domain D of f, then the following statements are equivalent.*
(a) *The function f is continuous at c.*
(b) *The deleted limit $\lim_c f$ exists and equals $f(c)$.*
(c) *The non-deleted limit $\text{Lim}_c f$ exists.*

PROOF. If (a) holds and V is a neighborhood of $f(c)$, then there exists a neighborhood U of c such that if x belongs to $U \cap D$, then $f(x)$ belongs to V. Clearly, this implies that $\text{Lim} f$ exists at c and equals $f(c)$. Similarly,

$f(x)$ belongs to V for all $x \neq c$ for which $x \in U \cap D$, in which case $\lim f$ exists and equals $f(c)$. Conversely, statements (b) and (c) are readily seen to imply (a).   Q.E.D.

If f and g are two functions which have deleted (respectively, non-deleted) limits at a cluster point c of $D(f + g) = D(f) \cap D(g)$, then their sum $f + g$ has a deleted (respectively, non-deleted) limit at c and

$$\lim_c (f + g) = \lim_c f + \lim_c g,$$

$$\left(\text{respectively, } \operatorname{Lim}_c (f + g) = \operatorname{Lim}_c f + \operatorname{Lim}_c g \right).$$

Similar results hold for other algebraic combinations of functions, as is easily seen. The following result, concerning the composition of two functions, is slightly deeper and is a place where the non-deleted limit is simpler than the deleted limit.

25.6 THEOREM. *Suppose that f has domain $D(f)$ in $\mathbf{R}^p$ and range in $\mathbf{R}^q$ and that g has domain $D(g)$ in $\mathbf{R}^q$ and range in $\mathbf{R}^r$. Let $g \circ f$ be the composition of g and f and let c be a cluster point of $D(g \circ f)$.*

(a) If the deleted limits $b = \lim_c f$ and $a = \lim_b g$ both exist and if either g is continuous at b or $f(x) \neq b$ for x in a neighborhood of c, then the deleted limit of $g \circ f$ exists at c and $a = \lim_c g \circ f$.

(b) If the non-deleted limits $b = \operatorname{Lim}_c f$ and $a = \operatorname{Lim}_b g$ both exist, then the non-deleted limit of $g \circ f$ exists at c and

$$a = \operatorname{Lim}_c g \circ f.$$

PROOF. (a) Let W be a neighborhood of a in $\mathbf{R}^r$; since $a = \lim g$ at b, there is a neighborhood V of b such that if y belongs to $V \cap D(g)$ and $y \neq b$, then $g(y) \in W$. Since $b = \lim f$ at c, there is a neighborhood U of c such that if x belongs to $U \cap D(f)$ and $x \neq c$, then $f(x) \in V$. Hence, if x belongs to the possibly smaller set $U \cap D(g \circ f)$, and $x \neq c$, then $f(x) \in V \cap D(g)$. If $f(x) \neq b$ on some neighborhood U_1 of c, it follows that for $x \neq c$ in $(U_1 \cap U) \cap D(g \circ f)$, then $(g \circ f)(x) \in W$, so that a is the deleted limit of $g \circ f$ at c. If g is continuous at b, then $(g \circ f)(x) \in W$ for x in $U \cap D(g \circ f)$ and $x \neq c$.

To prove part (b), we note that the exceptions made in the proof of (a) are no longer necessary. Hence if x belongs to $U \cap D(g \circ f)$, then $f(x) \in V \cap D(g)$ and, therefore, $(g \circ f)(x) \in W$.   Q.E.D.

The conclusion in part (a) of the preceding theorem may fail if we drop the condition that g is continuous at b or that $f(x) \neq b$ on a neighborhood of c. To substantiate this remark, let f be the function on $\mathbf{R}$ to $\mathbf{R}$ defined in formula (25.3) and let $g = f$ and $c = 0$. Then $g \circ f$ is given by

$$(g \circ f)(x) = 1, \qquad x \neq 0,$$
$$= 0, \qquad x = 0.$$

Furthermore, we have $\lim_{x \to 0} f(x) = 0$, and $\lim_{y \to 0} g(y) = 0$, whereas it is clear that $\lim_{x \to 0} (g \circ f)(x) = 1$. (Note that the non-deleted limits do not exist for these functions.)

Upper Limits at a Point

For the remainder of the present section, we shall consider the case where $q = 1$. Thus f is a function with domain D in $\boldsymbol{R}^p$ and values in $\boldsymbol{R}$ and the point c in $\boldsymbol{R}^p$ is a cluster point of D. We shall define the limit superior or the upper limit of f at c. Again there are two possibilities depending on whether deleted or non-deleted neighborhoods are considered, and we shall discuss both possibilities. It is clear that we can define the limit inferior in a similar fashion. One thing to be noted here is that, although the existence of the limit in $\boldsymbol{R}$ (deleted or not) is a relatively delicate matter, the limits superior to be defined have the virtue that if f is bounded, then their existence is guaranteed.

The ideas in this part are parallel to the notion of the limit superior of a sequence in $\boldsymbol{R}^p$ which was introduced in Section 19. However, we shall not assume familiarity with what was done there, except in some of the exercises.

25.7 DEFINITION. Suppose that f is bounded on a neighborhood of the point c. If $r > 0$, define $\varphi(r)$ and $\Phi(r)$ by

(a) $$\varphi(r) = \sup \{f(x) : 0 < \|x - c\| < r, x \in D\},$$

(b) $$\Phi(r) = \sup \{f(x) : \|x - c\| < r, x \in D\}$$

and set

(c) $$\lim_{x \to c} \sup f = \inf \{\varphi(r) : r > 0\},$$

(d) $$\text{Lim} \sup_{x \to c} f = \inf \{\Phi(r) : r > 0\}.$$

These quantities are called the **deleted limit superior** and the **non-deleted limit superior** of f at c, respectively.

Since these quantities are defined as the infima of the image under f of ever-decreasing neighborhoods of c, it is probably not clear that they deserve the terms "limit superior." The next lemma indicates a justification for the terminology.

25.8 LEMMA. *If φ, Φ are as defined as above, then*

(a) $$\limsup_{x \to c} f = \lim_{r \to 0} \varphi(r),$$

(b) $$\operatorname{Lim\,sup}_{x \to c} f = \lim_{r \to 0} \Phi(r).$$

PROOF. We observe that if $0 < r < s$, then

$$\limsup_{x \to c} f \le \varphi(r) \le \varphi(s).$$

Furthermore, by 25.7(c), if $\varepsilon > 0$ there exists an $r_\varepsilon > 0$ such that

$$\varphi(r_\varepsilon) < \limsup_{x \to c} f + \varepsilon.$$

Therefore, if r satisfies $0 < r < r_\varepsilon$, we have $|\varphi(r) - \limsup_{x \to c} f| < \varepsilon$, which proves (a). The proof of (b) is similar and will be omitted.   Q.E.D.

25.9 LEMMA. (a) *If $M > \limsup_{x \to c} f$, then there exists a neighborhood U of c such that*

$$f(x) < M \qquad \text{for} \qquad c \neq x \in D \cap U.$$

(b) *If $M > \operatorname{Lim\,sup}_{x \to c} f$, then there exists a neighborhood U of c such that*

$$f(x) < M \qquad \text{for} \qquad x \in D \cap U.$$

PROOF. (a) By 25.7(c), we have $\inf \{\varphi(r) : r > 0\} < M$. Hence there exists a real number $r_1 > 0$ such that $\varphi(r_1) < M$ and we can take $U = \{x \in \boldsymbol{R}^p : \|x - c\| < r_1\}$. The proof of (b) is similar.   Q.E.D.

25.10 LEMMA. *Let f and g be bounded on a neighborhood of c and suppose that c is a cluster point of $D(f + g)$. Then*

(a) $$\limsup_{x \to c} (f + g) \le \limsup_{x \to c} f + \limsup_{x \to c} g$$

(b) $$\operatorname{Lim\,sup}_{x \to c} (f + g) \le \operatorname{Lim\,sup}_{x \to c} f + \operatorname{Lim\,sup}_{x \to c} g.$$

PROOF. In view of the relation

$$\sup \{f(x) + g(x) : x \in A\} \le \sup \{f(x) : x \in A\} + \sup \{g(x) : x \in A\},$$

it is clear that, using notation as in Definition 25.7, we have

$$\varphi_{f+g}(r) \le \varphi_f(r) + \varphi_g(r).$$

Now use Lemma 25.8 and let $r \to 0$ to obtain (a).   Q.E.D.

Results concerning other algebraic combinations will be found in Exercise 25.F.

Although we shall have no occasion to pursue these matters, in some areas of analysis it is useful to have the following generalization of the notion of continuity.

25.11 DEFINITION. A function f on D to $\mathbf{R}$ is said to be **upper semi-continuous** at a point c in D in case

$$(25.4) \qquad\qquad f(c) = \operatorname*{Lim\,sup}_{x \to c} f.$$

It is said to be **upper semi-continuous** on D if it is upper semi-continuous at every point of D.

Instead of defining upper semi-continuity by means of equation (25.4) we could require the equivalent, but less elegant, condition

$$(25.5) \qquad\qquad f(c) \geq \operatorname*{lim\,sup}_{x \to c} f.$$

One of the keys to the importance and the utility of upper semi-continuous functions is suggested by the following lemma, which may be compared with the Global Continuity Theorem 22.1.

25.12 LEMMA. *Let f be an upper semi-continuous function with domain D in $\mathbf{R}^p$ and let k be an arbitrary real number. Then there exists an open set G and a closed set F such that*

$$(25.6) \quad G \cap D = \{x \in D : f(x) < k\}, \qquad F \cap D = \{x \in D : f(x) \geq k\}.$$

PROOF. Suppose that c is a point in D such that $f(c) < k$. According to Definition 25.11 and Lemma 25.9(b), there is a neighborhood $U(c)$ of c such that $f(x) < k$ for all x in $D \cap U(c)$. Without loss of generality we can select $U(c)$ to be an open neighborhood; setting

$$G = \bigcup \{U(c) : c \in D\},$$

we have an open set with the property stated in (25.6). If F is the complement of G, then F is closed in $\mathbf{R}^p$ and satisfies the stated condition.

Q.E.D.

It is possible to show, using the lemma just proved, (cf. Exercise 25.M) that if K is a compact subset of $\mathbf{R}^p$ and f is upper semi-continuous on K, then f is bounded above on K and *there exists a point in K where f attains its supremum.* Thus upper semi-continuous functions on compact sets possess some of the properties we have established for continuous functions, even though an upper semi-continuous function can have many points of discontinuity.

It will have occurred to the reader that it is possible to extend the notion of limit superior at a point to the case where the function is not bounded by using ideas along the lines given at the end of Section 18. Similarly one can define the limit superior as $x \to \pm\infty$. These ideas are useful, but we will leave them as exercises.

Exercises

25.A. Discuss the existence of both the deleted and the non-deleted limits of the following functions at the point $x = 0$.

(a) $f(x) = |x|$,

(b) $f(x) = 1/x$, $x \neq 0$,

(c) $f(x) = x \sin (1/x)$, $x \neq 0$,

(d) $f(x) = \sin (1/x)$, $x \neq 0$

(e) $f(x) = \begin{cases} x \sin (1/x), & x \neq 0, \\ 1, & x = 0, \end{cases}$

(f) $f(x) = \begin{cases} 0, & x \leq 0, \\ 1, & x > 0. \end{cases}$

25.B. Prove Lemma 25.2.

25.C. If f denotes the function defined in equation (25.3), show that the deleted limit at $x = 0$ equals 0 and that the non-deleted limit at $x = 0$ does not exist. Discuss the existence of these two limits for the composition $f \circ f$.

25.D. Prove Lemma 25.4.

25.E. Show that statements 25.5(b) and 25.5(c) imply statement 25.5(a).

25.F. Show that if f and g have deleted limits at a cluster point c of the set $D(f) \cap D(g)$, then the sum $f + g$ has a deleted limit at c and

$$\lim_{c} (f + g) = \lim_{c} f + \lim_{c} g.$$

Under the same hypotheses, the inner product $f \cdot g$ has a deleted limit at c and

$$\lim_{c} (f \cdot g) = \left(\lim_{c} f \right) \cdot \left(\lim_{c} g \right).$$

25.G. Let f be defined on a subset $D(f)$ of $\mathbf{R}$ into $\mathbf{R}^q$. If c is a cluster point of the set $V = \{x \in \mathbf{R} : x \in D(f), x > c\}$, and if f_1 is the restriction of f to V, then we define the **right-hand deleted limit of** f at c to be $\lim_{c^+} f_1$, whenever this limit exists. Sometimes the limit is denoted by $\lim_{c+} f$ or by $f(c + 0)$. Formulate and establish a result analogous to Theorem 25.3 for the right-hand deleted limit. (A similar definition can be given for the **right-hand non-deleted limit** and both left-hand limits at c.)

25.H. Let f be defined on $D = \{x \in \mathbf{R} : x \geq 0\}$ to $\mathbf{R}$. We say that a number L is the **limit of** f at $+\infty$ if for each $\varepsilon > 0$ there exists a real number $m(\varepsilon)$ such that if $x \geq m(\varepsilon)$, then $|f(x) - L| < \varepsilon$. In this case we write $L = \lim_{x \to +\infty} f$. Formulate and prove a result analogous to Theorem 25.3 for this limit.

25.I. If f is defined on a set $D(f)$ in $\mathbf{R}$ to $\mathbf{R}$ and if c is a cluster point of $D(f)$, then we say that $f(x) \to +\infty$ as $x \to c$, or that

$$\lim_{x \to c} f = +\infty$$

in case for each positive number M there exists a neighborhood U of c such that if $x \in U \cap D(f)$, $x \neq c$, then $f(x) > M$. Formulate and establish a result analogous to Theorem 25.3 for this limit.

25.J. In view of Exercises 25.H and 25.I, give a definition of what is meant by the expressions:

$$\lim_{x \to +\infty} f = +\infty, \qquad \lim_{x \to c} f = -\infty.$$

25.K. Establish Lemma 25.8 for the non-deleted limit superior. Give the proof of Lemma 25.9(b).

25.L. Define what is meant by $\lim \sup_{x \to +\infty} f = L$, and $\lim \inf_{x \to +\infty} f = -\infty$.

25.M. Show that if f is an upper semi-continuous function on a compact subset K of R^p with values in R, then f is bounded above and attains its supremum on K.

25.N. Show that an upper semi-continuous function on a compact set may not be bounded below and may not attain its infimum.

25.O. Show that if A is an open subset of R^p and if f is defined on R^p to R by $f(x) = 1$ for $x \in A$, and $f(x) = 0$ for $x \notin A$, then f is a lower semi-continuous function. If A is a closed subset of R^p, show that f is upper semi-continuous.

25.P. Give an example of an upper semi-continuous function which has an infinite number of points of discontinuity.

25.Q. Is it true that a function on R^p to R is continuous at a point if and only if it is both upper and lower semi-continuous at this point?

25.R. If (f_n) is a bounded sequence of continuous functions on R^p to R and if f^* is defined on R^p by $f^*(x) = \sup \{f_n(x) : n \in N\}$ for $x \in R^p$, then is it true that f^* is upper semi-continuous on R^p?

25.S. If (f_n) is a bounded sequence of continuous functions on R^p to R and if f_* is defined on R^p by $f_*(x) = \inf \{f_n(x) : n \in N\}$ for $x \in R^p$, then is it true that f_* is upper semi-continuous on R^p?

25.T. Let f be defined on a subset D of $R^p \times R^q$ and with values in R'. Let (a, b) be a cluster point of D. By analogy with Definition 19.4, define the double and the two iterated limits of f at (a, b). Show that the existence of the double and the iterated limits implies their equality. Show that the double limit can exist without either iterated limit existing and that both iterated limits can exist and be equal without the double limit existing.

25.U. Let f be as in the preceding exercise. By analogy with Definitions 17.4 and 19.8, define what it means to say that

$$g(y) = \lim_{x \to a} f(x, y)$$

uniformly for y in a set. Formulate and prove a result analogous to Theorem 19.10.

25.V. Let f be as in Definition 25.1 and suppose that the deleted limit at c exists and that for some element A in R^q and $r > 0$ the inequality $\|f(x) - A\| < r$ holds on some neighborhood of c. Prove that $\|\lim_c f - A\| \le r$. Does the same conclusion hold for the non-deleted limit?

25.W. Discuss the upper and lower semi-continuity of the functions in parts (g) and (h) of Example 20.5.

25.X. If $f : [0, +\infty) \to R$ is continuous on $[0, +\infty)$ and $\lim_{x \to +\infty} f(x) = 0$, show that f is uniformly continuous on $[0, +\infty)$.

Section 26   Some Further Results

We shall present some theorems in this section that will not be applied later in this book, but which are often useful in topology and analysis. The first results are far-reaching extensions of the Weierstrass Approximation Theorem, next is a theorem giving conditions under which a continuous function has a continuous extension, and the final result is analogous to the Bolzano-Weierstrass in the space $C_{pq}(K)$ of continuous functions on a compact set K.

The Stone-Weierstrass Theorem

To facilitate our discussion, we introduce the following terminology. If f and g are functions with domain D in $\mathbf{R}^p$ and with values in $\mathbf{R}$, then the functions h and k defined for x in D by

$$h(x) = \sup\{f(x), g(x)\}, \qquad k(x) = \inf\{f(x), g(x)\},$$

are called the **supremum** and **infimum,** respectively, of the functions f and g. If f and g are continuous on D, then both h and k are also continuous. This follows from Theorem 20.7 and the observation that if a, b are real numbers, then

$$\sup\{a, b\} = \tfrac{1}{2}\{a + b + |a - b|\},$$
$$\inf\{a, b\} = \tfrac{1}{2}\{a + b - |a - b|\}.$$

We now state one form of Stone's† generalization of the Weierstrass Approximation Theorem. Despite its recent discovery it has already become "classical" and should be a part of the background of every student of mathematics. The reader should refer to the article by Stone listed in the References for extensions, applications, and a much fuller discussion than is presented here.

26.1 STONE APPROXIMATION THEOREM. *Let K be a compact subset of $\mathbf{R}^p$ and let $\mathcal{L}$ be a collection of continuous functions on K to $\mathbf{R}$ with the properties:*

(a) *If f, g belong to $\mathcal{L}$, then $\sup\{f, g\}$ and $\inf\{f, g\}$ belong to $\mathcal{L}$.*

(b) *If a, $b \in \mathbf{R}$ and $x \neq y \in K$, then there exists a function f in $\mathcal{L}$ such that $f(x) = a$, $f(y) = b$.*

Then any continuous function on K to $\mathbf{R}$ can be uniformly approximated on K by functions in $\mathcal{L}$.

PROOF. Let F be a continuous function on K to $\mathbf{R}$. If x, y belong to K, let $g_{xy} \in \mathcal{L}$ be such that $g_{xy}(x) = F(x)$ and $g_{xy}(y) = F(y)$. Since the functions F, g_{xy} are continuous and have the same value at y; given $\varepsilon > 0$, there is an

† MARSHALL H. STONE (1903–) studied at Harvard and has taught at Harvard and the Universities of Chicago and Massachusetts. The son of a chief justice, he has made basic contributions to modern analysis, especially to the theories of Hilbert space and Boolean algebras.

open neighborhood $U(y)$ of y such that if z belongs to $K \cap U(y)$, then

$$(26.1) \qquad\qquad g_{xy}(z) > F(z) - \varepsilon.$$

Hold x fixed and for each $y \in K$, select an open neighborhood $U(y)$ with this property. From the compactness of K, it follows that K is contained in a finite number of such neighborhoods: $U(y_1), \ldots, U(y_n)$. If $h_x = \sup\{g_{xy_1}, \ldots, g_{xy_n}\}$, then it follows from relation (26.1) that

$$(26.2) \qquad\qquad h_x(z) > F(z) - \varepsilon \qquad \text{for } z \in K.$$

Since $g_{xy_i}(x) = F(x)$, it is seen that $h_x(x) = F(x)$ and hence there is an open neighborhood $V(x)$ of x such that if z belongs to $K \cap V(x)$, then

$$(26.3) \qquad\qquad h_x(z) < F(z) + \varepsilon.$$

Use the compactness of K once more to obtain a finite number of neighborhoods $V(x_1), \ldots, V(x_m)$ and set $h = \inf\{h_{x_1}, \ldots, h_{x_m}\}$. Then h belongs to $\mathscr{L}$ and it follows from (26.2) that

$$h(z) > F(z) - \varepsilon \qquad \text{for } z \in K.$$

and from (26.3) that

$$h(z) < F(z) + \varepsilon \qquad \text{for } z \in K.$$

Combining these results, we have $|h(z) - F(z)| < \varepsilon$, $z \in K$, which yields the desired approximation.   Q.E.D.

The reader will have observed that the preceding result made no use of the Weierstrass Approximation Theorem. In the next result, we replace condition (a) above by three algebraic conditions on the set of functions. Here we make use of the classical Weierstrass Theorem 24.8 for the special case of the absolute value function φ defined for t in $\mathbf{R}$ by $\varphi(t) = |t|$, to conclude that φ can be approximated by polynomials on every compact set of real numbers.

26.2 STONE-WEIERSTRASS THEOREM. *Let K be a compact subset of $\mathbf{R}^p$ and let $\mathscr{A}$ be a collection of continuous functions on K to $\mathbf{R}$ with the properties:*

(a) *The constant function $e(x) = 1$, $x \in K$, belongs to $\mathscr{A}$.*

(b) *If f, g belong to $\mathscr{A}$, then $\alpha f + \beta g$ belongs to $\mathscr{A}$ for all α, β in $\mathbf{R}$.*

(c) *If f, g belong to $\mathscr{A}$, then fg belongs to $\mathscr{A}$.*

(d) *If $x \neq y$ are two points of K, there exists a function f in $\mathscr{A}$ such that $f(x) \neq f(y)$.*

Then any continuous function on K to $\mathbf{R}$ can be uniformly approximated on K by functions in $\mathscr{A}$.

PROOF. Let $a, b \in \mathbf{R}$ and $x \neq y$ belong to K. According to (d), there is a function f in $\mathscr{A}$ such that $f(x) \neq f(y)$. Since $e(x) = 1 = e(y)$, it follows that there are real numbers α, β such that

$$\alpha f(x) + \beta e(x) = a, \qquad \alpha f(y) + \beta e(y) = b.$$

Therefore, by (b) there exists a function $g \in \mathscr{A}$ such that $g(x) = a$ and $g(y) = b$.

Now let $\mathscr{L}$ be the collection of all continuous functions on K which can be uniformly approximated by functions in $\mathscr{A}$. Obviously $\mathscr{A} \subseteq \mathscr{L}$, so $\mathscr{L}$ has property (b) of the Stone Approximation Theorem 26.1. We shall now show that if $h \in \mathscr{L}$, then $|h| \in \mathscr{L}$. Since

$$\sup \{f, g\} = \tfrac{1}{2}(f + g + |f - g|),$$
$$\inf \{f, g\} = \tfrac{1}{2}(f + g - |f - g|),$$

this will imply that $\mathscr{L}$ has property 26.1(a) and hence that every continuous function on K to $\boldsymbol{R}$ belongs to $\mathscr{L}$.

Since h is continuous and K is compact, it follows that there exists an $M > 0$ such that $\|h\|_K \leq M$. Since $h \in \mathscr{L}$, there is a sequence (h_n) of functions in $\mathscr{A}$ which converge uniformly to h on K and we may suppose that $\|h_n\|_K \leq M + 1$ for all $n \in \boldsymbol{N}$. (Why?) If $\varepsilon > 0$ is given we now apply the Weierstrass Approximation Theorem 24.8 to the absolute value function on the interval $[-(M + 1), M + 1]$ to get a polynomial p_ε such that

$$\big| |t| - p_\varepsilon(t) \big| \leq \tfrac{1}{2}\varepsilon \qquad \text{for } |t| \leq M + 1.$$

It therefore follows that

$$\big| |h_n(x)| - p_\varepsilon(h_n(x)) \big| \leq \tfrac{1}{2}\varepsilon \qquad \text{for } x \in K.$$

Now $p_\varepsilon \circ h_n$ belongs to $\mathscr{A}$ because of our hypotheses (a), (b), and (c). Since

$$\big| |h(x)| - |h_n(x)| \big| \leq \|h - h_n\|_K$$

it follows that if n is sufficiently large, then we have

$$\big| |h(x)| - p_\varepsilon \circ h(x) \big| \leq \varepsilon \qquad \text{for } x \in K.$$

Since $\varepsilon > 0$ is arbitrary, we infer that $|h| \in \mathscr{L}$ and the result now follows from the preceding theorem. Q.E.D.

We now obtain, as a special case of the Stone-Weierstrass Theorem, a stronger form of Theorem 24.8. This result strengthens the latter result in two ways: (i) it permits the domain to be an arbitrary compact subset of $\boldsymbol{R}^p$ and not just a compact cell in $\boldsymbol{R}$, and (ii) it permits the range to lie in any space $\boldsymbol{R}^q$, and not just $\boldsymbol{R}$. To understand the statement, we recall that a function f with domain D in $\boldsymbol{R}^p$ and range in $\boldsymbol{R}^q$ can be regarded as q functions on D to $\boldsymbol{R}$ by the coordinate representation:

$$(26.4) \qquad f(x) = (f_1(x), \ldots, f_q(x)) \qquad \text{for } x \in D.$$

If each coordinate function f_i is a polynomial in the p coordinates $(x_1, \ldots, x_p)$, then we say that f is a *polynomial function*.

26.3 POLYNOMIAL APPROXIMATION THEOREM. *Let f be a continuous function whose domain K is a compact subset of $\mathbf{R}^p$ and whose range belongs to $\mathbf{R}^q$ and let $\varepsilon > 0$. Then there exists a polynomial function p on $\mathbf{R}^p$ to $\mathbf{R}^q$ such that $\|f(x) - p(x)\| < \varepsilon$ for $x \in K$.*

PROOF.   Represent f by its q coordinate functions, as in (26.4). Since f is continuous on K, each of the coordinate functions f_i is continuous on K to $\mathbf{R}$. The polynomial functions defined on $\mathbf{R}^p$ to $\mathbf{R}$ evidently satisfy the properties of the Stone-Weierstrass Theorem. Hence the coordinate function f_i can be uniformly approximated on K within $\varepsilon/\sqrt{q}$ by a polynomial function p_i. Letting p be defined by

$$p(x) = (p_1(x), \ldots, p_q(x)),$$

we obtain a polynomial function from $\mathbf{R}^p$ to $\mathbf{R}^q$ which yields the desired approximation on K to the given function f.   Q.E.D.

Extension of Continuous Functions

Sometimes it is desirable to extend the domain of a continuous function to a larger set without changing the values on the original domain. This can always be done in a trivial way by defining the function to be 0 outside the original domain, but in general this method of extension does not yield a continuous function. After some reflection, the reader should see that it is not always possible to obtain a continuous extension. For example, if $D = \{x \in \mathbf{R} : x \neq 0\}$ and if f is defined for $x \in D$ to be $f(x) = 1/x$, then it is not possible to extend f in such a way as to obtain a continuous function on all of $\mathbf{R}$. However, it is important to know that an extension is always possible when the domain is a closed set. Furthermore, it is not necessary to increase the bound of the function (if it is bounded).

Before we prove this extension theorem, we observe that if A and B are two disjoint closed subsets of $\mathbf{R}^p$, then there exists a continuous function φ defined on $\mathbf{R}^p$ with values in $\mathbf{R}$ such that

$$\varphi(x) = 0, \quad x \in A; \qquad \varphi(x) = 1, \quad x \in B; \qquad 0 \le \varphi(x) \le 1, \quad x \in \mathbf{R}^p.$$

In fact, if $d(x, A) = \inf \{\|x - y\| : y \in A\}$ and $d(x, B) = \inf \{\|x - y\| : y \in B\}$, then we can define φ for $x \in \mathbf{R}^p$ by the equation

$$\varphi(x) = \frac{d(x, A)}{d(x, A) + d(x, B)}.$$

26.4 TIETZE† EXTENSION THEOREM. *Let f be a bounded continuous function defined on a closed subset D of $\mathbf{R}^p$ and with values in $\mathbf{R}$. Then*

---

† HEINRICH TIETZE (1880–1964) was professor at Munich and contributed to topology, geometry, and algebra. This extension theorem goes back to 1914.

there exists a continuous function g on $\boldsymbol{R}^p$ to $\boldsymbol{R}$ such that $g(x) = f(x)$ for x in D and such that $\sup\{\|g(x)\| : x \in \boldsymbol{R}^p\} = \sup\{\|f(x)\| : x \in \boldsymbol{D}\}$.

PROOF. Let $M = \sup\{\|f(x)\| : x \in D\}$ and consider $A_1 = \{x \in D : f(x) \le -M/3\}$ and $B_1 = \{x \in D : f(x) \ge M/3\}$. From the continuity of f and the fact that D is closed, it follows from Theorem 22.1(c) that A_1 and B_1 are closed subsets of $\boldsymbol{R}^p$. According to the observation preceding the statement of the theorem, there is a continuous function φ_1 on $\boldsymbol{R}^p$ to $\boldsymbol{R}$ such that

$$\varphi_1(x) = -\tfrac{1}{3}M, \quad x \in A_1; \qquad \varphi_1(x) = \tfrac{1}{3}M, \quad x \in B_1;$$
$$-\tfrac{1}{3}M \le \varphi_1(x) \le \tfrac{1}{3}M, \quad x \in \boldsymbol{R}^p.$$

We now set $f_2 = f - \varphi_1$ and note that f_2 is continuous on D and that $\sup\{\|f_2(x)\| : x \in D\} \le \tfrac{2}{3}M$.

Proceeding, we define $A_2 = \{x \in D : f_2(x) \le -\tfrac{1}{3}\tfrac{2}{3}M\}$ and $B_2 = \{x \in D : f_2(x) \ge \tfrac{1}{3}\tfrac{2}{3}M\}$ and obtain a continuous function φ_2 on $\boldsymbol{R}^p$ to $\boldsymbol{R}$ such that

$$\varphi_2(x) = -\tfrac{1}{3}\tfrac{2}{3}M, \quad x \in A_2; \qquad \varphi_2(x) = -\tfrac{1}{3}\tfrac{2}{3}M, \quad x \in B_2;$$
$$-\tfrac{1}{3}\tfrac{2}{3}M \le \varphi_2(x) \le \tfrac{1}{3}\tfrac{2}{3}M, \quad x \in \boldsymbol{R}^p.$$

Having done this, we set $f_3 = f_2 - \varphi_2$ and note that $f_3 = f - \varphi_1 - \varphi_2$ is continuous on D and that $\sup\{\|f_3(x)\| : x \in D\} \le (\tfrac{2}{3})^2 M$.

By proceeding in this manner, we obtain a sequence (φ_n) of functions defined on $\boldsymbol{R}^p$ to $\boldsymbol{R}$ such that, for each n,

$$(26.5) \qquad |f(x) - [\varphi_1(x) + \varphi_2(x) + \cdots + \varphi_n(x)]| \le (\tfrac{2}{3})^n M,$$

for all x in D and such that

$$(26.6) \qquad |\varphi_n(x)| \le (\tfrac{1}{3})(\tfrac{2}{3})^{n-1} M \qquad \text{for } x \in \boldsymbol{R}^p.$$

Let g_n be defined on $\boldsymbol{R}^p$ to $\boldsymbol{R}$ by $g_n = \varphi_1 + \varphi_2 + \cdots + \varphi_n$, whence it follows that g_n is continuous. From inequality (26.6) we infer that if $m \ge n$ and $x \in \boldsymbol{R}^p$, then

$$|g_m(x) - g_n(x)| = |\varphi_{n+1}(x) + \cdots + \varphi_m(x)| \le (\tfrac{1}{3})(\tfrac{2}{3})^n M[1 + \tfrac{2}{3} + (\tfrac{2}{3})^2 + \cdots] \le (\tfrac{2}{3})^n M,$$

which proves that the sequence (g_n) converges uniformly on $\boldsymbol{R}^p$ to a function we shall denote by g. Since each g_n is continuous on $\boldsymbol{R}^p$, then Theorem 24.1 implies that g is continuous at every point of $\boldsymbol{R}^p$. Also, it is seen from the inequality (26.5) that $|f(x) - g_n(x)| \le (\tfrac{2}{3})^n M$ for $x \in D$. We conclude, therefore, that $f(x) = g(x)$ for x in D. Finally, inequality (26.6) implies that for any x in $\boldsymbol{R}^p$ we have

$$|g_n(x)| \le \tfrac{1}{3}M[1 + \tfrac{2}{3} + \cdots + (\tfrac{2}{3})^{n-1}] \le M,$$

which establishes the final statement of the theorem. Q.E.D.

26.5 COROLLARY. *Let f be a bounded continuous function defined on a closed subset D of $\mathbf{R}^p$ and with values in $\mathbf{R}^q$. Then there exists a continuous function g on $\mathbf{R}^p$ to $\mathbf{R}^q$ with $g(x) = f(x)$ for x in D and such that*

$$\sup \{\|g(x)\| : x \in \mathbf{R}^p\} \le \sqrt{q} \sup \{\|f(x)\| : x \in D\}.$$

PROOF. This result has just been proved for $q = 1$. In the general case, we note that f defines q continuous real-valued coordinate functions on D, say,

$$f(x) = (f_1(x), f_2(x), \ldots, f_q(x)).$$

Since each of the f_j, $1 \le j \le q$, has a continuous extension g_j on $\mathbf{R}^p$ to $\mathbf{R}$, we define g on $\mathbf{R}^p$ to $\mathbf{R}^q$ by $g(x) = (g_1(x), g_2(x), \ldots, g_q(x))$. The function g is seen to have the required properties.   Q.E.D.

Equicontinuity

We have made frequent use of the Bolzano-Weierstrass Theorem 10.6 for sets (which asserts that every bounded infinite subset of $\mathbf{R}^p$ has a cluster point) and the corresponding Theorem 16.4 for sequences (which asserts that every bounded sequence in $\mathbf{R}^p$ has a convergent subsequence). We now present a theorem which is entirely analogous to the Bolzano-Weierstrass Theorem except that it pertains to *sets of continuous functions* and not sets of points. For the sake of brevity and simplicity, we shall present here only the sequential form of this theorem.

In what follows we let K be a fixed compact subset of $\mathbf{R}^p$, and we shall be concerned with functions which are continuous on K and have their range in $\mathbf{R}^q$. In view of Theorem 22.5, each such function is bounded, and hence $C_{pq}(K) = BC_{pq}(K)$. We say that a set $\mathcal{F}$ in $C_{pq}(K)$ is **bounded** (or **uniformly bounded**) on K if there exists a constant M such that $\|f\|_K \le M$, for all f in $\mathcal{F}$. It is clear that any finite set $\mathcal{F}$ of such functions is bounded; for if $\mathcal{F} = \{f_1, f_2, \ldots, f_n\}$, then we can set

$$M = \sup \{\|f_1\|_K, \|f_2\|_K, \ldots, \|f_n\|_K\}.$$

In general, an infinite set of continuous functions on K to $\mathbf{R}^q$ will not be bounded. However, a uniformly convergent sequence of continuous functions is bounded. (Cf. Exercise 26.M).

If f is a continuous function on the compact set K of $\mathbf{R}^p$, then Theorem 23.3 implies that it is uniformly continuous. Hence, if $\varepsilon > 0$ there exists $\delta(\varepsilon) > 0$, such that if x, y belong to K and $\|x - y\| < \delta(\varepsilon)$, then $\|f(x) - f(y)\| < \varepsilon$. Of course, the value of δ may depend on the function f as well as on ε and so we often write $\delta(\varepsilon, f)$. (When we are dealing with more than one function it is well to indicate this dependence explicitly.)

We notice that if $\mathscr{F} = \{f_1, \ldots, f_n\}$ is a finite set in $C_{pq}(K)$, then, by setting

$$\delta(\varepsilon, \mathscr{F}) = \inf\{\delta(\varepsilon, f_1), \ldots, \delta(\varepsilon, f_n)\},$$

we obtain a δ which "works" for all the functions in this finite set.

26.6 DEFINITION. A set $\mathscr{F}$ of functions on K to $\mathbf{R}^q$ is said to be **uniformly equicontinuous** on K if, for each real number $\varepsilon > 0$ there is a number $\delta(\varepsilon) > 0$ such that if x, y belong to K and $\|x - y\| < \delta(\varepsilon)$ and f is a function in $\mathscr{F}$, then $\|f(x) - f(y)\| < \varepsilon$.

It has been seen that a finite set of continuous functions on K is equicontinuous. It is also true that a sequence of continuous functions which converges uniformly on K is also equicontinuous. (Cf. Exercise 26.N.)

It follows that, in order for a sequence in $C_{pq}(K)$ to be uniformly convergent on K, it is necessary that the sequence be bounded and uniformly equicontinuous on K. We shall now show that these two properties are necessary and sufficient for a set $\mathscr{F}$ in $C_{pq}(K)$ to have the property that every sequence of functions from $\mathscr{F}$ has a subsequence which converges uniformly on K. This may be regarded as a generalization of the Bolzano-Weierstrass Theorem to sets of continuous functions and plays an important role in the theory of differential and integral equations.

26.7 ARZELÀ-ASCOLI† THEOREM. *Let K be a compact subset of $\mathbf{R}^p$ and let $\mathscr{F}$ be a collection of functions which are continuous on K and have values in $\mathbf{R}^q$. The following properties are equivalent:*

(a) *The family $\mathscr{F}$ is bounded and uniformly equicontinuous on K.*

(b) *Every sequence from $\mathscr{F}$ has a subsequence which is uniformly convergent on K.*

PROOF. First we shall show that if condition (a) is false, then so is condition (b). If $\mathscr{F}$ is not bounded, then there exists a sequence (f_n) in $\mathscr{F}$ such that $\|f_n\|_K \geq n$ for $n \in \mathbf{N}$. But then no subsequence of (f_n) can be uniformly convergent. Also if the set $\mathscr{F}$ is not uniformly equicontinuous, then for some $\varepsilon_0 > 0$ there exists (why?) a sequence (f_n) in $\mathscr{F}$ and sequences (x_n) and (y_n) in K with $\|x_n - y_n\| < 1/n$ but such that $\|f_n(x_n) - f_n(y_n)\| > \varepsilon_0$. But then no subsequence of (f_n) can be uniformly convergent on K.

We now show that, if the set $\mathscr{F}$ satisfies (a), then given any sequence (f_n) in $\mathscr{F}$ there is a subsequence which converges uniformly on K. To do this we notice that it follows from Exercise 10.H that there exists a countable

† CESARE ARZELÀ (1847–1912) was a professor at Bologna. He gave necessary and sufficient conditions for the limit of a sequence of continuous functions on a closed interval to be continuous, and he studied related topics.

GIULIO ASCOLI (1843–1896), a professor at Milan, formulated the definition of equicontinuity in a geometrical setting. He also made contributions to Fourier series.

set C in K such that if $y \in K$ and $\varepsilon > 0$, then there exists an element x in C such that $\|x - y\| < \varepsilon$. If $C = \{x_1, x_2, \ldots\}$, then the sequence $(f_n(x_1))$ is bounded in $\mathbf{R}^q$. It follows from the Bolzano-Weierstrass Theorem 16.4 that there is a subsequence

$$(f_1^1(x_1), f_2^1(x_1), \ldots, f_n^1(x_1), \ldots)$$

of $(f_n(x_1))$ which is convergent. Next we note that the sequence $(f_k^1(x_2) : k \in \mathbf{N})$ is bounded in $\mathbf{R}^q$; hence it has a subsequence

$$(f_1^2(x_2), f_2^2(x_2), \ldots, f_n^2(x_2), \ldots)$$

which is convergent. Again, the sequence $(f_n^2(x_3) : n \in \mathbf{N})$ is bounded in $\mathbf{R}^q$, so some subsequence

$$(f_1^3(x_3), f_2^3(x_3), \ldots, f_n^3(x_3), \ldots)$$

is convergent. We proceed in this way and then set $g_n = f_n^n$ so that g_n is the nth function in the nth subsequence. It is clear from the construction that the sequence (g_n) converges at each point of C.

We shall now prove that the sequence (g_n) converges at each point of K and that the convergence is uniform. To do this, let $\varepsilon > 0$ and let $\delta(\varepsilon)$ be as in Definition 26.6. Let $C_1 = \{y_1, \ldots, y_k\}$ be a finite subset of C such that every point in K is within $\delta(\varepsilon)$ of some point in C_1. Since the sequences

$$(g_n(y_1)), (g_n(y_2)), \ldots, (g_n(y_k))$$

converge, there exists a natural number M such that if $m, n \geq M$, then

$$\|g_m(y_i) - g_n(y_i)\| < \varepsilon \qquad \text{for } i = 1, 2, \ldots, k.$$

Given $x \in K$, there exists a $y_j \in C_1$ such that $\|x - y_j\| < \delta(\varepsilon)$. Hence, by the uniform equicontinuity, we have $\|g_n(x) - g_n(y_j)\| < \varepsilon$ for all $n \in \mathbf{N}$; in particular, this inequality holds for $n \geq M$. Therefore, we have

$$\|g_n(x) - g_m(x)\| \leq \|g_n(x) - g_n(y_j)\| + \|g_n(y_j) - g_m(y_j)\|$$
$$+ \|g_m(y_j) - g_m(x)\| < \varepsilon + \varepsilon + \varepsilon = 3\varepsilon,$$

provided $m, n \geq M$. This shows that

$$\|g_n - g_m\|_K \leq 3\varepsilon \qquad \text{for } m, n \geq M,$$

so the uniform convergence of the sequence (g_n) on K follows from the Cauchy Criterion for uniform convergence, given in 17.11. Q.E.D.

In the proof of this result, we constructed a sequence of subsequences of functions and then selected the "diagonal" sequence (g_n), where $g_n = f_n^n$. Such a construction is often called a "diagonal process" or "Cantor's diagonal method" and is frequently useful. The reader should recall that a similar type of argument was used in Section 3 to prove that the real numbers do not form a countable set.

Exercises

26.A. Show that condition (a) of Theorem 26.1 is equivalent to the condition:
(a') If f belongs to $\mathscr{L}$, then $|f|$ belongs to $\mathscr{L}$.

26.B. Show that every continuous real-valued function on the interval $[0, \pi]$ is the uniform limit of a sequence of "polynomials in cos x" (that is, of functions (P_n), where $P_n(x) = p_n(\cos x)$ for some polynomial p_n).

26.C. Show that every continuous real-valued function on $[0, \pi]$ is the uniform limit of a sequence of functions of the form

$$x \mapsto a_0 + a_1 \cos x + a_2 \cos 2x + \cdots + a_n \cos nx.$$

26.D. Explain why the result in Exercise 26.B fails if cos kx is replaced by sin kx, $k \in \mathbf{N}$.

26.E. Use Exercise 26.C to show that every continuous real-valued function f on $[0, \pi]$ with $f(0) = f(\pi)$ is the uniform limit of a sequence of functions of the form

$$x \mapsto b_0 + b_1 \sin x + b_2 \sin 2x + \cdots + b_n \sin nx.$$

26.F. Use Exercises 26.C and 26.E to show that every continuous real-valued function f on $[-\pi, \pi]$ with $f(-\pi) = f(\pi)$ is the uniform limit of a sequence of functions of the form

$$x \mapsto a_0 + a_1 \cos x + b_1 \sin x + \cdots + a_n \cos nx + b_n \sin nx.$$

[Hint: split f into the sum $f = f_e + f_o$ of an even function $f_e(x) = \frac{1}{2}(f(x) + f(-x))$ and an odd function $f_o(x) = \frac{1}{2}(f(x) - f(-x))$.]

26.G. Give a proof of the preceding exercise based upon Theorem 26.3 applied to the unit circle $T = \{(x, y) \in \mathbf{R}^2 : x^2 + y^2 = 1\}$ and the observations that there is a one-one correspondence between continuous functions on T to $\mathbf{R}$ and continuous functions on $[-\pi, \pi]$ to $\mathbf{R}$ which satisfy $f(-\pi) = f(\pi)$.

26.H. Let $J \subseteq \mathbf{R}$ be a compact interval and let $\mathscr{A}$ be a collection of continuous functions on $J \to \mathbf{R}$ which satisfy the properties of the Stone-Weierstrass Theorem 26.2. Show that any continuous function on $J \times J$ (in $\mathbf{R}^2$) to $\mathbf{R}$ can be uniformly approximated by functions of the form

$$f_1(x)g_1(y) + \cdots + f_n(x)g_n(y).$$

where f_i, g_i belong to $\mathscr{A}$.

26.I. Show that Tietze's Theorem 26.4 may fail if the domain is not closed.

26.J. Use Tietze's Theorem 26.4 to show that if $D \subseteq \mathbf{R}^p$ is closed and if f is an unbounded continuous function on $D \to \mathbf{R}$, then there exists a continuous extension of f to all of $\mathbf{R}^p$. [Hint: consider the composition $\phi \circ f$, where $\phi(x) = $ Arc tan x or $\phi(x) = x/(1 + x)$.]

26.K. Let $\mathscr{F}$ be a collection of functions on $D \subseteq \mathbf{R}^p$ to $\mathbf{R}^q$. Consider the property at the point $c \in D$: if $\varepsilon > 0$ there is a $\delta(c, \varepsilon) > 0$ such that if $x \in D$ and $\|x - c\| < \delta(c, \varepsilon)$ then $\|f(x) - f(c)\| < \varepsilon$ for all $f \in \mathscr{F}$. Show that $\mathscr{F}$ has this property at $c \in D$ if and only if for each sequence (x_n) in D with $c = \lim(x_n)$, then $f(c) = \lim(f(x_n))$ uniformly for $f \in \mathscr{F}$. (Sometimes we say that $\mathscr{F}$ is **equicontinuous at** $c \in D$ when this property is satisfied).

26.L. Let $\mathscr{F}$ be as in Exercise 26.K. If D is compact and the property in Exercise 26.K is satisfied for all $c \in D$, show that $\mathscr{F}$ is uniformly equicontinuous in the sense of Definition 26.6.

26.M. If $K \subseteq \mathbf{R}^p$ is compact and (f_n) is a sequence of continuous functions on K to $\mathbf{R}^q$ which is uniformly convergent on K, show that the family $\{f_n\}$ is bounded on K (in the sense that there exists $M > 0$ such that $\|f_n(x)\| \leq M$ for all $x \in K$, $n \in N$ (or $\|f_n\|_K \leq M$ for $n \in N$).

26.N. If $K \subseteq \mathbf{R}^p$ is compact and (f_n) is a sequence of continuous functions on K to $\mathbf{R}^q$ which is uniformly convergent on K, show that the family $\{f_n\}$ is uniformly equicontinuous on K in the sense of Definition 26.6.

26.O. Let $\mathscr{F}$ be a bounded and uniformly equicontinuous collection of functions on $D \subseteq \mathbf{R}^p$ to $\mathbf{R}$ and let f^* be defined on $D \to \mathbf{R}$ by

$$f^*(x) = \sup \{f(x) : f \in \mathscr{F}\}.$$

Show that f^* is continuous on D to $\mathbf{R}$.

26.P. Show that the conclusion of the preceding exercise may fail if it is not assumed that $\mathscr{F}$ is uniformly equicontinuous.

26.Q. Consider the following sequences of functions which show that the Arzelà-Ascoli Theorem 26.7 may fail if the various hypotheses are dropped.

(a) $f_n(x) = x + n \qquad$ for $x \in [0, 1]$;

(b) $f_n(x) = x^n \qquad$ for $x \in [0, 1]$;

(c) $f_n(x) = \dfrac{1}{1 + (x - n)^2} \qquad$ for $x \in [0, +\infty)$.

26.R. Let (f_n) be a sequence of continuous functions on $\mathbf{R}$ to $\mathbf{R}^q$ which converges at each point of the set $\mathbf{Q}$ of rationals. If the set $\{f_n\}$ is uniformly equicontinuous on $\mathbf{R}$, show that the sequence converges at every point of $\mathbf{R}$ and that the convergence is uniform on every compact set of $\mathbf{R}$, but not necessarily uniform on $\mathbf{R}$.

V
FUNCTIONS OF
ONE VARIABLE

We shall now commence the study of the differentiation and integration of functions. In doing so it will be convenient to treat first the case of functions of *one* variable; in Chapters VII and VIII we shall return to the study of functions of several variables. It will be seen, in comparing these chapters, that the case of functions of several variables is quite similar in outline to what we shall do here, but that certain complications arise. Furthermore, since the general theory makes use of results from the case of one variable, it is convenient to have made a study of this case previously.

In Sections 27 and 28 we introduce the derivative of a function defined on a real interval and establish the important Mean Value Theorem and some of its corollaries. In Section 29 we shall introduce the definition of the Riemann (and the Riemann-Stieltjes) integral of bounded functions on an interval $[a, b]$. The basic properties of the integral are established in this section and in Sections 30 and 31. In the last two sections, we discuss "improper" and infinite integrals. Although the results of these sections are used very little in the following portions of this book, they are important for many applications.

Section 27   The Mean Value Theorem

Since the reader is assumed to be already familiar with the connection between the derivative of a function on R to R and the slope of its graph, and with the notion of instantaneous rate of change, we shall focus our attention entirely on the mathematical aspects of the derivative and not go into its applications to physics, economics, *et cetera*. In this and the next section we shall consider a function with domain D and range contained in

R. Although we are primarily interested in the derivative at an interior point, we shall define the derivative somewhat more generally so that an end point of an interval, for example, can be considered. However, we do require that the point at which the derivative is being defined is a cluster point of D and belongs to D.

27.1 DEFINITION. If c is a cluster point of D and belongs to D, we say that a real number L is the **derivative of f at** c if for every number $\varepsilon > 0$ there is a number $\delta(\varepsilon) > 0$ such that if x belongs to D and if $0 < |x - c| < \delta(\varepsilon)$, then

$$(27.1) \qquad \left| \frac{f(x) - f(c)}{x - c} - L \right| < \varepsilon.$$

In this case we write $f'(c)$ for L.

Alternatively, we could define $f'(c)$ as the limit

$$\lim_{x \to c} \frac{f(x) - f(c)}{x - c} \qquad (x \in D, x \neq c).$$

It is to be noted that if c is an interior point of D, then in (27.1) we consider the points x both to the left and the right of the point c. On the other hand, if D is an interval and c is the left end point of D, then in relation (27.1) we can only take x to the right of c.

Whenever the derivative of f at c exists, we denote its value by $f'(c)$. In this way we obtain a function f' whose domain is a subset of the domain of f. We now show that continuity of f at c is a necessary condition for the existence of the derivative at c.

27.2 LEMMA. *If f has a derivative at c, then f is continuous there.*

PROOF. Let $\varepsilon = 1$ and take $\delta = \delta(1)$ such that

$$\left| \frac{f(x) - f(c)}{x - c} - f'(c) \right| < 1,$$

for all $x \in D$ satisfying $0 < |x - c| < \delta$. From the Triangle Inequality, we infer that for these values of x we have

$$|f(x) - f(c)| \le |x - c| \{|f'(c)| + 1\}.$$

The left side of this expression can be made less than ε if we take x in D with $|x - c| < \inf \{\delta, \varepsilon/(|f'(c)| + 1)\}$. Q.E.D.

It is easily seen that continuity at c is not a sufficient condition for the derivative to exist at c. For example, if $D = R$ and $f(x) = |x|$, then f is continuous at every point of R but has a derivative at a point c if and only if $c \neq 0$. By taking simple algebraic combinations, it is easy to construct continuous functions which do not have a derivative at a finite or even a countable number of points. In 1872,

Weierstrass shocked the mathematical world by giving an example of a *function which is continuous at every point but whose derivative does not exist anywhere.* (In fact, the function defined by the series

$$f(x) = \sum_{n=0}^{\infty} \frac{1}{2^n} \cos(3^n x),$$

can be proved to have this property. We shall not go through the details, but refer the reader to the books of Titchmarsh and Boas for further details and references.)

27.3 LEMMA. (a) *If f has a derivative at c and $f'(c) > 0$, there exists a number $\delta > 0$ such that if $x \in D$ and $c < x < c + \delta$, then $f(c) < f(x)$.*

(b) *If $f'(c) < 0$, there exists a number $\delta > 0$ such that if $x \in D$ and $c - \delta < x < c$, then $f(c) < f(x)$.*

PROOF. (a) Let ε_0 be such that $0 < \varepsilon_0 < f'(c)$ and let $\delta = \delta(\varepsilon_0)$ correspond to ε_0 as in Definition 27.1. If $x \in D$ and $c < x < c + \delta$, then we have

$$-\varepsilon_0 < \frac{f(x) - f(c)}{x - c} - f'(c).$$

Since $x - c > 0$, this relation implies that

$$0 < (f'(c) - \varepsilon_0)(x - c) < f(x) - f(c),$$

which proves the assertion in (a). The proof of (b) is similar. Q.E.D.

We recall that the function f is said to have a **relative maximum** at a point c in D if there exists a $\delta > 0$ such that $f(x) \le f(c)$ when $x \in D$ satisfies $|x - c| < \delta$. A similar definition applies to the term **relative mimimum.** The next result provides the theoretical justification for the familiar process of finding points at which f has relative maxima and minima by examining the zeros of the derivative. It is to be noted that this procedure applies only to interior points of the interval. In fact, if $f(x) = x$ on $D = [0, 1]$, then the end point $x = 0$ yields the unique relative minimum and the end point $x = 1$ yields the unique relative maximum of f, but neither is a root of the derivative. For simplicity, we shall state this result only for relative maxima, leaving the formulation of the corresponding result for relative minima to the reader.

27.4 INTERIOR MAXIMUM THEOREM. *Let c be an interior point of D at which f has a relative maximum. If the derivative of f at c exists, then it must be equal to zero.*

PROOF. If $f'(c) > 0$, then from Lemma 27.3(a) there is a $\delta > 0$ such that if $c < x < c + \delta$ and $x \in D$, then $f(c) < f(x)$. This contradicts the assumption that f has a relative maximum at c. If $f'(c) < 0$, we use Lemma 27.3(b). Q.E.D.

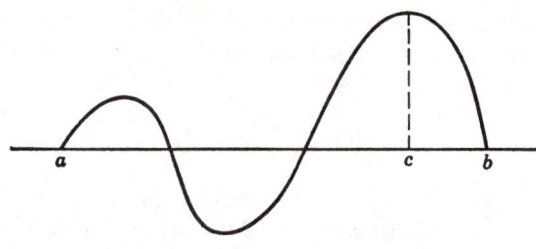

Figure 27.1

27.5 Rolle's Theorem.† *Suppose that f is continuous on a closed interval J = [a, b], that the derivative f′ exists in the open interval (a, b), and that f(a) = f(b) = 0. Then there exists a point c in (a, b) such that f′(c) = 0.*

PROOF. If f vanishes identically on J, we can take $c = (a+b)/2$. Hence we suppose that f does not vanish identically; replacing f by $-f$, if necessary, we may suppose that f assumes some positive values. By the Maximum Value Theorem 22.7, the function f attains the value $\sup \{f(x) : x \in J\}$ at some point c of J. Since $f(a) = f(b) = 0$, the point c satisfies $a < c < b$. By hypothesis $f′(c)$ exists and, since f has a relative maximum point at c, the Interior Maximum Theorem implies that $f′(c) = 0$.   Q.E.D.

As a consequence of Rolle's Theorem, we obtain the very fundamental Mean Value Theorem.

27.6 Mean Value Theorem. *Suppose that f is continuous on a closed interval J = [a, b] and has a derivative in the open interval (a, b). Then there exists a point c in (a, b) such that*

$$f(b) - f(a) = f′(c)(b - a).$$

PROOF. Consider the function φ defined on J by

$$\varphi(x) = f(x) - f(a) - \frac{f(b) - f(a)}{b - a} (x - a).$$

[It is easily seen that φ is the difference of f and the function whose graph consist of the line segment passing through the points $(a, f(a))$ and $(b, f(b))$; see Figure 27.2.] It follows from the hypotheses that φ is continuous on $J = [a, b]$ and it is easily checked that φ has a derivative in (a, b). Furthermore, we have $\varphi(a) = \varphi(b) = 0$. Applying Rolle's Theorem, there

† This theorem is generally attributed to Michel Rolle (1652–1719), a member of the French Academy, who made contributions to analytic geometry and the early work leading to calculus.

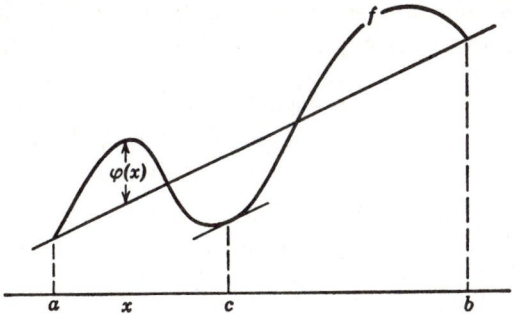

Figure 27.2. The mean value theorem.

exists a point c inside J such that

$$0 = \varphi'(c) = f'(c) - \frac{f(b) - f(a)}{b - a}$$

from which the result follows. Q.E.D.

27.7 COROLLARY. *If f has a derivative on $J = [a, b]$, then there exists a point c in (a, b) such that*

$$f(b) - f(a) = f'(c)(b - a).$$

Sometimes it is convenient to have a more general version of the Mean Value Theorem involving two functions.

27.8 CAUCHY MEAN VALUE THEOREM. *Let f, g be continuous on $J = [a, b]$ and have derivatives inside (a, b). Then there exists a point c in (a, b) such that*

$$f'(c)[g(b) - g(a)] = g'(c)[f(b) - f(a)].$$

PROOF. When $g(b) = g(a)$ the result is immediate if we take c so that $g'(c) = 0$. If $g(b) \neq g(a)$, consider the function φ defined on J by

$$\varphi(x) = f(x) - f(a) - \frac{f(b) - f(a)}{g(b) - g(a)} [g(x) - g(a)].$$

Applying Rolle's Theorem to φ, we obtain the desired result. Q.E.D.

Although the derivative of a function *need not* be continuous, there is an elementary but striking theorem due to Darboux† asserting that the derivative f' attains every value between $f'(a)$ and $f'(b)$ on the interval $[a, b]$. (See Exercise 27.H.)

† GASTON DARBOUX (1842–1917) was a student of Hermite and a professor at the Collège de France. Although he is known primarily as a geometer, he made important contributions to analysis as well.

It is easy to remember the statement of the Mean Value Theorem by drawing appropriate diagrams. While this should not be discouraged, it tends to suggest that its importance is geometrical in nature, which is quite misleading. In fact the Mean Value Theorem is a wolf in sheep's clothing and is *the* Fundamental Theorem of the Differential Calculus. We close this section with a few elementary consequences of this result. More will be given in the next section, and still others will appear later.

27.9   THEOREM. *Suppose that f is continuous on $J = [a, b]$ and that its derivative exists in (a, b).*

(i) *If $f'(x) = 0$ for $a < x < b$, then f is constant on J.*

(ii) *If $f'(x) = g'(x)$ for $a < x < b$, then f and g differ on J by a constant.*

(iii) *If $f'(x) \geq 0$ for $a < x < b$ and if $x_1 \leq x_2$ belong to J, then $f(x_1) \leq f(x_2)$.*

(iv) *If $f'(x) > 0$ for $a < x < b$ and if $x_1 < x_2$ belong to J, then $f(x_1) < f(x_2)$.*

(v) *If $f'(x) \geq 0$ for $a < x < a + \delta$, then a is a relative minimum point of f.*

(vi) *If $f'(x) \geq 0$ for $b - \delta < x < b$, then b is a relative maximum point of f.*

(vii) *If $|f'(x)| \leq M$ for $a < x < b$, then f satisfies the Lipschitz condition:*

$$|f(x_1) - f(x_2)| \leq M |x_1 - x_2| \qquad \text{for } x_1, x_2 \text{ in } J.$$

We leave the proof to the reader.

Exercises

27.A. Using the definition, calculate the derivative (when it exists) of the functions given by the expressions:

(a) $f(x) = x^2$ for $x \in \mathbf{R}$,

(b) $g(x) = x^n$ for $x \in \mathbf{R}$,

(c) $h(x) = \sqrt{x}$ for $x \geq 0$,

(d) $F(x) = 1/x$ for $x \neq 0$,

(e) $G(x) = |x|$ for $x \in \mathbf{R}$,

(f) $H(x) = 1/x^2$ for $x \neq 0$.

27.B. If f and g are real-valued functions defined on an interval J, and if they are differentiable at a point c, show that their product h, defined by $h(x) = f(x)g(x)$, for $x \in J$, is differentiable at c and

$$h'(c) = f'(c)g(c) + f(c)g'(c).$$

27.C. Show that the function defined for $x \neq 0$ by

$$f(x) = \sin (1/x)$$

is differentiable at each non-zero real number. Show that its derivative is not bounded on a neighborhood of $x = 0$. (You may make use of trigonometric identities, the continuity of the sine and cosine functions, and the elementary limiting relation $(\sin u)/u \to 1$ as $u \to 0$.)

27.D. Show that the function defined by

$$g(x) = x^2 \sin(1/x), \qquad x \neq 0,$$
$$= 0, \qquad x = 0,$$

is differentiable for all real numbers, but that g' is not continuous at $x = 0$.

27.E. The function $h: R \to R$ defined by $h(x) = x^2$ for $x \in Q$ and $h(x) = 0$ for $x \notin Q$ is continuous at exactly one point. Is it differentiable there?

27.F. Let $c \in D$ be a cluster point of D and let $f: D \to R$. Show that $f'(c)$ exists if and only if for every sequence (x_n) in D with $x_n \neq c$ for $n \in N$ such that $\lim (x_n) = c$, the limit of the sequence

$$\left(\frac{f(x_n) - f(c)}{x_n - c} \right)$$

exists. In this case the limit of all such sequences are equal to $f'(c)$.

27.G. If $f: D \to R$ is differentiable at $c \in D$ and if $c + 1/n \in D$ for all $n \in N$, show that

$$f'(c) = \lim \left(n\{f(c + 1/n) - f(c)\} \right).$$

However show that the existence of the limit of this sequence does not imply the existence of the derivative.

27.H. (Darboux) If f is differentiable on $[a, b]$, if $f'(a) = A$, $f'(b) = B$, and if C lies between A and B, then there exists a point c in (a, b) for which $f'(c) = C$. (Hint: consider the lower bound of the function $g(x) = f(x) - C(x - a)$.)

27.I. If $g(x) = 0$ for $x < 0$ and $g(x) = 1$ for $x \geq 0$, prove that there does not exist a function $f: R \to R$ such that $f'(x) = g(x)$ for all $x \in R$.

27.J. Give an example of a continuous function with a unique relative maximum point but such that the derivative does not exist at this point.

27.K. Give an example of a uniformly continuous function which is differentiable on $(0, 1)$ but is such that its derivative is not bounded on $(0, 1)$.

27.L. Let $f: [a, b] \to R$ be differentiable at $c \in [a, b]$. Show that for every $\varepsilon > 0$, there is a $\delta(\varepsilon) > 0$ such that if $0 < |x - y| < \delta(\varepsilon)$ and $a \leq x \leq c \leq y \leq b$, then

$$\left| \frac{f(x) - f(y)}{x - y} - f'(c) \right| < \varepsilon.$$

27.M. Let $f: [a, b] \to R$ be differentiable on $[a, b]$. Show that f' is continuous on $[a, b]$ if and only if for every $\varepsilon > 0$ there is a $\delta(\varepsilon) > 0$ such that if $0 < |x - y| < \delta(\varepsilon)$, $x, y \in [a, b]$, then

$$\left| \frac{f(x) - f(y)}{x - y} - f'(x) \right| < \varepsilon.$$

27.N. Let $f: [a, b] \to R$ be continuous on $[a, b]$ and differentiable in (a, b). If $\lim_a f'(x) = A$ show that $f'(a)$ exists and equals A.

27.O. If $f: R \to R$ and $f'(a)$ exists, show that

$$f'(a) = \lim_{h \to 0} \frac{f(a + h) - f(a - h)}{2h}.$$

However, give an example to show that the existence of this limit does not imply the existence of the derivative.

27.P. A function $f: R \to R$ is said to be **even** if $f(-x) = f(x)$ for all $x \in R$, and to be **odd** if $f(-x) = -f(x)$ for all $x \in R$. If f is differentiable on R and even (respectively, odd), show that f' is odd (respectively, even).

27.Q. Let $f:(a, b) \to R$ and $c \in (a, b)$. We put $f(c+) = \lim_{x \to c} f(x)$ (the right-hand limit of f at c). If the right-hand limit

$$A_r = \lim_{\substack{x \to c \\ x > c}} \frac{f(x) - f(c+)}{x - c}$$

exists in R, we say that f has a **right-hand derivative** at c and denote A_r by $f'_+(c)$. Similarly for left-hand derivatives

Show that if f is continuous at c, then $f'(c)$ exists if and only if $f'_+(c)$ and $f'_-(c)$ exist and are equal. Show that we can have $g'_-(c) = g'_+(c)$ without $g'(c)$ existing.

27.R. Let I and J be intervals in R and let $f: I \to R$ and $g: J \to R$ be such that g is differentiable at a point $b \in J$ and f is differentiable at an interior point $a = g(b)$ of I. Show that the composition $h = f \circ g$ defined for $\{x \in J : g(x) \in I\}$ is differentiable at b and that $h'(b) = f'(a)g'(b)$. [Hint: let H be defined on $D(h)$ by

$$H(x) = \frac{f(g(x)) - f(g(b))}{g(x) - g(b)} \qquad \text{if} \qquad g(x) \neq g(c),$$

$$= f'(a) \qquad \text{if} \qquad g(x) = g(c).$$

Show that $\lim_b H(x) = f'(a)$. Then use the fact that $(g(x) - g(b))H(x) = f(g(x)) - f(g(b))$ for all x in $D(h)$.]

27.S. Let $f: [0, +\infty) \to R$ be differentiable on $(0, +\infty)$.

(a) If $f'(x) \to b \in R$ as $x \to +\infty$, show that for any $h > 0$ we have

$$\lim_{x \to +\infty} \frac{f(x + h) - f(x)}{h} = b.$$

(b) If $f(x) \to a \in R$ and $f'(x) \to b \in R$ as $x \to +\infty$, then $b = 0$.

(c) If $f'(x) \to b \in R$ as $x \to +\infty$, then $f(x)/x \to b$ as $x \to +\infty$.

27.T. Let $f:[a, b] \to R$ be differentiable with $0 < m \le f'(x) \le M$ for $x \in [a, b]$ and let $f(a) < 0 < f(b)$. Given $x_1 \in [a, b]$, define the sequence (x_n) by

$$x_{n+1} = x_n - \frac{1}{M} f(x_n), \qquad n \in N.$$

Prove that this sequence is well-defined and converges to the unique root $\bar{x}$ of the equation $f(x) = 0$ in $[a, b]$ and that

$$|x_{n+1} - \bar{x}| \le \frac{|f(x_1)|}{m} \left(1 - \frac{m}{M}\right)^n$$

for $n \in N$. (Hint: let $\varphi: [a, b] \to R$ be defined by $\varphi(x) = x - f(x)/M$. Show that φ is increasing and a contraction (see 23.4) with constant $1 - m/M$.)

27.U. Let $f: R \to R$ have a continuous derivative and be such that $f(a) = b$ and $f'(a) \neq 0$. Let $\delta > 0$ be such that if $|x - a| \le \delta$ then $|f'(x) - f'(a)| \le \frac{1}{2}|f'(a)|$, and let

$\eta = \frac{1}{2}\delta \, |f'(a)|$. Prove that if $|\bar{y} - b| \leq \eta$, then the sequence (x_n) defined by $x_1 = a$ and

$$x_{n+1} = x_n - \frac{f(x_n) - \bar{y}}{f'(a)}, \qquad n \in \mathbf{N}$$

converges to the unique point $\bar{x}$ in $[a - \delta, a + \delta]$ such that $f(\bar{x}) = \bar{y}$. (Hint: show that the function defined by $\varphi(x) = x - (f(x) - \bar{y})/f'(a)$ is a contraction with constant $\frac{1}{2}$ on the interval $[a - \delta, a + \delta]$.)

Section 28 Further Applications of the Mean Value Theorem

It is hardly possible to overemphasize the importance of the Mean Value Theorem, for it plays a crucial role in many theoretical considerations. At the same time it is very useful in many practical matters. In 27.9 we indicated some immediate consequences of the Mean Value Theorem which are often useful. We shall now suggest some other areas in which it can be applied; in doing so we shall draw more freely than before on the past experience of the reader and his knowledge concerning the derivatives of certain well-known functions.

28.1 APPLICATION. Rolle's Theorem can be used for the location of roots of a function. For, if a function g can be identified as the derivative of a function f, then between any two roots of f there is at least one root of g. For example, let $g(x) = \cos x$; then g is known to be the derivative of $f(x) = \sin x$. Hence, between any two roots of sin x there is at least one root of cos x. On the other hand, $g'(x) = -\sin x = -f(x)$, so another application of Rolle's Theorem tells us that between any two roots of cos x there is at least one root of sin x. Therefore, we conclude that *the roots of* sin x *and* cos x *interlace each other*. This conclusion is probably not news to the reader; however, the same type of argument can be applied to *Bessel*† *functions* J_n of order $n = 0, 1, 2, \ldots$ by using the relations

$$[x^n J_n(x)]' = x^n J_{n-1}(x), [x^{-n} J_n(x)]' = -x^{-n} J_{n+1}(x) \qquad \text{for } x > 0.$$

The details of this argument should be supplied by the reader.

28.2 APPLICATION. We can apply the Mean Value Theorem for approximate calculations and to obtain error estimates. For example, suppose it is desired to evaluate $\sqrt{105}$. We employ the Mean Value Theorem with $f(x) = \sqrt{x}$, $a = 100$, $b = 105$ to obtain

$$\sqrt{105} - \sqrt{100} = \frac{5}{2\sqrt{c}},$$

† FRIEDRICH WILHELM BESSEL (1784–1846) was an astronomer and mathematician. A close friend of Gauss, he is best known for the differential equation which bears his name.

for some number c with $100 < c < 105$. Since $10 < \sqrt{c} < \sqrt{105} < \sqrt{121} = 11$, we can assert that

$$\frac{5}{2(11)} < \sqrt{105} - 10 < \frac{5}{2(10)},$$

whence it follows that $10.22 < \sqrt{105} < 10.25$. This estimate may not be as sharp as desired. It is clear that the estimate $\sqrt{c} < \sqrt{105} < \sqrt{121}$ was wasteful and can be improved by making use of our conclusion that $\sqrt{105} < 10.25$. Thus, $\sqrt{c} < 10.25$ and we easily determine that

$$0.243 < \frac{5}{2(10.25)} < \sqrt{105} - 10.$$

Our improved estimate is $10.243 < \sqrt{105} < 10.250$ and more accurate estimates can be obtained in this way.

28.3 APPLICATION. The Mean Value Theorem and its corollaries can be used to establish inequalities and to extend inequalities that are known for integral or rational values to real values.

For example, we recall that Bernoulli's Inequality 5.C asserts that if $1 + x > 0$ and $n \in \mathbf{N}$, then $(1 + x)^n \geq 1 + nx$. We shall show that this inequality holds for any real exponent $r \geq 1$, To do so, let $f(x) = (1 + x)^r$, so that $f'(x) = r(1 + x)^{r-1}$. If $-1 < x < 0$, then $f'(x) < r$, while if $x > 0$, then $f'(x) > r$. If we apply the Mean Value Theorem to both of these cases, we obtain the result

$$(1 + x)^r \geq 1 + rx,$$

when $1 + x > 0$ and $r \geqq 1$. Moreover, if $r > 1$, then the equality occurs if and only if $x = 0$.

As a similar result, let α be a real number satisfying $0 < \alpha < 1$ and let $g(x) = \alpha x - x^{\alpha}$ for $x \geq 0$. Then $g'(x) = \alpha(1 - x^{\alpha-1})$, so that $g'(x) < 0$ for $0 < x < 1$ and $g'(x) > 0$ for $x > 1$. Consequently, if $x \geq 0$, then $g(x) \geq g(1)$ and $g(x) = g(1)$ if and only if $x = 1$. Therefore, if $x \geq 0$ and $0 < \alpha < 1$, then we have

$$x^{\alpha} \leq \alpha x + (1 - \alpha).$$

If $a \geq 0$ and $b > 0$ and if we let $x = a/b$ and multiply by b, we obtain the inequality

$$a^{\alpha} b^{1-\alpha} \leq \alpha a + (1 - \alpha) b.$$

where equality holds if and only if $a = b$. This inequality is often the starting point in establishing the important Hölder Inequality (cf. Project 8.β).

28.4 APPLICATION. The familiar rules of L'Hospital[†] on the evaluation of "indeterminant forms" can be established by means of the Cauchy Mean Value Theorem. For example, suppose that f, g are continuous on $[a, b]$ and have derivatives in (a, b), that $f(a) = g(a) = 0$, but that g, g' do not vanish for $x \neq a$. Then there exists a point c with $a < c < b$ such that

$$\frac{f(b)}{g(b)} = \frac{f'(c)}{g'(c)}.$$

It follows that if $\lim_{x \to a} f'(x)/g'(x)$ exists, then

$$\lim_{x \to a} \frac{f(x)}{g(x)} = \lim_{x \to a} \frac{f'(x)}{g'(x)}.$$

The case where the functions become infinite at $x = a$, or where the point at which the limit is taken is infinite, or where we have an "indeterminant" of some other form, can often be treated by taking logarithms, exponentials or some similar manipulation.

For example, if $a = 0$ and we wish to evaluate the limit of $h(x) = x \log x$ as $x \to 0$, we cannot apply the above argument. We write $h(x)$ in the form $f(x)/g(x)$ where $f(x) = \log x$ and $g(x) = 1/x$, $x > 0$. It is seen that

$$\frac{f'(x)}{g'(x)} = \frac{\dfrac{1}{x}}{\dfrac{-1}{x^2}} = -x \to 0, \qquad \text{as} \qquad x \to 0.$$

Let $\varepsilon > 0$ and choose a fixed number $0 < x_1 < 1$ such that if $0 < x < x_1$, then $|f'(x)/g'(x)| < \varepsilon$. Applying the Cauchy Mean Value Theorem, we have

$$\left| \frac{f(x) - f(x_1)}{g(x) - g(x_1)} \right| = \left| \frac{f'(x_2)}{g'(x_2)} \right| < \varepsilon,$$

with x_2 satisfying $0 < x < x_2 < x_1$. Since $f(x) \neq 0$ and $g(x) \neq 0$ for $0 < x < x_1$, we can write the quantity appearing on the left side in the more convenient form

$$\frac{f(x)}{g(x)} \left\{ \frac{1 - \dfrac{f(x_1)}{f(x)}}{1 - \dfrac{g(x_1)}{g(x)}} \right\}.$$

Holding x_1 fixed, we let $x \to 0$. Since the quantity in braces converges to

† GUILLAUME FRANÇOIS L'HOSPITAL (1661–1704) was a student of Johann Bernoulli (1667–1748). The Marquis de L'Hospital published his teacher's lectures on differential calculus in 1696, thereby presenting the first textbook on calculus to the world.

1, it exceeds $\frac{1}{2}$ for x sufficiently small. We infer from the above that

$$|h(x)| = \left| \frac{f(x)}{g(x)} \right| < 2\varepsilon,$$

for x sufficiently near 0. Thus the limit at $x = 0$ of h is 0.

Interchange of Limit and Derivative

Let (f_n) be a sequence of functions defined on an interval J of $\boldsymbol{R}$ and with values in $\boldsymbol{R}$. It is easy to give an example of a sequence of functions which have derivatives at every point of J and which converges on J to a function f which does not have a derivative at some points of J. (Do so!) Moreover, the example of Weierstrass mentioned before can be used to give an example of a sequence of functions possessing derivatives at every point of $\boldsymbol{R}$ and converging uniformly on $\boldsymbol{R}$ to a continuous function which has a derivative at no point. Thus it is not permissible, in general, to differentiate the limit of a convergent sequence of functions possessing derivatives even when the convergence is uniform.

We shall now show that if the *sequence of derivatives* is uniformly convergent, then all is well. If one adds the hypothesis that the derivatives are continuous, then it is possible to give a short proof based on the Riemann integral. However, if the derivatives are not assumed to be continuous, a somewhat more delicate argument is required.

28.5   THEOREM.   *Let (f_n) be a sequence of functions defined on a finite interval J of $\boldsymbol{R}$ and with values on $\boldsymbol{R}$. Suppose that there is a point x_0 in J at which the sequence $(f_n(x_0))$ converges, that the derivatives f_n' exist on J, and that the sequence (f_n') converges uniformly on J to a function g. Then the sequence (f_n) converges uniformly on J to a function f which has a derivative at every point of J and $f' = g$.*

PROOF.   Suppose the end points of J are $a < b$ and let x be any point of J. If m, n are natural numbers, we apply the Mean Value Theorem to the difference $f_m - f_n$ on the interval with end points x_0, x to conclude that there exists a point y (depending on m, n) such that

$$f_m(x) - f_n(x) = f_m(x_0) - f_n(x_0) + (x - x_0)\{f_m'(y) - f_n'(y)\}.$$

Hence we infer that

$$\|f_m - f_n\|_J \le |f_m(x_0) - f_n(x_0)| + (b - a)\,\|f_m' - f_n'\|_J$$

so the sequence (f_n) converges uniformly on J to a function we shall denote by f. Since the f_n are continuous and the convergence of (f_n) to f is uniform, then f is continuous on J.

To establish the existence of the derivative of f at a point c in J, we apply the Mean Value Theorem to the difference $f_m - f_n$ on an interval with end points c, x to infer that there exists a point z (depending on m, n) such that

$$\{f_m(x) - f_n(x)\} - \{f_m(c) - f_n(c)\} = (x - c)\{f'_m(z) - f'_n(z)\}.$$

We infer that, when $c \neq x$, then

$$\left| \frac{f_m(x) - f_m(c)}{x - c} - \frac{f_n(x) - f_n(c)}{x - c} \right| \leq \|f'_m - f'_n\|_J.$$

In virtue of the uniform convergence of the sequence (f'_n), the right hand side is dominated by ε when $m, n \geq M(\varepsilon)$. Taking the limit with respect to m, we infer from Lemma 15.8 that

$$\left| \frac{f(x) - f(c)}{x - c} - \frac{f_n(x) - f_n(c)}{x - c} \right| \leq \varepsilon,$$

when $n \geq M(\varepsilon)$. Since $g(c) = \lim (f'_n(c))$, there exists an $N(\varepsilon)$ such that if $n \geq N(\varepsilon)$, then $|f'_n(c) - g(c)| < \varepsilon$. Now let $K = \sup \{M(\varepsilon), N(\varepsilon)\}$. In view of the existence of $f'_K(c)$, if $0 < |x - c| < \delta_K(\varepsilon)$, then

$$\left| \frac{f_K(x) - f_K(c)}{x - c} - f'_K(c) \right| < \varepsilon.$$

Therefore, it follows that if $0 < |x - c| < \delta_K(\varepsilon)$, then

$$\left| \frac{f(x) - f(c)}{x - c} - g(c) \right| < 3\varepsilon.$$

This shows that $f'(c)$ exists and equals $g(c)$.   Q.E.D.

Taylor's Theorem

If the derivative $f'(x)$ of f exists at every point x of a set D, we can consider the existence of the derivative of the function f' at a point $c \in D$. In case f' has a derivative of c, we refer to the resulting number as the **second derivative** of f at c and shall ordinarily denote this number by $f''(c)$, or by $f^{(2)}(c)$. In a similar fashion we define the third $f'''(c) = f^{(3)}(c), \ldots,$ and the nth derivative $f^{(n)}(c), \ldots,$ whenever these derivatives exist.

We shall now obtain the celebrated theorem attributed to Brook Taylor,† which plays an important role in many investigations and can be regarded as an extension of the Mean Value Theorem.

28.6 TAYLOR'S THEOREM. *Suppose that n is a natural number, that f and its derivatives $f', f'', \ldots, f^{(n-1)}$ are defined and continuous on $J = [a, b]$,*

---

† BROOK TAYLOR (1685–1731) was an early English mathematician. In 1715 he gave the infinite series expansion, but—true to the spirit of the time—did not discuss convergence. The remainder was supplied by Lagrange.

and that $f^{(n)}$ exists in (a, b). If α, β belong to J, then there exists a number γ between α and β such that

$$f(\beta) = f(\alpha) + \frac{f'(\alpha)}{1!}(\beta - \alpha) + \frac{f''(\alpha)}{2!}(\beta - \alpha)^2$$
$$+ \cdots + \frac{f^{(n-1)}(\alpha)}{(n-1)!}(\beta - \alpha)^{n-1} + \frac{f^{(n)}(\gamma)}{n!}(\beta - \alpha)^n.$$

PROOF.   Let P be the real number defined by the relation

$$(28.1) \quad \frac{(\beta - \alpha)^n}{n!} P = f(\beta) - \left\{ f(\alpha) + \frac{f'(\alpha)}{1!}(\beta - \alpha) \right.$$
$$\left. + \cdots + \frac{f^{(n-1)}(\alpha)}{(n-1)!}(\beta - \alpha)^{n-1} \right\}.$$

and consider the function φ defined on J by

$$\varphi(x) = f(\beta) - \left\{ f(x) + \frac{f'(x)}{1!}(\beta - x) + \cdots + \frac{f^{(n-1)}(x)}{(n-1)!}(\beta - x)^{n-1} + \frac{P}{n!}(\beta - x)^n \right\}.$$

Clearly, φ is continuous on J and has a derivative on (a, b). It is evident that $\varphi(\beta) = 0$ and it follows from the definition of P that $\varphi(\alpha) = 0$. By Rolle's Theorem, there exists a point γ between α and β such that $\varphi'(\gamma) = 0$. On calculating the derivative φ' (using the usual formula for the derivative of a sum and product of two functions), we obtain the telescoping sum

$$\varphi'(x) = - \left\{ f'(x) - f'(x) + \frac{f''(x)}{1!}(\beta - x) + \cdots + (-1)\frac{f^{(n-1)}(x)}{(n-2)!}(\beta - x)^{n-2} \right.$$
$$\left. + \frac{f^{(n)}(x)}{(n-1)!}(\beta - x)^{n-1} - \frac{P}{(n-1)!}(\beta - x)^{n-1} \right\}$$
$$= \frac{P - f^{(n)}(x)}{(n-1)!}(\beta - x)^{n-1}.$$

Since $\varphi'(\gamma) = 0$, then $P = f^{(n)}(\gamma)$, proving the assertion.   Q.E.D.

REMARK.   The remainder term

$$(28.2) \qquad\qquad R_n = \frac{f^{(n)}(\gamma)}{n!}(\beta - \alpha)^n$$

given above is often called the **Lagrange form** of the remainder. There are many other expressions for the remainder, but for the present, we mention only the **Cauchy form** which asserts that for some number θ with $0 < \theta < 1$, then

$$(28.3) \qquad\qquad R_n = (1 - \theta)^{n-1} \frac{f^{(n)}((1 - \theta)\alpha + \theta\beta)}{(n-1)!}(\beta - \alpha)^n.$$

This form can be established as above, except that on the left side of equation (28.1) we put $(\beta - \alpha)Q/(n-1)!$ and we define φ as above except its last term is $(\beta - x)Q/(n-1)!$ We leave the details as an exercise. (In Section 31 we shall obtain another form involving use of the integral to evaluate the remainder term.)

Exercises

28.A. Using the formulas in 28.1, show that if $n = 0, 1, 2, \ldots$, then the roots of the Bessel functions J_n and J_{n+1} on $(0, +\infty)$ interlace each other.

28.B. Show that if $x > 0$, then

$$1 + \frac{x}{2} - \frac{x^2}{8} \le \sqrt{1+x} \le 1 + \frac{x}{2}.$$

28.C. Calculate $\sqrt{1.2}$ and $\sqrt{2}$. What is the best accuracy you can be sure of?

28.D. Get estimates similar to those in Exercise 28.B for $(1+x)^{1/3}$ on the interval $[0, 7]$. Use these to calculate $\sqrt[3]{1.5}$ and $\sqrt[3]{2}$.

28.E. Suppose that $0 < r < 1$ and $-1 < x$. Show that we have $(1+x)^r \le 1 + rx$ and that the equality holds if and only if $x = 0$.

28.F. A root x_0 of a polynomial p is said to be **simple** (or have **multiplicity one**) if $p'(x_0) \neq 0$, and to have **multiplicity** n if $p(x_0) = p'(x_0) = \cdots = p^{(n-1)}(x_0) = 0$, but $p^{(n)}(x_0) \neq 0$.

If $a < b$ are consecutive roots of a polynomial, then there are an odd number (counting multiplicities) of roots of its derivative in (a, b).

28.G. Show that if the roots of the polynomial p are all real, then the roots of p' are all real. If, in addition, the roots of p are all simple, then the roots of p' are all simple.

28.H. If $f(x) = (x^2 - 1)^n$ and if p is the nth derivative of f, then p is a polynomial of degree n whose roots are simple and lie in the open interval $(-1, 1)$.

28.I. Establish the Cauchy form of the remainder term R_n in Taylor's Theorem given in formula (28.3).

28.J. A proof of Taylor's Theorem 28.6 using the Cauchy Mean Value Theorem can be given by letting

$$R(x) = f(x) - \left[f(\alpha) + \frac{x-\alpha}{1!} f'(\alpha) + \cdots + \frac{(x-\alpha)^{n-1}}{(n-1)!} f^{(n-1)}(\alpha) \right].$$

Show that $R(\alpha) = R'(\alpha) = \cdots = R^{(n-1)}(\alpha) = 0$ and $R^{(n)}(x) = f^{(n)}(x)$. Note that there exists γ_1 between α and β such that

$$\frac{R(\beta)}{(\beta - \alpha)^n} = \frac{R(\beta) - R(\alpha)}{(\beta - \alpha)^n - 0^n} = \frac{R'(\gamma_1)}{n(\gamma_1 - \alpha)^{n-1}}.$$

Continue this to find that $R(\beta) = (\beta - \alpha)^n f^{(n)}(\gamma_n)/n!$ for some γ_n between α and β.

28.K. If $f(x) = e^x$, show that the remainder term in Taylor's Theorem converges to zero as $n \to \infty$ for each fixed α, β.

28.L. If $f(x) = \sin x$, show that the remainder term in Taylor's Theorem converges to zero as $n \to \infty$ for each fixed α, β.

28.M. If $f(x) = (1+x)^m$ where $m \in Q$, $|x| < 1$, the usual differentiation formulas from calculus and Taylor's Theorem lead to the expression

$$(1+x)^m = 1 + \binom{m}{1}x + \binom{m}{2}x^2 + \cdots + \binom{m}{n-1}x^{n-1} + R_n,$$

where R_n can be given in Lagrange's form by $R_n = x^n f^{(n)}(\theta_n x)/n!$ where $0 < \theta_n < 1$. Show that if $0 \le x < 1$, then $\lim (R_n) = 0$. Show that if $-1 < x < 0$, then we cannot use the same argument to show that $\lim (R_n) = 0$.

28.N. In the preceding exercise, use Cauchy's form of the remainder to obtain

$$R_n = \frac{m(m-1)\cdots(m-n+1)}{1\cdot 2\cdots(n-1)} \frac{(1-\theta_n)^{n-1}x^n}{(1+\theta_n x)^{n-m}},$$

where $0 < \theta_n < 1$. When $|x| < 1$ show that $|(1-\theta_n)/(1+\theta_n x)| < 1$, and prove that $\lim (R_n) = 0$. (Hint: If $|x| < 1$, the set $\{(1 + \theta x)^{m-1}; \ 0 \le \theta \le 1\}$ is bounded.)

28.O. If $f: R \to R$, if $f'(x)$ exist for $x \in R$, and if $f''(a)$ exists, show that

$$f''(a) = \lim_{h \to 0} \frac{f(a+h) - 2f(a) + f(a-h)}{h^2}.$$

Give an example where this limit exists, but the function does not have a second derivative at a.

28.P. Let $f_n(x) = |x|^{1+1/n}$ for x in $[-1, 1]$. Show that each f_n is differentiable on $[-1, 1]$ and that (f_n) converges uniformly on $[-1, 1]$ to $f(x) = |x|$.

Projects

28.α. In this project we consider the exponential function from the point of view of differential calculus.

(a) Suppose that a function E on $J = (a, b)$ to R has a derivative at every point of J and that $E'(x) = E(x)$ for all $x \in J$. Observe that E has derivatives of all orders on J and they all equal E.

(b) If $E(\alpha) = 0$ for some $\alpha \in J$, apply Taylor's Theorem 28.6 and Exercise 14.L to show that $E(x) = 0$ for all $x \in J$.

(c) Show that there exists at most one function E on R to R which satisfies

$$E'(x) = E(x) \qquad \text{for } x \in R, \qquad E(0) = 1.$$

(d) Prove that if E satisfies the conditions in part (c), then it also satisfies the functional equation

$$E(x + y) = E(x)E(y) \qquad \text{for } x, y \in R.$$

(Hint: if $f(x) = E(x+y)/E(y)$, then $f'(x) = f(x)$ and $f(0) = 1$.)

(e) Let (E_n) be the sequence of functions defined on R by

$$E_1(x) = 1 + x, \qquad E_n(x) = E_{n-1}(x) + x^n/n!.$$

Let A be any positive number; if $|x| \le A$ and if $m \ge n > 2A$, then

$$|E_m(x) - E_n(x)| \le \frac{A^{n+1}}{(n+1)!}\left[1 + \frac{A}{n} + \cdots + \left(\frac{A}{n}\right)^{m-n}\right] < \frac{2A^{n+1}}{(n+1)!}.$$

Hence the sequence (E_n) converges uniformly for $|x| \le A$.

(f) If (E_n) is the sequence of functions defined in part (e), then

$$E'_n(x) = E_{n-1}(x), \qquad \text{for } x \in \mathbf{R}.$$

Show that the sequence (E_n) converges on $\mathbf{R}$ to a function E with the properties displayed in part (c). Therefore, E is the unique function with these properties.

(g) Let E be the function with $E' = E$ and $E(0) = 1$. If we define e to be the number

$$e = E(1),$$

then e lies between $2\frac{2}{3}$ and $2\frac{3}{4}$. (Hint: $1 + 1 + \frac{1}{2} + \frac{1}{6} < e < 1 + 1 + \frac{1}{2} + \frac{1}{6} + \frac{1}{12}$. More precisely, we can show that $2.708 < 2 + \frac{17}{24} < e < 2 + \frac{13}{18} < 2.723$.)

28.β. In this project, you may use the results of the preceding one. Let E denote the unique function on $\mathbf{R}$ such that

$$E' = E \qquad \text{and} \qquad E(0) = 1$$

and let $e = E(1)$.

(a) Show that E is strictly increasing and has range $P = \{x \in \mathbf{R} : x > 0\}$.

(b) Let L be the inverse function of E, so that the domain of L is P and its range is all of $\mathbf{R}$. Prove that L is strictly increasing on P, that $L(1) = 0$, and that $L(e) = 1$.

(c) Show that $L(xy) = L(x) + L(y)$ for all x, y in P.

(d) If $0 < x < y$, then

$$\frac{1}{y}(y - x) < L(y) - L(x) < \frac{1}{x}(y - x).$$

(Hint: apply the Mean Value Theorem to E.)

(e) The function L has a derivative for $x > 0$ and $L'(x) = 1/x$.

(f) The number e satisfies

$$e = \lim\left(\left(1 + \frac{1}{n}\right)^n\right).$$

(Hint: evaluate $L'(1)$ by using the sequence $((1 + 1/n))$ and the continuity of E.)

28.γ. In this project we shall introduce the sine and cosine.

(a) Let h be defined on an interval $J = (a, b)$ to $\mathbf{R}$ and satisfy

$$h''(x) + h(x) = 0$$

for all x in J. Show that h has derivatives of all orders and that if there is a point α in J such that $h(\alpha) = 0$, $h'(\alpha) = 0$, then $h(x) = 0$ for all $x \in J$. (Hint: use Taylor's Theorem 28.6.)

(b) Show that there exists at most one function C on $\mathbf{R}$ satisfying the conditions

$$C'' + C = 0, \qquad C(0) = 1, \qquad C'(0) = 0,$$

and at most one function S on $\mathbf{R}$ satisfying

$$S'' + S = 0, \qquad S(0) = 0, \qquad S'(0) = 1.$$

(c) We define a sequence (C_n) by

$$C_1(x) = 1 - x^2/2, \qquad C_n(x) = C_{n-1}(x) + (-1)^n \frac{x^{2n}}{(2n)!}.$$

Let A be any positive number; if $|x| \leq A$ and if $m \geq n > A$, then

$$|C_m(x) - C_n(x)| \leq \frac{A^{2n+2}}{(2n+2)!}\left[1 + \left(\frac{A}{2n}\right)^2 + \cdots + \left(\frac{A}{2n}\right)^{2m-2n}\right]$$
$$< \left(\frac{4}{3}\right)\frac{A^{2n+2}}{(2n+2)!} .$$

Hence the sequence (C_n) converges uniformly for $|x| \leq A$. Show also that $C_n'' = -C_{n-1}$, and $C_n(0) = 1$ and $C_n'(0) = 0$. Prove that the limit C of the sequence (C_n) is the unique function with the properties in part (b). (Use Theorem 28.5.)

(d) Let (S_n) be defined by

$$S_1(x) = x, \qquad S_n(x) = S_{n-1}(x) + (-1)^{n-1}\frac{x^{2n-1}}{(2n-1)!} .$$

Show that (S_n) converges uniformly for $|x| \leq A$ to the unique function S with the properties in part (b).

(e) Prove that $S' = C$ and $C' = -S$.

(f) Establish the Pythagorean Identity $S^2 + C^2 = 1$. (Hint: calculate the derivative of $S^2 + C^2$.)

28.δ. The project continues the discussion of the sine and cosine functions. Free use may be made of the properties established in the preceding project.

(a) Suppose that h is a function on $\mathbf{R}$ which satisfies the equation

$$h'' + h = 0.$$

Show that there exist constants α, β such that $h = \alpha C + \beta S$. (Hint: $\alpha = h(0)$, $\beta = h'(0)$.)

(b) The function C is even and S is odd in the sense that

$$C(-x) = C(x) \qquad \text{and} \qquad S(-x) = -S(x) \qquad \text{for all } x \text{ in } \mathbf{R}.$$

(c) Show that the "addition formulas"

$$C(x + y) = C(x)C(y) - S(x)S(y),$$
$$S(x + y) = S(x)C(y) + C(x)S(y),$$

hold for all x, y in $\mathbf{R}$. (Hint: let y be fixed, define $h(x) = C(x + y)$, and show that $h'' + h = 0$, $h(0) = C(y)$, $h'(0) = -S(y)$.)

(d) Show that the "duplication formulas"

$$C(2x) = 2[C(x)]^2 - 1 = 1 - 2[S(x)]^2,$$
$$S(2x) = 2S(x)C(x),$$

hold for all x in $\mathbf{R}$.

(e) Prove that C satisfies the inequality

$$C_1(x) = 1 - \frac{x^2}{2} \leq C(x) \leq 1 - \frac{x^2}{2} + \frac{x^4}{24} = C_2(x).$$

Therefore, the smallest positive root γ of C lies between the positive root of $x^2 - 2 = 0$ and the smallest positive root of $x^4 - 12x^2 + 24 = 0$. Using this, prove that $\sqrt{2} < \gamma < \sqrt{3}$.

(f) We define π to be the smallest strictly positive root of S. Then $\pi = 2\gamma$ and hence that $2\sqrt{2} < \pi < 2\sqrt{3}$.

(g) Prove that both C and S are periodic functions with period 2π in the sense that $C(x + 2\pi) = C(x)$ and $S(x + 2\pi) = S(x)$ for all x in **R**. Also show that

$$S(x) = C\left(\frac{\pi}{2} - x\right) = -C\left(x + \frac{\pi}{2}\right),$$
$$C(x) = S\left(\frac{\pi}{2} - x\right) = S\left(x + \frac{\pi}{2}\right),$$

for all x in **R**.

28.ε. Following the model of the preceding two projects, introduce the hyperbolic cosine and sine as functions satisfying

$$c'' = c, \qquad c(0) = 1, \qquad c'(0) = 0,$$
$$s'' = s, \qquad s(0) = 0, \qquad s'(0) = 1,$$

respectively. Establish the existence and the uniqueness of these functions and show that
$$c^2 - s^2 = 1.$$

Prove results similar to (a)–(d) of Project 28.δ and show that, if the exponential function is denoted by E, then

$$c(x) = \tfrac{1}{2}(E(x) + E(-x)), \qquad s(x) = \tfrac{1}{2}(E(x) - E(-x)).$$

28.ζ. A function φ on an interval J of **R** to **R** is said to be (**midpoint**) **convex** in case

$$\varphi\left(\frac{x+y}{2}\right) \le \tfrac{1}{2}(\varphi(x) + \varphi(y))$$

for each x, y in J. (In geometrical terms: the midpoint of any chord of the curve $y = \varphi(x)$, lies above or on the curve.) In this project we shall suppose that φ is a continuous convex function.

(a) If $n = 2^m$ and if $x_1, \ldots, x_n$ belong to J, then

$$\varphi\left(\frac{x_1 + x_2 + \cdots + x_n}{n}\right) \le \frac{1}{n}(\varphi(x_1) + \cdots + \varphi(x_n)).$$

(b) If $n < 2^m$ and if $x_1, \ldots, x_n$ belong to J, let x_j for $j = n+1, \ldots, 2^m$ be equal to

$$\bar{x} = \left(\frac{x_1 + x_2 + \cdots + x_n}{n}\right).$$

Show that the same inequality holds as in part (a).

(c) Since φ is continuous, show that if x, y belong to J and $t \in I$, then

$$\varphi((1-t)x + ty) \le (1-t)\varphi(x) + t\varphi(y).$$

(In geometrical terms: the entire chord lies above or on the curve.)

(d) Suppose that φ has a second derivative on J. Then a necessary and sufficient condition that φ be convex on J is that $\varphi''(x) \ge 0$ for $x \in J$. (Hint: to prove the necessity, use Exercise 28.O. To prove the sufficiency, use Taylor's Theorem and expand about $\bar{x} = (x + y)/2$.)

(e) If φ is a continuous convex function on J and if $x < y < z$ belong to J, show that

$$\frac{\varphi(y) - \varphi(x)}{y - x} \leq \frac{\varphi(z) - \varphi(x)}{z - x} .$$

Therefore, if $w < x < y < z$ belong to J, then

$$\frac{\varphi(x) - \varphi(w)}{x - w} \leq \frac{\varphi(z) - \varphi(y)}{z - y} .$$

(f) Prove that a continuous convex function φ on J has a left-hand derivative and a right-hand derivative at every point interior to J. Furthermore, the subset where φ' does not exist is countable.

Section 29   The Riemann-Stieltjes Integral

In this section we shall define the Riemann-Stieltjes† integral of bounded functions on a compact interval of $\mathbf{R}$. Since we assume that the reader is acquainted, at least informally, with the integral from a calculus course we will not provide an extensive motivation for it.

The reader who continues his study of mathematical analysis will want to become familiar with the more general Lebesgue integral at an early date. However, since the Riemann and the Riemann-Stieltjes integrals are adequate for many purposes and are more familiar to the reader, we prefer to treat them here and leave the more advanced Lebesgue theory for a later course.

We shall consider bounded real-valued functions on closed intervals of the real number system, define the integral of one such function with respect to another, and derive the main properties of this integral. The type of integration considered here is somewhat more general than that considered in earlier courses and the added generality makes it very useful in certain applications, especially in statistics. At the same time, there is little additional complication to the theoretical machinery that a rigorous discussion of the ordinary Riemann integral requires. Therefore, it is worthwhile to develop this type of integration theory as far as its most frequent applications require.

---

† (George Friedrich) Bernhard Riemann (1826–1866) was the son of a poor country minister and was born near Hanover. He studied at Göttingen and Berlin and taught at Göttingen. He was one of the founders of the theory of analytic functions, but also made fundamental contributions to geometry, number theory, and mathematical physics.

Thomas Joannes Stieltjes (1856–1894) was a Dutch astronomer and mathematician. He studied in Paris with Hermite and obtained a professorship at Toulouse. His most famous work was a memoir on continued fractions, the moment problem, and the Stieltjes integral, which was published in the last year of his short life.

Let f and g denote real-valued functions defined on a closed interval $J = [a, b]$ of the real line. *We shall suppose that both f and g are bounded on J*; this standing hypothesis will not be repeated. A **partition** of J is a finite collection of non-overlapping intervals whose union is J. Usually, we describe a partition P by specifying a finite set of real numbers $(x_0, x_1, \ldots, x_n)$ such that

$$a = x_0 \leq x_1 \leq \cdots \leq x_n = b$$

and such that the subintervals occurring in the partition P are the intervals $[x_{k-1}, x_k]$, $k = 1, 2, \ldots, n$. More properly, we refer to the end points x_k, $k = 0, 1, \ldots, n$ as the **partition points** corresponding to P. However, in practice it is often convenient and can cause no confusion to use the word "partition" to denote either the collection of subintervals or the collection of end points of these subintervals. Hence we write $P = (x_0, x_1, \ldots, x_n)$.

If P and Q are partitions of J, we say that Q is a **refinement of** P or that Q is **finer than** P in case every subinterval in Q is contained in some subinterval in P. This is equivalent to the requirement that every partition point in P is also a partition point in Q. For this reason, we write $P \subseteq Q$ when Q is a refinement of P.

29.1 Definition. If P is a partition of J, then a **Riemann-Stieltjes sum of** f with respect to g and corresponding to $P = (x_0, x_1, \ldots, x_n)$ is a real number $S(P; f, g)$ of the form

$$(29.1) \qquad S(P; f, g) = \sum_{k=1}^{n} f(\xi_k)\{g(x_k) - g(x_{k-1})\}.$$

Here we have selected numbers ξ_k satisfying

$$x_{k-1} \leq \xi_k \leq x_k \qquad \text{for} \qquad k = 1, 2, \ldots, n.$$

Note that if the function g is given by $g(x) = x$, then the expression in equation (29.1) reduces to

$$(29.2) \qquad \sum_{k=1}^{n} f(\xi_k)(x_k - x_{k-1}).$$

The sum (29.2) is usually called a **Riemann sum** of f corresponding to the partition P and can be interpreted as the area of the union of rectangles with sides $[x_{k-1}, x_k]$ and heights $f(\xi_k)$. (See Figure 29.1.) Thus if the partition P is very fine, it is expected that the Riemann sum (29.2) yields an approximation to the "area under the graph of f." For a general function g, the reader should interpret the Riemann-Stieltjes sum (29.1) as being similar to the Riemann sum (29.2)—except that, instead of considering the *length* $x_k - x_{k-1}$ of the subinterval $[x_{k-1}, x_k]$, we are considering some other measure of magnitude of this subinterval, namely the difference $g(x_k) - g(x_{k-1})$. Thus if $g(x)$ is the total "mass" or "charge" on the interval $[a, x]$, then $g(x_k) - g(x_{k-1})$ denotes the "mass" or "charge" on the subinterval $[x_{k-1}, x_k]$. The idea is that we want to be able to consider measures of magnitude of an interval other than length, so we allow for the slightly more general sums (29.1).

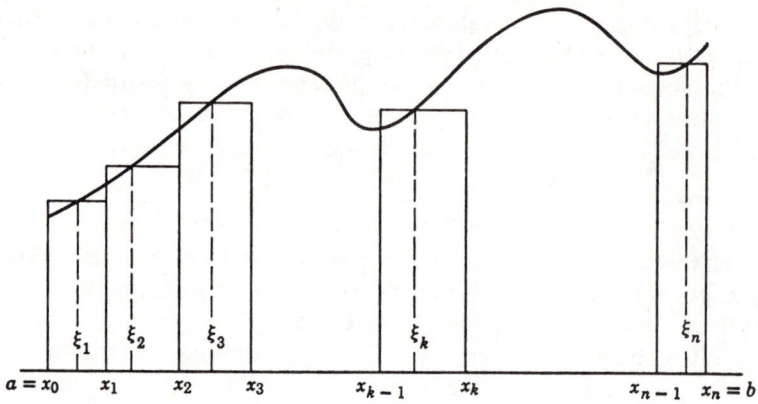

Figure 29.1.   The Riemann sum as an area.

It will be noted that both of the sums (29.1) and (29.2) depend upon the choice of the "intermediate points"; that is, upon the numbers ξ_k, $1 \le k \le n$. Thus it might be thought advisable to introduce a notation displaying the choice of these numbers. However, by introducing a finer partition, it can always be assumed that the intermediate points ξ_k are partition points. In fact, if we introduce the partition $Q = (x_0, \xi_1, x_1, \xi_2, \ldots, \xi_n, x_n)$ and the sum $S(Q; f, g)$ where we take the intermediate points to be alternately the right and the left end points of the subinterval, then the sum $S(Q; f, g)$ yields the same value as the sum in (29.1). We could always assume that the partition divides the interval into an even number of subintervals and the intermediate points are alternately the right and left end points of these subintervals. However, we shall not find it necessary to require this "standard" partitioning process, nor shall we find it necessary to display these intermediate points.

29.2   DEFINITION.   We say that f is **integrable** with respect to g on J if there exists a real number I such that for every number $\varepsilon > 0$ there is a partition P_ε of J such that if P is any refinement of P_ε and $S(P; f, g)$ is any Riemann-Stieltjes sum corresponding to P, then

(29.3) $$|S(P; f, g) - I| < \varepsilon.$$

In this case the number I is uniquely determined and is denoted by

$$I = \int_a^b f \, dg = \int_a^b f(t) \, dg(t);$$

it is called the **Riemann-Stieltjes integral** of f with respect to g over $J = [a, b]$. We call the function f the **integrand**, and g the **integrator**. Sometimes we say that f is g-**integrable** if f is integrable with respect to g. In the special case $g(x) = x$, if f is integrable with respect to g, we usually say that f is **Riemann integrable**.

Before we develop any of the properties of the Riemann-Stieltjes integral, we shall consider some examples. In order to keep the calculations simple, some of these examples are chosen to be extreme cases; more typical examples are found by combining the ones given below.

29.3 EXAMPLES. (a) We have already noted that if $g(x) = x$, then the integral reduces to the ordinary Riemann integral of elementary calculus.

(b) If g is constant on the interval $[a, b]$, then any function f is integrable with respect to g and the value of the integral is 0.

(c) Let g be defined on $J = [a, b]$ by

$$g(x) = 0, \qquad x = a,$$
$$= 1, \qquad a < x \le b.$$

We leave it as an exercise to show that a function f is integrable with respect to g if and only if f is continuous at a and that in this case the value of the integral is $f(a)$.

(d) Let c be an interior point of the interval $J = [a, b]$ and let g be defined by

$$g(x) = 0, \qquad a \le x \le c,$$
$$= 1, \qquad c < x \le b.$$

It is an exercise to show that a function f is integrable with respect to g if and only if it is **continuous at c from the right** (in the sense that for every $\varepsilon > 0$ there exists $\delta(\varepsilon) > 0$ such that if $c \le x < c + \delta(\varepsilon)$ and $x \in J$, then we have $|f(x) - f(c)| < \varepsilon$). If f satisfies this condition, then the value of the integral is $f(c)$. (Observe that the integrator function g is continuous at c from the left.)

(e) Modifying the preceding example, let h be defined by

$$h(x) = 0, \qquad a \le x < c,$$
$$= 1, \qquad c \le x \le b.$$

Then h is continuous at c from the right and a function f is integrable with respect to h if and only if f is continuous at c from the left. In this case the value of the integral is $f(c)$.

(f) Let $c_1 < c_2$ be interior points of $J = [a, b]$ and let g be defined by

$$g(x) = \alpha_1, \qquad a \le x \le c_1,$$
$$= \alpha_2, \qquad c_1 < x \le c_2,$$
$$= \alpha_3, \qquad c_2 < x \le b.$$

If f is continuous at the points c_1, c_2, then f is integrable with respect to g and

$$\int_a^b f \, dg = (\alpha_2 - \alpha_1)f(c_1) + (\alpha_3 - \alpha_2)f(c_2).$$

By taking more points we can obtain a sum involving the values of f at points in J, weighted by the values of the jumps of g at these points.

(g) Let the function f be Dirichlet's discontinuous function [cf. Example 20.5(g)] defined by

$$f(x) = 1, \qquad \text{if } x \text{ is rational,}$$
$$= 0, \qquad \text{if } x \text{ is irrational,}$$

and let $g(x) = x$. Consider these functions on $I = [0, 1]$. If a partition P consists of n equal subintervals, then by selecting k of the intermediate points in the sum $S(P; f, g)$ to be rational and the remaining to be irrational, $S(P; f, g) = k/n$. It follows that f is *not* Riemann integrable.

(h) Let f be the function defined on I by $f(0) = 1$, $f(x) = 0$ for x irrational, and $f(m/n) = 1/n$ when m and n are natural numbers with no common factors except 1. It was seen in Example 20.5(h) that f is continuous at every irrational number and discontinuous at every rational number. If $g(x) = x$, then it is an exercise to show that f is integrable with respect to g and that the value of the integral is 0.

29.4 CAUCHY CRITERION FOR INTEGRABILITY. *The function f is integrable with respect to g over $J = [a, b]$ if and only if for each number $\varepsilon > 0$ there is a partition Q_ε of J such that if P and Q are refinements of Q_ε and if $S(P; f, g)$ and $S(Q; f, g)$ are any corresponding Riemann-Stieltjes sums, then*

$$(29.4) \qquad\qquad |S(P; f, g) - S(Q; f, g)| < \varepsilon.$$

PROOF. If f is integrable, there is a partition P_ε such that if P, Q are refinements of P_ε, then any corresponding Riemann-Stieltjes sums satisfy $|S(P; f, g) - I| < \varepsilon/2$ and $|S(Q; f, g) - I| < \varepsilon/2$. By using the Triangle Inequality, we obtain (29.4).

Conversely, suppose the criterion is satisfied. To show that f is integrable with respect to g, we need to produce the value of its integral and use Definition 29.2. Let Q_1 be a partition of J such that if P and Q are refinements of Q_1, then $|S(P; f, g) - S(Q; f, g)| < 1$. Inductively, we choose Q_n to be a refinement of Q_{n-1} such that if P and Q are refinements of Q_n, then

$$(29.5) \qquad\qquad |S(P; f, g) - S(Q; f, g)| < 1/n.$$

Consider a sequence $(S(Q_n; f, g))$ of real numbers obtained in this way. Since Q_n is a refinement of Q_m when $n \geq m$, this sequence of sums is a Cauchy sequence of real numbers, regardless of how the intermediate points are chosen. By Theorem 16.10, the sequence converges to some real number L. Hence, if $\varepsilon > 0$, there is an integer N such that $2/N < \varepsilon$ and

$$|S(Q_N; f, g) - L| < \varepsilon/2.$$

If P is a refinement of Q_N, then it follows from the construction of Q_N that

$$|S(P; f, g) - S(Q_N; f, g)| < 1/N < \varepsilon/2.$$

Hence, for any refinement P of Q_N and any corresponding Riemann-Stieltjes sum, we have

$$(29.6) \qquad |S(P; f, g) - L| < \varepsilon,$$

This shows that f is integrable with respect to g over J and that the value of this integral is L. Q.E.D.

Some Properties of the Integral

The next property is sometimes referred to as the **bilinearity** of the Riemann-Stieltjes integral.

29.5 THEOREM. (a) *If f_1, f_2 are integrable with respect to g on J and α, β are real numbers, then $\alpha f_1 + \beta f_2$ is integrable with respect to g on J and*

$$(29.7) \qquad \int_a^b (\alpha f_1 + \beta f_2)\, dg = \alpha \int_a^b f_1\, dg + \beta \int_a^b f_2\, dg.$$

(b) *If f is integrable with respect to g_1 and g_2 on J and α, β are real numbers, then f is integrable with respect to $g = \alpha g_1 + \beta g_2$ on J and*

$$(29.8) \qquad \int_a^b f\, dg = \alpha \int_a^b f\, dg_1 + \beta \int_a^b f\, dg_2.$$

PROOF. (a) Let $\varepsilon > 0$ and let $P_1 = (x_0, x_1, \ldots, x_n)$ and $P_2 = (y_0, y_1, \ldots, y_m)$ be partitions of $J = [a, b]$ such that if Q is a refinement of both P_1 and P_2, then for any corresponding Riemann-Stieltjes sums, we have

$$|I_1 - S(Q; f_1, g)| < \varepsilon, \qquad |I_2 - S(Q; f_2, g)| < \varepsilon.$$

Let P_ε be a partition of J which is a refinement of both P_1 and P_2 (for example, all the partition points in P_1 and P_2 are combined to form P_ε). If Q is a partition of J such that $P_\varepsilon \subseteq Q$, then both of the relations above still hold. When the same intermediate points are used, we evidently have

$$S(Q; \alpha f_1 + \beta f_2, g) = \alpha S(Q; f_1, g) + \beta S(Q; f_2, g).$$

It follows from this and the preceding inequalities that

$$|\alpha I_1 + \beta I_2 - S(Q; \alpha f_1 + \beta f_2, g)| = |\alpha \{I_1 - S(Q; f_1, g)\} + \beta \{I_2 - S(Q; f_2, g)\}|$$
$$\leq (|\alpha| + |\beta|)\varepsilon.$$

This proves that $\alpha I_1 + \beta I_2$ is the integral of $\alpha f_1 + \beta f_2$ with respect to g. This establishes part (a); the proof of part (b) is similar and will be left to the reader. Q.E.D.

There is another useful additivity property possessed by the Riemann-Stieltjes integral; namely, with respect to the interval over which the integral is extended. It is in order to obtain the next result that we employed the type of limiting introduced in Definition 29.2. A more restrictive type of limiting would be to require inequality (29.3) for any Riemann-Stieltjes sum corresponding to a partition $P = (x_0, x_1, \ldots, x_n)$ which is such that

$$\|P\| = \sup \{x_1 - x_0, x_2 - x_1, \ldots, x_n - x_{n-1}\} < \delta(\varepsilon).$$

This type of limiting is generally used in defining the Riemann integral and sometimes used in defining the Riemann-Stieltjes integral. However, many authors employ the definition we introduced, which is due to S. Pollard, for it enlarges slightly the class of integrable functions. As a result of this enlargement, the next result is valid without any additional restriction. See Exercises 29. P–R.

29.6   THEOREM.   (a) *Suppose that $a \le c \le b$ and that f is integrable with respect to g over both of the subintervals $[a, c]$ and $[c, b]$. Then f is integrable with respect to g on the interval $[a, b]$ and*

(29.9) $$\int_a^b f \, dg = \int_a^c f \, dg + \int_c^b f \, dg.$$

(b) *Let f be integrable with respect to g on the interval $[a, b]$ and let c satisfy $a \le c \le b$. Then f is integrable with respect to g on the subintervals $[a, c]$ and $[c, b]$ and formula (29.9) holds.*

PROOF.   (a) If $\varepsilon > 0$, let P_ε' be a partition of $[a, c]$ such that if P' is a refinement of P_ε', then inequality (29.3) holds for any Riemann-Stieltjes sum. Let P_ε'' be a corresponding partition of $[c, b]$. If P_ε is the partition of $[a, b]$ formed by using the partition points in both P_ε' and P_ε'', and if P is a refinement of P_ε, then

$$S(P; f, g) = S(P'; f, g) + S(P''; f, g),$$

where P', P'' denote the partitions of $[a, c]$, $[c, b]$ induced by P and where the corresponding intermediate points are used. Therefore, we have

$$\left| \int_a^c f \, dg + \int_c^b f \, dg - S(P; f, g) \right|$$

$$\le \left| \int_a^c f \, dg - S(P'; f, g) \right| + \left| \int_c^b f \, dg - S(P''; f, g) \right| < 2\varepsilon.$$

It follows that f is integrable with respect to g over $[a, b]$ and that the value of its integral is

$$\int_a^c f \, dg + \int_c^b f \, dg.$$

(b) We shall use the Cauchy Criterion 29.4 to prove that f is integrable over $[a, c]$. Since f is integrable over $[a, b]$, given $\varepsilon > 0$ there is a partition

Q_ε of $[a, b]$ such that if P, Q are refinements of Q_ε, then relation (29.4) holds for any corresponding Riemann-Stieltjes sums. It is clear that we may suppose that the point c belongs to Q_ε, and we let Q_ε' be the partition of $[a, c]$ consisting of those points of Q_ε which belong to $[a, c]$. Suppose that P' and Q' are partitions of $[a, c]$ which are refinements of Q_ε' and extend them to partitions P and Q of $[a, b]$ by using the points in Q_ε which belong to $[c, b]$. Since P, Q are refinements of Q_ε, then relation (29.4) holds. However, it is clear from the fact that P, Q are identical on $[c, b]$ that, if we use the same intermediate points, then

$$|S(P'; f, g) - S(Q'; f, g)| = |S(P; f, g) - S(Q; f, g)| < \varepsilon.$$

Therefore, the Cauchy Criterion establishes the integrability of f with respect to g over the subinterval $[a, c]$ and a similar argument also applies to the interval $[c, b]$. Once this integrability is known, part (a) yields the validity of formula (29.9).   Q.E.D.

Thus far we have not interchanged the roles of the integrand f and the integrator g, and it may not have occurred to the reader that it might be possible to do so. Although the next result is not exactly the same as the "integration by parts formula" of calculus, the relation is close and this result is usually referred to by that name.

29.7 INTEGRATION BY PARTS. *A function f is integrable with respect to g over $[a, b]$ if and only if g is integrable with respect to f over $[a, b]$. In this case,*

$$(29.10) \qquad \int_a^b f \, dg + \int_a^b g \, df = f(b)g(b) - f(a)g(a).$$

PROOF. We shall suppose that f is integrable with respect to g. Let $\varepsilon > 0$ and let P_ε be a partition of $[a, b]$ such that if Q is a refinement of P_ε and $S(Q; f, g)$ is any corresponding Riemann-Stieltjes sum, then

$$(29.11) \qquad \left| S(Q; f, g) - \int_a^b f \, dg \right| < \varepsilon.$$

Now let P be a refinement of P_ε and consider a Riemann-Stieltjes sum $S(P; g, f)$ given by

$$S(P; g, f) = \sum_{k=1}^n g(\xi_k)\{f(x_k) - f(x_{k-1})\},$$

where $x_{k-1} \le \xi_k \le x_k$. Let $Q = (y_0, y_1, \ldots, y_{2n})$ be the partition of $[a, b]$ obtained by using both the ξ_k and x_k as partition points; hence $y_{2k} = x_k$ and $y_{2k-1} = \xi_k$. Add and subtract the terms $f(y_{2k})g(y_{2k})$ for $k = 0, 1, \ldots, n$, to $S(P; g, f)$ and rearrange to obtain

$$S(P; g, f) = f(b)g(b) - f(a)g(a) - \sum_{k=1}^{2n} f(\eta_k)\{g(y_k) - g(y_{k-1})\},$$

where the intermediate points η_k are selected to be the points x_j. Thus we have

$$S(P; g, f) = f(b)g(b) - f(a)g(a) - S(Q; f, g),$$

where the partition $Q = (y_0, y_1, \ldots, y_{2n})$ is a refinement of P_ε. In view of formula (29.11),

$$\left| S(P; g, f) - \left\{ f(b)g(b) - f(a)g(a) - \int_a^b f \, dg \right\} \right| < \varepsilon$$

provided P is a refinement of P_ε. This proves that g is integrable with respect to f over $[a, b]$ and establishes formula (29.10). Q.E.D.

Modification of the Integral

When the integrator function g has a continuous derivative, it is possible and often convenient to replace the Riemann-Stieltjes integral by a Riemann integral. We now establish the validity of this reduction.

29.8 THEOREM. *If the derivative g' exists and is continuous on J and if f is integrable with respect to g, then the product fg' is Riemann integrable and*

$$(29.12) \qquad \int_a^b f \, dg = \int_a^b fg'.$$

PROOF. The hypothesis implies that g' is uniformly continuous on J. If $\varepsilon > 0$, let $P = (x_0, x_1, \ldots, x_n)$ be a partition of J such that if ξ_k and ζ_k belong to $[x_{k-1}, x_k]$ then $|g'(\xi_k) - g'(\zeta_k)| < \varepsilon$. We consider the difference of the Riemann-Stieltjes sum $S(P; f, g)$ and the Riemann sum $S(P; fg')$, using the same intermediate points ξ_k. In doing so we have a sum of terms of the form

$$f(\xi_k)\{g(x_k) - g(x_{k-1})\} - f(\xi_k)g'(\xi_k)\{x_k - x_{k-1}\}.$$

If we apply the Mean Value Theorem 27.6 to g, we can write this difference in the form

$$f(\xi_k)\{g'(\zeta_k) - g'(\xi_k)\}(x_k - x_{k-1}),$$

where ζ_k is some point in the interval $[x_{k-1}, x_k]$. Since this term is dominated by $\varepsilon \|f\| (x_k - x_{k-1})$, we conclude that

$$|S(P; f, g) - S(P; fg')| \le \varepsilon \|f\| (b - a),$$

provided the partition P is sufficiently fine. Since the integral on the left side of (29.12) exists and is the limit of the Riemann-Stieltjes sums $S(P; f, g)$, we infer that the integral on the right side of (29.12) also exists and that the equality holds. Q.E.D.

For an extension of this result, see Theorem 30.13.

29.9 EXAMPLES.   (a) It follows from results to be proved in Section 30 that $f(x) = x$ is integrable with respect to $g(x) = x^2$ on $J = [0, 1]$. Assuming this, Theorem 29.8 shows that

$$\int_0^1 x\, d(x^2) = \int_0^1 x \cdot 2x\, dx = \tfrac{2}{3}x^3 \Big|_0^1 = \frac{2}{3}.$$

(Here we have made use of results from calculus that will be proved in Section 30.)

(b) If we apply Theorem 29.7 on Integration by Parts to the functions in (a), we get

$$\int_0^1 x\, d(x^2) = x^3 \Big|_0^1 - \int_0^1 x^2\, dx = 1 - \tfrac{1}{3}x \Big|_0^1 = \frac{2}{3}.$$

(c) It follows from results to be proved in Section 30 that $f(x) = \sin x$ is integrable with respect to f on $J = [0, \pi/2]$. Assuming this, we have

$$\int_0^{\pi/2} \sin x\, d(\sin x) = \int_0^{\pi/2} \sin x \cos x\, dx = \tfrac{1}{2}(\sin x)^2 \Big|_0^{\pi/2} = \frac{1}{2}.$$

(d) If we apply Theorem 29.7 on Integration by Parts to part (c) we get

$$\int_0^{\pi/2} \sin x\, d(\sin x) = (\sin x)^2 \Big|_0^{\pi/2} - \int_0^{\pi/2} \sin x\, d(\sin x),$$

whence it follows that

$$\int_0^{\pi/2} \sin x\, d(\sin x) = \tfrac{1}{2}(\sin x)^2 \Big|_0^{\pi/2} = \frac{1}{2}.$$

(e) We introduce the **greatest integer function** on R to R, denoted by the special symbol $[\,\cdot\,]$ and defined by requiring that if $x \in R$, then $[x]$ is the largest integer less than or equal to x. Hence $[\pi] = 3$, $[e] = 2$, $[-2.5] = -3$. The reader should draw a graph of this function and note that it is continuous from the right, with jumps equal to 1 at the integers. It follows that if f is continuous on $[0, 5]$, then f is integrable with respect to $g(x) = [x]$, $x \in [0, 5]$, and that

$$\int_0^5 f(x)\, d([x]) = \sum_{j=1}^5 f(j).$$

(f) It follows from results in Section 30 that $f(x) = x^2$ is integrable with respect to both $g_1(x) = x$ and $g_2(x) = [x]$ on $[0, 5]$. Therefore it is integrable with respect to $g(x) = x + [x]$ and we have

$$\int_0^5 x^2\, d(x + [x]) = \int_0^5 x^2\, dx + \int_0^5 x^2\, d([x])$$
$$= \tfrac{1}{3}5^3 + 1^2 + 2^2 + 3^2 + 4^2 + 5^2.$$

Exercises

29.A. If f is constant on the interval $[a, b]$, then it is integrable with respect to any function g and

$$\int_a^b f \, dg = f(a)\{g(b) - g(a)\}.$$

29.B. If g is as in Example 29.3(c), show that f is integrable with respect to g if and only if f is continuous at a.

29.C. Let g be defined on $I = [0, 1]$ by $g(x) = 0$ for $0 \le x \le \frac{1}{2}$ and $g(x) = 1$ for $\frac{1}{2} < x \le 1$. Show that f is integrable with respect to g on I if and only if it is continuous at $\frac{1}{2}$ from the right. In this case the value of the integral is $f(\frac{1}{2})$.

29.D. Show that the function f, given in Example 29.3(h) is Riemann integrable on I and that the value of its integral is 0.

29.E. If f is integrable on $[a, b]$ with respect to f, then

$$\int_a^b f \, df = \frac{1}{2}\{(f(b))^2 - (f(a))^2\}.$$

(a) Prove this by examining the two Riemann-Stieltjes sums for a partition $P = (x_0, x_1, \ldots, x_n)$ obtained by taking $\xi_k = x_{k-1}$ and $\xi_k = x_k$.

(b) Prove this by using the Integration by Parts Theorem 29.7.

29.F. Show directly that if f is the greatest integer function $f(x) = [x]$ defined in Example 29.9(e), then f is *not* integrable with respect to f on the interval $[0, 2]$.

29.G. If f is Riemann integrable on $[0, 1]$, then

$$\int_0^1 f = \lim \left(\frac{1}{n} \sum_{k=1}^n f\left(\frac{k}{n}\right) \right).$$

29.H. Show that if g is *not* integrable on $[0, 1]$, then the sequence of averages

$$\left(\frac{1}{n} \sum_{k=1}^n g\left(\frac{k}{n}\right) \right)$$

may or may not be convergent.

29.I. Show that the function h, defined on I by $h(x) = x$ for x rational and $h(x) = 0$ for x irrational, is not Riemann integrable on I.

29.J. Suppose that f is Riemann integrable on $[a, b]$. If f_1 is a function on $[a, b]$ to R such that $f_1(x) = f(x)$ except for a *finite number* of points in $[a, b]$, show that f_1 is Riemann integrable and that

$$\int_a^b f_1 = \int_a^b f.$$

(Thus we can change the value of a Riemann integrable function—or leave it undefined—at a finite number of points.)

20.K. Give an example to show that the conclusion of the preceding exercise may fail if the number of exceptional points is infinite.

29.L. Let $c \in (a, b)$ and let k be defined on $[a, b]$ by $k(c) = 1$ and $k(x) = 0$ for $x \in [a, b]$, $x \ne c$. If $f : [a, b] \to R$ is continuous at c show directly that f is

k-integrable, that k is f-integrable, and that

$$\int_a^b f \, dk = \int_a^b k \, df = 0.$$

29.M. Suppose that f is g-integrable on $[a, b]$. If $g_1:[a, b] \to \mathbf{R}$ is such that $g_1(x) = g(x)$ except for a finite number of points in (a, b) at which f is continuous, then f is g_1-integrable and

$$\int_a^b f \, dg_1 = \int_a^b f \, dg.$$

29.N. Suppose that g is continuous on $[a, b]$, that $x \mapsto g'(x)$ exists and is continuous on $[a, b] \setminus \{c\}$, and that the one-sided limits

$$g'(c-) = \lim_{\substack{x \to c \\ x < c}} g'(x), \qquad g'(c+) = \lim_{\substack{x \to c \\ x > c}} g'(x)$$

exist. If f is integrable with respect to g on $[a, b]$, then fg' can be defined at c to be Riemann integrable on $[a, b]$ and such that

$$\int_a^b f \, dg = \int_a^b fg'.$$

(Hint: consider Exercise 27.N.)

29.O If f is Riemann integrable on $[-5, 5]$, show that f is integrable with respect to $g(x) = |x|$ and

$$\int_{-5}^5 f \, dg = \int_0^5 f - \int_{-5}^0 f.$$

29.P. If $P = (x_0, x_1, \ldots, x_n)$ is a partition of $J = [a, b]$, let $\|P\|$ be defined to be

$$\|P\| = \sup\{x_j - x_{j-1} : j = 1, 2, \ldots, n\};$$

we call $\|P\|$ the **norm** of the partition P. Define f to be $(*)$-**integrable** with respect to g on J in case there exists a number A with the property: if $\varepsilon > 0$ then there is a $\delta(\varepsilon) > 0$ such that if $\|P\| < \delta(\varepsilon)$ and if $S(P; f, g)$ is any corresponding Riemann-Stieltjes sum, then $|S(P; f, g) - A| < \varepsilon$. If this is satisfied the number A is called the $(*)$-**integral** of f with respect to g on J. Show that if f is $(*)$-integrable with respect to g on J, then f is integrable with respect to g (in the sense of Definition 29.2) and that the values of these integrals are equal.

29.Q. Let g be defined on I as in Exercise 29.C. Show that a bounded function f is $(*)$-integrable with respect to g in the sense of the preceding exercise if and only if f is continuous at $\frac{1}{2}$ when the value of the $(*)$-integral is $f(\frac{1}{2})$. If h is defined by

$$h(x) = 0, \qquad 0 \le x < \tfrac{1}{2},$$
$$= 1, \qquad \tfrac{1}{2} \le x \le 1,$$

then h is $(*)$-integrable with respect to g on $[0, \frac{1}{2}]$ and on $[\frac{1}{2}, 1]$ but it is not $(*)$-integrable with respect to g on $[0, 1]$. Hence Theorem 29.6(a) may fail for the $(*)$-integral.

29.R. Let $g(x) = x$ for $x \in J$. Show that for this integrator, a function f is integrable in the sense of Definition 29.2 if and only if it is $(*)$-integrable in the sense of Exercise 29.P.

29.S. Let f be Riemann integrable on J and let $f(x) \geq 0$ for $x \in J$. If f is continuous at a point $c \in J$ and if $f(c) > 0$, then

$$\int_a^b f > 0.$$

29.T. Let f be Riemann integrable on J and let $f(x) > 0$ for $x \in J$. Show that

$$\int_a^b f > 0.$$

(Hint: for each $n \in N$, let H_n be the closure of the set of points $x \in J$ such that $f(x) > 1/n$ and apply Exercise 11.N.)

Projects

29.α. The following outline is sometimes used as an approach to the Riemann-Stieltjes integral when the integrator function g is monotone increasing on the interval J. [This development has the advantage that it permits the definition of upper and lower integrals which always exists for a bounded function f. However, it has the disadvantage that it puts an additional restriction on g and tends to blemish somewhat the symmetry of the Riemann-Stieltjes integral given by the Integration of Parts Theorem 29.7.] If $P = (x_0, x_1, \ldots, x_n)$ is a partition of $J = [a, b]$ and f is a bounded function on J, let m_j, M_j be defined to be the infimum and the supremum of $\{f(x) : x_{j-1} \leq x \leq x_j\}$, respectively. Corresponding to the partition P, define the **lower** and the **upper sums** of f with respect to g to be

$$L(P; f, g) = \sum_{j=1}^n m_j \{g(x_j) - g(x_{j-1})\},$$

$$U(P; f, g) = \sum_{j=1}^n M_j \{g(x_j) - g(x_{j-1})\}.$$

(a) If $S(P; f, g)$ is any Riemann-Stieltjes sum corresponding to P, then

$$L(P; f, g) \leq S(P; f, g) \leq U(P; f, g).$$

(b) If $\varepsilon > 0$ then there exists a Riemann-Stieltjes sum $S_1(P; f, g)$ corresponding to P such that

$$S_1(P; f, g) \leq L(P; f, g) + \varepsilon,$$

and there exists a Riemann-Stieltjes sum $S_2(P; f, g)$ corresponding to P such that

$$U(P; f, g) - \varepsilon \leq S_2(P; f, g).$$

(c) If P and Q are partitions of J and if Q is a refinement of P (that is, $P \subseteq Q$), then

$$L(P; f, g) \leq L(Q; f, g) \leq U(Q; f, g) \leq U(P; f, g).$$

(d) If P_1 and P_2 are any partitions of J, then $L(P_1; f, g) \le U(P_2; f, g)$. [Hint: let Q be a partition which is a refinement of both P_1 and P_2 and apply (c).]

(e) Define the **lower** and the **upper integral** of f with respect to g to be, respectively

$$L(f, g) = \sup\{L(P; f, g)\},$$

$$U(f, g) = \inf\{U(P; f, g)\};$$

here the supremum and the infimum are taken over all partitions P of J. Show that $L(f, g) \le U(f, g)$.

(f) Prove that f is integrable with respect to the increasing function g if and only if the lower and upper integrals introduced in (e) are equal. In this case the common value of these integrals equals

$$\int_a^b f \, dg.$$

Show that f is integrable with respect to g if and only if the following Riemann condition is satisfied: for every $\varepsilon > 0$ there exists a partition P such that $U(P; f, g) - L(P; f, g) < \varepsilon$.

(g) If f_1 and f_2 are bounded on J, then the lower and upper integrals of $f_1 + f_2$ satisfy

$$L(f_1 + f_2, g) \ge L(f_1, g) + L(f_2, g),$$

$$U(f_1 + f_2, g) \le U(f_1, g) + U(f_2, g).$$

Show that strict inequality can hold in these relations.

29.β. This (and the next two) projects introduce and study the important class of functions which have "bounded variation" on a compact interval. Let $f : [a, b] \to \mathbf{R}$ be given; if $P = (a = x_0 < x_1 < \cdots < x_n = b)$ is a partition of $[a, b]$, let $v_f(P)$ be defined by

$$v_f(P) = \sum_{k=1}^n |f(x_k) - f(x_{k-1})|.$$

If the set $\{v_f(P) : P \text{ a partition of } [a, b]\}$ is bounded, we say that f has **bounded variation** on $[a, b]$. The collection of all functions having bounded variation on $[a, b]$ is denoted by $BV([a, b])$ or by $BV[a, b]$. If $f \in BV[a, b]$, then we define

$$V_f[a, b] = \sup\{v_f(P) : P \text{ a partition of } [a, b]\}.$$

We call the number $V_f[a, b]$ the **total variation** of f on $[a, b]$. Show that $V_f[a, b] = 0$ if and only if f is a constant function on $[a, b]$.

(a) If $f : [a, b] \to \mathbf{R}$, if P and Q are partitions of $[a, b]$, and if $P \supseteq Q$, show that $v_f(P) \ge v_f(Q)$. If $f \in BV[a, b]$, show that there exists a sequence (P_n) of partitions of $[a, b]$ such that $V_f[a, b] = \lim (v_f(P_n))$.

(b) If f is monotone increasing on $[a, b]$, show that $f \in BV[a, b]$ and that $V_f[a, b] = f(b) - f(a)$. What if f is monotone decreasing on $[a, b]$?

(c) If $g : [a, b] \to \mathbf{R}$ satisfies the Lipschitz condition $|g(x) - g(y)| \le M |x - y|$ for all x, y in $[a, b]$, show that $g \in BV[a, b]$ and that $V_g[a, b] \le M(b - a)$. If $|h'(x)| \le M$ for all $x \in [a, b]$, then $h \in BV[a, b]$ and $V_h[a, b] \le M(b - a)$. However, consider $k(x) = \sqrt{x}$ on $[0, 1]$.

(d) Let $f:[0, 1] \to \mathbf{R}$ be defined by $f(x) = 0$ for $x = 0$ and $f(x) = \sin(1/x)$ for $0 < x \le 1$. Show that f does not have bounded variation on $[0, 1]$. If g is defined by $g(x) = xf(x)$ for $x \in [0, 1]$, show that g is continuous but does not have bounded variation on $[0, 1]$. However, if h is defined by $h(x) = x^2 f(x)$ for $x \in [0, 1]$, show that h does have bounded variation on $[0, 1]$.

(e) If $f \in BV[a, b]$, show that $|f(x)| \le |f(a)| + V_f[a, b]$ for all $x \in [a, b]$, so that f is bounded on $I = [a, b]$ and $\|f\|_I \le |f(a)| + V_f[a, b]$.

(f) If $f, g \in BV[a, b]$ and $\alpha \in \mathbf{R}$, show that αf and $f + g$ belong to $BV[a, b]$ and that

$$V_{\alpha f}[a, b] = |\alpha| \, V_f[a, b],$$

$$V_{f+g}[a, b] \le V_f[a, b] + V_g[a, b].$$

Hence $BV[a, b]$ is a vector space of functions.

(g) If $f, g \in BV[a, b]$, show that the product fg belongs to $BV[a, b]$ and that

$$V_{fg}[a, b] \le \|f\|_I \, V_g[a, b] + \|g\|_I \, V_f[a, b].$$

Show that the quotient of two functions in $BV[a, b]$ may not belong to $BV[a, b]$.

(h) Show that the mapping $f \mapsto V_f[a, b]$ is not a norm on the vector space $BV[a, b]$, but that the mapping

$$f \mapsto \|f\|_{BV} = |f(a)| + V_f[a, b]$$

is a norm on this space.

29.γ. We continue our study of functions of bounded variation on an interval $[a, b] \subseteq \mathbf{R}$.

(a) If $f \in BV[a, b]$ and if $c \in (a, b)$, show that the restrictions of f to $[a, c]$ and $[c, b]$ have bounded variation on these intervals and that

$$V_f[a, b] = V_f[a, c] + V_f[c, b].$$

Conversely, if $g : [a, b] \to \mathbf{R}$ is such that for some $c \in (a, b)$ we have $g \in BV[a, c]$ and $g \in BV[c, b]$, then $g \in BV[a, b]$.

(b) If $f \in BV[a, b]$, then we define $p_f(x) = V_f[a, x]$ for $x \in (a, b]$, and $p_f(a) = 0$. Show that p_f is an increasing function on $[a, b]$.

(c) Note that if $a \le x \le y \le b$, then

$$f(y) - f(x) \le V_f[x, y].$$

Show that if we define $n_f(x) = p_f(x) - f(x)$ for $x \in [a, b]$, then n_f is an increasing function.

(d) Show that a function $f : [a, b] \to \mathbf{R}$ belongs to $BV[a, b]$ if and only if it is the difference of two increasing functions.

(e) If $f \in BV[a, b]$ is continuous from the right at a point $c \in [a, b)$, and if $\varepsilon > 0$, show that there exists $\delta > 0$ and a partition such that if $Q = (c < x_1 < \cdots < x_n = b)$ is a sufficiently fine partition of $[c, b]$ with $x_1 - c < \delta$, then

$$V_f[a, b] - \tfrac{1}{2}\varepsilon \le \tfrac{1}{2}\varepsilon + \sum_{k=2}^{n} |f(x_k) - f(x_{k-1})| \le \tfrac{1}{2}\varepsilon + V_f[x_1, b],$$

whence it follows that

$$V_f[c, x_1] = V_f[c, b] - V_f[x_1, b] < \varepsilon.$$

Show that f is continuous at $c \in [a, b]$ if and only if p_f is continuous at c.

(f) Deduce that a continuous function $f : [a, b] \to R$ belongs to $BV[a, b]$ if and only if it is the difference of two increasing continuous functions.

29.δ. It was proved by Lebesgue that a function with bounded variation has a derivative at every point except possibly for a set of "measure zero." The proof of this result is rather difficult and will not be outlined here, but we shall obtain some further properties of such functions.

(a) If $f \in BV[a, b]$ and if $c \in (a, b)$, then the left- and right-hand limits of f at c exist. These limits are equal except possibly for a countable collection of points in $[a, b]$.

(b) If (f_n) is a sequence of continuous functions in $BV[a, b]$ which is uniformly convergent on $[a, b]$ to a function f, show that it does not follow that f belongs to $BV[a, b]$.

(c) Let (f_n) be a sequence in $BV[a, b]$ which converges at every point of $[a, b]$ to a function f, and suppose that for some $M > 0$ we have $V_{f_n}[a, b] \le M$ for all $n \in N$. Show that f belongs to $BV[a, b]$ and that $V_f[a, b] \le M$.

(d) Let (f_n) be a sequence in $BV[a, b]$ such that $\|f_n - f_m\|_{BV} \to 0$ as $m, n \to \infty$. Show that there exists a function $f \in BV[a, b]$ such that $\|f_n - f\|_{BV} \to 0$ as $n \to \infty$.

(e) Let (f_n) be a sequence of monotone increasing functions defined on $I = [a, b]$ such that $\|f_n\|_I \le M$ for all $n \in N$. Use the diagonal process to obtain a subsequence (g_k) of (f_n) which converges for each rational number r in $[a, b]$. Define $g(r) = \lim (g_k(r))$ for $r \in Q \cap [a, b]$. Show that g is increasing on $Q \cap [a, b]$. We define g for $x \in [a, b)$ as the right-hand limit $g(x) = \lim_{r \to x+} g(r)$. Show that if $c \in [a, b)$ is a point of continuity of g, then $g(c) = \lim_k g_k(c)$. Since g has at most a countable collection of points of discontinuity, a further application of the diagonal process can be applied to get a subsequence (h_m) of (g_k) which converges everywhere on $[a, b]$.

(f) Making use of part (e), establish the following result, called the **Helly Selection Theorem:** Let (f_n) be a sequence of functions in $BV[a, b]$ such that $\|f_n\|_{BV} \le M$ for all $n \in N$. Then there exists a subsequence of (f_n) which converges at every point of $[a, b]$ to a function $f \in BV[a, b]$ for which $\|f\|_{BV} \le M$.

Section 30 Existence of the Integral

In the preceding section we established some useful properties of the Riemann-Stieltjes integral. However, we have not yet shown the existence of the integral for very many functions.

In this section we shall focus our attention on monotone increasing integrator functions, although much of what we do can be extended to functions g which have **bounded variation** on an interval $J = [a, b]$ in the sense that there exists a constant $M > 0$ such that if $P = (x_0, x_1, \ldots, x_n)$ is any partition of J, then

(30.1)
$$\sum_{j=1}^{n} |g(x_j) - g(x_{j-1})| \le M.$$

It is clear that, if g is monotone increasing, then the sum in (30.1) telescopes and one can take $M = g(b) - g(a)$. Hence a monotone increasing function has bounded variation. Conversely, it can be shown that every function with bounded variation is the difference of two increasing functions. (See Project 29.γ.)

We shall first establish a very powerful result.

30.1 RIEMANN CRITERION FOR INTEGRABILITY. *Let $J = [a, b]$ and let g be monotone increasing on J. A function $f : J \to R$ is integrable with respect to g on J if and only if for every $\varepsilon > 0$ there exists a partition P_ε of J such that if $P = (x_0, x_1, \ldots, x_n)$ is a refinement of P_ε, then*

$$(30.2) \qquad \sum_{j=1}^{n} (M_j - m_j)\{g(x_j) - g(x_{j-1})\} < \varepsilon,$$

where $M_j = \sup \{f(x) : x \in [x_{j-1}, x_j]\}$ and $m_j = \inf \{f(x) : x \in [x_{j-1}, x_j]\}$ for $j = 1, \ldots, n$.

PROOF.   If f is integrable with respect to g and $\varepsilon > 0$ is given, let P_ε be a partition of J such that if $P = (x_0, x_1, \ldots, x_n)$ is a refinement of P_ε, then

$$\left| S(P; f, g) - \int_a^b f \, dg \right| < \varepsilon$$

for any Riemann-Stieltjes sum corresponding to P. Now choose y_j and z_j in $[x_{j-1}, x_j]$ such that

$$M_j - \varepsilon < f(y_j), \qquad f(z_j) < m_j + \varepsilon.$$

This implies that $M_j - m_j < f(y_j) - f(z_j) + 2\varepsilon$ and hence

$$\sum_{j=1}^{n} (M_j - m_j)\{g(x_j) - g(x_{j-1})\} \le \sum_{j=1}^{n} f(y_j)\{g(x_j) - g(x_{j-1})\}$$
$$- \sum_{j=1}^{n} f(z_j)\{g(x_j) - g(x_{j-1})\} + 2\varepsilon\{g(b) - g(a)\}.$$

Now the right side of this inequality contains two Riemann-Stieltjes sums corresponding to P, which cannot differ by more than 2ε. Hence the condition (30.2) is satisfied with ε replaced by $2\varepsilon\{1 + g(b) - g(a)\}$.

Conversely suppose $\varepsilon > 0$ is given and P_ε is a partition such that (30.2) holds for any partition $P = (x_0, x_1, \ldots, x_n)$ refining P_ε. Let $Q = (y_0, y_1, \ldots, y_m)$ be a refinement of P; we shall estimate the difference $S(P; f, g) - S(Q; f, g)$ of two corresponding sums. Since every point in P belongs to Q, we can express both of these sums in the form

$$S(P; f, g) = \sum_{k=1}^{m} f(u_k)\{g(y_k) - g(y_{k-1})\},$$

$$S(Q; f, g) = \sum_{k=1}^{m} f(v_k)\{g(y_k) - g(y_{k-1})\}.$$

However, to write $S(P; f, g)$ in terms of the points in Q, we must permit repetitions of the points u_k and do not require u_k to belong to $[y_{k-1}, y_k]$. However, both u_k and v_k do belong to some interval $[x_{j-1}, x_j]$ and in this case $|f(u_k) - f(v_k)| \leq M_j - m_j$. Multiplying by $g(y_k) - g(y_{k-1}) \geq 0$ and adding, we obtain

$$|S(P; f, g) - S(Q; f, g)| \leq \sum (M_j - m_j)\{g(x_j) - g(x_{j-1})\} < \varepsilon.$$

Finally, let P and P' be arbitrary refinements of P_ε and let Q be a common refinement of both P and P'. Since the preceding argument applies to both P and P', we deduce that any sums $S(P; f, g)$ and $S(P'; f, g)$ could differ by at most 2ε. Hence the Cauchy Criterion 29.4 applies to yield the integrability of f. · \hfill Q.E.D.

30.2 INTEGRABILITY THEOREM. *If f is continuous and g is monotone increasing on J, then f is integrable with respect to g on J.*

PROOF. Since f is uniformly continuous on J, given $\varepsilon > 0$, there exists a $\delta(\varepsilon) > 0$ such that if $x, y \in J$ and $|x - y| < \delta(\varepsilon)$, then $|f(x) - f(y)| < \varepsilon$. Let $P_\varepsilon = (z_0, z_1, \ldots, z_r)$ be a partition such that $\sup \{z_k - z_{k-1}\} < \delta(\varepsilon)$. If $P = (x_0, x_1, \ldots, x_n)$ is a refinement of P_ε, then also $\sup \{x_j - x_{j-1}\} < \delta(\varepsilon)$ and so $M_j - m_j < \varepsilon$, whence it follows that

$$\sum_{j=1}^{n} (M_j - m_j)\{g(x_j) - g(x_{j-1})\} \leq \varepsilon (g(b) - g(a)).$$

Since $\varepsilon > 0$ is arbitrary, the Riemann Criterion applies. \hfill Q.E.D.

30.3 COROLLARY. *If f is monotone and g is continuous on J, then f is integrable with respect to g on J.*

PROOF. Apply the preceding theorem and Theorem 29.7 to $\pm f$. \hfill Q.E.D.

The Riemann Criterion enables us to show that the absolute value and the product of integrable functions are integrable.

30.4 THEOREM. *Let g be monotone increasing on $J = [a, b]$.*
(a) *If $f : J \to R$ is integrable, then $|f|$ is integrable with respect to g on J.*
(b) *If f_1 and f_2 are integrable, then the product $f_1 f_2$ is integrable with respect to g on J.*

PROOF. Let M_j and m_j have the meaning given in the Riemann Criterion and observe that

$$M_j - m_j = \sup \{f(x) - f(y) : x, y \in [x_{j-1}, x_j]\}.$$

To prove (a), note that $||f(x)| - |f(y)|| \leq |f(x) - f(y)|$, so the Riemann Criterion implies that $|f|$ is integrable when f is.

We also observe that if $|f(x)| \leq K$ for $x \in J$, then $|(f(x))^2 - (f(y))^2| \leq 2K |f(x) - f(y)|$, so the Riemann Criterion implies that f^2 is integrable when f

is. To prove that $f_1 f_2$ is integrable when f_1 and f_2 are, note that

$$2f_1 f_2 = (f_1 + f_2)^2 - f_1{}^2 - f_2{}^2. \qquad \text{Q.E.D.}$$

30.5 LEMMA. *Let g be monotone increasing on $J = [a, b]$ and suppose that f is integrable with respect to g on J. Then*

$$(30.3) \qquad \left| \int_a^b f \, dg \right| \le \int_a^b |f| \, dg \le \|f\|_J (g(b) - g(a)).$$

If $m \le f(x) \le M$ for all $x \in J$, then

$$(30.4) \qquad m(g(b) - g(a)) \le \int_a^b f \, dg \le M(g(b) - g(a)).$$

PROOF. It follows from Theorem 30.4 that $|f|$ is integrable with respect to g. If $P = (x_0, x_1, \ldots, x_n)$ is a partition of J and (z_j) is a set of intermediate points, then for $j = 1, 2, \ldots, n$, we have

$$-\|f\|_J \le -|f(z_j)| \le f(z_j) \le |f(z_j)| \le \|f\|_J.$$

Multiply by $g(x_j) - g(x_{j-1}) \ge 0$ and sum to obtain the estimate

$$-\|f\|_J (g(b) - g(a)) \le -S(P; |f|, g) \le S(P; f, g) \le S(P; |f|, g)$$
$$\le \|f\|_J (g(b) - g(a)),$$

whence it follows that $|S(P; f, g)| \le S(P; |f|, g) \le \|f\|_J (g(b) - g(a))$, which implies the validity of (30.3). The proof of (30.4) is similar and will be omitted.   Q.E.D.

Evaluation of the Integral

The next two results are useful in their own right, but also lead to the Fundamental Theorem which is the primary tool for evaluating Riemann integrals.

30.6 FIRST MEAN VALUE THEOREM. *If g is increasing on $J = [a, b]$ and f is continuous on J to $\mathbf{R}$, then there exists a number c in J such that*

$$(30.5) \qquad \int_a^b f \, dg = f(c) \int_a^b dg = f(c)\{g(b) - g(a)\}.$$

PROOF. If $m = \inf \{f(x) : x \in J\}$ and $M = \sup \{f(x) : x \in J\}$, it was seen in the preceding lemma that

$$m\{g(b) - g(a)\} \le \int_a^b f \, dg \le M\{g(b) - g(a)\}.$$

If $g(b) = g(a)$, then the relation (30.5) is trivial; if $g(b) > g(a)$, then it follows from Bolzano's Intermediate Value Theorem 22.4 that there exists a number c in J such that

$$f(c) = \left\{ \int_a^b f\,dg \right\} / \{g(b) - g(a)\}. \qquad \text{Q.E.D.}$$

30.7 DIFFERENTIATION THEOREM. *Suppose that f is continuous on J and that g is increasing on J and has a derivative at a point c in J. Then the function F, defined for x in J by*

$$F(x) = \int_a^x f\,dg,$$

has a derivative at c and $F'(c) = f(c)g'(c)$.

PROOF. If $h > 0$ is such that $c + h$ belongs to J, then it follows from Theorem 29.6 and the preceding result that

$$F(c + h) - F(c) = \int_a^{c+h} f\,dg - \int_a^c f\,dg = \int_c^{c+h} f\,dg$$
$$= f(c_1)\{g(c + h) - g(c)\},$$

for some c_1 with $c \le c_1 \le c + h$. A similar relation holds if $h < 0$. Since f is continuous and g has a derivative at c, then $F'(c)$ exists and equals $f(c)g'(c)$.

Q.E.D.

Specializing this theorem to the Riemann case, we obtain the result which provides the basis for the familiar method of evaluating integrals in calculus.

30.8 FUNDAMENTAL THEOREM OF INTEGRAL CALCULUS. *Let f be continuous on $J = [a, b]$. A function F on J satisfies*

$$(30.6) \qquad F(x) - F(a) = \int_a^x f \qquad \text{for } x \in J,$$

if and only if $F' = f$ on J.

PROOF. If relation (30.6) holds and $c \in J$, then it is seen from the preceding theorem that $F'(c) = f(c)$.

Conversely, let F_a be defined for x in J by

$$F_a(x) = \int_a^x f.$$

The preceding theorem asserts that $F_a' = f$ on J. If F is such that $F' = f$, then it follows from Theorem 27.9(ii), that there exists a constant C such that $F(x) = F_a(x) + C$, for $x \in J$. Since $F_a(a) = 0$, then $C = F(a)$ whence it follows that if $F' = f$ on J, then

$$F(x) - F(a) = \int_a^x f. \qquad \text{Q.E.D.}$$

NOTE. If F is a function defined on J such that $F' = f$ on J, then we sometimes say that F is an **indefinite integral,** an **anti-derivative,** or a **primitive** of f. In this terminology, the Differentiation Theorem 30.7 asserts that every continuous function has a primitive. Sometimes the Fundamental Theorem of Integral Calculus is formulated in ways differing from that given in 30.8, but it always includes the assertion that, under suitable hypotheses, the Riemann integral of f can be calculated by evaluating any primitive of f at the end points of the interval of integration. We have given the above formulation, which yields a necessary and sufficient condition for a function to be a primitive of a continuous function. A somewhat more general result, not requiring the continuity of the integrand, will be found in Exercise 30.J.

It should *not* be supposed that the Fundamental Theorem asserts that if the derivative f of a function F exists at every point of J, then f is integrable and (30.6) holds. In fact, it may happen that f is not Riemann integrable (see Exercise 30.K). Similarly, a function f may be Riemann integrable but not have a primitive (see Exercise 30.L).

As a consequence of the Fundamental Theorem and Theorem 29.8, we obtain the following variant of the First Mean Value Theorem 30.6, here stated for Riemann integrals.

30.9 FIRST MEAN VALUE THEOREM. *If f and p are continuous on $J = [a, b]$ and $p(x) \geq 0$ for all $x \in J$, then there exists a point $c \in J$ such that*

$$(30.7) \qquad \int_a^b f(x)p(x)\, dx = f(c)\int_a^b p(x)\, dx.$$

PROOF. Let $g : J \to R$ be defined for $x \in J$ by

$$g(x) = \int_a^x p(t)\, dt.$$

Since $p(x) \geq 0$, it is seen that g is increasing and it follows from the Differentiation Theorem 30.7 that $g' = p$. By Theorem 29.8, we conclude that

$$\int_a^b f\, dg = \int_a^b fp,$$

and from the First Mean Value Theorem 30.6, we infer that for some c in J, then

$$\int_a^b f\, dg = f(c)\int_a^b p. \qquad \text{Q.E.D.}$$

As a second application of Theorem 29.8 we shall reformulate Theorem 29.7, which is concerned with integration by parts, in a more traditional form. The proof will be left to the reader.

30.10 INTEGRATION BY PARTS. *If f and g have continuous derivatives on* $[a, b]$, *then*

$$\int_a^b fg' = f(b)g(b) - f(a)g(a) - \int_a^b f'g.$$

The next result is often useful.

30.11 SECOND MEAN VALUE THEOREM. (a) *If f is increasing and g is continuous on* $J = [a, b]$, *then there exists a point c in J such that*

(30.8)
$$\int_a^b f\,dg = f(a)\int_a^c dg + f(b)\int_c^b dg.$$

(b) *If f is increasing and h is continuous on J, then there exists a point c in J such that*

(30.9)
$$\int_a^b fh = f(a)\int_a^c h + f(b)\int_c^b h.$$

(c) *If φ is non-negative and increasing and h is continuous on J, then there exists a point c in J such that*

$$\int_a^b \varphi h = \varphi(b)\int_c^b h.$$

PROOF. The hypotheses, together with the Integrability Theorem 30.2 imply that g is integrable with respect to f on J. Furthermore, by the First Mean Value Theorem 30.6,

$$\int_a^b g\,df = g(c)\{f(b) - f(a)\}.$$

After using Theorem 29.7 concerning integration by parts, we conclude that f is integrable with respect to g and

$$\begin{aligned}
\int_a^b f\,dg &= \{f(b)g(b) - f(a)g(a)\} - g(c)\{f(b) - f(a)\} \\
&= f(a)\{g(c) - g(a)\} + f(b)\{g(b) - g(c)\} \\
&= f(a)\int_a^c dg + f(b)\int_c^b dg,
\end{aligned}$$

which establishes part (a). To prove (b) let g be defined on J by

$$g(x) = \int_a^x h,$$

so that $g' = h$. The conclusion then follows from part (a) by using Theorem 29.8. To prove (c) define F to be equal to φ for x in $(a, b]$ and define $F(a) = 0$. We now apply part (b) to F. Q.E.D.

Part (c) of the preceding theorem is frequently called the Bonnet† form of the second Mean Value Theorem. It is evident that there is a corresponding result for a decreasing function (cf. Exercise 30.N.)

Change of Variable

We shall now establish a theorem justifying the familiar formula relating to the "change of variable" in a Riemann integral.

30.12 CHANGE OF VARIABLE THEOREM. *Let φ be defined on an interval $[\alpha, \beta]$ to $\mathbf{R}$ with a continuous derivative and suppose that $a = \varphi(\alpha)$ and $b = \varphi(\beta)$. If f is continuous on the range of φ, then*

$$(30.10) \qquad \int_a^b f(x)\,dx = \int_\alpha^\beta f(\varphi(t))\varphi'(t)\,dt.$$

PROOF. Let $I = \varphi([\alpha, \beta])$ and let F be defined by

$$F(\xi) = \int_a^\xi f(x)\,dx \qquad \text{for } \xi \in I$$

and consider the function H defined by $H(t) = F(\varphi(t))$ for $\alpha \le t \le \beta$. Observe that $H(\alpha) = F(a) = 0$. If we differentiate with respect to t and use the fact that $F' = f$ (why?), we obtain

$$H'(t) = F'(\varphi(t))\varphi'(t) = f(\varphi(t))\varphi'(t).$$

We now apply the Fundamental Theorem to infer that

$$\int_a^b f(x)\,dx = F(b) = H(\beta) = \int_\alpha^\beta f(\varphi(t))\varphi'(t)\,dt. \qquad\qquad \text{Q.E.D.}$$

Modification of the Integral

The next result is often useful in reducing a Riemann-Stieltjes integral to a Riemann integral.

30.13 THEOREM. *If g' exists and f and g' are Riemann integrable on $[a, b]$, then f is Riemann-Stieltjes integrable with respect to g and*

$$(30.11) \qquad \int_a^b f\,dg = \int_a^b fg'.$$

PROOF. Let $M > 0$ be such that $|f(x)| \le M$ for $x \in [a, b]$ and let $\varepsilon > 0$. It follows from Theorem 30.4 that fg' is Riemann integrable. Therefore there exists a partition P_ε of $[a, b]$ such that if $P = (x_0, x_1, \ldots, x_n)$ is any

---

† OSSIAN BONNET (1819–1892) is primarily known for his work in differential geometry.

refinement of P_ε and if $\xi_j \in [x_{j-1}, x_j]$ for $j = 1, \ldots, n$, then

$$(30.12) \qquad \left| \sum_{j=1}^{n} f(\xi_j) g'(\xi_j)(x_j - x_{j-1}) - \int_a^b fg' \right| < \varepsilon.$$

Since g' is Riemann integrable we may also suppose (in virtue of the Riemann Criterion 30.1) that P_ε has been chosen so that

$$(30.13) \qquad \sum_{j=1}^{n} (M_j - m_j)(x_j - x_{j-1}) < \varepsilon,$$

where $M_j = \sup \{g'(x) : x \in [x_{j-1}, x_j]\}$ and $m_j = \inf \{g'(x) : x \in [x_{j-1}, x_j]\}$. If we use the Mean Value Theorem 27.6, we obtain points $\zeta_j \in (x_{j-1}, x_j)$ such that

$$\left| \sum_{j=1}^{n} f(\xi_j)\{g(x_j) - g(x_{j-1})\} - \int_a^b fg' \right|$$

$$= \left| \sum_{j=1}^{n} f(\xi_j) g'(\zeta_j)(x_j - x_{j-1}) - \int_a^b fg' \right|$$

$$\leq \left| \sum_{j=1}^{n} f(\xi_j)\{g'(\zeta_j) - g'(\xi_j)\}(x_j - x_{j-1}) \right|$$

$$+ \left| \sum_{j=1}^{n} f(\xi_j) g'(\xi_j)(x_j - x_{j-1}) - \int_a^b fg' \right|.$$

Now since $|g'(\zeta_j) - g'(\xi_j)| \leq M_j - m_j$, it follows from (30.12) and (30.13) that the preceding expression is dominated by

$$M \sum_{j=1}^{n} (M_j - m_j)(x_j - x_{j-1}) + \varepsilon \leq (M+1)\varepsilon.$$

Since $\varepsilon > 0$ and the choice of $\xi_j \in [x_{j-1}, x_j]$ are arbitrary, it follows that f is integrable with respect to g and that (30.11) holds. Q.E.D.

REMARK. The proof can be modified to apply to the case where f is bounded and g is continuous on $[a, b]$, and where g has a derivative except at a finite number of points at which g' can be defined so that g' and fg' are Riemann integrable on $[a, b]$.

Exercises

30.A. Show that a bounded function which has at most a finite number of discontinuities is Riemann integrable.

30.B. If $f : [a, b] \to R$ is discontinuous at some point of the interval, then there exists a monotone increasing function g such that f is not g-integrable.

30.C. Show that the Integrability Theorem 30.2 holds when g is a function of bounded variation on J.

30.D. Give an example of a function f which is not Riemann integrable over J but such that $|f|$ and f^2 are Riemann integrable over J.

30.E. Let f be positive and continuous on $J = [a, b]$ and let $M = \sup \{f(x) : x \in J\}$. Show that

$$M = \lim_n \left(\int_a^b (f(x))^n \, dx \right)^{1/n}.$$

30.F. Show that the First Mean Value Theorem 30.6 may fail if f is not continuous.

30.G. Show that the Differentiation Theorem 30.7 holds if it is assumed that f is integrable on J with respect to an increasing function g, that f is continuous at c, and that g is differentiable at c.

30.H. Suppose that f is integrable with respect to an increasing function g on $J = [a, b]$ and let F be defined for $x \in J$ by

$$F(x) = \int_a^x f \, dg.$$

Prove that (a) if g is continuous at c, then F is continuous at c, and (b) if f is positive, then F is increasing.

30.I. Give an example of a Riemann integrable function f on J such that the function F, defined for $x \in J$ by

$$F(x) = \int_a^x f,$$

does not have a derivative at some points of J. Can you find an integrable function f such that F is not continuous on J?

30.J. If f is Riemann integrable on $J = [a, b]$ and if $F' = f$ on J, then

$$F(b) - F(a) = \int_a^b f.$$

Hint: if $P = (x_0, x_1, \ldots, x_n)$ is a partition of J, write

$$F(b) - F(a) = \sum_{j=1}^n \{F(x_j) - F(x_{j-1})\}.$$

30.K. Let F be defined by

$$F(x) = x^2 \sin (1/x^2), \qquad 0 < x \le 1,$$
$$= 0, \qquad\qquad\quad x = 0.$$

Then F has a derivative at every point of I. However F' is not integrable on I and so F is not the integral of its derivative.

30.L. Let f be defined by $f(x) = [x]$ for $x \in [0, 2]$. Then f is Riemann integrable on $[0, 2]$, but it is not the derivative of any function.

30.M. In the First Mean Value Theorem 30.9, assume that p is Riemann integrable (instead of continuous). Show that the conclusion still holds.

30.N. If φ is non-negative and decreasing and h is continuous on $[a, b]$, then there exists a point $\xi \in [a, b]$ such that

$$\int_a^b \varphi h = \varphi(a) \int_a^\xi h.$$

30.O. Let f be continuous on $I = [0, 1]$, let $f_0 = f$, and let f_{n+1} be defined by

$$f_{n+1}(x) = \int_0^x f_n(t)\, dt \qquad \text{for } n \in \mathbf{N}, x \in \mathbf{I}.$$

By induction, show that $|f_n(x)| \le (M/n!)x^n \le M/n!$, where $M = \sup\{|f(x)| : x \in \mathbf{I}\}$. It follows that the sequence (f_n) converges uniformly on $\mathbf{I}$ to the zero function.

30.P. If f is integrable with respect to g on $J = [a, b]$, if φ is continuous and strictly increasing on $[c, d]$, and if $\varphi(c) = a$, $\varphi(d) = b$, then $f \circ \varphi$ is integrable with respect to $g \circ \varphi$ and

$$\int_a^b f\, dg = \int_c^d (f \circ \varphi)\, d(g \circ \varphi).$$

30.Q. If f is continuous on $[a, b]$ and if

$$\int_a^b fh = 0$$

for all continuous functions h, then $f(x) = 0$ for all x.

30.R. If f is integrable on $[a, b]$ and if

$$\int_a^b fh = 0$$

for all continuous functions h, then $f(x) = 0$ for all points of continuity of f.

30.S. Let p be continuous and positive on $[a, b]$ and let $c > 0$. If

$$p(x) \le c \int_a^x p(t)\, dt$$

for all $x \in [a, b]$, show that $p(x) = 0$ for all x.

30.T.   Let f be continuous and such that $f(x) \ge 0$ for all $x \in [a, b]$. If g is strictly increasing on $[a, b]$ show that

$$\int_a^b f\, dg = 0$$

if and only if $f(x) = 0$ for all $x \in [a, b]$.

30.U.   Show that if g is strictly increasing on $[a, b]$, then in the First Mean Value Theorem 30.6 one can take $c \in (a, b)$. Make a similar modification of the parts (a) and (b) of the Second Mean Value Theorem 30.11.

30.V.   Evaluate the following Riemann-Stieltjes integrals. (Here $x \mapsto [x]$ denotes the greatest integer function.)

(a) $\displaystyle\int_0^{1} x\, d(x^3)$,   (b) $\displaystyle\int_{-2}^{2} x\, d(|x|)$,

(c) $\displaystyle\int_0^{2} x^3\, d([x])$,   (d) $\displaystyle\int_0^{4} x^2\, d([x^2])$,

(e) $\displaystyle\int_0^{\pi} \cos x\, d(\sin x)$,   (f) $\displaystyle\int_{-\pi}^{\pi} \cos x\, d(|\sin x|)$.

Projects

30.α. The purpose of this project is to develop the logarithm by using an integral as its definition. Let $P = \{x \in \mathbf{R} : x > 0\}$.

(a) If $x \in P$, define $L(x)$ to be

$$L(x) = \int_1^x \frac{1}{t}\, dt.$$

Hence $L(1) = 0$. Prove that L is differentiable and that $L'(x) = 1/x$.

(b) Show that $L(x) < 0$ for $0 < x < 1$ and $L(x) > 0$ for $x > 1$. In fact,

$$1 - 1/x \le L(x) \le x - 1 \qquad \text{for } x > 0.$$

(c) Prove that $L(xy) = L(x) + L(y)$ for x, y in P. Hence $L(1/x) = -L(x)$ for x in P. (Hint: if $y \in P$, let L_1 be defined on P by $L_1(x) = L(xy)$ and show that $L_1' = L'$.)

(d) Show that if $n \in \mathbf{N}$, $n \ge 3$, then

$$\frac{1}{2} + \frac{1}{3} + \cdots + \frac{1}{n} < L(n) < 1 + \frac{1}{2} + \cdots + \frac{1}{n-1}.$$

(e) Prove that L is a one-one function mapping P onto all of $\mathbf{R}$. Letting e denote the unique number such that $L(e) = 1$, and using the fact that $L'(1) = 1$, show that $e = \lim ((1 + 1/n)^n)$.

(f) Let r be any positive rational number, then $\lim_{x \to +\infty} L(x)/x^r = 0$.

(g) Observe that

$$L(1 + x) = \int_1^{1+x} \frac{dt}{t} = \int_0^x \frac{dt}{1+t}.$$

Write $(1 + t)^{-1}$ as a finite geometric series to obtain

$$L(1 + x) = \sum_{k=1}^n \frac{(-1)^{k-1}}{k} x^k + R_n(x).$$

Show that $|R_n(x)| \le x^{n+1}/(n + 1)$ for $0 \le x \le 1$ and

$$|R_n(x)| \le \frac{|x|^{n+1}}{(n+1)(1+x)} \qquad \text{for } -1 < x < 0.$$

30.β. This project develops the trigonometric functions starting with an integral.

(a) Let A be defined for x in $\mathbf{R}$ by

$$A(x) = \int_0^x \frac{dt}{1+t^2}.$$

Then A is an odd function (that is, $A(-x) = -A(x)$), it is strictly increasing, and it is bounded by 2. Define π by $\pi/2 = \sup \{A(x) : x \in \mathbf{R}\}$.

(b) Let T be the inverse of A, so that T is a strictly increasing function with domain $(-\pi/2, \pi/2)$. Show that T has a derivative and that $T' = 1 + T^2$.

(c) Define C and S on $(-\pi/2, \pi/2)$ by the formulas

$$C = \frac{1}{(1 + T^2)^{1/2}}, \qquad S = \frac{T}{(1 + T^2)^{1/2}}.$$

Hence C is even and S is odd on $(-\pi/2, \pi/2)$. Show that $C(0) = 1$ and $S(0) = 0$ and $C(x) \to 0$ and $S(x) \to 1$ as $x \to \pi/2$.

(d) Prove that $C'(x) = -S(x)$ and $S'(x) = C(x)$ for x in $(-\pi/2, \pi/2)$. Therefore, both C and S satisfy the differential equation

$$h'' + h = 0$$

on the interval $(-\pi/2, \pi/2)$.

(e) Define $C(\pi/2) = 0$ and $S(\pi/2) = 0$ and define C, S, T outside the interval $(-\pi/2, \pi/2)$ by the equations

$$C(x + \pi) = -C(x), \qquad S(x + \pi) = -S(x),$$

$$T(x + \pi) = T(x).$$

If this is done successively, then C and S are defined for all R and have period 2π. Similarly, T is defined except at odd multiples of $\pi/2$ and has period π.

(f) Show that the functions C and S, as defined on R in the preceding part, are differentiable at every point of R and that they continue to satisfy the relations

$$C' = -S, \qquad S' = C$$

everywhere on R.

30.γ. This project develops the well-known Wallis† product formula. Throughout it we shall let

$$S_n = \int_0^{\pi/2} (\sin x)^n \, dx.$$

(a) If $n > 2$, then $S_n = [(n-1)/n]S_{n-2}$. (Hint: integrate by parts.)

(b) Establish the formulas

$$S_{2n} = \frac{1 \cdot 3 \cdot 5 \cdots (2n-1)}{2 \cdot 4 \cdot 6 \cdots (2n)} \frac{\pi}{2}, \qquad S_{2n+1} = \frac{2 \cdot 4 \cdots (2n)}{1 \cdot 3 \cdot 5 \cdots (2n+1)}.$$

(c) Show that the sequence (S_n) is monotone decreasing. (Hint: $0 \le \sin x \le 1$.)

(d) Let W_n be defined by

$$W_n = \frac{2 \cdot 2 \cdot 4 \cdot 4 \cdot 6 \cdot 6 \cdots (2n)(2n)}{1 \cdot 3 \cdot 3 \cdot 5 \cdot 5 \cdot 7 \cdots (2n-1)(2n+1)}.$$

Prove that $\lim (W_n) = \pi/2$. (This is Wallis's product.)

(e) Prove that $\lim ((n!)^2 2^{2n}/(2n)! \sqrt{n}) = \sqrt{\pi}$.

30.δ. This project develops the important Stirling‡ formula, which estimates the magnitude of $n!$.

---

† JOHN WALLIS (1616–1703), the Savilian professor of geometry at Oxford for sixty years, was a precurser of Newton. He helped to lay the groundwork for the development of calculus.
‡ JAMES STIRLING (1692–1770) was an English mathematician of the Newtonian school. The formula attributed to Stirling was actually established earlier by ABRAHAM DE MOIVRE (1667–1754), a French Huguenot who settled in London and was a friend of Newton.

(a) By comparing the area under the hyperbola $y = 1/x$ and the area of a trapezoid inscribed in it, show that

$$\frac{2}{2n+1} < \log\left(1+\frac{1}{n}\right).$$

From this, show that $e < (1+1/n)^{n+1/2}$.

(b) Show that

$$\int_1^n \log x \, dx = n \log n - n + 1 = \log(n/e)^n + 1.$$

Consider the figure F made up of rectangles with bases $[1,\frac{3}{2}], [n-\frac{1}{2}, n]$ and heights 2, $\log n$, respectively, and with trapezoids with bases $[k-\frac{1}{2}, k+\frac{1}{2}]$, $k = 2, 3, \ldots, n-1$, and with slant heights passing through the points $(k, \log k)$. Show that the area of F is

$$1 + \log 2 + \cdots + \log(n-1) + \tfrac{1}{2}\log n = 1 + \log(n!) - \log\sqrt{n}.$$

(c) Comparing the two areas in part (b), show that

$$u_n = \frac{(n/e)^n \sqrt{n}}{n!} < 1, \qquad n \in N.$$

(d) Show that the sequence (u_n) is monotone increasing. (Hint: consider u_{n+1}/u_n.)

(e) By considering u_n^2/u_{2n}, and making use of the result of part (e) of the preceding project, show that $\lim(u_n) = (2\pi)^{-1/2}$.

(f) Obtain Stirling's formula

$$\lim\left(\frac{(n/e)^n \sqrt{2\pi n}}{n!}\right) = 1.$$

Section 31   Further Properties of the Integral

In this section we shall present some further properties of the Riemann-Stieltjes (and the Riemann) integral that are often useful.

We first consider the possibility of "taking the limit under the integral sign"; that is, the integrability of the limit of a sequence of integrable functions.

Suppose that g is monotone increasing on an interval $J = [a, b]$ and that (f_n) is a sequence of functions which are integrable with respect to g and which converges at every point of J to a function f. It is quite natural to expect that the limit function f is integrable and that

(31.1) $$\int_a^b f \, dg = \lim \int_a^b f_n \, dg.$$

However, this need not be the case even for very nice functions.

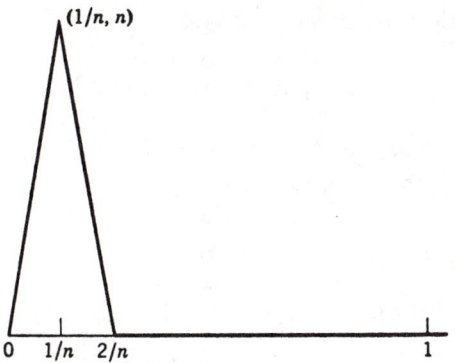

Figure 31.1. Graph of f_n.

31.1 EXAMPLE. Let $J = [0, 1]$, let $g(x) = x$, and let f_n be defined for $n \geq 2$ by

$$f_n(x) = n^2 x, \qquad 0 \leq x \leq 1/n,$$
$$= -n^2(x - 2/n), \qquad 1/n \leq x \leq 2/n,$$
$$= 0, \qquad 2/n \leq x \leq 1.$$

It is clear that for each n the function f_n is continuous on J, and hence it is integrable with respect to g. (See Figure 31.1.) Either by means of a direct calculation or referring to the significance of the integral as an area, we obtain

$$\int_0^1 f_n(x)\, dx = 1, \qquad n \geq 2.$$

In addition, the sequence (f_n) converges at every point of J to 0; hence the limit function f vanishes identically, is integrable, and

$$\int_0^1 f(x)\, dx = 0.$$

Therefore, equation (31.1) does not hold in this case even though both sides have a meaning.

Since equation (31.1) is very convenient, we inquire if there are any simple additional conditions that will imply it. We now show that, if the convergence is uniform, then this relation holds.

31.2 THEOREM. *Let g be a monotone increasing function on J and let (f_n) be a sequence of functions which are integrable with respect to g over J. Suppose that the sequence (f_n) converges uniformly on J to a limit function f.*

Then f is integrable with respect to g and

$$(31.1) \qquad \int_a^b f \, dg = \lim \int_a^b f_n \, dg.$$

PROOF. Let $\varepsilon > 0$ and let N be such that $\|f_N - f\|_J < \varepsilon$. Now let P_N be a partition of J such that if P, Q are refinements of P_N, then $|S(P; f_N, g) - S(Q; f_N, g)| < \varepsilon$, for any choice of the intermediate points. If we use the same intermediate points for f and f_N, then

$$|S(P; f_N, g) - S(P; f, g)| \le \sum_{k=1}^n \|f_N - f\|_J \{g(x_k) - g(x_{k-1})\}$$

$$= \|f_N - f\|_J \{g(b) - g(a)\} < \varepsilon \{g(b) - g(a)\}.$$

Since a similar estimate holds for the partition Q, then for refinements P, Q of P_N and corresponding Riemann-Stieltjes sums, we have

$$|S(P; f, g) - S(Q; f, g)| \le |S(P; f, g) - S(P; f_N, g)|$$
$$+ |S(P; f_N, g) - S(Q; f_N, g)|$$
$$+ |S(Q; f_N, g) - S(Q; f, g)|$$
$$\le \varepsilon (1 + 2\{g(b) - g(a)\}).$$

According to the Cauchy Criterion 29.4, the limit function f is integrable with respect to g.

To establish (31.1), we employ Lemma 30.5:

$$\left| \int_a^b f \, dg - \int_a^b f_n \, dg \right| = \left| \int_a^b (f - f_n) \, dg \right| \le \|f - f_n\|_J \{g(b) - g(a)\}.$$

Since $\lim \|f - f_n\|_J = 0$, the desired conclusion follows. Q.E.D.

The hypothesis made in Theorem 31.2 that the convergence of the sequence (f_n) is uniform is rather severe and restricts the utility of this result. We shall now state a result which does not restrict the convergence so heavily, but requires the integrability of the limit function. We shall not prove this result here, since the most natural proof would require an excursion into "measure theory." (However, the reader may consult the article of Luxemberg listed in the References.)

31.3 BOUNDED CONVERGENCE THEOREM. *Let (f_n) be a sequence of functions which are integrable with respect to a monotone increasing function g on $J = [a, b]$ to $\mathbf{R}$. Suppose that there exists $B > 0$ such that $|f_n(x)| \le B$ for all $n \in N$, $x \in J$. If the function $f(x) = \lim (f_n(x))$, $x \in J$, exists and is integrable with respect to g on J, then*

$$(31.1) \qquad \int_a^b f \, dg = \lim \int_a^b f_n \, dg.$$

The following consequence of the Bounded Convergence Theorem is frequently useful, so we shall state it formally.

31.4 Monotone Convergence Theorem. *Let (f_n) be a monotone sequence of functions which are integrable with respect to a monotone increasing function g on $J = [a, b]$ to $\mathbf{R}$. If the function $f(x) = \lim (f_n(x))$, $x \in J$, is integrable with respect to g on J, then*

$$(31.1) \qquad \int_a^b f \, dg = \lim \int_a^b f_n \, dg.$$

PROOF. Suppose that $f_1(x) \le f_2(x) \le \cdots \le f(x)$ for all $x \in J$. Then $|f_n(x)| \le B$, where $B = \|f_1\|_J + \|f\|_J$, so we can apply 31.3.   Q.E.D.

The main source of power of the Lebesgue (and Lebesgue-Stieltjes) theory of integration is that it enlarges the class of integrable functions so that equation (31.1) holds under weaker assumptions than given in the preceding theorems. See the author's *Elements of Integration*, listed in the References.

Integral Form for the Remainder

The reader will recall Taylor's Theorem 28.6, which enables one to calculate the value $f(b)$ in terms of the values $f(a), f'(a), \ldots, f^{(n-1)}(a)$ and a remainder term which involves $f^{(n)}$ evaluated at a point between a and b. For many applications it is more convenient to be able to express the remainder term as an integral involving $f^{(n)}$.

31.5 Taylor's Theorem. *Suppose that f and its derivatives f', $f'', \ldots, f^{(n)}$ are continuous on $[a, b]$ to $\mathbf{R}$. Then*

$$f(b) = f(a) + \frac{f'(a)}{1!} (b - a) + \cdots + \frac{f^{(n-1)}(a)}{(n-1)!} (b - a)^{n-1} + R_n,$$

where the remainder is given by

$$(31.2) \qquad R_n = \frac{1}{(n-1)!} \int_a^b (b - t)^{n-1} f^{(n)}(t) \, dt.$$

PROOF. Integrate R_n by parts to obtain

$$R_n = \frac{1}{(n-1)!} \left\{ (b - t)^{n-1} f^{(n-1)}(t) \Big|_{t=a}^{t=b} + (n-1) \int_a^b (b - t)^{n-2} f^{(n-1)}(t) \, dt \right\}$$

$$= -\frac{f^{(n-1)}(a)}{(n-1)!} (b - a)^{n-1} + \frac{1}{(n-2)!} \int_a^b (b - t)^{n-2} f^{(n-1)}(t) \, dt.$$

Continuing to integrate by parts in this way, we obtain the stated formula.
Q.E.D.

Instead of the formula (31.2), it is often convenient to make the change of variable $t = (1 - s)a + sb$, for s in $[0, 1]$, and to obtain the formula

$$(31.3) \qquad R_n = \frac{(b-a)^{n-1}}{(n-1)!} \int_0^1 (1-s)^{n-1} f^{(n)}[a + (b-a)s] \, ds.$$

This form of the remainder can be extended to the case where f has domain in $\mathbf{R}^p$ and range in $\mathbf{R}^q$.

Integrals Depending on a Parameter

It is often important to consider integrals in which the integrands depend on a parameter. In such cases one desires to have conditions assuring the continuity, the differentiability, and the integrability of the resulting function. The next few results are useful in this connection.

Let D be the rectangle in $\mathbf{R} \times \mathbf{R}$ given by

$$D = \{(x, t) : a \le x \le b, c \le t \le d\},$$

and suppose that f is continuous on D to $\mathbf{R}$. Then it is easily seen (cf. Exercise 22.G) that, for each fixed t in $[c, d]$, the function which sends x into $f(x, t)$ is continuous on $[a, b]$ and, therefore, Riemann integrable. We define F for t in $[c, d]$ by the formula

$$(31.4) \qquad F(t) = \int_a^b f(x, t) \, dx.$$

It will first be proved that F is continuous.

31.6 THEOREM. *If f is continuous on D to $\mathbf{R}$ and if F is defined by (31.4), then F is continuous on $[c, d]$ to $\mathbf{R}$.*

PROOF.   The Uniform Continuity Theorem 23.3 implies that if $\varepsilon > 0$, then there exists a $\delta(\varepsilon) > 0$ such that if t and t_0 belong to $[c, d]$ and $|t - t_0| < \delta(\varepsilon)$, then

$$|f(x, t) - f(x, t_0)| < \varepsilon,$$

for all x in $[a, b]$. It follows from Lemma 30.5 that

$$|F(t) - F(t_0)| = \left| \int_a^b \{f(x, t) - f(x, t_0)\} \, dx \right|$$

$$\le \int_a^b |f(x, t) - f(x, t_0)| \, dx \le \varepsilon (b - a),$$

which establishes the continuity of F.   Q.E.D.

In the next two results, we shall make use of the notion of the partial derivative of a function of two real variables. This concept, familiar to the reader from calculus, will be discussed further in Chapter VII.

31.7 **THEOREM.** *If f and its partial derivative f_t are continuous on D to $\mathbf{R}$, then the function F defined by (31.4) has a derivative on $[c, d]$ and*

$$(31.5) \qquad F'(t) = \int_a^b f_t(x, t)\, dx.$$

PROOF. From the uniform continuity of f_t on D we infer that if $\varepsilon > 0$, then there is a $\delta(\varepsilon) > 0$ such that if $|t - t_0| < \delta(\varepsilon)$, then

$$|f_t(x, t) - f_t(x, t_0)| < \varepsilon$$

for all x in $[a, b]$. Let t, t_0 satisfy this condition and apply the Mean Value Theorem 27.6 to obtain a t_1 (which may depend on x and lies between t and t_0) such that

$$f(x, t) - f(x, t_0) = (t - t_0) f_t(x, t_1).$$

Combining these two relations, we infer that if $0 < |t - t_0| < \delta(\varepsilon)$, then

$$\left| \frac{f(x, t) - f(x, t_0)}{t - t_0} - f_t(x, t_0) \right| < \varepsilon,$$

for all x in $[a, b]$. By applying Lemma 30.5, we obtain the estimate

$$\left| \frac{F(t) - F(t_0)}{t - t_0} - \int_a^b f_t(x, t_0)\, dx \right| \le \int_a^b \left| \frac{f(x, t) - f(x, t_0)}{t - t_0} - f_t(x, t_0) \right| dx$$

$$\le \varepsilon(b - a),$$

which establishes equation (31.5). Q.E.D.

Sometimes the parameter t enters in the limits of integration as well as in the integrand. The next result considers this possibility. In its proof we shall make use of a very special case of the Chain Rule (to be discussed in Chapter VII) which will be familiar to the reader.

31.8 **LEIBNIZ'S FORMULA.** *Suppose that f and f_t are continuous on D to $\mathbf{R}$ and that α and β are functions which are differentiable on the interval $[c, d]$ and have values in $[a, b]$. If φ is defined on $[c, d]$ by*

$$(31.6) \qquad \varphi(t) = \int_{\alpha(t)}^{\beta(t)} f(x, t)\, dx,$$

then φ has a derivative for each t in $[c, d]$ which is given by

$$(31.7) \qquad \varphi'(t) = f(\beta(t), t)\beta'(t) - f(\alpha(t), t)\alpha'(t) + \int_{\alpha(t)}^{\beta(t)} f_t(x, t)\, dx.$$

PROOF. Let H be defined for (u, v, t) by

$$H(u, v, t) = \int_v^u f(x, t)\, dx,$$

when u, v belong to $[a, b]$ and t belongs to $[c, d]$. The function φ defined in (31.6) is the composition given by $\varphi(t) = H(\beta(t), \alpha(t), t)$. Applying the Chain Rule, we have

$$\varphi'(t) = H_u(\beta(t), \alpha(t), t)\beta'(t) + H_v(\beta(t), \alpha(t), t)\alpha'(t) + H_t(\beta(t), \alpha(t), t).$$

According to the Differentiation Theorem 30.7,

$$H_u(u, v, t) = f(u, t), \qquad H_v(u, v, t) = -f(v, t),$$

and from the preceding theorem, we have

$$H_t(u, v, t) = \int_v^u f_t(x, t)\, dx.$$

If we substitute $u = \beta(t)$ and $v = \alpha(t)$, then we obtain the formula (31.7).

Q.E.D.

If f is continuous on D to $\mathbf{R}$ and if F is defined by formula (31.4), then it was proved in Theorem 31.6 that F is continuous and hence Riemann integrable on the interval $[c, d]$. We now show that this hypothesis of continuity is sufficient to insure that we may *interchange the order of integration*. In formulas, this may be expressed as

$$(31.8) \qquad \int_c^d \left\{ \int_a^b f(x, t)\, dx \right\} dt = \int_a^b \left\{ \int_c^d f(x, t)\, dt \right\} dx.$$

31.9 INTERCHANGE THEOREM. *If f is continuous on D with values in $\mathbf{R}$, then formula (31.8) is valid.*

PROOF. Theorem 31.6 and the Integrability Theorem 30.2 imply that both of the iterated integrals appearing in (31.8) exist; it remains only to establish their equality. Since f is uniformly continuous on D, if $\varepsilon > 0$ there exists a $\delta(\varepsilon) > 0$ such that if $|x - x'| < \delta(\varepsilon)$ and $|t - t'| < \delta(\varepsilon)$, then $|f(x, t) - f(x', t')| < \varepsilon$. Let n be chosen so large that $(b - a)/n < \delta(\varepsilon)$ and $(d - c)/n < \delta(\varepsilon)$ and divide D into n^2 equal rectangles by dividing $[a, b]$ and $[c, d]$ each into n equal parts. For $j = 0, 1, \ldots, n$, we let

$$x_j = a + (b - a)j/n, \qquad t_j = c + (d - c)j/n.$$

We can write the integral on the left of (31.8) in the form of the sum

$$\sum_{k=1}^n \sum_{j=1}^n \int_{t_{k-1}}^{t_k} \left\{ \int_{x_{j-1}}^{x_j} f(x, t)\, dx \right\} dt.$$

Applying the First Mean Value Theorem 30.6 twice, we infer that there exists a number x_j' in $[x_{j-1}, x_j]$ and numbers t_{jk}' in $[t_{k-1}, t_k]$ such that

$$\int_{t_{k-1}}^{t_k} \left\{ \int_{x_{j-1}}^{x_j} f(x, t)\, dx \right\} dt = f(x_j', t_{jk}')(x_j - x_{j-1})(t_k - t_{k-1}).$$

Hence we have

$$\int_c^d \left\{ \int_a^b f(x, t)\, dx \right\} dt = \sum_{k=1}^{n} \sum_{j=1}^{n} f(x_j', t_{jk}')(x_j - x_{j-1})(t_k - t_{k-1}).$$

The same line of reasoning, applied to the integral on the right of (31.8), yields the existence of numbers x_{jk}'' in $[x_{j-1}, x_j]$ and t_k'' in $[t_{k-1}, t_k]$ such that

$$\int_a^b \left\{ \int_c^d f(x, t)\, dt \right\} dx = \sum_{k=1}^{n} \sum_{j=1}^{n} f(x_{jk}'', t_k'')(x_j - x_{j-1})(t_k - t_{k-1}).$$

Since both x_j' and x_{jk}'' belong to $[x_{j-1}, x_j]$ and t_{jk}', t_k'' belong to $[t_{k-1}, t_k]$, we conclude from the uniform continuity of f that the two double sums, and therefore the two iterated integrals, differ by at most $\varepsilon (b-a)(d-c)$. Since $\varepsilon > 0$ is arbitrary, the equality of these integrals is confirmed. Q.E.D.

The Riesz Representation Theorem†

We shall conclude this section with a deep theorem which, although it will not be applied below, plays a very important role in functional analysis.

First it will be convenient to collect some results which have already been established or are direct consequences of what we had done.

Let $J = [a, b]$ be a closed cell in $\mathbf{R}$, let $C(J)$ denote the vector space of all continuous functions on J to $\mathbf{R}$, and let $\|f\|_J$ be the norm on $C(J)$ defined by

$$\|f\|_J = \sup \{|f(x)| : x \in J\}.$$

A **linear functional on** $C(J)$ is a linear function $G : C(J) \to \mathbf{R}$ defined on the vector space $C(J)$; hence

$$G(\alpha f_1 + \beta f_2) = \alpha G(f_1) + \beta G(f_2)$$

for all α, β in $\mathbf{R}$ and f_1, f_2 in $C(J)$. A linear functional G on $C(J)$ is said to be **positive** if for each $f \in C(J)$ with $f(x) \ge 0$, $x \in J$, we have

$$G(f) \ge 0.$$

A linear functional G on $C(J)$ is said to be **bounded** if there exists $M \ge 0$ such that

$$|G(f)| \le M \|f\|_J$$

for all $f \in C(J)$.

31.10 LEMMA. *If g is a monotone increasing function on J and if G is defined for f in $C(J)$ by*

$$(31.9) \qquad G(f) = \int_a^b f\, dg,$$

then G is a bounded positive linear functional on $C(J)$.

† The rest of this section may be omitted on a first reading.

PROOF.   It follows from Theorem 29.5(a) and Theorem 30.2 that G is a linear function on $C(J)$ and from Lemma 30.5 that G is bounded with $M = g(b) - g(a)$. If f belongs to $C(J)$ and $f(x) \geq 0$ for $x \in J$, then taking $m = 0$ in formula (30.4) we conclude that $G(f) \geq 0$.   Q.E.D.

We shall now show that, conversely, every bounded positive linear functional on $C(J)$ is generated by the Riemann-Stieltjes integral with respect to some monotone increasing function g. This is a form of the celebrated "Riesz Representation Theorem," which is one of the keystones for the subject of "functional analysis" and has many far-reaching generalizations and applications. The theorem was proved by the great Hungarian mathematician Frederic Riesz.[†]

31.11   RIESZ REPRESENTATION THEOREM.   *If G is a bounded positive linear functional on $C(J)$, then there exists a monotone increasing function g on J such that*

$$(31.9) \qquad\qquad G(f) = \int_a^b f \, dg,$$

for every f in $C(J)$.

PROOF.   We shall first define a monotone increasing function g and then show that (31.9) holds.

There exists a constant M such that if $0 \leq f_1(x) \leq f_2(x)$ for all x in J, then $0 \leq G(f_1) \leq G(f_2) \leq M \, \|f_2\|_J$. If t is any real number such that $a < t < b$, and if n is a sufficiently large natural number, we let $\varphi_{t,n}$ be the function (see Figure 31.2) in $C(J)$ defined by

$$
\begin{aligned}
(31.10) \qquad \varphi_{t,n}(x) &= 1, & a &\leq x \leq t, \\
&= 1 - n(x - t), & t &< x \leq t + 1/n, \\
&= 0, & t + 1/n &< x \leq b.
\end{aligned}
$$

It is readily seen that if $n \leq m$, then for each t with $a < t < b$,

$$0 \leq \varphi_{t,m}(x) \leq \varphi_{t,n}(x) \leq 1,$$

so that the sequence $(G(\varphi_{t,n}) : n \in N)$ is a bounded decreasing sequence of real numbers which converges to a real number. We define $g(t)$ to be equal to this limit. If $a < t \leq s < b$ and $n \in N$, then

$$0 \leq \varphi_{t,n}(x) \leq \varphi_{s,n}(x) \leq 1,$$

† FREDERIC RIESZ (1880–1955), a brilliant Hungarian mathematician, was one of the founders of topology and functional analysis. He also made beautiful contributions to potential, ergodic, and integration theory.

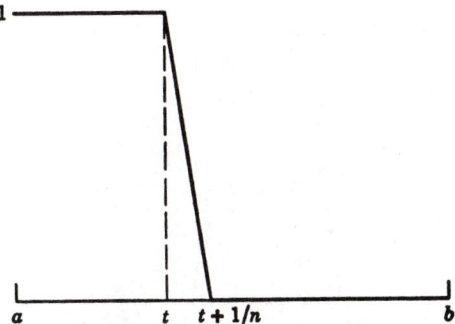

Figure 31.2.   Graph of $\varphi_{t,n}$.

whence it follows that $g(t) \le g(s)$. We define $g(a) = 0$ and if $\varphi_{b,n}$ denotes the function $\varphi_{b,n}(x) = 1$, $x \in J$, then we set $g(b) = G(\varphi_{b,n})$. If $a < t < b$ and n is sufficiently large, then for all x in J we have

$$0 \le \varphi_{t,n}(x) \le \varphi_{b,n}(x) = 1,$$

so that $g(a) = 0 \le G(\varphi_{t,n}) \le G(\varphi_{b,n}) = g(b)$. This shows that $g(a) \le g(t) \le g(b)$ and completes the construction of the monotone increasing function g.

If f is continuous on J and $\varepsilon > 0$, there is a $\delta(\varepsilon) > 0$ such that if $|x - y| < \delta(\varepsilon)$ and $x, y \in J$, then $|f(x) - f(y)| < \varepsilon$. Since f is integrable with respect to g, there exists a partition P_ε of J such that if Q is a refinement of P_ε, then for any Riemann-Stieltjes sum, we have

$$\left| \int_a^b f \, dg - S(Q; f, g) \right| < \varepsilon.$$

Now let $P = (t_0, t_1, \ldots, t_m)$ be a partition of J into distinct points which is a refinement of P_ε such that $\sup\{t_k - t_{k-1}\} < \frac{1}{2}\delta(\varepsilon)$ and let n be a natural number so large that

$$2/n < \inf\{t_k - t_{k-1}\}.$$

Then only consecutive intervals

(31.11) $[t_0, t_1 + 1/n], \ldots, [t_{k-1}, t_k + 1/n], \ldots, [t_{m-1}, t_m]$

have any points in common. (See Figure 31.3.) For each $k = 1, \ldots, m$, the decreasing sequence $(G(\varphi_{t_k,n}))$ converges to $g(t_k)$ and hence we may suppose that n is so large that

(31.12) $g(t_k) \le G(\varphi_{t_k,n}) \le g(t_k) + (\varepsilon/m \, \|f\|_J).$

We now consider the function f^* defined on J by

(31.13) $f^*(x) = f(t_1)\varphi_{t_1,n}(x) + \displaystyle\sum_{k=2}^m f(t_k)\{\varphi_{t_k,n}(x) - \varphi_{t_{k-1},n}(x)\}.$

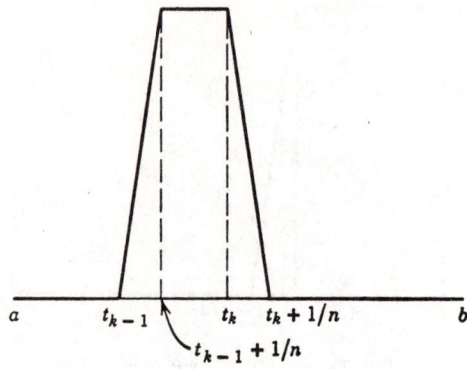

Figure 31.3.   Graph of $\varphi_{t_k,n} - \varphi_{t_{k-1},n}$.

An element x in J either belongs to one or two intervals in (31.11).   If it belongs to one interval, then we must have $t_0 \le x < t_1$ and $f^*(x) = f(t_1)$ or we have $t_{k-1} + (1/n) < x \le t_k$ for some $k = 1, 2, \ldots, m$ in which case $f^*(x) = f(t_k)$.   (See Figure 31.4.)   Hence

$$|f(x) - f^*(x)| < \varepsilon.$$

If the x belongs to two intervals in (31.11), then $t_k \le x \le t_k + 1/n$ for some

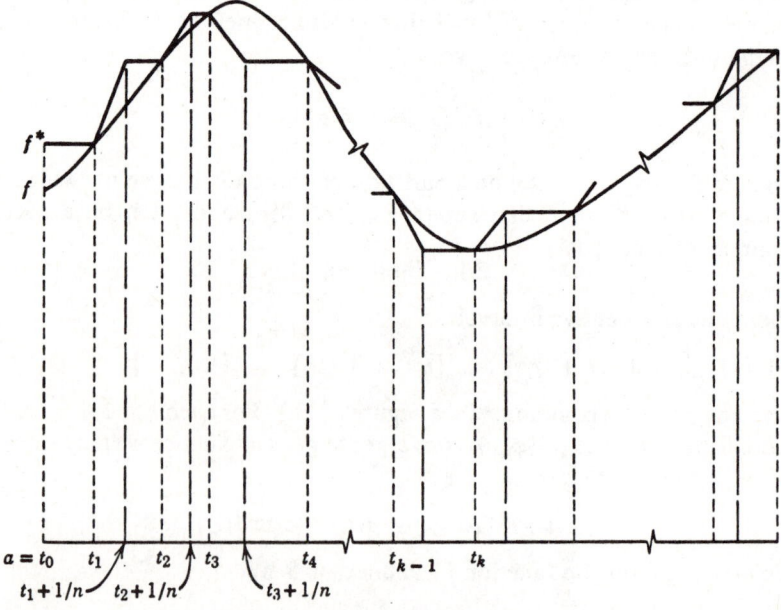

Figure 31.4.   Graphs of f and f^*.

$k = 1, \ldots, m-1$ and we infer that

$$f^*(x) = f(t_k)\varphi_{t_k,n}(x) + f(t_{k+1})\{1 - \varphi_{t_k,n}(x)\}.$$

If we refer to the definition of the φ's in (31.10), we have

$$f^*(x) = f(t_k)(1 - n(x - t_k)) + f(t_{k+1})n(x - t_k).$$

Since $|x - t_k| < \delta(\varepsilon)$ and $|x - t_{k+1}| < \delta(\varepsilon)$, we conclude that

$$|f(x) - f^*(x)| \leq |f(x) - f(t_k)|\,(1 - n(x - t_k)) + |f(x) - f(t_{k+1})|\,n(x - t_k)$$
$$< \varepsilon\{1 - n(x - t_k) + n(x - t_k)\} = \varepsilon.$$

Consequently, we have the estimate

$$\|f - f^*\|_J = \sup\{|f(x) - f^*(x)| : x \in J\} \leq \varepsilon.$$

Since G is a bounded linear functional on $C(J)$, it follows that

(31.14) $$|G(f) - G(f^*)| \leq M\varepsilon.$$

In view of relation (31.12) we see that

$$|\{G(\varphi_{t_k,n}) - G(\varphi_{t_{k-1},n})\} - \{g(t_k) - g(t_{k-1})\}| < \varepsilon/2m\,\|f\|_J$$

for $k = 2, 3, \ldots, m$. Applying G to the function f^* defined by equation (31.13) and recalling that $g(t_0) = 0$, we obtain

$$\left| G(f^*) - \sum_{k=1}^{m} f(t_k)\{g(t_k) - g(t_{k-1})\} \right| < \varepsilon.$$

But the second term on the left side is a Riemann-Stieltjes sum $S(P; f, g)$ for f with respect to g corresponding to the partition P which is a refinement of P_ε. Hence we have

$$\left| \int_a^b f\,dg - G(f^*) \right| \leq \left| \int_a^b f\,dg - S(P; f, g) \right| + |S(P; f, g) - G(f^*)| < 2\varepsilon.$$

Finally, using relation (31.14), we find that

(31.15) $$\left| \int_a^b f\,dg - G(f) \right| < (M+2)\varepsilon.$$

Since $\varepsilon > 0$ is arbitrary and the left side of (31.15) does not depend on it, we conclude that

$$G(f) = \int_a^b f\,dg. \qquad \text{Q.E.D.}$$

For some purposes it is important to know that there is a one-one correspondence between bounded positive linear functionals on $C(J)$ and certain normalized monotone increasing functions. Our construction can be checked to show that it yields an increasing function g such that $g(a) = 0$

and g is continuous from the right at every interior point of J. With these additional conditions, there is a one-one correspondence between positive functionals and increasing functions.

Exercises

31.A. If $a > 0$, show directly that

$$\lim_n \int_0^a e^{-nx} \, dx = 0.$$

Which of the results of this section apply?

31.B. If $0 < a < 2$, show that

$$\lim_n \int_a^2 e^{-nx^2} \, dx = 0.$$

What happens if $a = 0$?

31.C. Discuss $\lim_n \int_0^1 nx(1-x)^n \, dx$.

31.D. If $a > 0$, show that

$$\lim_n \int_a^\pi \frac{\sin nx}{nx} \, dx = 0.$$

What happens if $a = 0$?

31.E. Let $f_n(x) = nx(1+nx)^{-1}$ for $x \in [0, 1]$, and let $f(x) = 0$ for $x = 0$ and $f(x) = 1$ for $x \in (0, 1]$. Show that $f_n(x) \to f(x)$ for all $x \in [0, 1]$ and that

$$\int_0^1 f_n(x) \, dx \to \int_0^1 f(x) \, dx.$$

31.F. Let $h_n(x) = nx \, e^{-nx^2}$ for $x \in [0, 1]$ and let $h(x) = 0$. Show that

$$0 = \int_0^1 h(x) \, dx \neq \lim_n \int_0^1 h_n(x) \, dx = \frac{1}{2}.$$

31.G. Let (g_n) be a sequence of increasing functions on $[a, b]$ which converges uniformly to a function g on $[a, b]$. If an increasing function f is integrable with respect to g_n for all $n \in \mathbf{N}$, show that f is integrable with respect to g and

$$\int_a^b f \, dg = \lim_n \int_a^b f \, dg_n.$$

31.H. Give example to show that the conclusion in the preceding exercise may fail if the convergence is not uniform.

31.I. If $\alpha > 0$, show that $\int_0^1 t^\alpha (\log t)^2 \, dt = 2/(\alpha + 1)^3$.

31.J. Suppose that f and its partial derivative f_t are continuous for (x, t) in $[a, b] \times [c, d]$. Apply the Interchange Theorem 31.9 to

$$\int_c^t \left\{ \int_a^b f_t(x, t) \, dx \right\} dt, \qquad c \leq t \leq d,$$

and differentiate to obtain another proof of Theorem 31.7.

31.K. Use the Fundamental Theorem 30.8 to show that if a sequence (f_n) of functions converges on J to a function f and if the derivatives (f'_n) are continuous and converge uniformly on J to a function g, then f' exists and equals g. (This result is less general than Theorem 28.5, but it is easier to establish.)

31.L. Let $\{r_1, r_2, \ldots, r_n, \ldots\}$ be an enumeration of the rational numbers in I. Let f_n be defined to be 1 if $x \in \{r_1, \ldots, r_n\}$ and to be 0 otherwise. Then f_n is Riemann integrable on I and the sequence (f_n) converges monotonely to the Dirichlet discontinuous function f (which equals 1 on $I \cap Q$ and equals 0 on $I \setminus Q$). Hence the monotone limit of a sequence of Riemann integrable functions does not need to be Riemann integrable.

31.M. Let g be a fixed monotone increasing function on $J = [a, b]$. If f is any function which is integrable with respect to g on J, then we define $\|f\|_1$ by

$$\|f\|_1 = \int_a^b |f| \, dg.$$

Show that the following "norm properties" are satisfied:

  (a) $\|f\|_1 \geq 0$;
  (b) If $f(x) = 0$ for all $x \in J$, then $\|f\|_1 = 0$;
  (c) If $c \in R$, then $\|cf\|_1 = |c| \, \|f\|_1$;
  (d) $|\, \|f\|_1 - \|h\|_1 \,| \leq \|f \pm h\|_1 \leq \|f\|_1 + \|h\|_1$.

However, it is possible to have $\|f\|_1 = 0$ without having $f(x) = 0$ for all $x \in J$. (Can this occur when $g(x) = x$?)

31.N. If g is monotone increasing on J, and if f and f_n, $n \in N$, are functions which are integrable with respect to g, then we say that the sequence (f_n) **converges in mean** (with respect to g) in case

$$\|f_n - f\|_1 \to 0.$$

(The notation here is the same as in the preceding exercise.) Show that if (f_n) converges in mean to f, then

$$\int_a^b f_n \, dg \to \int_a^b f \, dg.$$

Prove that if a sequence (f_n) of integrable functions converges uniformly on J to f, then it also converges in mean to f. In fact,

$$\|f_n - f\|_1 \leq \{g(b) - g(a)\} \|f_n - f\|_J.$$

However, if f_n denotes the function in Example 31.1, and if $g_n = (1/n)f_n$, then the sequence (g_n) converges in mean [with respect to $g(x) = x$] to the zero function, but the convergence is not uniform on I.

31.O. Let $g(x) = x$ on $J = [0, 2]$ and let (I_n) be a sequence of closed intervals in J such that (i) the length of I_n is $1/n$, (ii) $I_n \cap I_{n+1} = \emptyset$, and (iii) every point x in J belongs to infinitely many of the I_n. Let f_n be defined by

$$f_n(x) = 1, \qquad x \in I_n,$$
$$= 0, \qquad x \notin I_n.$$

Prove that the sequence (f_n) converges in mean [with respect to $g(x) = x$] to the zero function on J, but that the sequence (f_n) does not converge uniformly. Indeed, the sequence (f_n) does not converge at *any* point!

31.P. Let g be monotone increasing on $J = [a, b]$. If f and h are integrable with respect to g on J to $\mathbf{R}$, we define the **inner product** (f, h) of f and h by the formula

$$(f, h) = \int_a^b f(x)h(x)\, dg(x).$$

Verify that all of the properties of Definition 8.3 are satisfied except (ii). If $f = h$ is the zero function on J, then $(f, f) = 0$; however, it may happen that $(f, f) = 0$ for a function f which does not vanish everywhere on J.

31.Q. Define $\|f\|_2$ to be

$$\|f\|_2 = \left\{ \int_a^b |f(x)|^2\, dg(x) \right\}^{1/2},$$

so that $\|f\|_2 = (f, f)^{1/2}$. Establish the Schwarz Inequality

$$|(f, h)| \le \|f\|_2\, \|h\|_2$$

(Theorem 8.7 and 8.8). Show that the Norm Properties 8.5 hold, except that $\|f\|_2 = 0$ does not imply that $f(x) = 0$ for all x in J. Show that $\|f\|_1 \le \{g(b) - g(a)\}^{1/2}\, \|f\|_2$.

31.R. Let f and f_n, $n \in \mathbf{N}$, be integrable on J with respect to an increasing function g. We say that the sequence (f_n) **converges in mean square** (with respect to g on J) to f if $\|f_n - f\|_2 \to 0$.

(a) Show that if the sequence is uniformly convergent on J, then it also converges in mean square to the same function.

(b) Show that if the sequence converges in mean square, then it converges in mean to the same function.

(c) Show that Exercise 31.O proves that convergence in mean square does not imply convergence at *any* point of J.

(d) If, in Exercise 31.O, we take I_n to have length $1/n^2$ and if we set $h_n = nf_n$, then the sequence (h_n) converges in mean, but does not converge in mean square, to the zero function.

31.S. Show that, if the nth derivative $f^{(n)}$ is continuous on $[a, b]$, then the Integral Form of Taylor's Theorem 31.5 and the First Mean Value Theorem 30.9 can be used to obtain the Lagrange form of the remainder given in 28.6.

31.T. If $J_1 = [a, b]$, $J_2 = [c, d]$, and if f is continuous on $J_1 \times J_2$ to $\mathbf{R}$ and g is Riemann integrable on J_1, then the function F, defined on J_2 by

$$F(t) = \int_a^b f(x, t)g(x)\, dx,$$

is continuous on J_2.

31.U. Let g be an increasing function on $J_1 = [a, b]$ to $\mathbf{R}$ and for each fixed t in $J_2 = [c, d]$, suppose that the integral

$$F(t) = \int_a^b f(x, t)\, dg(x)$$

exists. If the partial derivative f_t is continuous on $J_1 \times J_2$, then the derivative F'

exists on J_2 and is given by

$$F'(t) = \int_a^b f_t(x, t) \, dg(x).$$

31.V. Let $J_1 = [a, b]$ and $J_2 = [c, d]$.   Assume that the real valued function g is monotone on J_1, that h is monotone on J_2, and that f is continuous on $J_1 \times J_2$. Define G on J_2 and H on J_1 by

$$G(t) = \int_a^b f(x, t) \, dg(x), \qquad H(x) = \int_c^d f(x, t) \, dh(t).$$

Show that G is integrable with respect to h on J_2, that H is integrable with respect to g on J_1 and that

$$\int_c^d G(t) \, dh(t) = \int_a^b H(x) \, dg(x).$$

We can write this last equation in the form

$$\int_c^d \left\{ \int_a^b f(x, t) \, dg(x) \right\} dh(t) = \int_a^b \left\{ \int_c^d f(x, t) \, dh(t) \right\} dg(x).$$

31.W. Let f, J_1, and J_2 be as in Exercise 31.V. If φ is in $C(J_1)$ (that is, φ is a continuous function on J_1 to $\mathbf{R}$), let $T(\varphi)$ be the function defined on J_2 by the formula

$$T(\varphi)(t) = \int_a^b f(x, t)\varphi(x) \, dx.$$

Show that T is a **linear transformation** of $C(J_1)$ into $C(J_2)$ in the sense that if φ, ψ belong to $C(J_1)$, then
  (a) $T(\varphi)$ belongs to $C(J_2)$,
  (b) $T(\varphi + \psi) = T(\varphi) + T(\psi)$,
  (c) $T(c\varphi) = cT(\varphi)$ for $c \in \mathbf{R}$.
If $M = \sup \{|f(x, t)| : (x, t) \in J_1 \times J_2\}$, then T is **bounded** in the sense that
  (d) $\|T(\varphi)\|_{J_2} \leq M \|\varphi\|_{J_1}$ for $\varphi \in C(J_1)$.
  31.X. Continuing the notation of the preceding exercise, show that if $r > 0$, then T sends the collection

$$B_r = \{\varphi \in C(J_1) : \|\varphi\|_{J_1} \leq r\}$$

into a uniformly equicontinuous set of functions in $C(J_2)$ (see Definition 28.6). Therefore, if (φ_n) is any sequence of functions in B_r, there is a subsequence (φ_{n_k}) such that the sequence $(T(\varphi_{n_k}))$ converges uniformly on J_2.
  31.Y. Let J_1 and J_2 be as before and let f be continuous on $\mathbf{R} \times J_2$ into $\mathbf{R}$. If φ is in $C(J_1)$, let $S(\varphi)$ be the function defined on J_2 by the formula

$$S(\varphi)(t) = \int_a^b f(\varphi(x), t) \, dx.$$

Show that $S(\varphi)$ belongs to $C(J_2)$, but that, in general, S is not a linear transformation in the sense of Exercise 31.W. However, show that S sends the collection B_r

of Exercise 31.X into a uniformly equicontinuous set of functions in $C(J_2)$. Also, if (φ_n) is any sequence in B_r, there is a subsequence such that $(S(\varphi_{n_k}))$ converges uniformly on J_2. (This result is important in the theory of non-linear integral equations.)

31.Z. Show that if we define G_0, G_1, G_2 for f in $C(I)$ by

$$G_0(f) = f(0), \qquad G_1(f) = 2\int_0^{1/2} f(x)\, dx,$$

$$G_2(f) = \tfrac{1}{2}\{f(0) + f(1)\};$$

then G_0, G_1, and G_2 are bounded positive linear functionals on $C(I)$. Give monotone increasing functions g_0, g_1, g_2 which represent these linear functionals as Riemann-Stieltjes integrals. Show that the choice of these g_j is not uniquely determined unless one requires that $g_j(0) = 0$ and that g_j is continuous from the right at each interior point of I.

Project

31.α. This project establishes the existence of a unique solution of a first order differential equation under the presence of a Lipschitz condition. Let $\Omega \subseteq \mathbf{R}^2$ be open and let $f : \Omega \to \mathbf{R}$ be continuous and satisfy the Lipschitz condition: $|f(x, y) - f(x, y')| \le K |y - y'|$ for all points $(x, y), (x, y')$ in Ω. Let I be a closed cell

$$I = \{(x, y) : |x - a| \le \alpha,\ |y - b| \le \beta\}$$

contained in Ω and suppose that $M\alpha \le \beta$, where $|f(x, y)| \le M$ for $(x, y) \in I$.

(a) If $J = [a - \alpha, a + \alpha]$, define $\varphi_0(x) = b$ for $x \in J$ and, if $n \in N$, define

$$\varphi_n(x) = b + \int_a^x f(t, \varphi_{n-1}(t))\, dt$$

for $x \in J$. Prove by induction that the sequence (φ_n) is well-defined on J and that

(i) $|\varphi_n(x) - b| \le \beta$,

(ii) $|\varphi_n(x) - \varphi_{n-1}(x)| \le \dfrac{MK^{n-1}}{(n-1)!} |x - a|^{n-1}$,

for all $x \in J$.

(b) Show that each of the functions φ_n is continuous on J and that the sequence (φ_n) converges uniformly on J to a function φ.

(c) Conclude that the function φ is continuous on J, satisfies $\varphi(a) = b$, and

$$\varphi(x) = b + \int_a^x f(t, \varphi(t))\, dt,$$

for all $x \in J$. Deduce that φ is differentiable on J and satisfies

$$\varphi'(x) = f(x, \varphi(x)) \qquad \text{for } x \in J.$$

(d) If ψ is continuous on J and satisfies

$$\psi(a) = b, \qquad \psi'(t) = f(x, \psi(x))$$

for all $x \in J$, show that

$$\psi(x) = b + \int_a^b f(t, \psi(t)) \, dt \qquad \text{for } x \in J.$$

(e) If φ is as in (c) and ψ is as in (d), show by induction that

$$|\varphi(x) - \psi(x)| \le K \left| \int_a^x |\varphi(t) - \psi(t)| \, dt \right|$$

$$\le \frac{K^n}{n!} \|\varphi - \psi\|_J \, |x - a|^n.$$

Hence $\|\varphi - \psi\|_J \le \|\varphi - \psi\|_J \, K^n \alpha^n/n!$, whence it follows that $\varphi(x) = \psi(x)$ for all $x \in J$.

Section 32   Improper and Infinite Integrals

In the preceding three sections we have had two standing assumptions: we required the functions to be bounded and we required the domain of integration to be compact. If either of these hypotheses is dropped, the foregoing integration theory does not apply without some change. Since there are a number of important applications where it is desirable to permit one or both of these new phenomena, we shall indicate here the changes that are to be made.

Unbounded Functions

Let $J = [a, b]$ be an interval in $\mathbf{R}$ and let f be a real-valued function which is defined at least for x satisfying $a < x \le b$. If f is Riemann integrable on the interval $[c, b]$ for each c satisfying $a < c \le b$, let

$$(32.1) \qquad\qquad I_c = \int_c^b f.$$

We shall define the improper integral of f over $J = [a, b]$ to be the limit of I_c as $c \to a$.

32.1 DEFINITION. Suppose that the Riemann integral in (32.1) exists for each c in $(a, b]$. Suppose that there exists a real number I such that for every $\varepsilon > 0$ there is a $\delta(\varepsilon) > 0$ such that if $a < c < a + \delta(\varepsilon)$ then $|I_c - I| < \varepsilon$. In this case we say that I is the **improper integral** of f over $J = [a, b]$ and we sometimes denote the value I of this improper integral by

$$(32.2) \qquad\qquad \int_{a+}^b f \quad \text{or by} \quad \int_{a+}^b f(x) \, dx,$$

although it is more usual not to write the plus signs in the lower limit.

32.2   EXAMPLES.   (a) Suppose the function f is defined on $(a, b]$ and is bounded on this interval. If f is Riemann integrable on every interval $[c, b]$ with $a < c \le b$, then it is easily seen (Exercise 32.A) that the improper integral (32.2) exists. Thus the function $f(x) = \sin(1/x)$ has an improper integral on the interval $[0, 1]$.

(b) If $f(x) = 1/x$ for x on $(0, 1]$ and if c is in $(0, 1]$ then it follows from the Fundamental Theorem 30.8 and the fact that f is the derivative of the logarithm that

$$I_c = \int_c^1 f = \log 1 - \log c = -\log c.$$

Since $\log c$ becomes unbounded as $c \to 0$, the improper integral of f on $[0, 1]$ does not exist.

(c) Let $f(x) = x^\alpha$ for x in $(0, 1]$. If $\alpha < 0$, the function is continuous but not bounded on $(0, 1]$. If $\alpha \ne -1$, then f is the derivative of

$$g(x) = \frac{1}{\alpha + 1} x^{\alpha + 1}.$$

It follows from the Fundamental Theorem 30.8 that

$$\int_c^1 x^\alpha \, dx = \frac{1}{\alpha + 1} (1 - c^{\alpha + 1}).$$

If α satisfies $-1 < \alpha < 0$, then $c^{\alpha + 1} \to 0$ as $c \to 0$, and f has an improper integral. On the other hand, if $\alpha < -1$, then $c^{\alpha + 1}$ does not have a (finite) limit as $c \to 0$, and hence f does not have an improper integral.

The preceding discussion pertained to a function which is not defined or not bounded at the left end point of the interval. It is obvious how to treat analogous behavior at the right end point. Somewhat more interesting is the case where the function is not defined or not bounded at an interior point of the interval. Suppose that p is an interior point of $[a, b]$ and that f is defined at every point of $[a, b]$ except perhaps p. If both of the improper integrals

$$\int_a^{p-} f, \qquad \int_{p+}^b f$$

exist, then we define the **improper integral** of f over $[a, b]$ to be their sum. In the limit notation, we define the improper integral of f over $[a, b]$ to be

(32.3) $$\lim_{\varepsilon \to 0+} \int_a^{p-\varepsilon} f(x) \, dx + \lim_{\delta \to 0+} \int_{p+\delta}^b f(x) \, dx.$$

It is clear that if those two limits exist, then the single limit

(32.4) $$\lim_{\varepsilon \to 0+} \left\{ \int_a^{p-\varepsilon} f(x) \, dx + \int_{p+\varepsilon}^b f(x) \, dx \right\}$$

also exists and has the same value. However, the existence of the limit (32.4) does not imply the existence of (32.3). For example, if f is defined for $x \in [-1, 1]$, $x \neq 0$, by $f(x) = 1/x^3$, then it is easily seen that

$$\int_{-1}^{-\varepsilon} \left(\frac{1}{x^3}\right) dx + \int_{\varepsilon}^{1} \left(\frac{1}{x^3}\right) dx = \left(\frac{-1}{2}\right)\left(\frac{1}{\varepsilon^2} - 1\right) + \left(\frac{-1}{2}\right)\left(1 - \frac{1}{\varepsilon^2}\right) = 0$$

for all ε satisfying $0 < \varepsilon < 1$. However, we have seen in Example 32.2(c) that if $\alpha = -3$, then the improper integrals

$$\int_{-1}^{0-} \frac{1}{x^3}\, dx, \qquad \int_{0+}^{1} \frac{1}{x^3}\, dx$$

do not exist.

The preceding comments show that the limit in (32.4) may exist without the limit in (32.3) existing. We defined the improper integral (which is sometimes called the **Cauchy integral**) of f to be given by (32.3). The limit in (32.4) is also of interest and is called the **Cauchy principal value** of the integral and denoted by

$$(\text{CPV}) \int_a^b f(x)\, dx.$$

It is clear that a function which has a finite number of points where it is not defined or bounded can be treated by breaking the interval into subintervals with these points as end points.

Infinite Integrals

It is important to extend the integral to certain functions which are defined on unbounded sets. For example, if f is defined on $\{x \in \mathbf{R} : x \geq a\}$ to $\mathbf{R}$ and is Riemann integrable over $[a, c]$ for every $c > a$, we let I_c be the partial integral given by

$$(32.5) \qquad I_c = \int_a^c f.$$

We shall now define the "infinite integral" of f for $x \geq a$ to be the limit of I_c as c increases.

32.3 Definition. If f is Riemann integrable over $[a, c]$ for each $c > a$, let I_c be the **partial integral** given by (32.5). A real number I is said to be the **infinite integral** of f over $\{x : x \geq a\}$ if for every $\varepsilon > 0$, there exists a real number $M(\varepsilon)$ such that if $c > M(\varepsilon)$ then $|I - I_c| < \varepsilon$. In this case we denote I by

$$(32.6) \qquad \int_a^{+\infty} f \quad \text{or} \quad \int_a^{+\infty} f(x)\, dx.$$

It should be remarked that infinite integrals are sometimes called "improper integrals of the first kind." We prefer the present terminology, which is due to Hardy,† for it is both simpler and parallel to the terminology used in connection with infinite series.

32.4 EXAMPLES. (a) If $f(x) = 1/x$ for $x > a > 0$, then the partial integrals are

$$I_c = \int_a^c \frac{1}{x}\, dx = \log c - \log a.$$

Since $\log c$ becomes unbounded as $c \to +\infty$, the infinite integral of f does not exist.

(b) Let $f(x) = x^\alpha$ for $x \ge a > 0$ and $\alpha \neq -1$. Then

$$I_c = \int_a^c x^\alpha\, dx = \frac{1}{\alpha + 1}(c^{\alpha+1} - a^{\alpha+1}).$$

If $\alpha > -1$, then $\alpha + 1 > 0$ and the infinite integral does not exist. However, if $\alpha < -1$, then

$$\int_a^{+\infty} x^\alpha\, dx = -\frac{a^{\alpha+1}}{\alpha + 1}.$$

(c) Let $f(x) = e^{-x}$ for $x \ge 0$. Then

$$\int_0^c e^{-x}\, dx = -(e^{-c} - 1);$$

hence the infinite integral of f over $\{x : x \ge 0\}$ exists and equals 1.

It is also possible to consider the integral of a function defined on all of **R**. In this case we require that f be Riemann integrable over every finite interval in **R** and consider the limits

(32.7a) $$\int_{-\infty}^a f(x)\, dx = \lim_{b \to -\infty} \int_b^a f(x)\, dx,$$

(32.7b) $$\int_a^{+\infty} f(x)\, dx = \lim_{c \to +\infty} \int_a^c f(x)\, dx.$$

It is easily seen that if both of these limits exist for one value of a, then they both exist for all values of a. In this case we define the **infinite integral of f over R** to be the sum of these two infinite integrals:

(32.8) $$\int_{-\infty}^{+\infty} f(x)\, dx = \lim_{b \to -\infty} \int_b^a f(x)\, dx + \lim_{c \to +\infty} \int_a^c f(x)\, dx.$$

---

† GEOFFREY H. HARDY (1877–1947) was professor at Cambridge and long-time dean of British mathematics. He made many and deep contributions to mathematical analysis.

As in the case of the improper integral, the existence of both of the limits in (32.8) implies the existence of the limit

$$(32.9) \qquad \lim_{c \to +\infty} \left\{ \int_{-c}^{a} f(x)\, dx + \int_{a}^{c} f(x)\, dx \right\} = \lim_{c \to +\infty} \int_{-c}^{c} f(x)\, dx,$$

and the equality of (32.8) and (32.9). The limit in (32.9), when it exists, is often called the **Cauchy principal value** of the infinite integral over R and is denoted by

$$(32.10) \qquad \text{(CPV)} \int_{-\infty}^{+\infty} f(x)\, dx.$$

However, the existence of the Cauchy principal value does not imply the existence of the infinite integral (32.8). This is seen by considering $f(x) = x$, whence

$$\int_{-c}^{c} x\, dx = \tfrac{1}{2}(c^2 - c^2) = 0$$

for all c. Thus the Cauchy principal value of the infinite integral for $f(x) = x$ exists and equals 0, but the infinite integral of this function does not exist, since neither of the infinite integrals in (32.7) exists.

Existence of the Infinite Integral

We now obtain a few conditions for the existence of the infinite integral over the set $\{x : x \geq a\}$. These results can also be applied to give conditions for the infinite integral over R, since the latter involves consideration of infinite integrals over the sets $\{x : x \leq a\}$ and $\{x : x \geq a\}$. First we state the Cauchy Criterion.

32.5 CAUCHY CRITERION. *Suppose that f is integrable over $[a, c]$ for all $c \geq a$. Then the infinite integral*

$$\int_{a}^{+\infty} f$$

exists if and only if for every $\varepsilon > 0$ there exists a $K(\varepsilon)$ such that if $b \geq c \geq K(\varepsilon)$, then

$$(32.11) \qquad \left| \int_{c}^{b} f \right| < \varepsilon.$$

PROOF. The necessity of the condition is established in the usual manner. Suppose that the condition is satisfied and let I_n be the partial integral defined for $n \in N$ by

$$I_n = \int_{a}^{a+n} f.$$

It is seen that (I_n) is a Cauchy sequence of real numbers. If $I = \lim (I_n)$ and $\varepsilon > 0$, then there exists $N(\varepsilon)$ such that if $n \geq N(\varepsilon)$, then $|I - I_n| < \varepsilon$. Let $M(\varepsilon) = \sup \{K(\varepsilon), a + N(\varepsilon)\} + 1$ and let $c > M(\varepsilon)$. Then there exists a natural number $n \geq N(\varepsilon)$ such that $K(\varepsilon) \leq a + n < c$. Therefore the partial integral I_c is given by

$$I_c = \int_a^c f = \int_a^{a+n} f + \int_{a+n}^c f,$$

whence it follows that $|I - I_c| < 2\varepsilon$.   Q.E.D.

In the important case where $f(x) \geq 0$ for all $x \geq a$, the next result provides a useful test.

32.6 THEOREM. *Suppose that $f(x) \geq 0$ for all $x \geq a$ and that f is integrable over $[a, c]$ for all $c \geq a$. Then the infinite integral of f exists if and only if the set $\{I_c : c \geq a\}$ is bounded. In this case*

$$\int_a^{+\infty} f = \sup \left\{ \int_a^c f : c \geq a \right\}.$$

PROOF. The hypothesis that $f(x) \geq 0$ implies that I_c is a monotone increasing function of c. Therefore, the existence of $\lim I_c$ is equivalent to the boundedness of $\{I_c : c \geq a\}$.   Q.E.D.

32.7 COMPARISON TEST. *Suppose that f and g are integrable over $[a, c]$ for all $c \geq a$ and that $|f(x)| \leq g(x)$ for all $x \geq a$. If the infinite integral of g exists, then the infinite integral of f exists and*

$$\left| \int_a^{+\infty} f \right| \leq \int_a^{+\infty} g.$$

PROOF. If $a \leq c < b$, then it follows from Lemma 30.5 that $|f|$ is integrable on $[c, b]$ and that

$$\left| \int_c^b f \right| \leq \int_c^b |f| \leq \int_c^b g.$$

It follows from the Cauchy Criterion 32.5 that the infinite integrals of f and $|f|$ exist. Moreover, we have

$$\left| \int_a^{+\infty} f \right| \leq \int_a^{+\infty} |f| \leq \int_a^{+\infty} g.$$ Q.E.D.

32.8 LIMIT COMPARISON TEST. *Suppose that f and g are positive and integrable over $[a, c]$ for all $c \geq a$ and that*

(32.12) $$\lim_{x \to +\infty} \frac{f(x)}{g(x)} \neq 0.$$

Then both or neither of the infinite integrals $\int_a^{+\infty} f$, $\int_a^{+\infty} g$ exist.

PROOF.   In view of the relation (32.12) we infer that there exist positive numbers $A < B$ and $K \geq a$ such that

$$Ag(x) \leq f(x) \leq Bg(x) \qquad \text{for } x \geq K.$$

The Comparison Test 32.7 and this relation show that both or neither of the infinite integrals $\int_K^{+\infty} f$, $\int_K^{+\infty} g$ exist. Since both f and g are integrable on $[a, K]$, the statement follows.   Q.E.D.

32.9   DIRICHLET'S TEST.   *Suppose that f is continuous for $x \geq a$, that the partial integrals*

$$I_c = \int_a^c f, \qquad c \geq a,$$

are bounded, and that φ is monotone decreasing to zero as $x \to +\infty$. Then the infinite integral $\int_a^{+\infty} f\varphi$ exists.

PROOF.   Let A be a bound for the set $\{|I_c| : c \geq a\}$. If $\varepsilon > 0$, let $K(\varepsilon)$ be such that if $x \geq K(\varepsilon)$, then $0 \leq \varphi(x) \leq \varepsilon/2A$. If $b \geq c \geq K(\varepsilon)$, then it follows from Exercise 30.N that there exists a number ξ in $[c, b]$ such that

$$\int_c^b f\varphi = \varphi(c) \int_c^{\xi} f.$$

In view of the estimate

$$\left| \int_c^{\xi} f \right| = |I_\xi - I_c| \leq 2A,$$

it follows that

$$\left| \int_c^b f\varphi \right| < \varepsilon$$

when $b \geq c$ both exceed $K(\varepsilon)$. We can then apply the Cauchy Criterion 32.5.   Q.E.D.

32.10   EXAMPLES.   (a) If $f(x) = 1/(1 + x^2)$ and $g(x) = 1/x^2$ for $x \geq a > 0$, then $0 \leq f(x) \leq g(x)$. Since we have already seen in Example 32.4(b) that the infinite integral $\int_1^{+\infty} (1/x^2)\, dx$ exists, it follows from the Comparison Test 32.7 that the infinite integral $\int_1^{+\infty} (1/(1 + x^2))\, dx$ also exists. (This could be shown directly by noting that

$$\int_1^c \frac{1}{1 + x^2}\, dx = \text{Arc tan } c - \text{Arc tan } 1$$

and that Arc tan $c \to \pi/2$ as $c \to +\infty$.)

(b) If $h(x) = e^{-x^2}$ and $g(x) = e^{-x}$ then $0 \leq h(x) \leq g(x)$ for $x \geq 1$. It was seen in Example 32.4(c) that the infinite integral $\int_0^{+\infty} e^{-x}\, dx$ exists, whence it follows from the Comparison Test 32.7 that the infinite integral $\int_0^{+\infty} e^{-x^2}\, dx$

also exists. This time, a direct evaluation of the partial integrals is not possible, using elementary functions. However, it will be seen later that this infinite integral equals $\frac{1}{2}\sqrt{\pi}$.

(c) Let $p > 0$ and consider the existence of the infinite integral

$$\int_1^{+\infty} \frac{\sin x}{x^p}\, dx.$$

If $p > 1$, then the integrand is dominated by $1/x^p$, which was seen in Example 32.4(b) to be convergent. In this case the Comparison Test implies that the infinite integral converges. If $0 < p \le 1$, this argument fails; however, if we set $f(x) = \sin x$ and $\varphi(x) = 1/x^p$, then Dirichlet's Test 32.9 shows that the infinite integral exists.

(d) Let $f(x) = \sin x^2$ for $x \ge 0$ and consider the **Fresnel† Integral**

$$\int_0^{+\infty} \sin x^2\, dx.$$

It is clear that the integral over $[0, 1]$ exists, so we shall examine only the integral over $\{x : x \ge 1\}$. If we make the substitution $t = x^2$ and apply the Change of Variable Theorem 30.12, we obtain

$$\int_1^c \sin x^2\, dx = \frac{1}{2} \int_1^{c^2} \frac{\sin t}{\sqrt{t}}\, dt.$$

The preceding example shows that the integral on the right converges when $c \to +\infty$; hence it follows that $\int_1^{+\infty} \sin x^2\, dx$ exists. (It should be observed that the integrand does not converge to 0 as $x \to +\infty$.)

(e) Suppose that $\alpha \ge 1$ and let $\Gamma(\alpha)$ be defined by the integral

(32.13)
$$\Gamma(\alpha) = \int_0^{+\infty} e^{-x} x^{\alpha-1}\, dx.$$

In order to see that this infinite integral exists, consider the function $g(x) = 1/x^2$ for $x \ge 1$. Since

$$\lim_{x \to +\infty} \frac{e^{-x} x^{\alpha-1}}{x^{-2}} = \lim_{x \to +\infty} \frac{x^{\alpha+1}}{e^x} = 0,$$

it follows that if $\varepsilon > 0$ then there exists $K(\varepsilon)$ such that

$$0 < e^{-x} x^{\alpha-1} \le \varepsilon x^{-2} \qquad \text{for } x \ge K(\varepsilon).$$

Since the infinite integral $\int_K^{+\infty} x^{-2}\, dx$ exists, we infer that the integral (32.13) also converges. The important function defined for $\alpha \ge 1$ by formula (32.13) is called the **Gamma function**. It will be quickly seen that if $\alpha < 1$,

---

† AUGUSTIN FRESNEL (1788–1827), a French mathematical physicist, helped to reestablish the undulatory theory of light which was introduced earlier by Huygens.

then the integrand $e^{-x}x^{\alpha-1}$ becomes unbounded near $x = 0$. However, if α satisfies $0 < \alpha < 1$, then we have seen in Example 32.2(c) that the function $x^{\alpha-1}$ has an improper integral over the interval $[0, 1]$. Since $0 < e^{-x} \le 1$ for all $x \ge 0$, it is readily established that the improper integral

$$\int_{0+}^{1} e^{-x}x^{\alpha-1}\,dx$$

exists when $0 < \alpha < 1$. Hence we can extend the definition of the Gamma function to be given for all $\alpha > 0$ by an integral of the form of (32.13) provided it is interpreted as a sum

$$\int_{0+}^{a} e^{-x}x^{\alpha-1}\,dx + \int_{a}^{+\infty} e^{-x}x^{\alpha-1}\,dx$$

of an improper integral and an infinite integral.

Absolute Convergence

If f is Riemann integrable on $[a, c]$ for every $c \ge a$, then it follows from Theorem 30.4(a) that $|f|$, the absolute value of f, is also Riemann integrable on $[a, c]$ for $c \ge a$. It follows from the Comparison Test 32.7 that if the infinite integral

$$(32.14) \qquad \int_{a}^{+\infty} |f(x)|\,dx$$

exists, then the infinite integral

$$(32.15) \qquad \int_{a}^{+\infty} f(x)\,dx$$

also exists and is bounded in absolute value by (32.14).

32.11 DEFINITION. If the infinite integral (32.14) exists, then we say that f is **absolutely integrable** over $\{x : x \ge a\}$, or that the infinite integral (32.15) is **absolutely convergent.**

We have remarked that if f is absolutely integrable over $\{x : x \ge a\}$, then the infinite integral (32.15) exists. The converse is not true, however, as may be seen by considering the integral

$$\int_{\pi}^{+\infty} \frac{\sin x}{x}\,dx.$$

The convergence of this integral was established in Example 32.10(c). However, it is easily seen that in each interval $[k\pi, (k+1)\pi]$, $k \in \mathbf{N}$, there is a subinterval of length $b > 0$ on which

$$|\sin x| \ge \tfrac{1}{2}.$$

(In fact, we can take $b = 2\pi/3$.) Therefore, we have

$$\int_\pi^{k\pi} \left|\frac{\sin x}{x}\right| dx = \int_\pi^{2\pi} + \cdots + \int_{(k-1)\pi}^{k\pi} \geq \frac{b}{2}\left\{\frac{1}{2\pi} + \frac{1}{3\pi} + \cdots + \frac{1}{k\pi}\right\},$$

whence it follows (see 16.11(c)) that the function $f(x) = \sin x/x$ is not absolutely integrable over $\{x : x \geq \pi\}$.

We observe that the Comparison Test 32.7 in fact establishes the *absolute convergence* of the infinite integral of f over the interval $[a, +\infty)$.

Exercises

32.A. Suppose that f is a bounded real-valued function on $J = [a, b]$ and that f is integrable over $[c, b]$ for all $c > a$. Prove that the improper integral of f over J exists.

32.B. Suppose that f is integrable over $[c, b]$ for all $c > a$ and that the improper integral $\int_{a+}^b |f|$ exists. Show that the improper integral $\int_{a+}^b f$ exists, but that the converse may not be true.

32.C. Suppose that f and g are integrable on $[c, b]$ for all $c \in (a, b)$. If $|f(x)| \leq g(x)$ for $x \in J = [a, b]$ and if g has an improper integral on J, then so does f.

32.D. Discuss the convergence or the divergence of the following improper integrals:

(a) $\displaystyle\int_0^1 \frac{dx}{(x+x^2)^{1/2}}$,

(b) $\displaystyle\int_0^1 \frac{dx}{(x-x^2)^{1/2}}$,

(c) $\displaystyle\int_0^1 \frac{x \, dx}{(1-x^3)}$,

(d) $\displaystyle\int_0^1 \frac{\log x}{\sqrt{x}} \, dx$,

(e) $\displaystyle\int_0^1 \frac{\log x}{1-x^2} \, dx$,

(f) $\displaystyle\int_0^1 \frac{x \, dx}{(1-x^3)^{1/2}}$.

32.E. Determine the values of p and q for which the following integrals converge:

(a) $\displaystyle\int_0^1 x^p (1-x)^q \, dx$,

(b) $\displaystyle\int_0^{\pi/2} x^p (\sin x)^q \, dx$,

(c) $\displaystyle\int_1^2 (\log x)^p \, dx$,

(d) $\displaystyle\int_0^1 x^p (-\log x)^q \, dx$.

32.F. Discuss the convergence or the divergence of the following integrals. Which are absolutely convergent?

(a) $\displaystyle\int_1^{+\infty} \frac{dx}{x(1+\sqrt{x})}$,

(b) $\displaystyle\int_1^{+\infty} \frac{x+2}{x^2+1} \, dx$,

(c) $\displaystyle\int_1^{+\infty} \frac{\sin (1/x)}{x} \, dx$,

(d) $\displaystyle\int_1^{+\infty} \frac{\cos x}{\sqrt{x}} \, dx$,

(e) $\displaystyle\int_0^{+\infty} \frac{x \sin x}{1+x^2} \, dx$,

(f) $\displaystyle\int_0^{+\infty} \frac{\sin x \sin 2x}{x} \, dx$.

32.G. For what values of p and q are the following integrals convergent? For what values are they absolutely convergent?

(a) $\displaystyle\int_{1}^{+\infty} \frac{x^{p}}{1+x^{q}}\, dx,$

(b) $\displaystyle\int_{1}^{+\infty} \frac{\sin x}{x^{q}}\, dx,$

(c) $\displaystyle\int_{1}^{+\infty} \frac{\sin x^{p}}{x}\, dx,$

(d) $\displaystyle\int_{1}^{+\infty} \frac{1-\cos x}{x^{q}}\, dx.$

32.H. If f is integrable on any interval $[0, c]$ for $c > 0$, show that the infinite integral $\int_{0}^{+\infty} f$ exists if and only if the infinite integral $\int_{5}^{+\infty} f$ exists.

32.I. Give an example where the infinite integral $\int_{0}^{+\infty} f$ exists but where f is not bounded on the set $\{x : x \geq 0\}$.

32.J. If f is monotone and the infinite integral $\int_{0}^{+\infty} f$ exists, then $xf(x) \to 0$ as $x \to +\infty$.

Section 33 Uniform Convergence and Infinite Integrals

In many applications it is important to consider infinite integrals in which the integrand depends on a parameter. In order to handle this situation easily, the notion of uniform convergence of the integral relative to the parameter is of prime importance. We shall first treat the case that the parameter belongs to an interval $J = [\alpha, \beta]$.

33.1 DEFINITION. Let f be a real-valued function, defined for (x, t) satisfying $x \geq a$ and $\alpha \leq t \leq \beta$. Suppose that for each t in $J = [\alpha, \beta]$ the infinite integral

$$(33.1) \qquad F(t) = \int_{a}^{+\infty} f(x, t)\, dx$$

exists. We say that this convergence is **uniform on** J if for every $\varepsilon > 0$ there exists a number $N(\varepsilon)$ such that if $c \geq N(\varepsilon)$ and $t \in J$, then

$$\left| F(t) - \int_{a}^{c} f(x, t)\, dx \right| < \varepsilon.$$

The distinction between ordinary convergence of the infinite integrals given in (33.1) and uniform convergence is that $M(\varepsilon)$ can be chosen to be independent of the value of t in J. We leave it to the reader to write out the definition of uniform convergence of the infinite integrals when the parameter t belongs to the set $\{t : t \geq \alpha\}$ or to the set $\mathbf{N}$.

It is useful to have some tests for uniform convergence of the infinite integral.

33.2 CAUCHY CRITERION. *Suppose that for each $t \in J$, the infinite integral (33.1) exists. Then the convergence is uniform on J if and only if for* *each $\varepsilon > 0$ there is a number $K(\varepsilon)$ such that if $b \geq c \geq K(\varepsilon)$ and $t \in J$, then*

(33.2)
$$\left| \int_c^b f(x, t) \, dx \right| < \varepsilon.$$

We leave the proof as an exercise.

33.3   WEIERSTRASS M-TEST.   *Suppose that f is Riemann integrable over $[a, c]$ for all $c \geq a$ and all $t \in J$. Suppose that there exists a positive function M defined for $x \geq a$ such that*

$$|f(x, t)| \leq M(x) \qquad for \ x \geq a, \ t \in J,$$

and such that the infinite integral $\int_a^{+\infty} M(x) \, dx$ exists. Then, for each $t \in J$, the integral in (33.1) is (absolutely) convergent and the convergence is uniform on J.

PROOF.   The convergence of

$$\int_a^{+\infty} |f(x, t)| \, dx \qquad for \ t \in J,$$

is an immediate consequence of the Comparison Test and the hypotheses. Therefore, the integral yielding $F(t)$ is absolutely convergent for $t \in J$. If we use the Cauchy Criterion together with the estimate

$$\left| \int_c^b f(x, t) \, dx \right| \leq \int_c^b |f(x, t)| \, dx \leq \int_c^b M(x) \, dx,$$

we can readily establish the uniform convergence on J.   Q.E.D.

The Weierstrass M-test is useful when the convergence is absolute as well as uniform, but it is not delicate enough to handle the case of non-absolute uniform convergence. For this, we turn to an analogue of Dirichlet's Test 32.9.

33.4   DIRICHLET'S TEST.   *Let f be continuous in (x, t) for $x \geq a$ and t in J and suppose that there exists a constant A such that*

$$\left| \int_a^c f(x, t) \, dx \right| \leq A \qquad for \ c \geq a, \quad t \in J.$$

Suppose that for each $t \in J$, the function $\varphi(x, t)$ is monotone decreasing for $x \geq a$, and converges to 0 as $x \to +\infty$ uniformly for $t \in J$. Then the integral

$$F(t) = \int_a^{+\infty} f(x, t) \varphi(x, t) \, dx$$

converges uniformly on J.

PROOF.   Let $\varepsilon > 0$ and choose $K(\varepsilon)$ such that if $x \geq K(\varepsilon)$ and $t \in J$, then $\varphi(x, t) < \varepsilon/2A$. If $b \geq c \geq K(\varepsilon)$, then it follows from Exercise 30.N that for each $t \in J$, there exists a number $\xi(t)$ in $[c, b]$ such that

$$\int_c^b f(x, t)\varphi(x, t)\, dx = \varphi(c, t)\int_c^{\xi(t)} f(x, t)\, dx.$$

Therefore, if $b \geq c \geq K(\varepsilon)$ and $t \in J$, we have

$$\left|\int_c^b f(x, t)\varphi(x, t)\, dx\right| \leq \varphi(c, t)2A < \varepsilon,$$

so the uniformity of the convergence follows from the Cauchy Criterion 33.2.
Q.E.D.

33.5 EXAMPLES. (a) If f is given by

$$f(x, t) = \frac{\cos tx}{1 + x^2}, \qquad x \geq 0, \quad t \in \mathbf{R},$$

and if we define $M(x) = (1 + x^2)^{-1}$, then $|f(x, t)| \leq M(x)$. Since the infinite integral of M on $[0, +\infty)$ exists, it follows from the Weierstrass M-test that the infinite integral

$$\int_0^{+\infty} \frac{\cos tx}{1 + x^2}\, dx$$

converges uniformly for $t \in \mathbf{R}$.

(b) Let $f(x, t) = e^{-x}x^t$ for $x \geq 0$, $t \geq 0$. It is seen that the integral

$$\int_0^{+\infty} e^{-x}x^t\, dx$$

converges uniformly for t in an interval $[0, \beta]$ for any $\beta > 0$. However, it does not converge uniformly on $\{t \in \mathbf{R} : t \geq 0\}$. (See Exercise 33.A.)

(c) If $f(x, t) = e^{-tx}\sin x$ for $x \geq 0$ and $t \geq \gamma > 0$, then

$$|f(x, t)| \leq e^{-tx} \leq e^{-\gamma x}.$$

If we set $M(x) = e^{-\gamma x}$, then the Weierstrass M-test implies that the integral

$$\int_0^{+\infty} e^{-tx}\sin x\, dx$$

converges uniformly for $t \geq \gamma > 0$ and an elementary calculation shows that it converges to $(1 + t^2)^{-1}$. (Note that if $t = 0$, then the integral no longer converges.)

(d) Consider the infinite integral

$$\int_0^{+\infty} e^{-tx}\frac{\sin x}{x}\, dx \qquad \text{for } t \geq 0,$$

where we interpret the integrand to be 1 for $x = 0$. Since the integrand is dominated by 1, it suffices to show that the integral over $\varepsilon \leq x$ converges uniformly for $t \geq 0$. The Weierstrass M-test does not apply to this integrand. However, if we take $f(x, t) = \sin x$ and $\varphi(x, t) = e^{-tx}/x$, then the hypotheses of Dirichlet's Test are satisfied.

Infinite Integrals Depending on a Parameter

Suppose that f is a continuous function of (x, t) defined for $x \geq a$ and for t in $J = [\alpha, \beta]$. Furthermore, suppose that the infinite integral

$$(33.1) \qquad F(t) = \int_a^{+\infty} f(x, t)\, dx$$

exists for each $t \in J$. We shall now show that if this convergence is uniform, then F is continuous on J and its integral can be calculated by interchanging the order of integration. A similar result will be established for the derivative.

33.6 THEOREM. *Suppose that f is continuous in (x, t) for $x \geq a$ and t in $J = [\alpha, \beta]$ and that the convergence in (33.1) is uniform on J. Then F is continuous on J.*

PROOF. If $n \in N$, let F_n be defined on J by

$$F_n(t) = \int_a^{a+n} f(x, t)\, dx.$$

It follows from Theorem 31.6 that F_n is continuous on J. Since the sequence (F_n) converges to F uniformly on J, it follows from Theorem 24.1 that F is continuous on J. Q.E.D.

33.7 THEOREM. *Under the hypotheses of the preceding theorem, then*

$$\int_\alpha^\beta F(t)\, dt = \int_a^{+\infty} \left\{ \int_\alpha^\beta f(x, t)\, dt \right\} dx,$$

which can be written in the form

$$(33.3) \qquad \int_\alpha^\beta \left\{ \int_a^{+\infty} f(x, t)\, dx \right\} dt = \int_a^{+\infty} \left\{ \int_\alpha^\beta f(x, t)\, dt \right\} dx.$$

PROOF. If F_n is defined as in the preceding proof, then it follows from Theorem 31.9 that

$$\int_\alpha^\beta F_n(t)\, dt = \int_a^{a+n} \left\{ \int_\alpha^\beta f(x, t)\, dt \right\} dx.$$

Since (F_n) converges to F uniformly on J, then Theorem 31.2 implies that

$$\int_\alpha^\beta F(t)\, dt = \lim_n \int_\alpha^\beta F_n(t)\, dt.$$

Combining the last two relations, we obtain (33.3). Q.E.D.

33.8 THEOREM. *Suppose that f and its partial derivative f_t are continuous in (x, t) for $x \geq a$ and t in $J = [\alpha, \beta]$. Suppose that (33.1) exists for all $t \in J$ and that*

$$G(t) = \int_a^{+\infty} f_t(x, t)\, dx$$

is uniformly convergent on J. Then F is differentiable on J and $F' = G$. In symbols:

$$\frac{d}{dt} \int_a^{+\infty} f(x, t)\, dx = \int_a^{+\infty} \frac{\partial f}{\partial t}(x, t)\, dx.$$

PROOF. If F_n is defined for $t \in J$ to be

$$F_n(t) = \int_a^{a+n} f(x, t)\, dx,$$

then it follows from Theorem 31.7 that F_n is differentiable and that

$$F_n'(t) = \int_a^{a+n} f_t(x, t)\, dx.$$

By hypothesis, the sequence (F_n) converges on J to F and the sequence (F_n') converges uniformly on J to G. It follows from Theorem 28.5 that F is differentiable on J and that $F' = G$. Q.E.D.

33.9 EXAMPLES. (a) We observe that if $t > 0$, then

$$\frac{1}{t} = \int_0^{+\infty} e^{-tx}\, dx$$

and that the convergence is uniform for $t \geq t_0 > 0$. If we integrate both sides of this relation with respect to t over an interval $[\alpha, \beta]$ where $0 < \alpha < \beta$, and use Theorem 33.7, we obtain the formula

$$\log(\beta/\alpha) = \int_\alpha^\beta \frac{1}{t}\, dt = \int_0^{+\infty} \left\{ \int_\alpha^\beta e^{-tx}\, dt \right\} dx$$

$$= \int_0^{+\infty} \frac{e^{-\alpha x} - e^{-\beta x}}{x}\, dx.$$

(Observe that the last integrand can be defined to be continuous at $x = 0$.)

(b) Instead of integrating with respect to t, we differentiate and formally obtain

$$\frac{1}{t^2} = \int_0^{+\infty} x e^{-tx} \, dx.$$

Since this latter integral converges uniformly with respect to t, provided $t \geq t_0 > 0$, the formula holds for $t > 0$. By induction we obtain

$$\frac{n!}{t^{n+1}} = \int_0^{+\infty} x^n e^{-tx} \, dx \qquad \text{for } t > 0.$$

Referring to the definition of the Gamma function, given in Example 32.10(e), we see that $\Gamma(n+1) = n!$.

(c) If $\alpha > 1$ is a real number and $x > 0$, then $x^{\alpha-1} = e^{(\alpha-1)\log x}$. Hence $f(\alpha) = x^{\alpha-1}$ is a continuous function of (α, x). Moreover, it is seen that there exists a neighborhood of α on which the integral

$$\Gamma(\alpha) = \int_0^{+\infty} x^{\alpha-1} e^{-x} \, dx$$

is uniformly convergent. It follows from Theorem 33.6 that the Gamma function is continuous at least for $\alpha > 1$. (If $0 < \alpha \leq 1$, the same conclusion can be drawn, but the fact that the integral is improper at $x = 0$ must be considered.)

(d) Let $t \geq 0$ and $u \geq 0$ and let F be defined by

$$F(u) = \int_0^{+\infty} e^{-tx} \frac{\sin ux}{x} \, dx.$$

If $t > 0$, then this integral is uniformly convergent for $u \geq 0$ and so is the integral

$$F'(u) = \int_0^{+\infty} e^{-tx} \cos ux \, dx.$$

Moreover, integration by parts shows that

$$\int_0^A e^{-tx} \cos ux \, dx = \left[\frac{e^{-tx}[u \sin ux - t \cos ux]}{t^2 + u^2} \right]_{x=0}^{x=A}.$$

If we let $A \to +\infty$, we obtain the formula

$$F'(u) = \int_0^{+\infty} e^{-tx} \cos ux \, dx = \frac{t}{t^2 + u^2}, \qquad u \geq 0.$$

Therefore, there exists a constant C such that

$$F(u) = \text{Arc tan } (u/t) + C \qquad \text{for} \qquad u \geq 0.$$

In order to evaluate the constant C, we use the fact that $F(0) = 0$ and

Arc tan $(0) = 0$ and infer that $C = 0$. Hence, if $t > 0$ and $u \geq 0$, then

$$\text{Arc tan } (u/t) = \int_0^{+\infty} e^{-tx} \frac{\sin ux}{x} \, dx.$$

(e) Now hold $u > 0$ fixed in the last formula and observe, as in Example 33.5(d) that the integral converges uniformly for $t \geq 0$ so that the limit is continuous for $t \geq 0$. If we let $t \to 0+$, we obtain the important formula

(33.4) $$\frac{\pi}{2} = \int_0^{+\infty} \frac{\sin ux}{x} \, dx, \qquad u > 0.$$

Infinite Integrals of Sequences

Let (f_n) be a sequence of real-valued functions which are defined for $x \geq a$. We shall suppose that the infinite integrals $\int_a^{+\infty} f_n$ all exist and that the limit $f(x) = \lim (f_n(x))$ exists for each $x \geq a$. We would like to be able to conclude that the infinite integral of f exists and that

(33.5) $$\int_a^{+\infty} f = \lim \int_a^{+\infty} f_n.$$

In Theorem 31.2 it was proved that if a sequence (f_n) of Riemann integrable functions converges uniformly on an interval $[a, c]$ to a function f, then f is Riemann integrable and the integral of f is the limit of the integrals of the f_n. The corresponding result is not necessarily true for infinite integrals; it will be seen in Exercise 33.J that the limit function need not possess an infinite integral. Moreover, even if the infinite integral does exist and both sides of (33.5) have a meaning, the equality may fail (cf. Exercise 33.K). Similarly, the obvious extension of the Bounded Convergence Theorem 31.3 may fail for infinite integrals. However, there are two important and useful results which give conditions under which equation (33.5) holds. In proving them we shall make use of the Bounded Convergence Theorem 31.3. The first result is a special case of a celebrated theorem due to Lebesgue. (Since we are dealing with infinite Riemann integrals, we need to add the hypothesis that the limit function is integrable. In the more general Lebesgue theory of integration, this additional hypothesis is not required.)

33.10 DOMINATED CONVERGENCE THEOREM. *Suppose that (f_n) is a bounded sequence of real-valued functions, that $f(x) = \lim (f_n(x))$ for all $x \geq a$, and that f and f_n, $n \in \mathbf{N}$, are Riemann integrable over $[a, c]$ for all $c > a$. Suppose that there exists a function M which has an integral over $x \geq a$ and that*

$$|f_n(x)| \leq M(x) \qquad \text{for } x \geq a, \quad n \in \mathbf{N}.$$

Then f has an integral over x ≥ a and

(33.5)
$$\int_a^{+\infty} f = \lim \int_a^{+\infty} f_n.$$

PROOF. It follows from the Comparison Test 32.7 that the infinite integrals

$$\int_a^{+\infty} f, \quad \int_a^{+\infty} f_n, \quad n \in \mathbf{N},$$

exist. If $\varepsilon > 0$, let K be chosen such that $\int_K^{+\infty} M < \varepsilon$, from which it follows that

$$\left| \int_K^{+\infty} f \right| < \varepsilon \quad \text{and} \quad \left| \int_K^{+\infty} f_n \right| < \varepsilon, \quad n \in \mathbf{N}.$$

Since $f(x) = \lim (f_n(x))$ for all $x \in [a, K]$ it follows from the Bounded Convergence Theorem 31.3 that $\int_a^K f = \lim_n \int_a^K f_n$. Therefore, we have

$$\left| \int_a^{+\infty} f - \int_a^{+\infty} f_n \right| \le \left| \int_a^K f - \int_a^K f_n \right| + 2\varepsilon,$$

which is less than 3ε for sufficiently large n.   Q.E.D.

33.11 MONOTONE CONVERGENCE THEOREM. *Suppose that (f_n) is a bounded sequence of positive functions on $\{x : x \ge a\}$ which is monotone increasing in the sense that $f_n(x) \le f_{n+1}(x)$ for $n \in \mathbf{N}$ and $x \ge a$, and such that f and each f_n has an integral over $[a, c]$ for all $c > a$. Then the limit function f has an integral over $\{x : x \ge a\}$ if and only if the set $\{\int_a^{+\infty} f_n : n \in \mathbf{N}\}$ is bounded. In this case*

$$\int_a^{+\infty} f = \sup_n \left\{ \int_a^{+\infty} f_n \right\} = \lim_n \int_a^{+\infty} f_n.$$

PROOF. Since the sequence (f_n) is monotone increasing, we infer that the sequence $(\int_a^{+\infty} f_n : n \in \mathbf{N})$ is also monotone increasing. If f has an integral over $\{x : x \ge a\}$, then the Dominated Convergence Theorem (with $M = f$) shows that

$$\int_a^{+\infty} f = \lim \int_a^{+\infty} f_n.$$

Conversely, suppose that the set of infinite integrals is bounded and let S be the supremum of this set. If $c > a$, then the Monotone Convergence Theorem 31.4 implies that

$$\int_a^c f = \lim_n \int_a^c f_n = \sup_n \left\{ \int_a^c f_n \right\}.$$

Since $f_n \ge 0$, it follows that $\int_a^c f_n \le \int_a^{+\infty} f_n \le S$, and hence that $\int_a^c f \le S$. By

Theorem 32.6 the infinite integral of f exists and

$$\int_a^{+\infty} f = \sup_c \int_a^c f = \sup_c \left\{ \sup_n \int_a^c f_n \right\}$$

$$= \sup_n \left\{ \sup_c \int_a^c f_n \right\} = \sup_n \int_a^{+\infty} f_n. \qquad \text{Q.E.D.}$$

Iterated Infinite Integrals

In Theorem 33.7 we obtained a result which justifies the interchange of the order of integration over the region $\{(x, t): a \le x, \alpha \le t \le \beta\}$. It is also desirable to be able to interchange the order of integration of an iterated infinite integral. That is, we wish to establish the equality

(33.6)
$$\int_\alpha^{+\infty} \left\{ \int_a^{+\infty} f(x, t) \, dx \right\} dt = \int_a^{+\infty} \left\{ \int_\alpha^{+\infty} f(x, t) \, dt \right\} dx,$$

under suitable hypotheses. It turns out that a simple condition can be given which will also imply absolute convergence of the integrals. However, in order to treat iterated infinite integrals which are not necessarily absolutely convergent, a more complicated set of conditions is required.

33.12 THEOREM. *Suppose that f is a positive function defined for (x, t) satisfying $x \ge a$, $t \ge \alpha$. Suppose that*

(33.7)
$$\int_a^b \left\{ \int_\alpha^{+\infty} f(x, t) \, dt \right\} dx = \int_\alpha^{+\infty} \left\{ \int_a^b f(x, t) \, dx \right\} dt$$

for each $b \ge a$ and that

(33.7′)
$$\int_\alpha^\beta \left\{ \int_a^{+\infty} f(x, t) \, dx \right\} dt = \int_a^{+\infty} \left\{ \int_\alpha^\beta f(x, t) \, dt \right\} dx$$

for each $\beta \ge \alpha$. Then, if one of the iterated integrals in equation (33.6) exists, the other also exists and they are equal.

PROOF. Suppose that the integral on the left side of (33.6) exists. Since f is positive,

$$\int_a^b f(x, t) \, dx \le \int_a^{+\infty} f(x, t) \, dx$$

for each $b \ge a$ and $t \ge \alpha$. Therefore, it follows from the Comparison Test 32.7, that

$$\int_\alpha^{+\infty} \left\{ \int_a^b f(x, t) \, dx \right\} dt \le \int_\alpha^{+\infty} \left\{ \int_a^{+\infty} f(x, t) \, dx \right\} dt.$$

Employing relation (33.7), we conclude that

$$\int_a^b \left\{ \int_\alpha^{+\infty} f(x, t) \, dt \right\} dx \le \int_\alpha^{+\infty} \left\{ \int_a^{+\infty} f(x, t) \, dx \right\} dt$$

for each $b \ge a$. An application of Theorem 32.6 shows that we can take the limit as $b \to +\infty$, so the other iterated integral exists and

$$\int_a^{+\infty} \left\{ \int_\alpha^{+\infty} f(x, t) \, dt \right\} dx \le \int_\alpha^{+\infty} \left\{ \int_a^{+\infty} f(x, t) \, dx \right\} dt.$$

If we repeat this argument and apply equation (33.7'), we obtain the reverse inequality. Therefore, the equality must hold.   Q.E.D.

33.13 THEOREM. *Suppose f is continuous for $x \ge a$, $t \ge \alpha$, and that there exist positive functions M and N such that the infinite integrals $\int_a^{+\infty} M$ and $\int_\alpha^{+\infty} N$ exist. If the inequality*

$$(33.8) \qquad |f(x, t)| \le M(x)N(t), \qquad x \ge a, \quad t \ge \alpha,$$

holds, then the iterated integrals in (33.6) both exist and are equal.

PROOF. Let g be defined for $x \ge a$, $t \ge \alpha$ by $g(x, t) = f(x, t) + M(x)N(t)$ so that

$$0 \le g(x, t) \le 2M(x)N(t).$$

Since N is bounded on each interval $[\alpha, \beta]$, it follows from the inequality (33.8) and the Weierstrass M-test 33.3 that the integral

$$\int_a^{+\infty} g(x, t) \, dx$$

exists uniformly for $t \in [\alpha, \beta]$. By applying Theorem 33.7, we observe that equation (33.7') holds (with f replaced by g) for each $\beta \ge \alpha$. Similarly, (33.7) holds (with f replaced by g) for each $b \ge a$. Also the Comparison Test 32.7 implies that the iterated integrals in (33.6) exist (with f replaced by g). We deduce from Theorem 33.12 that these iterated integrals of g are equal. But this implies that the iterated integrals of f exist and are equal.
Q.E.D.

The preceding results deal with the case that the iterated integrals are absolutely convergent. We now present a result which treats the case of non-absolute convergence.

33.14 THEOREM. *Suppose that the real-valued function f is continuous in (x, t) for $x \ge a$ and $t \ge \alpha$ and that the infinite integrals*

$$(33.9) \qquad \int_a^{+\infty} f(x, t) \, dx, \qquad \int_\alpha^{+\infty} f(x, t) \, dt$$

*converge uniformly for $t \geq \alpha$ and $x \geq a$, respectively. In addition, let F be
defined for $x \geq a$, $\beta \geq \alpha$, by*

$$F(x, \beta) = \int_{\alpha}^{\beta} f(x, t)\, dt$$

and suppose that the infinite integral

(33.10)
$$\int_{a}^{+\infty} F(x, \beta)\, dx$$

*converges uniformly for $\beta \geq \alpha$. Then both iterated infinite integrals exist and
are equal.*

PROOF. Since the infinite integral (33.10) is uniformly convergent for
$\beta \geq \alpha$, if $\varepsilon > 0$ there exists a number $A_\varepsilon \geq a$ such that if $A \geq A_\varepsilon$, then

(33.11)
$$\left| \int_{a}^{A} F(x, \beta)\, dx - \int_{a}^{+\infty} F(x, \beta)\, dx \right| < \varepsilon$$

for all $\beta \geq \alpha$. Also we observe that

$$\int_{a}^{A} F(x, \beta)\, dx = \int_{a}^{A} \left\{ \int_{\alpha}^{\beta} f(x, t)\, dt \right\} dx$$

$$= \int_{\alpha}^{\beta} \left\{ \int_{a}^{A} f(x, t)\, dx \right\} dt.$$

From Theorem 33.7 and the uniform convergence of the second integral in
(33.9), we infer that

$$\lim_{\beta \to +\infty} \int_{a}^{A} F(x, \beta)\, dx = \int_{a}^{A} \left\{ \int_{\alpha}^{+\infty} f(x, t)\, dt \right\} dx.$$

Hence there exists a number $B \geq \alpha$ such that if $\beta_2 \geq \beta_1 \geq B$, then

(33.12)
$$\left| \int_{a}^{A} F(x, \beta_2)\, dx - \int_{a}^{A} F(x, \beta_1)\, dx \right| < \varepsilon.$$

By combining (33.11) and (33.12), it is seen that if $\beta_2 \geq \beta_1 \geq B$, then

$$\left| \int_{a}^{+\infty} F(x, \beta_2)\, dx - \int_{a}^{+\infty} F(x, \beta_1)\, dx \right| < 3\varepsilon,$$

whence it follows that the limit of $\int_{a}^{+\infty} F(x, \beta)\, dx$ exists as $\beta \to +\infty$. After
applying Theorem 33.7 to the uniform convergence of the first integral in
(33.9), we have

$$\lim_{\beta \to +\infty} \int_{a}^{+\infty} F(x, \beta)\, dx = \lim_{\beta \to +\infty} \int_{a}^{+\infty} \left\{ \int_{\alpha}^{\beta} f(x, t)\, dt \right\} dx$$

$$= \lim_{\beta \to +\infty} \int_{\alpha}^{\beta} \left\{ \int_{a}^{+\infty} f(x, t)\, dx \right\} dt$$

$$= \int_{\alpha}^{+\infty} \left\{ \int_{a}^{+\infty} f(x, t)\, dx \right\} dt.$$

Since both terms on the left side of (33.11) have limits as $\beta \to +\infty$ we conclude, on passing to the limit, that

$$\left| \int_a^A \left\{ \int_\alpha^{+\infty} f(x, t) \, dt \right\} dx - \int_\alpha^{+\infty} \left\{ \int_a^{+\infty} f(x, t) \, dx \right\} dt \right| \leq \varepsilon.$$

If we let $A \to +\infty$, we obtain the equality of the iterated improper integrals.

Q.E.D.

The theorems given above justifying the interchange of the order of integration are often useful, but they still leave ample foom for ingenuity. Frequently they are used in conjunction with the Dominated or Monotone Convergence Theorems 33.10 and 33.11.

33.15 EXAMPLES. (a) If $f(x, t) = e^{-(x+t)} \sin xt$, then we can take $M(x) = e^{-x}$ and $N(t) = e^{-t}$ and apply Theorem 33.13 to infer that

$$\int_0^{+\infty} \left\{ \int_0^{+\infty} e^{-(x+t)} \sin xt \, dx \right\} dt = \int_0^{+\infty} \left\{ \int_0^{+\infty} e^{-(x+t)} \sin xt \, dt \right\} dx.$$

(b) If $g(x, t) = e^{-xt}$, for $x \geq 0$ and $t \geq 0$, then we are in trouble on the lines $x = 0$ and $t = 0$. However, if $a > 0$, $\alpha > 0$, and $x \geq a$ and $t \geq \alpha$, then we observe that

$$e^{-xt} = e^{-xt/2} e^{-xt/2} \leq e^{-\alpha x/2} e^{-at/2},$$

If we set $M(x) = e^{-\alpha x/2}$ and $N(t) = e^{-at/2}$, then Theorem 33.13 implies that

$$\int_\alpha^{+\infty} \left\{ \int_a^{+\infty} e^{-xt} \, dx \right\} dt = \int_a^{+\infty} \left\{ \int_\alpha^{+\infty} e^{-xt} \, dt \right\} dx.$$

(c) Consider the function $f(x, y) = xe^{-x^2(1+y^2)}$ for $x \geq a > 0$ and $y \geq 0$. If we put $M(x) = xe^{-x^2}$ and $N(y) = e^{-a^2y^2}$, then we can invert the order of integration over $a \leq x$ and $0 \leq y$. Since we have

$$\int_a^{+\infty} xe^{-(1+y^2)x^2} \, dx = \frac{-e^{-(1+y^2)x^2}}{2(1+y^2)} \Big|_{x=a}^{x \to +\infty} = \frac{e^{-a^2(1+y^2)}}{2(1+y^2)},$$

it follows that

$$\tfrac{1}{2} e^{-a^2} \int_0^{+\infty} \frac{e^{-a^2y^2}}{1+y^2} \, dy = \int_a^{+\infty} e^{-x^2} \left\{ \int_0^{+\infty} xe^{-x^2y^2} \, dy \right\} dx.$$

If we introduce the change of variable $t = xy$, we find that

$$\int_0^{+\infty} xe^{-x^2y^2} \, dy = \int_0^{+\infty} e^{-t^2} \, dt = I.$$

It follows that

$$\int_0^{+\infty} \frac{e^{-a^2y^2}}{1+y^2} \, dy = 2e^{a^2} I \int_a^{+\infty} e^{-x^2} \, dx.$$

If we let $a \to 0$, the expression on the right side converges to $2I^2$. On the left hand side, we observe that the integrand is dominated by the integrable function $(1+y^2)^{-1}$. Applying the Dominated Convergence Theorem, we have

$$\tfrac{1}{2}\pi = \int_0^{+\infty} \frac{dy}{1+y^2} = \lim_{a \to 0} \int_0^{+\infty} \frac{e^{-a^2 y^2}}{1+y^2}\, dy = 2I^2.$$

Therefore $I^2 = \pi/4$, which yields a derivation of the formula

$$\int_0^{+\infty} e^{-x^2}\, dx = \tfrac{1}{2}\sqrt{\pi}.$$

(d) If we integrate by parts twice, we obtain the formula

(33.13)
$$\int_a^{+\infty} e^{-xy} \sin x\, dx = \frac{e^{-ay}}{1+y^2} \cos a + \frac{ye^{-ay}}{1+y^2} \sin a.$$

If $x \geq a > 0$ and $y \geq \alpha > 0$, we can argue as in Example (b) to show that

$$\int_\alpha^{+\infty} \frac{e^{-ay} \cos a}{1+y^2}\, dy + \int_\alpha^{+\infty} \frac{ye^{-ay} \sin a}{1+y^2}\, dy$$

$$= \int_a^{+\infty} \left\{ \int_\alpha^{+\infty} e^{-xy} \sin x\, dy \right\} dx = \int_a^{+\infty} \frac{e^{-\alpha x} \sin x}{x}\, dx.$$

We want to take the limit as $a \to 0$. In the last integral this can evidently be done, and we obtain $\int_0^{+\infty} (e^{-\alpha x} \sin x/x)\, dx$. In view of the fact that $e^{-ay} \cos a$ is dominated by 1 for $y \geq 0$, and the integral $\int_\alpha^{+\infty} (1/(1+y^2))\, dy$ exists, we can use the Dominated Convergence Theorem 33.10 to conclude that

$$\lim_{a \to 0} \int_\alpha^{+\infty} \frac{e^{-ay} \cos a}{1+y^2}\, dy = \int_\alpha^{+\infty} \frac{dy}{1+y^2}.$$

The second integral is a bit more troublesome as the same type of estimate shows that

$$\left| \frac{ye^{-ay} \sin a}{1+y^2} \right| \leq \frac{y}{1+y^2},$$

and the dominant function is not integrable; hence we must do better. Since $u \leq e^u$ and $|\sin u| \leq u$ for $u \geq 0$, we infer that $|e^{-ay} \sin a| \leq 1/y$, whence we obtain the sharper estimate

$$\left| \frac{ye^{-ay} \sin a}{1+y^2} \right| \leq \frac{1}{1+y^2}.$$

We can now employ the Dominated Convergence Theorem to take the limit under the integral sign, to obtain

$$\lim_{a \to 0} \int_{\alpha}^{+\infty} \frac{ye^{-ay} \sin a}{1+y^2} \, dy = 0.$$

We have arrived at the formula

$$\tfrac{1}{2}\pi - \text{Arc tan } \alpha = \int_{\alpha}^{+\infty} \frac{dy}{1+y^2} = \int_{0}^{+\infty} \frac{e^{-ax} \sin x}{x} \, dx.$$

We now want to take the limit as $\alpha \to 0$. This time we cannot use the Dominated Convergence Theorem, since $\int_{0}^{+\infty} x^{-1} \sin x \, dx$ is not absolutely convergent. Although the convergence of e^{-ax} to 1 as $\alpha \to 0$ is monotone, the fact that $\sin x$ takes both signs implies that the convergence of the entire integrand is not monotone. Fortunately, we have already seen in Example 33.5(d) that the convergence of the integral is uniform for $\alpha \geq 0$. According to Theorem 33.6, the integral is continuous for $\alpha \geq 0$ and hence we once more obtain the formula

(33.14)
$$\int_{0}^{+\infty} \frac{\sin x}{x} \, dx = \tfrac{1}{2}\pi.$$

Exercises

33.A. Show that the integral $\int_{0}^{+\infty} x^t e^{-x} \, dx$ converges uniformly for t in an interval $[0, \beta]$ but that it does not converge uniformly for $t \geq 0$.

33.B. Show that the integral

$$\int_{0}^{+\infty} \frac{\sin (tx)}{x} \, dx$$

is uniformly convergent for $t \geq 1$, but that it is not absolutely convergent for any of these values of t.

33.C. For what values of t do the following infinite integrals converge uniformly?

(a) $\int_{0}^{+\infty} \frac{dx}{x^2 + t^2}$,

(b) $\int_{0}^{+\infty} \frac{dx}{x^2 + t}$,

(c) $\int_{0}^{+\infty} e^{-x} \cos tx \, dx$,

(d) $\int_{0}^{+\infty} x^n e^{-x^2} \cos tx \, dx$,

(e) $\int_{0}^{+\infty} e^{-x^2 - t^2/x^2} \, dx$,

(f) $\int_{0}^{+\infty} \frac{t}{x^2} e^{-x^2 - t^2/x^2} \, dx$.

33.D. Use formula (33.14) to show that $\Gamma(\tfrac{1}{2}) = \sqrt{\pi}$.

33.E. Use formula (33.14) to show that $\int_{0}^{+\infty} e^{-tx^2} \, dx = \tfrac{1}{2}\sqrt{\pi/t}$ for $t > 0$. Justify the differentiation and show that

$$\int_{0}^{+\infty} x^{2n} e^{-x^2} \, dx = \frac{1 \cdot 3 \cdots (2n-1)}{2^{n+1}} \sqrt{\pi}.$$

33.F. Establish the existence of the integral $\int_{0}^{+\infty} (1 - e^{-x^2}) x^{-2} \, dx$. (Note that the

integrand can be defined to be continuous at $x = 0$.)   Evaluate this integral by
- (a) replacing e^{-x^2} by e^{-tx^2} and differentiating with respect to t;
- (b) integrating $\int_1^{+\infty} e^{-tx^2} \, dx$ with respect to t.   Justify all of the steps.

33.G. Let F be given for $t \in \mathbf{R}$ by

$$F(t) = \int_0^{+\infty} e^{-x^2} \cos tx \, dx.$$

Differentiate with respect to t and integrate by parts to prove that $F'(t) = (-1/2)tF(t)$.
Then find $F(t)$ and, after a change of variable, establish the formula

$$\int_0^{+\infty} e^{-cx^2} \cos tx \, dx = \tfrac{1}{2}\sqrt{\pi/c} \, e^{-t^2/4c}, \qquad c > 0.$$

33.H. Let G be defined for $t > 0$ by

$$G(t) = \int_0^{+\infty} e^{-x^2 - t^2/x^2} \, dx.$$

Differentiate and change variables to show that $G'(t) = -2G(t)$.   Then find $G(t)$ and
establish the formula

$$\int_0^{+\infty} e^{-x^2 - t^2/x^2} \, dx = \tfrac{1}{2}\sqrt{\pi} \, e^{-2|t|}.$$

33.I. Use formula (33.4), elementary trigonometric formulas, and manipulations
to show that

(a) $\dfrac{2}{\pi} \displaystyle\int_0^{+\infty} \dfrac{\sin ax}{x} \, dx = 1, \qquad a > 0,$

$\qquad\qquad\qquad\qquad = 0, \qquad a = 0,$

$\qquad\qquad\qquad\qquad = -1, \qquad a < 0.$

(b) $\dfrac{2}{\pi} \displaystyle\int_0^{+\infty} \dfrac{\sin x \cos ax}{x} \, dx = 1, \qquad |a| < 1,$

$\qquad\qquad\qquad\qquad\qquad = \tfrac{1}{2}, \qquad |a| = 1,$

$\qquad\qquad\qquad\qquad\qquad = 0, \qquad |a| > 1.$

(c) $\dfrac{2}{\pi} \displaystyle\int_0^{+\infty} \dfrac{\sin x \sin ax}{x} \, dx = \dfrac{1}{\pi} \log \dfrac{a+1}{1-a}, \qquad |a| < 1,$

$\qquad\qquad\qquad\qquad\qquad = \dfrac{1}{\pi} \log \dfrac{a+1}{a-1}, \qquad |a| > 1,$

(d) $\dfrac{2}{\pi} \displaystyle\int_0^{+\infty} \left[\dfrac{\sin x}{x}\right]^2 dx = 1.$

33.J. For $n \in \mathbf{N}$ let f_n be defined by

$$f_n(x) = 1/x, \qquad 1 \le x \le n,$$
$$= 0, \qquad\quad x > n.$$

Each f_n has an integral for $x \ge 1$ and the sequence (f_n) is bounded, monotone
increasing, and converges uniformly to a continuous function which is not integrable
over $\{x \in \mathbf{R} : x \ge 1\}$.

33.K. Let g_n be defined by

$$g_n(x) = 1/n, \qquad 0 \le x \le n^2,$$
$$= 0, \qquad\qquad x > n^2.$$

Each g_n has an integral over $x \ge 0$ and the sequence (g_n) is bounded and converges to a function g which has an integral over $x \ge 0$, but it is *not* true that

$$\lim \int_0^{+\infty} g_n = \int_0^{+\infty} g.$$

Is the convergence monotone?

33.L. If $f(x, t) = (x - t)/(x + t)^3$, show that

$$\int_1^A \left\{ \int_1^{+\infty} f(x, t)\, dx \right\} dt > 0 \qquad \text{for each } A \ge 1;$$

$$\int_1^B \left\{ \int_1^{+\infty} f(x, t)\, dt \right\} dx < 0 \qquad \text{for each } B \ge 1.$$

Hence, show that

$$\int_1^{+\infty} \left\{ \int_1^{+\infty} f(x, t)\, dx \right\} dt \ne \int_1^{+\infty} \left\{ \int_1^{+\infty} f(x, t)\, dt \right\} dx.$$

33.M. Using an argument similar to that in Example 33.15(c) and formulas from Exercises 33.G and 33.H, show that

$$\int_0^{+\infty} \frac{\cos ty}{1 + y^2}\, dy = \frac{\pi}{2} e^{-|t|}.$$

33.N. By considering the iterated integrals of $e^{-(a+y)x} \sin y$ over the quadrant $x \ge 0$, $y \ge 0$, establish the formula

$$\int_0^{+\infty} \frac{e^{-ax}}{1 + x^2}\, dx = \int_0^{+\infty} \frac{\sin y}{a + y}\, dy, \qquad a > 0.$$

Projects

33.α. This project treats the **Gamma function,** which was introduced in Example 32.10(e). Recall that Γ is defined for x in $P = \{x \in \mathbf{R} : x > 0\}$ by the integral

$$\Gamma(x) = \int_{0+}^{+\infty} e^{-t} t^{x-1}\, dt.$$

We have already seen that this integral converges for $x \in P$ and that $\Gamma(\tfrac{1}{2}) = \sqrt{\pi}$.

(a) Show that Γ is continuous on P.

(b) Prove that $\Gamma(x + 1) = x\Gamma(x)$ for $x \in P$. (Hint: integrate by parts on the interval $[\varepsilon, c]$.)

(c) Show that $\Gamma(n + 1) = n!$ for $n \in \mathbf{N}$.

(d) Show that $\lim_{x \to 0+} x\Gamma(x) = 1$. Hence it follows that Γ is not bounded to the right of $x = 0$.

(e) Show that Γ is differentiable on P and that the second derivative is always positive. (Hence Γ is a convex function on P.)

(f) By changing the variable t, show that

$$\Gamma(x) = 2\int_{0+}^{+\infty} e^{-s^2}s^{2x-1}\,ds = u^x\int_{0+}^{+\infty} e^{-us}s^{x-1}\,ds.$$

33.β. We introduce the **Beta function** of Euler. Let $B(x, y)$ be defined for x, y in $P = \{x \in \mathbf{R} : x > 0\}$ by

$$B(x, y) = \int_{0+}^{1-} t^{x-1}(1-t)^{y-1}\,dt.$$

If $x \geq 1$ and $y \geq 1$, this integral is proper, but if $0 < x < 1$ or $0 < y < 1$, the integral is improper.

(a) Establish the convergence of the integral for x, y in P.

(b) Prove that $B(x, y) = B(y, x)$.

(c) Show that if x, y belong to P, then

$$B(x, y) = 2\int_{0+}^{(\pi/2)-} (\sin t)^{2x-1}(\cos t)^{2y-1}\,dt$$

and

$$B(x, y) = \int_{0+}^{+\infty} \frac{u^{x-1}}{(1+u)^{x+y}}\,du.$$

(d) By integrating the positive function

$$f(t, u) = e^{-t^2-u^2}t^{2x-1}u^{2y-1}$$

over $\{(t, u) : t^2 + u^2 = R^2, t \geq 0, u \geq 0\}$ and comparing this integral with the integral over inscribed and circumscribed squares, derive the important formula

$$B(x, y) = \frac{\Gamma(x)\Gamma(y)}{\Gamma(x+y)}.$$

(e) Establish the integration formulas

$$\int_0^{\pi/2} (\sin x)^{2n}\,dx = \frac{\sqrt{\pi}\,\Gamma(n+\frac{1}{2})}{2\Gamma(n+1)} = \frac{1\cdot3\cdot5\cdots(2n-1)}{2\cdot4\cdot6\cdots(2n)}\,\frac{\pi}{2},$$

$$\int_0^{\pi/2} (\sin x)^{2n+1}\,dx = \frac{\sqrt{\pi}\,\Gamma(n+1)}{2\Gamma(n+\frac{3}{2})} = \frac{2\cdot4\cdot6\cdots(2n)}{1\cdot3\cdot5\cdot7\cdots(2n+1)}.$$

33.γ. This and the next project present a few of the properties of the Laplace† transform, which is important both for theoretical and applied mathematics. To simplify the discussion, we shall restrict our attention to continuous functions f defined on $\{t \in \mathbf{R} : t \geq 0\}$ to $\mathbf{R}$. The **Laplace transform** of f is the function $\hat{f}$ defined at the real number s by the formula

$$\hat{f}(s) = \int_0^{+\infty} e^{-st}f(t)\,dt,$$

whenever this integral converges. Sometimes we denote $\hat{f}$ by $\mathscr{L}(f)$.

† PIERRE-SIMON LAPLACE (1749–1827), the son of a Norman farmer, became professor at the Military School in Paris and was elected to the Academy of Sciences. He is famous for his work on celestial mechanics and probability.

(a) Suppose there exists a real number c such that $|f(t)| \le e^{ct}$ for sufficiently large t. Then the integral defining the Laplace transform $\hat{f}$ converges for $s > c$. Moreover, it converges uniformly for $s \ge c + \delta$ if $\delta > 0$.

(b) If f satisfies the boundedness condition in part (a), then $\hat{f}$ is continuous and has a derivative for $s > c$ given by the formula

$$\hat{f}'(s) = \int_0^{+\infty} e^{-st}(-t)f(t)\, dt.$$

[Thus the derivative of the Laplace transform of f is the Laplace transform of the function $g(t) = -tf(t)$.]

(c) By induction, show that under the boundedness condition in (a), then $\hat{f}$ has derivatives of all orders for $s > c$ and that

$$\hat{f}^{(n)}(s) = \int_0^{+\infty} e^{-st}(-t)^n f(t)\, dt.$$

(d) Suppose f and g are continuous functions whose Laplace transforms $\hat{f}$ and $\hat{g}$ converge for $s > s_0$, and if a and b are real numbers then the function $af + bg$ has a Laplace transform converging for $s > s_0$ and which equals $a\hat{f} + b\hat{g}$.

(e) If $a > 0$ and $g(t) = f(at)$, then $\hat{g}$ converges for $s > as_0$ and

$$\hat{g}(s) = \frac{1}{a}\hat{f}(s/a).$$

Similarly, if $h(t) = (1/a)f(t/a)$, then $\hat{h}$ converges for $s > s_0/a$ and

$$\hat{h}(s) = \hat{f}(as).$$

(f) Suppose that the Laplace transform $\hat{f}$ of f exists for $s > s_0$ and let f be defined for $t < 0$ to be equal to 0. If $b > 0$ and if $g(t) = f(t - b)$, then $\hat{g}$ converges for $s > s_0$ and

$$\hat{g}(s) = e^{-bs}\hat{f}(s).$$

Similarly, if $h(t) = e^{bt}f(t)$ for any real b, then $\hat{h}$ converges for $s > s_0 + b$ and

$$\hat{h}(s) = \hat{f}(s - b).$$

33.δ. This project continues the preceding one and makes use of its results.
(a) Establish the following short table of Laplace transforms.

$f(t)$	$\hat{f}(s)$	Interval of Convergence
1	$1/s$	$s > 0$,
t^n	$n!/s^{n+1}$	$s > 0$,
e^{at}	$(s-a)^{-1}$	$s > a$,
$t^n e^{at}$	$n!/(s-a)^{n+1}$	$s > a$,
$\sin at$	$\dfrac{a}{s^2 + a^2}$	all s,
$\cos at$	$\dfrac{s}{s^2 + a^2}$	all s,
$\sinh at$	$\dfrac{a}{s^2 - a^2}$	$s > a$,
$\cosh at$	$\dfrac{s}{s^2 - a^2}$	$s > a$,
$\dfrac{\sin t}{t}$	Arc tan $(1/s)$	$s > 0$.

(b) Suppose that f and f' are continuous for $t \geq 0$, that $\hat{f}$ converges for $s > s_0$ and that $e^{-st}f(t) \to 0$ as $t \to +\infty$ for all $s > s_0$. Then the Laplace transform of f' exists for $s > s_0$ and

$$\widehat{f'}(s) = s\hat{f}(s) - f(0).$$

(Hint: integrate by parts.)

(c) Suppose that f, f' and f'' are continuous for $t \geq 0$ and that $\hat{f}$ converges for $s > s_0$. In addition, suppose that $e^{-st}f(t)$ and $e^{-st}f'(t)$ approach 0 as $t \to +\infty$ for all $s > s_0$. Then the Laplace transform of f'' exists for $s > s_0$ and

$$\widehat{f''}(s) = s^2\hat{f}(s) - sf(0) - f'(0).$$

(d) When all or part of an integrand is seen to be a Laplace transform, the integral can sometimes be evaluated by changing the order of integration. Use this method to evaluate the integral

$$\int_0^{+\infty} \frac{\sin s}{s}\, ds = \tfrac{1}{2}\pi.$$

(e) It is desired to solve the differential equation

$$y'(t) + 2y(t) = 3 \sin t, \qquad y(0) = 1.$$

Assume that this equation has a solution y such that the Laplace transforms of y and y' exist for sufficiently large s. In this case the transform of y must satisfy the equation

$$s\hat{y}(s) - y(0) + 2\hat{y}(s) = 4/(s-1), \qquad s > 1,$$

from which it follows that

$$\hat{y}(s) = \frac{s+3}{(s+2)(s-1)}.$$

Use partial fractions and the table in (a) to obtain $y(t) = \tfrac{4}{3}e^t - \tfrac{1}{3}e^{-2t}$, which can be directly verified to be a solution.

(f) Find the solution of the equation

$$y'' + y' = 0, \qquad y(0) = a, \qquad y'(0) = b,$$

by using the Laplace transform.

(g) Show that a linear homogeneous differential equation with constant coefficients can be solved by using the Laplace transform and the technique of decomposing a rational function into partial fractions.

VI
INFINITE SERIES

This chapter is concerned with establishing the most important theorems in the theory of infinite series. Although a few peripheral results are included here, our attention is directed to the basic propositions. The reader is referred to more extensive treatises for advanced results and applications.

In the first section we shall present the main theorems concerning the convergence of infinite series in R^p. We shall obtain some results of a general nature which serve to establish the convergence of series and justify certain manipulations with series.

In Section 35 we shall give some familiar "tests" for absolute convergence of series. In addition to guaranteeing the convergence of the series to which the tests are applicable, each of these tests yields a quantitative estimate concerning the *rapidity* of the convergence.

The next section provides some useful tests for conditional convergence, and gives a brief discussion of double series and the multiplication of series.

In Section 37 we introduce the study of series of *functions* and establish the basic properties of power series. In the final section of this chapter we shall establish some of the main results from the theory of Fourier series.

Section 34 Convergence of Infinite Series

In elementary texts, an infinite series is sometimes "defined" to be "an expression of the form

$$(34.1) \qquad\qquad x_1 + x_2 + \cdots + x_n + \cdots."$$

This "definition" lacks clarity, however, since there is no particular value that we can attach *a priori* to this array of symbols which calls for an infinite

number of additions to be performed. Although there are other definitions that are suitable, we shall take an infinite series to be the same as the
sequence of partial sums.

34.1 DEFINITION. If $X = (x_n)$ is a sequence in $\mathbf{R}^p$, then the **infinite
series** (or simply the series) generated by X is the sequence $S = (s_k)$ defined
by

$$s_1 = x_1,$$
$$s_2 = s_1 + x_2 \quad (= x_1 + x_2),$$
$$\cdots\cdots\cdots\cdots$$
$$s_k = s_{k-1} + x_k \quad (= x_1 + x_2 + \cdots + x_k),$$
$$\cdots\cdots\cdots\cdots\cdots$$

If S converges, we refer to $\lim S$ as the **sum** of the infinite series. The
elements x_n are called the **terms** and the elements s_k are called the **partial
sums** of this infinite series.

It is conventional to use the expression (34.1) or one of the symbols

$$\sum (x_n), \qquad \sum_{n=1}^{\infty} (x_n), \qquad \sum_{n=1}^{\infty} x_n$$

both to denote the infinite series generated by the sequence $X = (x_n)$ and also to
denote $\lim S$ in the case that this infinite series is convergent. In actual practice,
the double use of these notations does not lead to confusion, provided it is
understood that the convergence of the series must be established.

The reader should guard against confusing the words "sequence" and "series."
In non-mathematical language, these words are interchangeable; in mathematics,
however, they are not synonyms. According to our definition, an infinite series is a
sequence S obtained from a given sequence X according to a special procedure that
was stated above. There are many other ways of generating new sequences and
attaching "sums" to the given sequence X. The reader should consult books on
divergent series, asymptotic series, and the **summability of series** for examples of
such theories.

A final word on notational matters. Although we generally index the elements
of the series by natural numbers, it is sometimes more convenient to start with
$n = 0$, with $n = 5$, or with $n = k$. When such is the case, we shall denote the
resulting series or their sums by notations such as

$$\sum_{n=0}^{\infty} x_n, \qquad \sum_{n=5}^{\infty} x_n, \qquad \sum_{n=k}^{\infty} x_n.$$

In Definition 14.2, we defined the sum and difference of two sequences
X, Y in $\mathbf{R}^p$. Similarly, if c is a real number and if w is an element in $\mathbf{R}^p$, we
defined the sequences $cX = (cx_n)$ and $(w \cdot x_n)$ in $\mathbf{R}^p$ and $\mathbf{R}$, respectively.
We now examine the series generated by these sequences.

34.2 THEOREM.   (a) *If the series $\sum (x_n)$ and $\sum (y_n)$ converge, then the series $\sum (x_n + y_n)$ converges and the sums are related by the formula*

$$\sum (x_n + y_n) = \sum (x_n) + \sum (y_n).$$

A similar result holds for the series generated by $X - Y$.

(b) *If the series $\sum (x_n)$ is convergent, c is a real number, and w is a fixed element of $\mathbf{R}^p$, then the series $\sum (cx_n)$ and $\sum (w \cdot x_n)$ converge and*

$$\sum (cx_n) = c \sum (x_n), \qquad \sum (w \cdot x_n) = w \cdot \sum (x_n).$$

PROOF.   This result follows directly from Theorem 15.6 and Definition 34.1.   Q.E.D.

It might be expected that if the sequences $X = (x_n)$ and $Y = (y_n)$ generate convergent series, then the sequence $X \cdot Y = (x_n \cdot y_n)$ also generates a convergent series.   That this is not always true may be seen by taking $X = Y = ((-1)^n/\sqrt{n})$ in $\mathbf{R}$.

We now present a very simple necessary condition for convergence of a series.   It is far from sufficient, however.

34.3 LEMMA.   *If $\sum (x_n)$ converges in $\mathbf{R}^p$, then $\lim (x_n) = 0$.*

PROOF.   By definition, the convergence of $\sum (x_n)$ means that $\lim (s_k)$ exists.   But, since $x_k = s_k - s_{k-1}$, then $\lim (x_k) = \lim (s_k) - \lim (s_{k-1}) = 0$.
                                                                  Q.E.D.

The next result, although limited in scope, is of great importance.

34.4 THEOREM.   *Let (x_n) be a sequence of positive real numbers.   Then $\sum (x_n)$ converges if and only if the sequence $S = (s_k)$ of partial sums is bounded.   In this case,*

$$\sum x_n = \lim (s_k) = \sup \{s_k\}.$$

PROOF.   Since $x_n \geq 0$, the sequence of partial sums is monotone increasing:

$$s_1 \leq s_2 \leq \cdots \leq s_k \leq \cdots$$

According to the Monotone Convergence Theorem 16.1, the sequence S converges if and only if it is bounded.   Q.E.D.

Since the following Cauchy Criterion is precisely a reformulation of Theorem 16.10, we shall omit its proof.

34.5 CAUCHY CRITERION FOR SERIES. *The series $\sum (x_n)$ in $\mathbf{R}^p$ converges if and only if for each number $\varepsilon > 0$ there is a natural number $M(\varepsilon)$ such that if $m \geq n \geq M(\varepsilon)$, then*

$$\|s_m - s_n\| = \|x_{n+1} + x_{n+2} + \cdots + x_m\| < \varepsilon.$$

The notion of absolute convergence is often of great importance in treating series, as we shall show later.

34.6 DEFINITION. Let $x = (x_n)$ be a sequence in $\mathbf{R}^p$. We say that the series $\sum (x_n)$ is **absolutely convergent** if the series $\sum (\|x_n\|)$ is convergent in $\mathbf{R}$. A series is said to be **conditionally convergent** if it is convergent but not absolutely convergent.

It is stressed that for series whose elements are positive real numbers, there is no distinction between ordinary convergence and absolute convergence. However, for other series there may be a difference.

34.7 THEOREM. *If a series in $\mathbf{R}^p$ is absolutely convergent, then it is convergent.*

PROOF. By hypothesis, the series $\sum (\|x_n\|)$ converges. Therefore, it follows from the necessity of the Cauchy Criterion 34.5 that given $\varepsilon > 0$ there is a natural number $M(\varepsilon)$ such that if $m \geq n \geq M(\varepsilon)$, then

$$\|x_{n+1}\| + \|x_{n+2}\| + \cdots + \|x_m\| < \varepsilon.$$

According to the Triangle Inequality, the left-hand side of this relation dominates

$$\|x_{n+1} + x_{n+2} + \cdots + x_m\|.$$

We apply the sufficiency of the Cauchy Criterion to conclude that the $\sum (x_n)$ must converge. Q.E.D.

34.8 EXAMPLES. (a) We consider the real sequence $X = (a^n)$, which generates the **geometric series**

$$(34.2) \qquad\qquad a + a^2 + \cdots + a^n + \cdots$$

A necessary condition for convergence is that $\lim (a^n) = 0$, which requires that $|a| < 1$. If $m \geq n$, then

$$(34.3) \qquad a^{n+1} + a^{n+2} + \cdots + a^m = \frac{a^{n+1} - a^{m+1}}{1 - a},$$

as can be verified by multiplying both sides by $1 - a$ and noticing the telescoping on the left side. Hence the partial sums satisfy

$$|s_m - s_n| = |a^{n+1} + \cdots + a^m| \leq \frac{|a^{n+1}| + |a^{m+1}|}{|1 - a|}, \qquad m \geq n.$$

If $|a|<1$, then $|a^{n+1}|\to 0$ so the Cauchy Criterion implies that the geometric series (34.2) converges if and only if $|a|<1$. Letting $n=0$ in (34.3) and passing to the limit with respect to m we find that (34.2) converges to the limit $a/(1-a)$ when $|a|<1$.

(b) Consider the **harmonic series** $\sum(1/n)$, which is well-known to diverge. Since $\lim(1/n)=0$, we cannot use Lemma 34.3 to establish this divergence, but must carry out a more delicate argument, which we shall base on Theorem 34.4. We shall show that a subsequence of the partial sums is not bounded. In fact, if $k_1=2$, then

$$s_{k_1}=\frac{1}{1}+\frac{1}{2},$$

and if $k_2=2^2$, then

$$s_{k_2}=\frac{1}{1}+\frac{1}{2}+\frac{1}{3}+\frac{1}{4}=s_{k_1}+\frac{1}{3}+\frac{1}{4}>s_{k_1}+2\left(\frac{1}{4}\right)=1+\frac{2}{2}.$$

By mathematical induction, we establish that if $k_r=2^r$, then

$$s_{k_r}>s_{k_{r-1}}+2^{r-1}\left(\frac{1}{2^r}\right)=s_{k_{r-1}}+\frac{1}{2}\ge 1+\frac{r}{2}.$$

Therefore, the subsequence (s_{k_n}) is not bounded and the harmonic series does not converge.

(c) We now treat the **p-series** $\sum(1/n^p)$ where $0<p\le 1$ and use the elementary inequality $n^p\le n$, for $n\in\mathbf{N}$. From this it follows that, when $0<p\le 1$, then

$$\frac{1}{n}\le\frac{1}{n^p},\qquad n\in\mathbf{N}.$$

Since the partial sums of the harmonic series are not bounded, this inequality shows that the partial sums of $\sum(1/n^p)$ are not bounded for $0<p\le 1$. Hence the series diverges for these values of p.

(d) Consider the p-series for $p>1$. Since the partial sums are monotone, it is sufficient to show that some subsequence remains bounded in order to establish the convergence of the series. If $k_1=2^1-1=1$, then $s_{k_1}=1$. If $k_2=2^2-1=3$, we have

$$s_{k_2}=\frac{1}{1}+\left(\frac{1}{2^p}+\frac{1}{3^p}\right)<1+\frac{2}{2^p}=1+\frac{1}{2^{p-1}},$$

and if $k_3=2^3-1$, we have

$$s_{k_3}=s_{k_2}+\left(\frac{1}{4^p}+\frac{1}{5^p}+\frac{1}{6^p}+\frac{1}{7^p}\right)<s_{k_2}+\frac{4}{4^p}<1+\frac{1}{2^{p-1}}+\frac{1}{4^{p-1}}.$$

Let $a = 1/2^{p-1}$; since $p > 1$, it is seen that $0 < a < 1$. By mathematical induction, we find that if $k_r = 2^r - 1$, then

$$0 < s_{k_r} < 1 + a + a^2 + \cdots + a^{r-1}.$$

Hence the number $1/(1-a)$ is an upper bound for the partial sums of the p-series when $1 < p$. From Theorem 34.4 it follows that for such values of p, the p-series converges.

(e) Consider the series $\sum (1/(n^2 + n))$. By using partial fractions, we can write

$$\frac{1}{k^2 + k} = \frac{1}{k(k+1)} = \frac{1}{k} - \frac{1}{k+1}.$$

This expression shows that the partial sums are telescoping and hence

$$s_n = \frac{1}{1 \cdot 2} + \frac{1}{2 \cdot 3} + \cdots + \frac{1}{n(n+1)} = \frac{1}{1} - \frac{1}{n+1}.$$

It follows that the sequence (s_n) is convergent to 1.

Rearrangements of Series

Loosely speaking, a rearrangement of a series is another series which is obtained from the given one by using all of the terms exactly once, but scrambling the order in which the terms are taken. For example, the harmonic series

$$\frac{1}{1} + \frac{1}{2} + \frac{1}{3} + \cdots + \frac{1}{n} + \cdots$$

has rearrangements

$$\frac{1}{2} + \frac{1}{1} + \frac{1}{4} + \frac{1}{3} + \cdots + \frac{1}{2n} + \frac{1}{2n-1} + \cdots,$$

$$\frac{1}{1} + \frac{1}{2} + \frac{1}{4} + \frac{1}{3} + \frac{1}{5} + \frac{1}{7} + \cdots$$

The first rearrangement is obtained by interchanging the first and second terms, the third and fourth terms, and so forth. The second rearrangement is obtained from the harmonic series by taking one "odd term," two "even terms," three "odd terms," and so on. It is evident that there are infinitely many other possible rearrangements of the harmonic series.

34.9 DEFINITION. A series $\sum (y_m)$ in $\mathbf{R}^p$ is a **rearrangement** of a series $\sum (x_n)$ if there exists a bijection f of $\mathbf{N}$ onto $\mathbf{N}$ such that $y_m = x_{f(m)}$ for all $m \in \mathbf{N}$.

There is a remarkable observation due to Riemann, that if $\sum (x_n)$ is a series in $\mathbf{R}$ which is conditionally convergent (that is, it is convergent but not absolutely convergent) and if c is an arbitrary real number, then there exists a rearrangement of $\sum (x_n)$ which converges to c. The idea of the proof of this assertion is very elementary: we take positive terms until we obtain a partial sum exceeding c, then we take negative terms from the given series until we obtain a partial sum of terms less than c, etc. Since $\lim (x_n) = 0$, it is not difficult to see that a rearrangement which converges to c can be constructed.

In our manipulations with series, we generally find it convenient to be sure that rearrangements will not affect the convergence or the value of the limit.

34.10 REARRANGEMENT THEOREM *Let $\sum (x_n)$ be an absolutely convergent series in $\mathbf{R}^p$. Then any rearrangement of $\sum (x_n)$ converges absolutely to the same value.*

PROOF. Let $x = \sum (x_n)$, let $\sum (y_m)$ be a rearrangement of $\sum (x_n)$, and let K be an upper bound for the partial sums of $\sum (\|x_n\|)$. Clearly, if $t_r = y_1 + \cdots + y_r$ is a partial sum of $\sum (y_m)$, then

$$\|y_1\| + \cdots + \|y_r\| \le K,$$

whence it follows that $\sum (y_m)$ is absolutely convergent to some element y of $\mathbf{R}^p$. We wish to show that $x = y$. If $\varepsilon > 0$, let $N(\varepsilon)$ be such that if $m > n \ge N(\varepsilon)$, and $s_n = x_1 + \cdots + x_n$, then $\|x - s_n\| < \varepsilon$ and

$$\sum_{k=n+1}^{m} \|x_k\| < \varepsilon.$$

Choose a partial sum t_r of $\sum (y_m)$ such that $\|y - t_r\| < \varepsilon$ and such that each $x_1, x_2, \ldots, x_n$ occurs in t_r. Having done this, choose $m > n$ so large that every y_k appearing in t_r also appears in s_m. Therefore

$$\|x - y\| \le \|x - s_m\| + \|s_m - t_r\| + \|t_r - y\| < \varepsilon + \sum_{n+1}^{m} \|x_k\| + \varepsilon < 3\varepsilon.$$

Since $\varepsilon > 0$ is arbitrary, we infer that $x = y$.   Q.E.D.

Exercises

34.A. Let $\sum (a_n)$ be a given series and let $\sum (b_n)$ be one in which the terms are the same as those in $\sum (a_n)$, except those for which $a_n = 0$ have been omitted. Show that $\sum (a_n)$ converges to a number A if and only if $\sum (b_n)$ converges to A.

34.B. Show that the convergence of a series is not affected by changing a finite number of its terms. (Of course, the sum may well be changed.)

34.C. Show that grouping the terms of a convergent series by introducing parentheses containing a finite number of terms does not destroy the convergence or the value of the limit. However, grouping terms in a divergent series can produce convergence.

34.D. Show that if a convergent series of real numbers contains only a finite number of negative terms, then it is absolutely convergent.

34.E. Show that if a series of real numbers is conditionally convergent, then the series of positive terms is divergent and the series of negative terms is divergent.

34.F. By using partial fractions, show that

(a) $\displaystyle\sum_{n=0}^{\infty} \frac{1}{(\alpha+n)(\alpha+n+1)} = \frac{1}{\alpha}$ if $\alpha > 0$,

(b) $\displaystyle\sum_{n=1}^{\infty} \frac{1}{n(n+1)(n+2)} = \frac{1}{4}$.

34.G. If $\sum (a_n)$ is a convergent series of real numbers, then is $\sum (a_n^2)$ always convergent? If $a_n \geq 0$, then is it true that $\sum (\sqrt{a_n})$ is always convergent?

34.H. If $\sum (a_n)$ is convergent and $a_n \geq 0$, then is $\sum (\sqrt{a_n a_{n+1}})$ convergent?

34.I. Let $\sum (a_n)$ be a series of strictly positive numbers and let b_n, $n \in \mathbf{N}$, be defined to be $b_n = (a_1 + a_2 + \cdots + a_n)/n$. Show that $\sum (b_n)$ always diverges.

34.J. Let $\sum (a_n)$ be convergent and let c_n, $n \in \mathbf{N}$, be defined to be the weighted means

$$c_n = \frac{a_1 + 2a_2 + \cdots + na_n}{n(n+1)}.$$

Then $\sum (c_n)$ converges and equals $\sum (a_n)$.

34.K. Let $\sum (a_n)$ be a series of monotone decreasing positive numbers. Prove that $\sum_{n=1}^{\infty} (a_n)$ converges if and only if the series

$$\sum_{n=1}^{\infty} 2^n a_{2^n}$$

converges. This result is often called the **Cauchy Condensation Test**. (Hint: group the terms into blocks as in Examples 34.8(b, d).)

34.L. Use the Cauchy Condensation Test to discuss the convergence of the p-series $\sum (1/n^p)$.

34.M. Use the Cauchy Condensation Test to show that the series

$$\sum \frac{1}{n \log n}, \qquad \sum \frac{1}{n(\log n)(\log \log n)},$$

$$\sum \frac{1}{n(\log n)(\log \log n)(\log \log \log n)}$$

are divergent.

34.N. Show that if $c > 1$, the series

$$\sum \frac{1}{n(\log n)^c}, \qquad \sum \frac{1}{n(\log n)(\log \log n)^c}$$

are convergent.

34.O. Suppose that (a_n) is a monotone decreasing sequence of positive numbers. Show that if the series $\sum (a_n)$ converges, then $\lim (na_n) = 0$. Is the converse true?

34.P. If $\lim (a_n) = 0$, then $\sum (a_n)$ and $\sum (a_n + 2a_{n+1})$ are both convergent or both divergent.

Section 35   Tests for Absolute Convergence

In the preceding section we obtained some results concerning the manipulation of infinite series, especially in the important case where the series are absolutely convergent. However, except for the Cauchy Criterion and the fact that the terms of a convergent series converge to zero, we did not establish any necessary or sufficient conditions for convergence of infinite series.

We shall now give some results which can be used to establish the convergence or divergence of infinite series. In view of its importance, we shall pay special attention to absolute convergence. Since the absolute convergence of the series $\sum (x_n)$ in R^p is equivalent with the convergence of the series $\sum (\|x_n\|)$ of positive elements of R, it is clear that results establishing the convergence of positive real series have particular interest.

Our first test shows that if the terms of a positive real series are dominated by the corresponding terms of a convergent series, then the first series is convergent. It yields a test for absolute convergence that the reader should formulate.

35.1   COMPARISON TEST.   Let $X = (x_n)$ and $Y = (y_n)$ be positive real sequences and suppose that for some natural number K,

$$(35.1) \qquad\qquad x_n \leq y_n \qquad for\ n \geq K,$$

Then the convergence of $\sum (y_n)$ implies the convergence of $\sum (x_n)$.

PROOF.   If $m \geq n \geq \sup \{K, M(\varepsilon)\}$, then

$$x_{n+1} + \cdots + x_m \leq y_{n+1} + \cdots + y_m < \varepsilon,$$

from which the assertion is evident.   Q.E.D.

35.2   LIMIT COMPARISON TEST.   Suppose that $X = (x_n)$ and $Y = (y_n)$ are positive real sequences.
  (a) If the relation

$$(35.2) \qquad\qquad \lim (x_n/y_n) \neq 0$$

holds, then $\sum (x_n)$ is convergent if and only if $\sum (y_n)$ is convergent.

(b) *If the limit in (35.2) is zero and $\sum (y_n)$ is convergent, then $\sum (x_n)$ is convergent.*

PROOF. It follows from (35.2) that for some real number $c > 1$ and some natural number K, then

$$(1/c)y_n \le x_n \le cy_n \qquad \text{for} \qquad n \ge K.$$

If we apply the Comparison Test 35.1 twice, we obtain the assertion in part (a). The proof of (b) is similar and will be omitted. Q.E.D.

The Root and Ratio Tests

We now give an important test due to Cauchy.

35.3 ROOT TEST. (a) *If $X = (x_n)$ is a sequence in $\mathbf{R}^p$ and there exists a positive number $r < 1$ and a natural number K such that*

$$(35.3) \qquad \|x_n\|^{1/n} \le r \qquad \text{for } n \ge K,$$

then the series $\sum (x_n)$ is absolutely convergent.

(b) *If there exists a number $r > 1$ and a natural number K such that*

$$(35.4) \qquad \|x_n\|^{1/n} \ge r \qquad \text{for } n \ge K,$$

then the series $\sum (x_n)$ is divergent.

PROOF. (a) If (35.3) holds, then we have $\|x_n\| \le r^n$. Now for $0 \le r \le 1$, the series $\sum (r^n)$ is convergent, as was seen in Example 34.8(a). Hence it follows from the Comparison Test that $\sum (x_n)$ is absolutely convergent.

(b) If (35.4) holds, then $\|x_n\| \ge r^n$. However, since $r \ge 1$, it is false that $\lim (\|x_n\|) = 0$. Q.E.D.

In addition to establishing the convergence of $\sum (x_n)$, the root test can be used to obtain an estimate of the rapidity of convergence. This estimate is useful in numerical computations and in some theoretical estimates as well.

35.4 COROLLARY. *If r satisfies $0 < r < 1$ and if the sequence $X = (x_n)$ satisfies (35.3), then the partial sums s_n, $n \ge K$, approximate the sum $s = \sum (x_n)$ according to the estimate*

$$(35.5) \qquad \|s - s_n\| \le \frac{r^{n+1}}{1 - r} \qquad \text{for } n \ge K.$$

PROOF. If $m \ge n \ge K$, we have

$$\|s_m - s_n\| = \|x_{n+1} + \cdots + x_m\| \le \|x_{n+1}\| + \cdots + \|x_m\| \le r^{n+1} + \cdots + r^m < \frac{r^{n+1}}{1 - r}.$$

Now take the limit with respect to m to obtain (35.5). Q.E.D.

It is often convenient to make use of the following variant of the Root Test.

35.5   COROLLARY.   *Let $X = (x_n)$ be a sequence in $\mathbf{R}^p$ and set*

(35.6) $$r = \lim (\|x_n\|^{1/n}),$$

whenever this limit exists.   Then $\sum (x_n)$ is absolutely convergent when $r < 1$ and is divergent when $r > 1$.

PROOF.   It follows that if the limit in (35.6) exists and is less than 1, then there is a real number r_1 with $r < r_1 < 1$ and a natural number K such that

$$\|x_n\|^{1/n} \leq r_1 \qquad \text{for } n \geq K.$$

In this case the series is absolutely convergent.   If this limit exceeds 1, then there is a real number $r_2 > 1$ and a natural number K such that

$$\|x_n\|^{1/n} \geq r_2 \qquad \text{for } n \geq K,$$

in which case series is divergent.   Q.E.D.

This corollary can be generalized by using the limit superior instead of the limit.   We leave the details as an exercise.   The next test is due to D'Alembert.†

35.6   RATIO TEST.   (a) *If $X = (x_n)$ is a sequence of non-zero elements of $\mathbf{R}^p$ and there is a positive number $r < 1$ and a natural number K such that*

(35.7) $$\frac{\|x_{n+1}\|}{\|x_n\|} \leq r \qquad \text{for } n \geq K,$$

then the series $\sum (x_n)$ is absolutely convergent.

   (b) *If there exists a number $r \geq 1$ and a natural number K such that*

(35.8) $$\frac{\|x_{n+1}\|}{\|x_n\|} \geq r \qquad \text{for} \qquad n \geq K,$$

then the series $\sum (x_n)$ is divergent.

PROOF.   (a) If (35.7) holds, then an elementary induction argument shows that $\|x_{K+m}\| \leq r^m \|x_K\|$ for $m \geq 1$.   It follows that for $n \geq K$ the terms of $\sum (x_n)$ are dominated by a fixed multiple of the terms of the geometric series $\sum (r^n)$ with $0 \leq r < 1$.   From the Comparison Test 35.1, we infer that $\sum (x_n)$ is absolutely convergent.

† JEAN LE ROND D'ALEMBERT (1717–1783) was a son of the Chevalier Destouches.   He became the secretary of the French Academy and the leading mathematician of the Encyclopedists.   He contributed to dynamics and differential equations.

(b) If (35.8) holds, then an elementary induction argument shows that $\|x_{K+m}\| \geq r^m \|x_K\|$ for $m \geq 1$. Since $r \geq 1$, it is impossible to have $\lim(\|x_n\|) = 0$, so the series cannot converge. Q.E.D.

35.7 COROLLARY. *If r satisfies $0 \leq r < 1$ and if the sequence $X = (x_n)$ satisfies (35.7) for $n \geq K$, then the partial sums approximate the sum $s = \sum(x_n)$ according to the estimate*

$$(35.9) \qquad \|s - s_n\| \leq \frac{r}{1-r} \|x_n\| \qquad for \ n \geq K.$$

PROOF. The relation (35.7) implies that $\|x_{n+k}\| \leq r^k \|x_n\|$ when $n \geq K$. Therefore, if $m \geq n \geq K$, we have

$$\|s_m - s_n\| = \|x_{n+1} + \cdots + x_m\| \leq \|x_{n+1}\| + \cdots + \|x_m\|$$
$$\leq (r + r^2 + \cdots + r^{m-n}) \|x_n\| < \frac{r}{1-r} \|x_n\|.$$

Again we take the limit with respect to m to obtain (35.9). Q.E.D.

35.8 COROLLARY. *Let $X = (x_n)$ be a sequence in $\mathbf{R}^p$ and set*

$$r = \lim\left(\frac{\|x_{n+1}\|}{\|x_n\|}\right),$$

whenever the limit exists. Then the series $\sum(x_n)$ is absolutely convergent when $r < 1$ and divergent when $r > 1$.

PROOF. Suppose that the limit exists and $r < 1$. If r_1 satisfies $r < r_1 < 1$, then there is a natural number K such that

$$\frac{\|x_{n+1}\|}{\|x_n\|} < r_1 \qquad for \ n \geq K.$$

In this case Theorem 35.6 establishes the absolute convergence of the series. If $r > 1$, and if r_2 satisfies $1 < r_2 < r$, then there is a natural number K such that

$$\frac{\|x_{n+1}\|}{\|x_n\|} > r_2 \qquad for \ n \geq K,$$

and in this case there is divergence. Q.E.D.

Raabe's Test

If $r = 1$, both the Ratio and the Root Tests fail and either convergence or divergence may take place. (See Example 35.13(d)). For some purposes it is useful to have a more delicate form of the Ratio Test for the case when $r = 1$. The next result, which is attributed to Raabe†, is usually adequate.

35.9 RAABE'S TEST. (a) If $X = (x_n)$ is a sequence of non-zero elements of $\mathbf{R}^p$ and there is a real number $a > 1$ and a natural number K such that

$$(35.10) \qquad \frac{\|x_{n+1}\|}{\|x_n\|} \leq 1 - \frac{a}{n} \qquad \text{for } n \geq K,$$

then the series $\sum (x_n)$ is absolutely convergent.

(b) If there is a real number $a \leq 1$ and a natural number K such that

$$(35.11) \qquad \frac{\|x_{n+1}\|}{\|x_n\|} \geq 1 - \frac{a}{n} \qquad \text{for } n \geq K,$$

then the series $\sum (x_n)$ is not absolutely convergent.

PROOF. (a) Assuming that relation (35.10) holds, we have

$$k \|x_{k+1}\| \leq (k-1) \|x_k\| - (a-1) \|x_k\| \qquad \text{for } k \geq K.$$

It follows that

$$(35.12) \qquad (k-1) \|x_k\| - k \|x_{k+1}\| \geq (a-1) \|x_k\| > 0 \qquad \text{for } k \geq K,$$

from which it follows that the sequence $(k \|x_{k+1}\|)$ is decreasing for $k \geq K$. On adding the relation (35.12) for $k = K, \ldots, n$ and noting that the left side telescopes, we find that

$$(K-1) \|x_K\| - n \|x_{n+1}\| \geq (a-1)(\|x_K\| + \cdots + \|x_n\|).$$

This shows that the partial sums of $\sum (\|x_n\|)$ are bounded and establishes the absolute convergence of $\sum (x_n)$.

(b) If the relation (35.11) holds for $n \geq K$ then, since $a \leq 1$,

$$n \|x_{n+1}\| \geq (n-a) \|x_n\| \geq (n-1) \|x_n\|.$$

Therefore, the sequence $(n \|x_{n+1}\|)$ is increasing for $n \geq K$, and there exists a number $c > 0$ such that

$$\|x_{n+1}\| > c/n, \qquad n \geq K.$$

Since the harmonic series $\sum (1/n)$ diverges, then $\sum (x_n)$ cannot be absolutely convergent.   Q.E.D.

We can also use Raabe's Test to obtain information on the rapidity of the convergence.

† JOSEPH L. RAABE (1801–1859) was born in the Ukraine and taught at Zürich. He worked in both geometry and analysis.

35.10 COROLLARY. *If $a > 1$ and if the sequence $X = (x_n)$ satisfies (35.10), then the partial sums approximate the sum s of $\sum (x_k)$ according to the estimate*

$$(35.13) \qquad \|s - s_n\| \le \frac{n}{a-1} \|x_{n+1}\| \qquad \text{for } n \ge K.$$

PROOF. Let $m > n \ge K$ and add the inequalities obtained from (35.12) for $k = n+1, \ldots, m$ to obtain

$$n \|x_{n+1}\| - m \|x_{m+1}\| \ge (a-1)(\|x_{n+1}\| + \cdots + \|x_m\|).$$

Hence we have

$$\|s_m - s_n\| \le \|x_{n+1}\| + \cdots + \|x_m\| \le \frac{n}{a-1} \|x_{n+1}\|;$$

taking the limit with respect to m, we obtain (35.13).   Q.E.D.

In the application of Raabe's Test, it may be convenient to use the following less sharp limiting form.

35.11 COROLLARY. *Let $X = (x_n)$ be a sequence of non-zero elements of $\mathbf{R}^p$ and set*

$$(35.14) \qquad a = \lim\left(n\left(1 - \frac{\|x_{n+1}\|}{\|x_n\|}\right)\right),$$

whenever this limit exists. Then $\sum (x_n)$ is absolutely convergent when $a > 1$ and is not absolutely convergent when $a < 1$.

PROOF. Suppose the limit (35.14) exists and satisfies $a > 1$. If a_1 is any number with $a > a_1 > 1$, then there exists a natural number K such that

$$a_1 < n\left(1 - \frac{\|x_{n+1}\|}{\|x_n\|}\right) \qquad \text{for } n \ge K.$$

Therefore, it follows that

$$\frac{\|x_{n+1}\|}{\|x_n\|} < 1 - \frac{a_1}{n} \qquad \text{for } n \ge K$$

and Theorem 35.9 assures the absolute convergence of the series. The case where $a < 1$ is handled similarly and will be omitted.   Q.E.D.

The Integral Test

We now present a powerful test, due to Maclaurin†, for a series of positive numbers.

† COLIN MACLAURIN (1698–1746) was a student of Newton's and professor at Edinburgh. He was the leading British mathematician of his time and contributed both to geometry and mathematical physics.

35.12 INTEGRAL TEST. *Let f be a positive, decreasing, continuous function on* $\{t : t \geq 1\}$. *Then the series* $\sum (f(n))$ *converges if and only if the infinite integral*

$$\int_1^{+\infty} f(t) \, dt = \lim_n \left(\int_1^n f(t) \, dt \right)$$

exists. In the case of convergence, the partial sum $s_n = \sum_{k=1}^n (f(k))$ *and the sum* $s = \sum_{k=1}^\infty (f(k))$ *satisfy the estimate*

(35.15) $$\int_{n+1}^{+\infty} f(t) \, dt \leq s - s_n \leq \int_n^{+\infty} f(t) \, dt.$$

PROOF. Since f is positive, continuous, and decreasing on the interval $[k-1, k]$, it follows that

(35.16) $$f(k) \leq \int_{k-1}^k f(t) \, dt \leq f(k-1).$$

By summing this inequality for $k = 2, 3, \ldots, n$, we obtain the relation

$$s_n - f(1) \leq \int_1^n f(t) \, dt \leq s_{n-1},$$

which shows that both or neither of the limits

$$\lim (s_n), \qquad \lim \left(\int_1^n f(t) \, dt \right)$$

exist. If they exist, we obtain on summing relation (35.16) for $k = n+1, \ldots, m$, that

$$s_m - s_n \leq \int_n^m f(t) \, dt \leq s_{m-1} - s_{n-1},$$

whence it follows that

$$\int_{n+1}^{m+1} f(t) \, dt \leq s_m - s_n \leq \int_n^m f(t) \, dt.$$

If we take the limit with respect to m in this last inequality, we obtain (35.15). Q.E.D.

We shall show how the results in Theorems 35.1 to 35.12 can be applied to the p-series, which were introduced in Example 34.8(c).

35.13 EXAMPLES. (a) First we shall apply the Comparison Test. Knowing that the harmonic series $\sum (1/n)$ diverges, it is seen that if $p \leq 1$, then $n^p \le n$ and hence

$$\frac{1}{n} \le \frac{1}{n^p}.$$

After using the Comparison Test 35.1, we conclude that the p-series $\sum (1/n^p)$ diverges for $p \le 1$.

(b) Now consider the case $p = 2$; that is, the series $\sum (1/n^2)$. We compare the series with the convergent series $\sum [1/n(n+1)]$ of Example 34.8(e). Since the relation

$$\frac{1}{n(n+1)} < \frac{1}{n^2}$$

holds and the terms on the left form a convergent series, we cannot apply the Comparison Theorem directly. However, we could apply this theorem if we compared the nth term of $\sum [1/n(n+1)]$ with the $(n+1)$st term of $\sum (1/n^2)$. Instead, we choose to apply the Limit Comparison Test 35.2 and note that

$$\frac{1}{n(n+1)} \div \frac{1}{n^2} = \frac{n^2}{n(n+1)} = \frac{n}{n+1}.$$

Since the limit of this quotient is 1 and $\sum [1/n(n+1)]$ converges, then so does the series $\sum (1/n^2)$.

(c) Now consider the case $p \ge 2$. If we note that $n^p \ge n^2$ for $p \ge 2$, then

$$\frac{1}{n^p} \le \frac{1}{n^2},$$

a direct application of the Comparison Test assures that $\sum (1/n^p)$ converges for $p \ge 2$. Alternatively, we could apply the Limit Comparison Test and note that

$$\frac{1}{n^p} \div \frac{1}{n^2} = \frac{n^2}{n^p} = \frac{1}{n^{p-2}}.$$

If $p > 2$, this expression converges to 0, whence it follows from Corollary 35.2(b) that the series $\sum (1/n^p)$ converges for $p \ge 2$.

By using the Comparison Test, we cannot gain any information concerning the p-series for $1 < p < 2$ unless we can find a series whose convergence character is known and which can be compared to the series in this range.

(d) We demonstrate the Root and the Ratio Tests as applied to the p-series. Note that

$$\left(\frac{1}{n^p}\right)^{1/n} = (n^{-p})^{1/n} = (n^{1/n})^{-p}.$$

Now it is known (see Example 14.8(e)) that the sequence $(n^{1/n})$ converges to 1. Hence we have

$$\lim\left(\left(\frac{1}{n^p}\right)^{1/n}\right) = 1,$$

so that the Root Test (in the form of Corollary 35.5) does not apply.

In the same way, since

$$\frac{1}{(n+1)^p} \div \frac{1}{n^p} = \frac{n^p}{(n+1)^p} = \frac{1}{(1+1/n)^p},$$

and since the sequence $((1+1/n)^p)$ converges to 1, the Ratio Test (in the form of Corollary 35.8) does not apply.

(e) In desperation, we apply Raabe's Test to the p-series for integral values of p. First, we attempt to use Corollary 35.11. Observe that

$$n\left(1 - \frac{(n+1)^{-p}}{n^{-p}}\right) = n\left(1 - \frac{n^p}{(n+1)^p}\right)$$
$$= n\left(1 - \frac{(n+1-1)^p}{(n+1)^p}\right) = n\left(1 - \left(1 - \frac{1}{n+1}\right)^p\right).$$

If p is an integer, then we can use the Binomial Theorem to obtain an estimate for the last term. In fact,

$$n\left(1 - \left(1 - \frac{1}{n+1}\right)^p\right) = n\left(1 - 1 + \frac{p}{n+1} - \frac{p(p-1)}{2(n+1)^2} + \cdots\right).$$

If we take the limit with respect to n, we obtain p. Hence this corollary to Raabe's Test shows that the series converges for integral values of $p \geq 2$ (and, if the Binomial Theorem is known for non-integral values of p, this could be improved).

(f) Finally, we apply the Integral Test to the p-series. Let $f(t) = t^{-p}$ and recall that

$$\int_1^n \frac{1}{t} \, dt = \log(n) - \log(1),$$

$$\int_1^n \frac{1}{t^p} \, dt = \frac{1}{1-p}(n^{1-p} - 1) \qquad \text{for } p \neq 1.$$

From these relations we see that the p-series converges if $p > 1$ and diverges if $p \leq 1$.

Exercises

35.A. Establish the convergence or the divergence of the series whose nth term is given by

(a) $\dfrac{1}{(n+1)(n+2)}$,

(b) $\dfrac{n}{(n+1)(n+2)}$,

(c) $2^{-1/n}$, (d) $n/2^n$,

(e) $[n(n+1)]^{-1/2}$, (f) $[n^2(n+1)]^{-1/2}$,

(g) $n!/n^n$, (h) $(-1)^n n/(n+1)$.

35.B. For each of the series in Exercise 35.A which converge, estimate the remainder if only four terms are taken. If we wish to determine the sum within $1/1000$, how many terms should we take?

35.C. Discuss the convergence or the divergence of the series with nth term (for sufficiently large n) given by

(a) $[\log n]^{-p}$, (b) $[\log n]^{-n}$,

(c) $[\log n]^{-\log n}$, (d) $[\log n]^{-\log \log n}$,

(e) $[n \log n]^{-1}$, (f) $[n(\log n)(\log \log n)^2]^{-1}$.

35.D. Discuss the convergence or the divergence of the series with nth term

(a) $2^n e^{-n}$, (b) $n^n e^{-n}$,

(c) $e^{-\log n}$, (d) $(\log n) e^{-\sqrt{n}}$,

(e) $n! e^{-n}$, (f) $n! e^{-n^2}$.

35.E. Show that the series

$$\frac{1}{1^2}+\frac{1}{2^3}+\frac{1}{3^2}+\frac{1}{4^3}+\cdots$$

is convergent, but that both the Ratio and the Root Tests fail to apply.

35.F. If a and b are positive numbers, then

$$\sum \frac{1}{(an+b)^p}$$

converges if $p>1$ and diverges if $p \le 1$.

35.G. Discuss the series whose nth term is

(a) $\dfrac{n!}{3 \cdot 5 \cdot 7 \cdots (2n+1)}$, (b) $\dfrac{(n!)^2}{(2n)!}$,

(c) $\dfrac{2 \cdot 4 \cdots (2n)}{3 \cdot 5 \cdots (2n+1)}$, (d) $\dfrac{2 \cdot 4 \cdots (2n)}{5 \cdot 7 \cdots (2n+3)}$.

35.H. The series given by

$$\left(\frac{1}{2}\right)^p + \left(\frac{1 \cdot 3}{2 \cdot 4}\right)^p + \left(\frac{1 \cdot 3 \cdot 5}{2 \cdot 4 \cdot 6}\right)^p + \cdots$$

converges for $p>2$ and diverges for $p \le 2$.

35.I. Let $X=(x_n)$ be a sequence in $\mathbf{R}^p$ and let r be given by

$$r = \lim \sup (\|x_n\|^{1/n}).$$

Then $\sum (x_n)$ is absolutely convergent if $r<1$ and divergent if $r>1$. [The limit superior $u = \lim \sup (b_n)$ of a bounded sequence of real numbers was defined in Section 18. It is the unique number u with the properties that (i) if $u<v$ then $b_n \le v$ for all sufficiently large $n \in \mathbf{N}$, and (ii) if $w<u$, then $w \le b_n$ for infinitely many $n \in \mathbf{N}$.]

35.J. Let $X = (x_n)$ be a sequence of non-zero elements of $\mathbf{R}^p$ and let r be given by $r = \lim \sup (\|x_{n+1}\|/\|x_n\|)$.

(a) Show that if $r < 1$, then the series $\sum (x_n)$ is absolutely convergent.

(b) Give an example of an absolutely convergent series with $r > 1$.

(c) If $\lim \inf (\|x_{n+1}\|/\|x_n\|) > 1$, show that the series $\sum (x_n)$ is not absolutely convergent.

35.K. Let $X = (x_n)$ be a sequence of non-zero elements of $\mathbf{R}^p$ and let a be given by $a = \lim \sup (n(1 - \|x_{n+1}\|/\|x_n\|))$.

(a) If $a < 1$, show that the series $\sum (x_n)$ is not absolutely convergent.

(b) Give an example of a divergent series with $a > 1$.

(c) If $\lim \inf (n(1 - \|x_{n+1}\|/\|x_n\|)) > 1$ show that the series $\sum (x_n)$ is absolutely convergent.

35.L. Let $X = (x_n)$ be such that $x_n > 0$ for $n \in N$. Show that the series $\sum (x_n)$ is divergent if

$$\lim \sup \left((\log n) \left[n \left(1 - \frac{x_{n+1}}{x_n} \right) - 1 \right] \right) < 1.$$

35.M. Let $x_n > 0$ for $n \in N$ and suppose that $n(1 - x_{n+1}/x_n) = a + k_n/n^p$, where $p > 0$ and (k_n) is bounded. Then the series $\sum (x_n)$ converges if $a > 1$ and diverges if $a \le 1$.

35.N. If $p > 0$, $q > 0$, then the series

$$\sum \frac{(p+1)(p+2) \cdots (p+n)}{(q+1)(q+2) \cdots (q+n)}$$

converges for $q > p + 1$ and diverges for $q \le p + 1$.

35.O. Show that the series $\sum (2^n n!)^2/(2n+1)!$ is divergent.

35.P. Let $x_n > 0$ and let $r = \lim \inf (-\log x_n/\log n)$. Show that $\sum (x_n)$ converges if $r > 1$ and diverges if $r < 1$.

25.Q. Suppose that none of the numbers a, b, c is a negative integer or zero. Prove that the **hypergeometric series**

$$\frac{ab}{1!c} + \frac{a(a+1)b(b+1)}{2!c(c+1)} + \frac{a(a+1)(a+2)b(b+1)(b+2)}{3!c(c+1)(c+2)} + \cdots$$

is absolutely convergent for $c > a + b$ and divergent for $c \le a + b$.

35.R. Let $a_n > 0$ and suppose that $\sum (a_n)$ converges. Construct a convergent series $\sum (b_n)$ with $b_n > 0$ such that $\lim (a_n/b_n) = 0$; hence $\sum (b_n)$ converges less rapidly than $\sum (a_n)$. (Hint: let (A_n) be the partial sums of $\sum (a_n)$ and A its limit. Define $r_0 = A$, $r_n = A - A_n$ and $b_n = \sqrt{r_{n-1}} - \sqrt{r_n}$.)

35.S. Let $a_n > 0$ and suppose that $\sum (a_n)$ diverges. Construct a divergent series $\sum (b_n)$ with $b_n > 0$ such that $\lim (b_n/a_n) = 0$; hence $\sum (b_n)$ diverges less rapidly than $\sum (a_n)$. (Hint: Let $b_1 = \sqrt{a_1}$ and $b_n = \sqrt{a_{n-1}} - \sqrt{a_n}$, $n > 1$.)

35.T. Let $\{n_1, n_2, \ldots\}$ denote the collection of natural numbers that do not use the digit 6 in their decimal expansion. Show that the series $\sum (1/n_k)$ converges to a number less than 90. If $\{m_1, m_2, \ldots\}$ is the collection that ends in 6, then $\sum (1/m_k)$ diverges.

Project

35.α. Although infinite products do not occur as frequently as infinite series, they are of importance in many investigations and applications. For simplicity, we shall restrict attention here to infinite products with terms $a_n > 0$. If $A = (a_n)$ is a sequence of strictly positive real numbers, then the **infinite product**, or the **sequence of partial products**, generated by A is the sequence $P = (p_n)$ defined by

$$p_1 = a_1, \ p_2 = p_1 a_2 (= a_1 a_2), \ldots,$$
$$p_n = p_{n-1} a_n (= a_1 a_2 \cdots a_{n-1} a_n), \ldots.$$

If the sequence P is convergent to *a non-zero number*, then we call $\lim P$ the **product** of the infinite product generated by A. In this case we say that the infinite product is **convergent** and write either

$$\prod_{n=1}^{\infty} a_n, \quad \prod (a_n), \quad \text{or} \quad a_1 a_2 a_3 \cdots a_n \cdots$$

to denote both P and $\lim P$.

(*Note:* the requirement that $\lim P \neq 0$ is not essential but is conventional, since it insures that certain properties of finite products carry over to infinite products.)

(a) Show that a necessary condition for the convergence of the infinite product is that $\lim (a_n) = 1$.

(b) Prove that a necessary and sufficient condition for the convergence of

$$\prod_{n=1}^{\infty} a_n, \quad a_n > 0, \quad \text{is the convergence of} \quad \sum_{n=1}^{\infty} \log a_n.$$

(c) Infinite products often have terms of the form $a_n = 1 + u_n$. In keeping with our standing restriction, we suppose $u_n > -1$ for all $n \in \mathbf{N}$. If $u_n \geq 0$, show that a necessary and sufficient condition for the convergence of the infinite product is the convergence of the infinite series $\sum (u_n)$. (Hint: use the Limit Comparison Test 35.2.)

(d) Let $u_n > -1$. Show that if the infinite series $\sum (u_n)$ is absolutely convergent, then the infinite product $\prod (1 + u_n)$ is convergent.

(e) Suppose that $u_n > -1$ and that the series $\sum (u_n)$ is convergent. Then a necessary and sufficient condition for the convergence of the infinite product $\prod (1 + u_n)$ is the convergence of the infinite series $\sum (u_n^2)$. (Hint: use Taylor's Theorem and show that there exist positive constants A and B such that if $|u| < \frac{1}{2}$, then $Au^2 \leq u - \log (1 + u) \leq Bu^2$.)

Section 36 Further Results for Series

The tests given in Section 35 all have the character that they guarantee that, if certain hypotheses are fulfilled, then the series $\sum (x_n)$ is absolutely convergent. Now it is known that absolute convergence implies ordinary convergence, but it is readily seen from an examination of special series, such as

$$\sum \frac{(-1)^n}{n}, \qquad \sum \frac{(-1)^n}{\sqrt{n}},$$

that convergence may take place even though absolute convergence fails. It is desired, therefore, to have a test which yields information about ordinary convergence. There are many such tests which apply to special types of series. Perhaps the ones with most general applicability are those due to Abel[†] and Dirichlet.

To establish these tests, we need a lemma which is sometimes called the **partial summation formula,** since it corresponds to the familiar integration by parts formula. In most applications, the sequences X and Y are both sequences in $\mathbf{R}$, but the results hold when X and Y are sequences in $\mathbf{R}^p$ and the inner product is used or when one of X and Y is a real sequence and the other is in $\mathbf{R}^p$.

36.1 ABEL'S LEMMA. *Let $X = (x_n)$ in $\mathbf{R}$ and $Y = (y_n)$ in $\mathbf{R}^p$ be sequences and let the partial sums of $\sum (y_n)$ be denoted by (s_k). If $m \geq n$, then*

$$(36.1) \qquad \sum_{j=n}^{m} x_j y_j = (x_{m+1} s_m - x_n s_{n-1}) + \sum_{j=n}^{m} (x_j - x_{j+1}) s_j.$$

PROOF. A proof of this result may be given by noting that $y_j = s_j - s_{j-1}$ and by matching the terms on each side of the equality. We shall leave the details to the reader.   Q.E.D.

We apply Abel's Lemma to conclude that the series $\sum (x_n y_n)$ is convergent in a case where both of the series $\sum (x_n)$ and $\sum (y_n)$ may be divergent.

36.2 DIRICHLET'S TEST. *Suppose the partial sums of $\sum (y_n)$ are bounded.* (a) *If the sequence $X = (x_n)$ converges to zero, and if*

$$(36.2) \qquad \qquad \sum |x_n - x_{n+1}|$$

is convergent, then the series $\sum (x_n y_n)$ is convergent.

(b) *In particular, if $X = (x_n)$ is a decreasing sequence of positive real numbers which converges to zero, then the series $\sum (x_n y_n)$ is convergent.*

---

† NIELS HENRIK ABEL (1802–1829) was the son of a poor Norwegian minister. When only twenty-two he proved the impossibility of solving the general quintic equation by radicals. This self-taught genius also did outstanding work on series and elliptic functions before his early death from tuberculosis.

PROOF. (a) Suppose that $\|s_j\| < B$ for all j. Using (36.1), we have the estimate

$$(36.3) \qquad \left\| \sum_{j=n}^{m} x_j y_j \right\| \le \{|x_{m+1}| + |x_n| + \sum_{j=n}^{m} |x_j - x_{j+1}|\} B.$$

If $\lim (x_n) = 0$, the first two terms on the right side can be made arbitrarily small by taking m and n sufficiently large. Also if the series (36.2) converges, then the Cauchy Criterion assures that the final term on this side can be made less than ε by taking $m \ge n \ge M(\varepsilon)$. Hence the Cauchy Criterion implies that the series $\sum (x_n y_n)$ is convergent.

(b) If $x_1 \ge x_2 \ge \ldots$, then the series in (36.2) is telescoping and convergent.
$\hfill$ Q.E.D.

36.3 COROLLARY. *In part (b), we have the error estimate*

$$\left\| \sum_{j=1}^{\infty} x_j y_j - \sum_{j=1}^{n} x_j y_j \right\| \le 2 x_{n+1} B,$$

where B is an upper bound for the partial sums $\sum (y_j)$.

PROOF. This is readily obtained from relation (36.3). $\hfill$ Q.E.D.

The next test strengthens the hypothesis on the series $\sum (y_n)$, but relaxes the hypothesis on $\sum (x_n)$.

36.4 ABEL'S TEST. *Suppose that the series $\sum (y_n)$ converges in $\mathbf{R}^p$.*
(a) *If the sequence $X = (x_n)$ in $\mathbf{R}$ is such that*

$$(36.2) \qquad \sum |x_n - x_{n+1}|$$

is convergent, then the series $\sum (x_n y_n)$ is convergent.
(b) *In particular, if the sequence $X = (x_n)$ is monotone and convergent to x in $\mathbf{R}$, then the series $\sum (x_n y_n)$ is convergent.*

PROOF. (a) By hypothesis, the partial sums s_k of $\sum (y_n)$ converge to some element s in $\mathbf{R}^p$. Hence there is a bound B for $\{\|s_k\| : k \in \mathbf{N}\}$ and, given $\varepsilon > 0$ there is $N_1(\varepsilon)$ such that if $n \ge N_1(\varepsilon)$, then $\|s_n - s\| < \varepsilon$.

Now the hypothesis that (36.2) is convergent implies that if $n \in \mathbf{N}$, then

$$|x_n| \le |x_1 + (x_2 - x_1) + \cdots + (x_n - x_{n-1})|$$
$$\le |x_1| + \sum_{k=1}^{n-1} |x_k - x_{k+1}|$$

so that $|x_n| < A$ for some $A > 0$. Moreover, there exists $N_2(\varepsilon)$ such that if $m > n \ge N_2(\varepsilon)$, then

$$(36.4) \qquad |x_{m+1} - x_n| \le \sum_{j=n}^{m} |x_{j+1} - x_j| < \varepsilon.$$

Now let $N_3(\varepsilon) = \sup \{N_1(\varepsilon), N_2(\varepsilon)\}$ so that if $m > n > N_3(\varepsilon)$, then we have

$$\|x_{m+1}s_m - x_n s_{n-1}\|$$
$$\leq \|x_{m+1}s_m - x_{m+1}s\| + \|x_{m+1}s - x_n s\| + \|x_n s - x_n s_{n-1}\|$$
$$\leq |x_{m+1}| \|s_m - s\| + |x_{m+1} - x_n| \|s\| + |x_n| \|s - s_{n-1}\|$$
$$\leq A\varepsilon + \varepsilon B + A\varepsilon = (2A + B)\varepsilon.$$

Therefore, by Abel's Lemma 36.1, if $m > n > N_3(\varepsilon)$, then we have

$$\left\|\sum_{j=n}^{m} x_j y_j\right\| \leq (2A + B)\varepsilon + \left\|\sum_{j=n}^{m} (x_j - x_{j+1})s_j\right\|$$
$$\leq (2A + B)\varepsilon + \left(\sum_{j=n}^{m} |x_j - x_{j+1}|\right)B$$
$$\leq 2(A + B)\varepsilon,$$

where we have used (36.4) in the last step. Since $\varepsilon > 0$ is arbitrary, the convergence of $\sum (x_j y_j)$ is established.

(b) If the sequence (x_n) is monotone and converges to x, then the series (36.2) is telescoping and converges either to $x - x_1$ or to $x_1 - x$.   Q.E.D.

If we use the same type of argument we can establish the following error estimate.

36.5 COROLLARY. *In part* (b), *we have the error estimate*

$$\left\|\sum_{j=1}^{\infty} x_j y_j - \sum_{j=1}^{n} x_j y_j\right\| \leq |x_{n+1}| \|s - s_n\| + 2B |x - x_{n+1}|.$$

Alternating Series

There is a particularly important class of conditionally convergent real series, namely those whose terms are alternately positive and negative.

36.6 DEFINITION. A sequence $X = (x_n)$ of non-zero real numbers is **alternating** if the terms $(-1)^n x_n$, $n = 1, 2, \ldots$, are all positive (or all negative) real numbers. If a sequence $X = (x_n)$ is alternating, we say that the series $\sum (x_n)$ it generates is an **alternating series.**

It is useful to set $x_n = (-1)^n z_n$ and require that $z_n > 0$ (or $z_n < 0$) for all $n = 1, 2, \ldots$. The convergence of alternating series is easily treated when the next result, proved by Leibniz, can be applied.

36.7 ALTERNATING SERIES TEST. *Let* $Z = (z_n)$ *be a decreasing sequence of strictly positive numbers with* $\lim (z_n) = 0$. *Then the alternating series* $\sum ((-1)^n z_n)$ *is convergent. Moreover, if s is the sum of this series and s_n*

is the nth partial sum, then we have the estimate

$$(36.5) \qquad |s - s_n| \le z_{n+1}$$

for the rapidity of convergence.

PROOF. This follows immediately from Dirichlet's Test 36.2(b) if we take $y_n = (-1)^n$, but the error estimate given in Corollary 36.3 is not as sharp as (36.5). We can also proceed directly and show by mathematical induction that if $m \ge n$, then

$$|s_m - s_n| = |z_{n+1} - z_{n+2} + \cdots + (-1)^{m-n-1} z_m| \le |z_{n+1}|.$$

This yields both the convergence and the estimate (36.5). Q.E.D.

36.8 EXAMPLES. (a) The series $\sum ((-1)^n/n)$, which is sometimes called the **alternating harmonic series,** is not absolutely convergent. However, it follows from the Alternating Series Test that it is convergent.

(b) Similarly, the series $\sum ((-1)^2/\sqrt{n})$ is convergent, but not absolutely convergent.

(c) Let $x \in R$ and let $k \in Z$. Then, since

$$2 \cos kx \sin \tfrac{1}{2}x = \sin (k + \tfrac{1}{2})x - \sin (k - \tfrac{1}{2})x,$$

it follows that

$$2 \sin \tfrac{1}{2}x [\cos x + \cdots + \cos nx] = \sin (n + \tfrac{1}{2})x - \sin \tfrac{1}{2}x.$$

Hence, if x is not an integer multiple of 2π, then

$$(36.6) \qquad \cos x + \cdots + \cos nx = \frac{\sin (n + \tfrac{1}{2})x - \sin \tfrac{1}{2}x}{2 \sin \tfrac{1}{2}x}$$

Therefore, if $x \notin \{2k\pi : k \in Z\}$, then

$$|\cos x + \cdots + \cos nx| \le \frac{1}{|\sin \tfrac{1}{2}x|}.$$

We can then apply Dirichlet's Test 36.2(b) to conclude that the series $\sum (1/n) \cos nx$ converges for all $x \notin \{2k\pi : k \in Z\}$. We note that this series diverges when $x = 2k\pi$ for some $k \in Z$.

(d) Let $x \in R$ and let $k \in Z$. Then since

$$2 \sin kx \sin \tfrac{1}{2}x = \cos (k - \tfrac{1}{2})x - \cos (k + \tfrac{1}{2})x,$$

it follows that

$$2 \sin \tfrac{1}{2}x [\sin x + \cdots + \sin nx] = \cos \tfrac{1}{2}x - \cos (n + \tfrac{1}{2})x.$$

Hence, if x is not an integer multiple of 2π, then

$$\sin x + \cdots + \sin nx = \frac{\cos \frac{1}{2}x - \cos (n + \frac{1}{2})x}{2 \sin \frac{1}{2}x}.$$

Therefore, if $x \notin \{2k\pi : k \in \mathbf{Z}\}$, then

$$|\sin x + \cdots + \sin nx| \le \frac{1}{|\sin \frac{1}{2}x|}.$$

As before, Dirichlet's Test implies the convergence of the series $\sum (1/n) \sin nx$ for all $x \notin \{2k\pi : k \in \mathbf{Z}\}$. We note that this series also converges when $x = 2k\pi$ for $k \in \mathbf{Z}$.

(e) Let $Y = (y_n)$ be the sequence in $\mathbf{R}^2$ whose elements are

$$y_1 = (1, 0), \qquad y_2 = (0, 1), \qquad y_3 = (-1, 0),$$

$$y_4 = (0, -1), \ldots, y_{n+4} = y_n, \ldots.$$

It is readily seen that the series $\sum (y_n)$ does not converge, but its partial sums s_n are bounded; in fact, we have $\|s_n\| \le \sqrt{2}$. Hence Dirichlet's Test shows that the series $\sum (1/n)y_n$ is convergent in $\mathbf{R}^2$.

Double Series

Sometimes it is necessary to consider infinite sums depending on two integral indices. The theory of such double series is developed by reducing them to double sequences; thus all of the results in Section 19 dealing with double sequences can be interpreted for double series. However, we shall not draw from the results of Section 19; instead, we shall restrict our attention to absolutely convergent double series, since those are the type of double series that arise most often,

Suppose that to every pair (i, j) in $\mathbf{N} \times \mathbf{N}$ one has an element x_{ij} in $\mathbf{R}^p$. One defines the (m, n)**th partial sum** s_{mn} to be

$$s_{mn} = \sum_{j=1}^{n} \sum_{i=1}^{m} x_{ij}.$$

By analogy with Definition 34.1, we shall say that the **double series** $\sum (x_{ij})$ **converges** to an element x in $\mathbf{R}^p$ if for every $\varepsilon > 0$ there exists a natural number $M(\varepsilon)$ such that if $m \ge M(\varepsilon)$ and $n \ge M(\varepsilon)$ then

$$\|x - s_{mn}\| < \varepsilon.$$

By analogy with Definition 34.6, we shall say that the double series $\sum (x_{ij})$ is **absolutely convergent** if the double series $\sum (\|x_{ij}\|)$ in $\mathbf{R}$ is convergent.

It is an exercise to show that if a double series is absolutely convergent, then it is convergent. Moreover, a double series is absolutely convergent if and only if the set

$$(36.7) \qquad \left\{ \sum_{j=1}^{n} \sum_{i=1}^{m} \|x_{ij}\| : m, n \in \mathbf{N} \right\}$$

is a bounded set of real numbers.

We wish to relate double series with iterated series, but we shall discuss only absolutely convergent series. The next result is very elementary, but it gives a useful criterion for the absolute convergence of the double series.

36.9 LEMMA. *Suppose that the iterated series* $\sum_{j=1}^{\infty} \sum_{i=1}^{\infty} \|x_{ij}\|$ *converges. Then the double series* $\sum (x_{ij})$ *is absolutely convergent.*

PROOF. By hypothesis each series $\sum_{i=1}^{\infty} \|x_{ij}\|$ converges to a positive number a_j, $j \in \mathbf{N}$. Moreover, the series $\sum (a_j)$ converges to a number A. It is clear that A is an upper bound for the set (36.7).   Q.E.D.

36.10 THEOREM. *Suppose that the double series* $\sum (x_{ij})$ *converges absolutely to* x *in* $\mathbf{R}^p$. *Then both of the iterated series*

$$(36.8) \qquad \sum_{j=1}^{\infty} \sum_{i=1}^{\infty} x_{ij}, \qquad \sum_{i=1}^{\infty} \sum_{j=1}^{\infty} x_{ij}$$

also converge to x.

PROOF. By hypothesis there exists a positive real number A which is an upper bound for the set in (36.7). If n is fixed, we observe that

$$\sum_{i=1}^{m} \|x_{in}\| \le \sum_{j=1}^{n} \sum_{i=1}^{m} \|x_{ij}\| \le A,$$

for each m in $\mathbf{N}$. It thus follows that, for each $n \in \mathbf{N}$, the single series $\sum_{i=1}^{\infty} (x_{in})$ is absolutely convergent to an element y_n in $\mathbf{R}^p$.

If $\varepsilon > 0$, let $M(\varepsilon)$ be such that if $m, n \ge M(\varepsilon)$, then

$$(36.9) \qquad \|s_{mn} - x\| < \varepsilon,$$

In view of the relation

$$s_{mn} = \sum_{i=1}^{m} x_{i1} + \sum_{i=1}^{m} x_{i2} + \cdots + \sum_{i=1}^{m} x_{in},$$

we infer that

$$\lim_{m} (s_{mn}) = \sum_{i=1}^{\infty} x_{i1} + \sum_{i=1}^{\infty} x_{i2} + \cdots + \sum_{i=1}^{\infty} x_{in}$$

$$= y_1 + y_2 + \cdots + y_n.$$

If we pass to the limit in (36.9) with respect to m, we obtain the relation

$$\left\| \sum_{j=1}^{n} y_j - x \right\| \leq \varepsilon, \qquad n \geq M(\varepsilon).$$

This proves that the first iterated sum in (36.8) exists and equals x. An analogous proof applies to the second iterated sum.   Q.E.D.

There is one additional method of summing double series that we shall consider, namely along the diagonals $i + j = n$.

36.11 Theorem. *Suppose that the double series $\sum (x_{ij})$ converges absolutely to x in $\mathbf{R}^p$. If we define*

$$t_k = \sum_{i+j=k} x_{ij} = x_{1,k-1} + x_{2,k-2} + \cdots + x_{k-1,1},$$

then the series $\sum (t_k)$ converges absolutely to x.

PROOF.   Let A be the supremum of the set in (36.7). We observe that

$$\sum_{k=2}^{n} \|t_k\| \leq \sum_{j=1}^{n} \sum_{i=1}^{n} \|x_{ij}\| \leq A.$$

Hence the series $\sum (t_k)$ is absolutely convergent; it remains to show that it converges to x. Let $\varepsilon > 0$ and let M be such that

$$A - \varepsilon < \sum_{j=1}^{M} \sum_{i=1}^{M} \|x_{ij}\| \leq A.$$

If $m, n \geq M$, then it follows that $\|s_{mn} - s_{MM}\|$ is no greater than the sum $\sum (\|x_{ij}\|)$ extended over all pairs (i, j) satisfying either $M < i \leq m$ or $M < j \leq n$. Hence $\|s_{mn} - s_{MM}\| < \varepsilon$, when $m, n \geq M$. It follows from this that $\|x - s_{MM}\| \leq \varepsilon$. A similar argument shows that if $n \geq 2M$, then

$$\left\| \sum_{k=2}^{n} t_k - s_{MM} \right\| < \varepsilon,$$

whence it follows that $x = \sum t_k$.   Q.E.D.

Cauchy Multiplication

In the process of multiplying two power series and collecting the terms according to the powers, there arises very naturally a new method of generating a series from two given ones. In this connection it is notationally useful to have the terms of the series indexed by $0, 1, 2, \ldots$.

36.12 Definition.   If $\sum_{i=0}^{\infty} (y_i)$ and $\sum_{j=0}^{\infty} (z_j)$ are infinite series in $\mathbf{R}^p$, their **Cauchy product** is the series $\sum_{k=0}^{\infty} (x_k)$, where

$$x_k = y_0 \cdot z_k + y_1 \cdot z_{k-1} + \cdots + y_k \cdot z_0.$$

Here the dot denotes the inner product in R^p. In like manner we can define the Cauchy product of a series in R and a series in R^p.

It is perhaps a bit surprising that the Cauchy product of two convergent series may fail to converge. However, it is seen that the series

$$\sum_{n=0}^{\infty} \frac{(-1)^n}{\sqrt{n+1}}$$

is convergent, but the nth term of the Cauchy product of this series with itself is

$$(-1)^n \left[\frac{1}{\sqrt{1}\sqrt{n+1}} + \frac{1}{\sqrt{2}\sqrt{n}} + \cdots + \frac{1}{\sqrt{n+1}\sqrt{1}} \right].$$

Since there are $n+1$ terms in the bracket and each term exceeds $1/(n+2)$, the terms in the Cauchy product do not converge to zero. Hence this Cauchy product cannot converge.

36.13 THEOREM. *If the series $\sum_{i=0}^{\infty} y_i$ and $\sum_{j=0}^{\infty} z_j$ converge absolutely to y, z in R^p, then their Cauchy product converges absolutely to y · z.*

PROOF. If $i, j = 0, 1, 2, \ldots$, let $x_{ij} = y_i \cdot z_j$. The hypotheses imply that the iterated series $\sum_{j=0}^{\infty} \sum_{i=0}^{\infty} \|x_{ij}\|$ converges. By Lemma 36.9, the double series $\sum (x_{ij})$ is absolutely convergent to a real number x. By applying Theorems 36.10 and 36.11, we infer that both of the series

$$\sum_{j=0}^{\infty} \sum_{i=0}^{\infty} x_{ij}, \qquad \sum_{k=0}^{\infty} \sum_{i+j=k} x_{ij}$$

converge to x. It is readily checked that the iterated series converges to y · z and that the diagonal series is the Cauchy product of $\sum (y_i)$ and $\sum (z_i)$.
Q.E.D.

In the case $p = 1$, it was proved by Mertens† that the absolute convergence of one of the series is sufficient to imply the convergence of the Cauchy product. In addition, Cesàro showed that the arithmetic means of the partial sums of the Cauchy product converge to yz. (See Exercises 37.O, P).

Exercises

36.A. Consider the series

$$1 - \frac{1}{2} - \frac{1}{3} + \frac{1}{4} + \frac{1}{5} - \frac{1}{6} - \frac{1}{7} + + - - \ldots,$$

where the signs come in pairs. Does it converge?

† FRANZ (C. J) MERTENS (1840–1927) studied at Berlin and taught at Cracow and Vienna. He contributed primarily to geometry, number theory, and algebra.

36.B. Let $a_n \in R$ for $n \in N$ and let $p < q$. If the series $\sum (a_n/n^p)$ is convergent, then the series $\sum (a_n/n^q)$ is also convergent.

36.C. If p and q are strictly positive numbers, then

$$\sum (-1)^n \frac{(\log n)^p}{n^q}$$

is a convergent series.

36.D. Discuss the series whose nth term is

(a) $(-1)^n \dfrac{n^n}{(n+1)^{n+1}}$,

(b) $\dfrac{n^n}{(n+1)^{n+1}}$,

(c) $(-1)^n \dfrac{(n+1)^n}{n^n}$,

(d) $\dfrac{(n+1)^n}{n^{n+1}}$.

36.E. Suppose that $\sum (a_n)$ is a convergent series of real numbers. Either prove that $\sum (b_n)$ converges or give a counter-example, when we define b_n by

(a) a_n/n,

(b) $\sqrt{a_n}/n$ $(a_n \ge 0)$,

(c) $a_n \sin n$,

(d) $\sqrt{a_n/n}$ $(a_n \ge 0)$,

(e) $n^{1/n} a_n$,

(f) $a_n/(1+|a_n|)$.

36.F. Show that the series

$$1 + \frac{1}{2} - \frac{1}{3} + \frac{1}{4} + \frac{1}{5} - \frac{1}{6} + + - \cdots$$

is divergent.

36.G. If the hypothesis that (z_n) is decreasing is dropped, show that the Alternating Series Test 36.7 may fail.

36.H. For $n \in N$, let c_n be defined by

$$c_n = \frac{1}{1} + \frac{1}{2} + \cdots + \frac{1}{n} - \log n.$$

Show that (c_n) is a decreasing sequence of positive numbers. The limit C of this sequence is called **Euler's Constant** and is approximately equal to 0.577. Show that if we put

$$b_n = \frac{1}{1} - \frac{1}{2} + \frac{1}{3} - \cdots - \frac{1}{2n},$$

then the sequence (b_n) converges to $\log 2$. (Hint: $b_n = c_{2n} - c_n + \log 2$.)

36.I. Let $\sum (a_{mn})$ be the double series given by

$$a_{mn} = +1, \quad \text{if } m - n = 1,$$
$$= -1, \quad \text{if } m - n = -1,$$
$$= 0, \quad \text{otherwise.}$$

Show that both iterated sums exist, but are unequal, and the double sum does not exist. However, if (s_{mn}) denote the partial sums, then $\lim (s_{nn})$ exists.

36.J. Show that if the double and the iterated series of $\sum (a_{mn})$ exist, then they are all equal. Show that the existence of the double series does not imply the existence of the iterated series; in fact the existence of the double series does not even imply that $\lim_n (a_{mn}) = 0$ for each m.

36.K. Show that if $p > 1$ and $q > 1$, then the double series

$$\sum \left(\frac{1}{m^p n^q}\right) \quad \text{and} \quad \sum \left(\frac{1}{(m^2 + n^2)^p}\right)$$

are convergent.

36.L. By separating $\sum (1/n^2)$ into odd and even parts, show that

$$\sum_{n=1}^{\infty} \frac{1}{n^2} = 4 \sum_{n=1}^{\infty} \frac{1}{(2n)^2} = \frac{4}{3} \sum_{n=1}^{\infty} \frac{1}{(2n-1)^2}.$$

36.M. If $|a| < 1$ and $|b| < 1$, prove that the series $a + b + a^2 + b^2 + a^3 + b^3 + \cdots$ converges. What is the limit?

36.N. If $\sum (a_n^2)$ and $\sum (b_n^2)$ are convergent, then $\sum (a_n b_n)$ is absolutely convergent and

$$\sum a_n b_n \leq \left\{ \sum a_n^2 \right\}^{1/2} \left\{ \sum b_n^2 \right\}^{1/2}.$$

In addition, $\sum (a_n + b_n)^2$ converges and

$$\left\{ \sum (a_n + b_n)^2 \right\}^{1/2} \leq \left\{ \sum a_n^2 \right\}^{1/2} + \left\{ \sum b_n^2 \right\}^{1/2}.$$

36.O. Prove Mertens' Theorem: If $\sum (a_n)$ converges absolutely to A and $\sum (b_n)$ converges to B, then their Cauchy product converges to AB. (Hint: Let the partial sums be denoted by A_n, B_n, C_n, respectively. Show that $\lim (C_{2n} - A_n B_n) = 0$ and $\lim (C_{2n+1} - A_n B_n) = 0$.)

36.P. Prove Cesàro's Theorem: Let $\sum (a_n)$ converge to A and $\sum (b_n)$ converge to B, and let $\sum (c_n)$ to their Cauchy product. If (C_n) is the sequence of partial sums of $\sum (c_n)$, then

$$\frac{1}{n} (C_1 + C_2 + \cdots + C_n) \to AB.$$

(Hint: write $C_1 + \cdots + C_n = A_1 B_n + \cdots + A_n B_1$; break this sum into three parts; and use the fact that $A_n \to A$ and $B_n \to B$.)

Section 37 Series of Functions

Because of their frequent appearance and importance, we now present a discussion of infinite series of functions. Since the convergence of an infinite series is handled by examining the sequence of partial sums, questions concerning series of functions are answered by examining corresponding questions for· sequences of functions. For this reason, a portion of the present section is merely a translation of facts already established for sequences of functions into series terminology. This is the case, for example, for the portion of the section dealing with series of general functions. However, in the second part of the section, where we discuss power series, some new features arise merely because of the special character of the functions involved.

37.1 DEFINITION. If (f_n) is a sequence of functions defined on a subset D of $\mathbf{R}^p$ with values in $\mathbf{R}^q$, the sequence of **partial sums** (s_n) of the infinite series $\sum (f_n)$ is defined for x in D by

$$s_1(x) = f_1(x),$$
$$s_2(x) = s_1(x) + f_2(x) \qquad [= f_1(x) + f_2(x)],$$
$$\cdots \cdots \cdots \cdots \cdots \cdots \cdots \cdots \cdots \cdots \cdots$$
$$s_{n+1}(x) = s_n(x) + f_{n+1}(x) \qquad [= f_1(x) + \cdots + f_n(x) + f_{n+1}(x)],$$
$$\cdots \cdots \cdots \cdots \cdots \cdots \cdots \cdots \cdots \cdots \cdots$$

In case the sequence (s_n) converges on D to a function f, we say that the infinite series of functions $\sum (f_n)$ **converges** to f on D. We shall often write

$$\sum (f_n), \qquad \sum_{n=1}^{\infty} (f_n), \qquad \text{or} \qquad \sum_{n=1}^{\infty} f_n$$

to denote either the series or the limit function, when it exists.

If the series $\sum (\|f_n(x)\|)$ converges for each x in D, then we say that $\sum (f_n)$ is **absolutely convergent** on D. If the sequence (s_n) is uniformly convergent on D to f, then we say that $\sum (f_n)$ is **uniformly convergent** on D, or that it **converges to f uniformly on D.**

One of the main reasons for the interest in uniformly convergent series of functions is the validity of the following results which give conditions justifying the change of order of the summation and other limiting operations.

37.2 THEOREM. *If f_n is continuous on $D \subseteq \mathbf{R}^p$ to $\mathbf{R}^q$ for each $n \in \mathbf{N}$ and if $\sum (f_n)$ converges to f uniformly on D, then f is continuous on D.*

This is a direct translation of Theorem 24.1 for series. The next result is a translation of Theorem 31.2.

37.3 THEOREM. *Suppose that the real-valued functions f_n, $n \in \mathbf{N}$, are Riemann-Stieltjes integrable with respect to a monotone function g on the interval $J = [a, b]$. If the series $\sum (f_n)$ converges to f uniformly on J, then f is Riemann-Stieltjes integrable with respect to g and*

$$(37.1) \qquad \int_a^b f \, dg = \sum_{n=1}^{\infty} \int_a^b f_n \, dg.$$

We now recast the Monotone Convergence Theorem 31.4 into series form.

37.4 THEOREM. *If the f_n are positive Riemann integrable functions on $J = [a, b]$ and if their sum $f = \sum (f_n)$ is Riemann integrable, then*

$$(37.2) \qquad \int_a^b f = \sum_{n=1}^{\infty} \int_a^b f_n.$$

Next we turn to the corresponding theorem pertaining to differentiation. Here we assume the uniform convergence of the series obtained after term-by-term differentiation of the given series. This result is an immediate consequence of Theorem 28.5.

37.5 THEOREM. *For each $n \in N$, let f_n be a real-valued function on $J = [a, b]$ which has a derivative f_n' on J. Suppose that the infinite series $\sum (f_n)$ converges for at least one point of J and that the series of derivatives $\sum (f_n')$ converges uniformly on J. Then there exists a real-valued function f on J such that $\sum (f_n)$ converges uniformly on J to f. In addition, f has a derivative on J and*

$$(37.3) \qquad f' = \sum f_n'.$$

Tests for Uniform Convergence

Since we have stated some consequences of uniform convergence of series, we shall now present a few tests which can be used to establish uniform convergence.

37.6 CAUCHY CRITERION. *Let (f_n) be a sequence of functions on $D \subseteq R^p$ to R^q. The infinite series $\sum (f_n)$ is uniformly convergent on D if and only if for every $\varepsilon > 0$ there exists an $M(\varepsilon)$ such that if $m \geq n \geq M(\varepsilon)$, then*

$$(37.4) \qquad \|f_n + f_{n+1} + \cdots + f_m\|_D < \varepsilon.$$

The proof of this result is immediate from 17.11, which is the corresponding Cauchy Criterion for the uniform convergence of sequences.

37.7 WEIERSTRASS M-TEST. *Let (M_n) be a sequence of non-negative real numbers such that $\|f_n\|_D \leq M_n$ for each $n \in N$. If the infinite series $\sum (M_n)$ is convergent, then $\sum (f_n)$ is uniformly convergent on D.*

PROOF. If $m > n$, we have the relation

$$\|f_n + \cdots + f_m\|_D \leq \|f_n\|_D + \cdots + \|f_m\|_D \leq M_n + \cdots + M_m.$$

The assertion follows from the Cauchy Criteria 34.5 and 37.6 and the convergence of $\sum (M_n)$.   Q.E.D.

The next two results are very useful in establishing uniform convergence when the convergence is not absolute. Their proofs are obtained by modifying the proofs of 36.2 and 36.4 and will be left as exercises.

37.8 DIRICHLET'S TEST. *Let (f_n) be a sequence of functions on $D \subseteq \mathbf{R}^p$ to $\mathbf{R}^q$ such that the partial sums*

$$s_n = \sum_{j=1}^{n} f_j, \qquad n \in \mathbf{N},$$

are all bounded in D-norm. Let (φ_n) be a decreasing sequence of functions on D to $\mathbf{R}$ which converges uniformly on D to zero. Then the series $\sum (\varphi_n f_n)$ converges uniformly on D.

37.9 ABEL'S TEST. *Let $\sum (f_n)$ be a series of functions on $D \subseteq \mathbf{R}^p$ to $\mathbf{R}^q$ which is uniformly convergent on D. Let (φ_n) be a monotone sequence of real-valued functions on D which is bounded in D-norm. Then the series $\sum (\varphi_n f_n)$ converges uniformly on D.*

37.10 EXAMPLES. (a) Consider the series $\sum_{n=1}^{\infty} (x^n/n^2)$. If $|x| \leq 1$, then $|x^n/n^2| \leq 1/n^2$. Since the series $\sum (1/n^2)$ is convergent, it follows from the Weierstrass M-test that the given series is uniformly convergent on the interval $[-1, 1]$.

(b) The series obtained after term-by-term differentiation of the series in (a) is $\sum_{n=1}^{\infty} (x^{n-1}/n)$. The Weierstrass M-test does not apply on the interval $[-1, 1]$ so we cannot apply Theorem 37.5. In fact, it is clear that this series of derivatives is not convergent for $x = 1$. However, if $0 < r < 1$, then the geometric series $\sum (r^{n-1})$ converges. Since

$$\left| \frac{x^{n-1}}{n} \right| \leq r^{n-1}$$

for $|x| \leq r$, it follows from the M-test that the differentiated series is uniformly convergent on the interval $[-r, r]$.

(c) A direct application of the M-test (with $M_n = 1/n^2$) shows that $\sum_{n=1}^{\infty} (1/n^2) \sin nx$ is uniformly convergent for all x in $\mathbf{R}$.

(d) Since the harmonic series $\sum (1/n)$ diverges, we cannot apply the M-test to

(37.5) $\displaystyle\sum_{n=1}^{\infty} (1/n) \sin nx.$

However, it follows from the discussion in Example 36.8(d) that if the interval $J = [a, b]$ is contained in the open interval $(0, 2\pi)$, then the partial sums $s_n(x) = \sum_{k=1}^{n} \sin kx$ are uniformly bounded on J. Since the sequence

$(1/n)$ decreases to zero, Dirichlet's Test 37.8 implies that the series (37.5) is uniformly convergent on J.

(e) Consider $\sum_{n=1}^{\infty} ((-1)^n/n)e^{-nx}$ on the interval $I = [0, 1]$. Since the norm of the nth term on I is $1/n$, we cannot apply the Weierstrass Test. Dirichlet's Test can be applied if we can show that the partial sums of $\sum ((-1)^n e^{-nx})$ are bounded. Alternatively, Abel's Test applies since $\sum ((-1)^n/n))$ is convergent and the bounded sequence (e^{-nx}) is monotone decreasing on I (but not uniformly convergent to zero).

Power Series

We shall now turn to a discussion of power series. This is an important class of series of functions and enjoys properties that are not valid for general series of functions.

37.11 Definition. A series of real functions $\sum (f_n)$ is said to be a **power series around** $x = c$ if the function f_n has the form

$$f_n(x) = a_n(x - c)^n,$$

where a_n and c belong to $\mathbf{R}$ and where $n = 0, 1, 2, \ldots$.

For the sake of simplicity of our notation, we shall treat only the case where $c = 0$. This is no loss of generality, however, since the translation $x' = x - c$ reduces a power series around c to a power series around 0. Thus whenever we refer to a power series, we shall mean a series of the form

$$(37.6) \qquad \sum_{n=0}^{\infty} a_n x^n = a_0 + a_1 x + \cdots + a_n x^n + \cdots .$$

Even though the functions appearing in (37.6) are defined over all of $\mathbf{R}$, it is not to be expected that the series (37.6) will converge for all x in $\mathbf{R}$. For example, by using the Ratio Test 35.8, we can show that the series

$$\sum_{n=0}^{\infty} n! x^n, \qquad \sum_{n=0}^{\infty} x^n, \qquad \sum_{n=0}^{\infty} x^n/n!,$$

converge for x in the sets

$$\{0\}, \qquad \{x \in \mathbf{R} : |x| < 1\}, \qquad \mathbf{R},$$

respectively. Thus the set on which a power series converges may be small, medium, or large. However, an arbitrary subset of $\mathbf{R}$ cannot be the precise set on which a power series converges, as we shall show.

If (b_n) is a bounded sequence of non-negative real numbers, then we define the **limit superior** of (b_n) to be the infimum of those numbers v such that $b_n \leq v$ for all sufficiently large $n \in \mathbf{N}$. This infimum is uniquely determined and is denoted by

$$\lim \sup (b_n).$$

Some other characterizations and properties of the limit superior of a sequence were given in Section 18, but the only thing we need to know is (i) that if $v > \lim \sup (b_n)$, then $b_n \leq v$ for all sufficiently large $n \in \mathbf{N}$, and (ii) that if $w < \lim \sup (b_n)$, then $w \leq b_n$ for infinitely many $n \in \mathbf{N}$.

37.12 DEFINITION. Let $\sum (a_n x^n)$ be a power series. If the sequence $(|a_n|^{1/n})$ is bounded, we set $\rho = \lim \sup (|a_n|^{1/n})$; if this sequence is not bounded we set $\rho = +\infty$. We define the **radius of convergence** of $\sum (a_n x^n)$ to be given by

$$
\begin{aligned}
R &= 0, &\quad \text{if} \quad & \rho = +\infty, \\
&= 1/\rho, &\quad \text{if} \quad & 0 < \rho < +\infty, \\
&= +\infty, &\quad \text{if} \quad & \rho = 0.
\end{aligned}
$$

The **interval of convergence** is the open interval $(-R, R)$.

We shall now justify the term "radius of convergence."

37.13 CAUCHY-HADAMARD† THEOREM. *If R is the radius of convergence of the power series $\sum (a_n x^n)$, then the series is absolutely convergent if $|x| < R$ and divergent if $|x| > R$.*

PROOF. We shall treat only the case where $0 < R < +\infty$, leaving the cases $R = 0$, $R = +\infty$, as exercises. If $0 < |x| < R$, then there exists a positive number $c < 1$ such that $|x| < cR$. Therefore $\rho < c/|x|$ and so it follows that if n is sufficiently large, then $|a_n|^{1/n} \leq c/|x|$. This is equivalent to the statement that

$$(37.7) \qquad\qquad |a_n x^n| \leq c^n$$

for all sufficiently large n. Since $c < 1$, the absolute convergence of $\sum (a_n x^n)$ follows from the Comparison Test 35.1.

If $|x| > R = 1/\rho$, then there are infinitely many $n \in \mathbf{N}$ for which we have $|a_n|^{1/n} > 1/|x|$. Therefore, $|a_n x^n| > 1$ for infinitely many n, so that the sequence $(a_n x^n)$ does not converge to zero.   Q.E.D.

---

† JACQUES HADAMARD (1865–1963), long-time dean of French mathematicians, was admitted to the École Polytechnique with the highest score attained during its first century. He was Henri Poincaré's successor in the Academy of Sciences and proved the Prime Number Theorem in 1896, although this theorem had been conjectured by Gauss many years before. Hadamard made other contributions to number theory, complex analysis, partial differential equations, and even psychology.

It will be noted that the Cauchy-Hadamard Theorem makes no statement as to whether the power series converges when $|x| = R$. Indeed, anything can happen, as the examples

$$(37.8) \qquad \sum x^n, \quad \sum \frac{1}{n} x^n, \quad \sum \frac{1}{n^2} x^n,$$

show. Since $\lim (n^{1/n}) = 1$ (cf. 14.8(e)), each of these power series has radius of convergence equal to 1. The first power series converges at neither of the points $x = -1$ and $x = +1$; the second series converges at $x = -1$ but diverges at $x = +1$; and the third power series converges at both $x = -1$ and $x = +1$. (Find a power series with $R = 1$ which converges at $x = +1$ but diverges at $x = -1$.)

It is an exercise to show that the radius of convergence of $\sum (a_n x^n)$ is also given by

$$(37.9) \qquad \lim \left(\frac{|a_n|}{|a_{n+1}|} \right),$$

provided this limit exists. Frequently, it is more convenient to use (37.9) than Definition 37.12.

The argument used in the proof of the Cauchy-Hadamard Theorem yields the uniform convergence of the power series on any fixed compact subset in the interval of convergence $(-R, R)$.

37.14 **THEOREM.** *Let R be the radius of convergence of $\sum (a_n x^n)$ and let K be a compact subset of the interval of convergence $(-R, R)$. Then the power series converges uniformly on K.*

PROOF. The compactness of $K \subseteq (-R, R)$ implies that there exists a positive constant $c < 1$ such that $|x| < cR$ for all $x \in K$. (Why?) By the argument in 37.13, we infer that for sufficiently large n, the estimate (37.7) holds for all $x \in K$. Since $c < 1$, the uniform convergence of $\sum (a_n x^n)$ on K is a direct consequence of the Weierstrass M-test with $M_n = c^n$. Q.E.D.

37.15 **THEOREM.** *The limit of a power series is continuous on the interval of convergence. A power series can be integrated term-by-term over any compact interval contained in the interval of convergence.*

PROOF. If $|x_0| < R$, then the preceding result asserts that $\sum (a_n x^n)$ converges uniformly on any compact neighborhood of x_0 contained in $(-R, R)$. The continuity at x_0 then follows from Theorem 37.2, and the term-by-term integration is justified by Theorem 37.3. Q.E.D.

We now show that a power series can be differentiated term-by-term. Unlike the situation for general series, we do not need to assume that the differentiated series is uniformly convergent. Hence this result is stronger than the corresponding result for the differentiation of infinite series.

37.16 DIFFERENTIATION THEOREM. *A power series can be differen-tiated term-by-term within the interval of convergence. In fact, if*

$$f(x) = \sum_{n=0}^{\infty} (a_n x^n), \qquad \text{then} \qquad f'(x) = \sum_{n=1}^{\infty} (n a_n x^{n-1}).$$

Both series have the same radius of convergence.

PROOF. Since $\lim (n^{1/n}) = 1$, the sequence $(|na_n|^{1/n})$ is bounded if and only if the sequence $(|a_n|^{1/n})$ is bounded. Moreover, it is easily seen that

$$\lim \sup (|na_n|^{1/n}) = \lim \sup (|a_n|^{1/n}).$$

Therefore, the radius of convergence of the two series is the same, so the formally differentiated series is uniformly convergent on each compact subset of the interval of convergence. We can then apply Theorem 37.5 to conclude that the formally differentiated series converges to the derivative of the given series.   Q.E.D.

It is to be observed that the theorem makes no assertion about the end points of the interval of convergence. If a series is convergent at an end point, then the differentiated series may or may not be convergent at this point. For example, the series $\sum_{n=0}^{\infty} (x^n/n^2)$ converges at both end points $x = -1$ and $x = +1$. However, the differentiated series

$$\sum_{n=1}^{\infty} \frac{x^{n-1}}{n} = \sum_{m=0}^{\infty} \frac{x^m}{m+1}$$

converges at $x = -1$ but diverges at $x = +1$.

By repeated application of the preceding result, we conclude that if k is any natural number, then the power series $\sum_{n=0}^{\infty} (a_n x^n)$ can be differentiated term-by-term k times to obtain

(37.10) $$\sum_{n=k}^{\infty} \frac{n!}{(n-k)!} a_n x^{n-k}.$$

Moreover, this series converges absolutely to $f^{(k)}$ for $|x| < R$ and uniformly over any compact subset of the interval of convergence.

If we substitute $x = 0$ in (37.10), we obtain the important formula

(37.11) $$f^{(k)}(0) = k! a_k.$$

37.17 UNIQUENESS THEOREM. *If $\sum (a_n x^n)$ and $\sum (b_n x^n)$ converge on some interval $(-r, r)$, $r > 0$, to the same function f, then*

$$a_n = b_n \qquad \text{for all } n \in \mathbf{N}.$$

PROOF. Our preceding remarks show that $n! a_n = f^{(n)}(0) = n! b_n$ for $n \in \mathbf{N}$.   Q.E.D.

Some Additional Results†

There are a number of results concerning various algebraic combinations of power series, but those involving substitution and inversion are more naturally proved using arguments from complex analysis. For this reason we shall not go into these questions but content ourselves with one result in this direction. Fortunately, it is one of the most useful.

37.18 MULTIPLICATION THEOREM. *If f and g are given on the interval $(-r, r)$ by the power series*

$$f(x) = \sum_{n=0}^{\infty} a_n x^n, \qquad g(x) = \sum_{n=0}^{\infty} b_n x^n,$$

then their product is given on this interval by the series $\sum (c_n x^n)$, where the coefficients (c_n) are

$$c_n = \sum_{k=0}^{n} a_k b_{n-k} \qquad for\ n = 0, 1, 2, \ldots.$$

PROOF. We have seen in 37.13 that if $|x| < r$, then the series giving $f(x)$ and $g(x)$ are absolutely converent. If we apply Theorem 36.13, we obtain the desired conclusion. Q.E.D.

The Multiplication Theorem asserts that the radius of convergence of the product is at least r. It can be larger, however, as is easily shown.

We have seen that, in order for a function f to be given by a power series on an interval $(-r, r)$, $r > 0$, it is necessary that all of the derivatives of f exist on this interval. It might be suspected that this condition is also sufficient; however, things are not quite so simple. For example, the function f, given by

(37.12)
$$f(x) = e^{-1/x^2}, \qquad x \neq 0,$$
$$= 0, \qquad x = 0,$$

can be shown (see Exercise 37.N) to possess derivatives of all orders and $f^{(n)}(0) = 0$ for $n = 0, 1, 2, \ldots$. If f can be given on an interval $(-r, r)$ by a power series around $x = 0$, then it follows from the Uniqueness Theorem 37.17 that the series must vanish identically, contrary to the fact that $f(x) \neq 0$ for $x \neq 0$.

Nevertheless, there are some useful sufficient conditions that can be given in order to guarantee that f can be given by a power series. As an example, we observe that it follows from Taylor's Theorem 28.6 that if there exists a constant $B > 0$ such that

(37.13)
$$|f^{(n)}(x)| \leq B$$

† The rest of this section can be omitted on the first reading.

for all $|x| < r$ and $n = 0, 1, 2, \ldots$, then $\sum_{n=0}^{\infty} f^{(n)}(0)x^n/n!$ converges to $f(x)$ for $|x| < r$. Similar (but less stringent) conditions on the magnitude of the derivatives can be given which yield the same conclusion.

As an example, we present an elegant and useful result due to Serge Bernstein concerning the one-sided expansion of a function in a power series.

37.19 BERNSTEÏN'S THEOREM. *Let f be defined and possess derivatives of all orders on an interval $[0, r]$ and suppose that f and all of its derivatives are positive on the interval $[0, r]$. If $0 \leq x < r$, then $f(x)$ is given by the expansion*

$$f(x) = \sum_{n=0}^{\infty} \frac{f^{(n)}(0)}{n!} x^n.$$

PROOF. We shall make use of the integral form for the remainder in Taylor's Theorem given by the relation (31.3). If $0 \leq x \leq r$, then

(37.14) $$f(x) = \sum_{k=0}^{n-1} \frac{f^{(k)}(0)}{k!} x^k + R_n,$$

where we have the formula

$$R_n = \frac{x^{n-1}}{(n-1)!} \int_0^1 (1-s)^{n-1} f^{(n)}(sx) \, ds.$$

Since all the terms in the sum in (37.14) are positive,

(37.15) $$f(r) \geq \frac{r^{n-1}}{(n-1)!} \int_0^1 (1-s)^{n-1} f^{(n)}(sr) \, ds.$$

Since $f^{(n+1)}$ is positive, $f^{(n)}$ is increasing on $[0, r]$; therefore, if x is in this interval, then

(37.16) $$0 \leq R_n \leq \frac{x^{n-1}}{(n-1)!} \int_0^1 (1-s)^{n-1} f^{(n)}(sr) \, ds.$$

By combining (37.15) and (37.16), we have $0 \leq R_n \leq (x/r)^{n-1} f(r)$. Hence, if $0 \leq x < r$, then $\lim (R_n) = 0$.   Q.E.D.

We have seen in Theorem 37.14 that a power series converges uniformly on every compact subset of its interval of convergence. However, there is no *a priori* reason to believe that this result can be extended to the end points of the interval of convergence. However, there is a theorem of Abel that, if convergence does take place at one of the end points, then the series converges uniformly out to this end point.

In order to simplify our notation, we shall suppose that the radius of convergence of the series is equal to 1. This is no loss of generality and can always be attained by letting $x' = x/R$, which is merely a change of scale.

37.20 ABEL'S THEOREM. *Suppose that the power series $\sum_{n=0}^{\infty} (a_n x^n)$ converges to $f(x)$ for $|x| < 1$ and that $\sum_{n=0}^{\infty} (a_k)$ converges to A. Then the power series converges uniformly in $I = [0, 1]$ and*

$$(37.17) \qquad\qquad \lim_{x \to 1-} f(x) = A.$$

PROOF. Abel's Test 37.9, with $f_n(x) = a_n$ and $\varphi_n(x) = x^n$, applies to give the uniform convergence of $\sum (a_n x^n)$ on I. Hence the limit is continuous on I; since it agrees with $f(x)$ for $0 \le x < 1$, the limit relation (37.17) follows. Q.E.D.

One of the most interesting things about this result is that it suggests a method of attaching a limit to series which may not be convergent. Thus, if $\sum_{n=1}^{\infty} (b_n)$ is an infinite series, we can form the corresponding power series $\sum (b_n x^n)$. If the b_n do not increase too rapidly, this power series converges to a function $B(x)$ for $|x| < 1$. If $B(x) \to \beta$ as $x \to 1-$, we say that the series $\sum (b_n)$ is **Abel summable** to β. This type of summation is similar to (but more powerful than) the Cesàro method of arithmetic means mentioned in Section 19 and has deep and interesting consequences. The content of Abel's Theorem 37.20 is similar to Theorem 19.3; it asserts that if a series is already convergent, then it is Abel summable to the same limit. The converse is not true, however, for the series $\sum_{n=0}^{\infty} (-1)^n$ is not convergent but since

$$\frac{1}{1+x} = \sum_{n=0}^{\infty} (-1)^n x^n,$$

it follows that $\sum (-1)^n$ is Abel summable to $\frac{1}{2}$.

It sometimes happens that if a series is known to be Abel summable, and if certain other conditions are satisfied, then it can be proved that the series is actually convergent. Theorems of this nature are called **Tauberian theorems** and are often very deep and difficult to prove. They are also useful because they enable one to go from a weaker type of convergence to a stronger type, provided certain additional hypotheses are satisfied.

Our final theorem is the first result of this type and was proved by A. Tauber† in 1897. It provides a partial converse to Abel's Theorem.

37.21 TAUBER'S THEOREM. *Suppose that the power series $\sum (a_n x^n)$ converges to $f(x)$ for $|x| < 1$ and that $\lim (n a_n) = 0$. If $\lim f(x) = A$ as $x \to 1-$, then the series $\sum (a_n)$ converges to A.*

---

† ALFRED TAUBER (1866-circa 1947) was a professor at Vienna. He contributed primarily to analysis.

PROOF.   It is desired to estimate differences such as $\sum^N (a_n) - A$.   To do this, we write

$$(37.18) \qquad \sum_{n=0}^{N} a_n - A = \left\{ \sum_{n=0}^{N} a_n - f(x) \right\} + \{f(x) - A\}$$

$$= \sum_{n=0}^{N} a_n(1 - x^n) - \sum_{N+1}^{\infty} a_n x^n + \{f(x) - A\}.$$

Since $0 \le x < 1$, we have $1 - x^n = (1 - x)(1 + x + \cdots + x^{n-1}) < n(1 - x)$, so we can dominate the first term on the right side by the expression $(1 - x) \sum_{n=0}^{N} na_n$.

By hypothesis $\lim (na_n) = 0$; hence Theorem 19.3 implies that

$$\lim\left(\frac{1}{m+1} \sum_{n=0}^{m} na_n \right) = 0.$$

In addition, we have the relation $A = \lim f(x)$.

Now let $\varepsilon > 0$ be given and choose a fixed natural number N which is so large that

(i) $\left| \sum_{n=0}^{N} na_n \right| < (N+1)\varepsilon$;

(ii) $|a_n| < \dfrac{\varepsilon}{N+1}$ for all $n \ge N$;

(iii) $|f(x_0) - A| < \varepsilon$ for $x_0 = 1 - \dfrac{1}{N+1}$.

We shall assess the magnitude of (37.18) for this value of N and x_0. From (i), (ii), (iii) and the fact that $(1 - x_0)(N+1) = 1$, we derive the estimate

$$\left| \sum_{n=0}^{N} a_n - A \right| \le (1 - x_0)(N+1)\varepsilon + \frac{\varepsilon}{N+1} \frac{x_0^{N+1}}{1 - x_0} + \varepsilon < 3\varepsilon.$$

Since this can be done for each $\varepsilon > 0$, the convergence of $\sum (a_n)$ to A is established.   Q.E.D.

Exercises

37.A. Discuss the convergence and the uniform convergence of the series $\sum (f_n)$, where $f_n(x)$ is given by

(a) $(x^2 + n^2)^{-1}$,   (b) $(nx)^{-2}$, $x \ne 0$,
(c) $\sin(x/n^2)$,   (d) $(x^n + 1)^{-1}$, $x \ge 0$,
(e) $x^n(x^n + 1)^{-1}$, $x \ge 0$,   (f) $(-1)^n(n + x)^{-1}$, $x \ge 0$.

37.B. If $\sum (a_n)$ is an absolutely convergent series, then the series $\sum (a_n \sin nx)$ is absolutely and uniformly convergent.

37.C. Let (c_n) be a decreasing sequence of positive numbers. If $\sum (c_n \sin nx)$ is uniformly convergent, then $\lim (nc_n) = 0$.

37.D. Give the details of the proof of Dirichlet's Test 37.8.

37.E. Give the details of the proof of Abel's Test 37.9.

37.F. Discuss the cases $R = 0$, $R = +\infty$ in the Cauchy-Hadamard Theorem 37.13.

37.G. Show that the radius of convergence R of the power series $\sum (a_n x^n)$ is given by $\lim (|a_n|/|a_{n+1}|)$ whenever this limit exists. Give an example of a power series where this limit does not exist.

37.H. Determine the radius of convergence of the series $\sum (a_n x^n)$, where a_n is given by

(a) $1/n^n$, (b) $n^\alpha/n!$,
(c) $n^n/n!$, (d) $(\log n)^{-1}$, $n \ge 2$.
(e) $(n!)^2/(2n)!$, (f) $n^{-\sqrt{n}}$.

37.I. If $a_n = 1$ when n is the square of a natural number and $a_n = 0$ otherwise, find the radius of convergence of $\sum (a_n x^n)$. If $b_n = 1$ when $n = m!$ for $m \in \mathbf{N}$ and $b_n = 0$ otherwise, find the radius of convergence of $\sum (b_n x^n)$.

37.J. Prove in detail that $\limsup (|na_n|^{1/n}) = \limsup (|a_n|^{1/n})$.

37.K. If $0 < p \le |a_n| \le q$ for all $n \in \mathbf{N}$, find the radius of convergence of $\sum (a_n x^n)$.

37.L. Let $f(x) = \sum (a_n x^n)$ for $|x| < R$. If $f(x) = f(-x)$ for all $|x| < R$, show that $a_n = 0$ for all odd n.

37.M. Prove that if f is defined for $|x| < r$ and if there exists a constant B such that $|f^{(n)}(x)| \le B$ for all $|x| < r$ and $n \in \mathbf{N}$, then the Taylor series expansion

$$\sum_{n=0}^{\infty} \frac{f^{(n)}(0)}{n!} x^n$$

converges to $f(x)$ for $|x| < r$.

37.N. Prove by induction that the function given in formula (37.12) has derivatives of all orders at every point and that all of these derivatives vanish at $x = 0$. Hence this function is not given by its Taylor expansion about $x = 0$.

37.O. Give an example of a function which is equal to its Taylor series expansion about $x = 0$ for $x \ge 0$, but which is not equal to this expansion for $x < 0$.

37.P. The argument outlined in Exercise 28.M shows that the Lagrange form of the remainder can be used to justify the general Binomial Expansion

$$(1+x)^m = \sum_{n=0}^{\infty} \binom{m}{n} x^n$$

when x is in the interval $0 \le x < 1$. Similarly Exercise 28.N validates this expansion for $-1 < x \le 0$, but the argument is based on the Cauchy form of the remainder and is somewhat more involved. To obtain an alternative proof of this second case, apply Bernstein's Theorem to $g(x) = (1-x)^m$ for $0 \le x < 1$.

37.Q. Consider the Binomial Expansion at the end points $x = \pm 1$. Show that if $x = -1$, then the series converges absolutely for $m \ge 0$ and diverges for $m < 0$. At $x = +1$, the series converges absolutely for $m \ge 0$, converges conditionally for $-1 < m < 0$, and diverges for $m \le -1$.

37.R. Let $f(x) = \tan x$ for $|x| < \pi/2$. Use the fact that f is odd and Bernstein's Theorem to show that f is given on this interval by its Taylor series expansion about $x = 0$.

37.S. Use Abel's Theorem to prove that if $f(x) = \sum (a_n x^n)$ for $|x| < R$, then

$$\int_0^R f(x)\, dx = \sum_{n=0}^{\infty} \frac{a_n}{n+1} R^{n+1},$$

provided that the series on the right side is convergent even though the original series may not converge at $x = R$. Hence it follows that

$$\log 2 = \sum_{n=1}^{\infty} \frac{(-1)^{n+1}}{n}, \qquad \frac{\pi}{4} = \sum_{n=0}^{\infty} \frac{(-1)^n}{2n+1}.$$

37.T. By using Abel's Theorem, prove that if the series $\sum (a_n)$ and $\sum (b_n)$ converge and if their Cauchy product $\sum (c_n)$ converges, then we have $\sum (c_n) = \sum (a_n) \cdot \sum (b_n)$.

37.U. Suppose that $a_n \geq 0$ and that $f(x) = \sum (a_n x^n)$ has radius of convergence 1. If $\sum (a_n)$ diverges, prove that $f(x) \to +\infty$ as $x \to 1-$. Use this result to prove the elementary Tauberian theorem: If $a_n \geq 0$ and if

$$A = \lim_{x \to 1-} \sum a_n x^n,$$

then $\sum (a_n)$ converges to A.

37.V. Let $\sum_{n=0}^{\infty} (p_n)$ be a divergent series of positive numbers such that the radius of convergence of $\sum (p_n x^n)$ is 1. Prove Appell's† Theorem: If $s = \lim (a_n/p_n)$, then the radius of convergence of $\sum (a_n x^n)$ is also 1 and

$$\lim_{x \to 1-} \frac{\sum a_n x^n}{\sum p_n x^n} = s.$$

(Hint: it is sufficient to treat the case $s = 0$. Also use the fact that $\lim_{x \to 1-} [\sum (p_n x^n)]^{-1} = 0$.)

37.W. Apply Appell's Theorem with $p(x) = \sum_{n=0}^{\infty} (x^n)$ to obtain Abel's Theorem.

37.X. If (a_n) is a sequence of real numbers and $a_0 = 0$, let $s_n = a_1 + \cdots + a_n$ and let $\sigma_n = (s_1 + \cdots + s_n)/n$. Prove Frobenius'‡ Theorem: If $s = \lim (\sigma_n)$ then

$$s = \lim_{x \to 1-} \sum_{n=0}^{\infty} a_n x^n.$$

REMARK. In the terminology of summability theory, this result says that if a sequence (a_n) is Cesàro summable to s, then it is also Abel summable to s. (Hint: apply Appell's Theorem to $p(x) = (1-x)^{-2} = \sum_{n=0}^{\infty} (nx^{n-1})$ and note that $\sum (n \cdot \sigma_n x^n) = p(x) \sum (a_n x^n)$.)

---

† PAUL APPELL (1855–1930) was a student of Hermite at the Sorbonne. He did research in complex analysis.

‡ GEORG FROBENIUS (1849–1917) was professor at Berlin. He is known for his work both in algebra and analysis.

Projects

37.α. The theory of power series presented in the text entends to complex power series.

(a) In view of the observations in Section 13, all of the definitions and theorems that are meaningful and valid for series in $\boldsymbol{R}^2$ are also valid for series with elements in $\boldsymbol{C}$. In particular the results pertaining to absolute convergence extend readily.

(b) Examine the results pertaining to rearrangements and the Cauchy product to see if they extend to $\boldsymbol{C}$.

(c) Show that the Comparison, Root, and Ratio Tests extended to $\boldsymbol{C}$.

(d) Let R be the radius of convergence of a complex power series

$$\sum_{n=0}^{\infty} a_n z^n.$$

Prove that the series converges absolutely for $|z| < R$ and uniformly on any compact subset of $\{z \in \boldsymbol{C} : |z| < R\}$.

(e) Let f and g be functions defined for $D = \{z \in \boldsymbol{C} : |z| < r\}$ with values in $\boldsymbol{C}$ which are the limits on D of two power series. Show that if f and g agree on $D \cap \boldsymbol{R}$, then they agree on all of D.

(f) Show that two power series in $\boldsymbol{C}$ can be multiplied together within their common circle of convergence.

37.β. In this project we define the exponential function in terms of power series. In doing so, we shall define it for complex numbers as well as real.

(a) Let E be defined for $z \in \boldsymbol{C}$ by the series

$$E(z) = \sum_{n=0}^{\infty} \frac{z^n}{n!}.$$

Show that the series is absolutely convergent for all $z \in \boldsymbol{C}$ and that it is uniformly convergent on any bounded subset of $\boldsymbol{C}$.

(b) Prove that E is a continuous function on $\boldsymbol{C}$ to $\boldsymbol{C}$, that $E(0) = 1$, and that

$$E(z + w) = E(z)E(w)$$

for z, w in $\boldsymbol{C}$. (Hint: the Binomial Theorem for $(z + w)^n$ holds when z, $w \in \boldsymbol{C}$ and $n \in \boldsymbol{N}$.)

(c) If x and y are real numbers, define E_1 and E_2 by $E_1(x) = E(x)$, $E_2(y) = E(iy)$; hence $E(x + iy) = E_1(x)E_2(y)$. Show that E_1 takes on only real values but that E_2 has some non-real values. Define C and S on $\boldsymbol{R}$ to $\boldsymbol{R}$ by

$$C(y) = \operatorname{Re} E_2(y), \qquad S(y) = \operatorname{Im} E_2(y)$$

for $y \in \boldsymbol{R}$, and show that

$$C(y_1 + y_2) = C(y_1)C(y_2) - S(y_1)S(y_2),$$
$$S(y_1 + y_2) = S(y_1)C(y_2) + C(y_1)S(y_2).$$

(d) Prove that C and S, as defined in (c), have the series expansions

$$C(y) = \sum_{n=0}^{\infty} \frac{(-1)^n y^{2n}}{(2n)!}, \qquad S(y) = \sum_{n=0}^{\infty} \frac{(-1)^n y^{2n+1}}{(2n+1)!}.$$

(e) Show that $C' = -S$ and $S' = C$. Hence $(C^2 + S^2)' = 2CC' + 2SS' = 0$ which implies that $C^2 + S^2$ is identically equal to 1. In particular, this implies that both C and S are bounded in absolute value by 1.

(f) Infer that the function E_2 on R to C satisfies $E_2(0) = 1$, $E_2(y_1 + y_2) = E_2(y_1)E_2(y_2)$. Hence $E_2(-y) = 1/E_2(y)$ and $|E_2(y)| = 1$ for all y in R.

Section 38   Fourier Series

We shall now give the definition of the Fourier† series of a piecewise continuous function with period 2π. Although our discussion will be brief, we shall present the main convergence theorems relating to Fourier series. These theorems are of considerable importance in analysis, and its applications to physics.

In the following we shall suppose that $f : R \to R$ has **period** 2π; that is, that $f(x + 2\pi) = f(x)$ for all $x \in R$. We also suppose that f is **piecewise continuous**; that is, f is continuous except possibly for a finite number of points $x_1, \ldots, x_r$ in any interval of length 2π, at which f has left and right hand limits:

$$f(x_j -) = \lim_{\substack{h \to 0 \\ h > 0}} f(x_j - h), \qquad f(x_j +) = \lim_{\substack{h \to 0 \\ h > 0}} f(x_j + h).$$

The set of all functions $f : R \to R$ which have period 2π and are piecewise continuous will be denoted by $PC(2\pi)$. It is readily seen that this set is a vector space under the operations:

$$(f + g)(x) = f(x) + g(x), \qquad (cf)(x) = cf(x), \qquad x \in R.$$

Because of the periodicity of $f \in PC(2\pi)$ it is only necessary to investigate f on an interval of length 2π; for example, we have

$$\int_{-\pi}^{\pi} f(x) \, dx = \int_{c}^{c+2\pi} f(x) \, dx$$

for any $c \in R$.

On the space $PC(2\pi)$ we shall be interested in the two norms

$$\|f\|_\infty = \sup\{|f(x)| : x \in [-\pi, \pi]\}, \qquad \|f\|_2 = \left(\int_{-\pi}^{\pi} (f(x))^2 \, dx \right)^{1/2},$$

---

† (J.-B.) Joseph Fourier (1768–1830) was the son of a French tailor. Educated in a monastery, he left to engage in mathematical and revolutionary activities. He accompanied Napoleon to Egypt in 1798 and was later appointed prefect of the Department of Isère in southern France. During this time he worked on his most famous accomplishment: the mathematical theory of heat. His work was a landmark in mathematical physics and has had a towering influence on both subjects to the present day.

which are well defined because a function in $PC(2\pi)$ is bounded and Riemann integrable. It is an elementary exercise to show that if $f \in PC(2\pi)$, then

(38.1) $$\|f\|_2 \leq \sqrt{2\pi}\, \|f\|_\infty.$$

It follows from this inequality that convergence in the norm $\|\cdot\|_\infty$ (that is, *uniform convergence*) implies convergence in the norm $\|\cdot\|_2$ (that is, *mean square convergence*). However, the converse is not true. (See Exercises 31.H and 38.L.)

38.1 DEFINITION. If $f \in PC(2\pi)$, then the **Fourier coefficients** of f are the numbers $a_0, a_1, a_2, \ldots, b_1, b_2, \ldots$ defined by

(38.2) $$a_n = \frac{1}{\pi} \int_{-\pi}^{\pi} f(t) \cos nt \, dt, \qquad b_n = \frac{1}{\pi} \int_{-\pi}^{\pi} f(t) \sin nt \, dt.$$

By the **Fourier series** of f we mean the series

(38.3) $$\tfrac{1}{2}a_0 + \sum_{n=1}^{\infty} (a_n \cos nx + b_n \sin nx).$$

To indicate the association of the Fourier series (38.3) to the function f, we often write

$$f(x) \sim \tfrac{1}{2}a_0 + \sum_{n=1}^{\infty} (a_n \cos nx + b_n \sin nx).$$

However it is to be emphasized that this notation is not intended to suggest that the Fourier series converges to $f(x)$ at any particular point x. Indeed, there exist continuous functions with period 2π whose Fourier series are divergent at infinitely many points. (See Burkhill and Burkhill, page 317, and Hewitt and Ross, page 300.)

38.2 EXAMPLES. (a) Let $f_1 \in PC(2\pi)$ be defined on $(-\pi, \pi]$ by $f_1(x) = -1$ for $-\pi < x < 0$ and $f_1(x) = +1$ for $0 \leq x \leq +\pi$. It is an exercise to show that the Fourier series for f_1 is given by

$$\frac{4}{\pi}\left[\frac{\sin x}{1} + \frac{\sin 3x}{3} + \frac{\sin 5x}{5} + \cdots\right].$$

It will be proved below that this Fourier series does indeed converge to f_1 for $0 < |x| < \pi$, but it does *not* converge to f_1 at $x = 0, \pm\pi$ (why?). Note that f_1 is piecewise continuous, but is not continuous at the points in $\{n\pi : n \in \mathbf{Z}\}$.

(b) Let $f_2 \in PC(2\pi)$ be defined on $(-\pi, \pi]$ by $f_2(x) = |x|$. It is an exercise to show that the Fourier series for f_2 is given by

$$\frac{\pi}{2} - \frac{4}{\pi}\left[\frac{\cos x}{1^2} + \frac{\cos 3x}{3^2} + \frac{\cos 5x}{5^2} + \cdots\right].$$

It is clear that this series converges uniformly on $\mathbf{R}$ and it will be proved below that it converges to f_2.

(c) Let $f \in PC(2\pi)$ be **even**; that is, $f(-x) = f(x)$ for all $x \in \mathbf{R}$. For such a function the Fourier coefficients $b_n = 0$ for $n = 1, 2, \ldots$, while

$$a_n = \frac{2}{\pi} \int_0^\pi f(t) \cos nt \, dt, \qquad n = 0, 1, 2, \ldots.$$

(Note that the function in (b) is even.)

(d) Let $g \in PC(2\pi)$ be **odd**; that is, $g(-x) = -g(x)$ for all $x \in \mathbf{R}$. For such a function the Fourier coefficients $a_n = 0$ for $n = 0, 1, 2, \ldots$, while

$$b_n = \frac{2}{\pi} \int_0^\pi g(t) \sin nt \, dt, \qquad n = 1, 2, \ldots.$$

(Note that the function in (a) is odd.)

(e) Let f be continuous on $\mathbf{R}$ with period 2π and let its derivative f' be piecewise continuous on $\mathbf{R}$ (and with period 2π). We shall relate the Fourier coefficients a_n, b_n of f with the Fourier coefficients a'_n, b'_n of f' for $n = 1, 2, \ldots$. In fact, integrating by parts, we have

$$a'_n = \frac{1}{\pi} \int_{-\pi}^{\pi} f'(t) \cos nt \, dt$$

$$= \frac{1}{\pi} \left[f(t) \cos nt \, \Big|_{-\pi}^{\pi} - \int_{-\pi}^{\pi} f(t)(-n) \sin nt \, dt \right].$$

If we use the fact that $t \mapsto f(t) \cos nt$ has period 2π the first term is seen to vanish and so $a'_n = nb_n$ for $n = 1, 2, \ldots$. Similarly it is shown that $b'_n = -na_n$ for $n = 1, 2, \ldots$, (We note that if f_1, f_2 are the functions in (a) and (b), then $f_1(x) = f'_2(x)$ for $x \notin \{n\pi : n \in \mathbf{Z}\}$, and that the Fourier coefficients for f_1 and f_2 for $n = 1, 2, \ldots$ satisfy the above relationships.)

In the next lemma, we shall calculate the square of the distance relative to the norm $\|\cdot\|_2$ from f in $PC(2\pi)$ to an arbitrary function T_n of the form

(38.4) $$T_n(x) = \tfrac{1}{2}\alpha_0 + \sum_{k=1}^n (\alpha_k \cos kx + \beta_k \sin kx);$$

such a function is sometimes called a **trigonometric polynomial of degree** n. In making this calculation it is useful to have the relations

$$\int_{-\pi}^{\pi} (\cos kx)^2 \, dx = \int_{-\pi}^{\pi} (\sin kx)^2 \, dx = \pi, \qquad k \in \mathbf{N},$$

$$\int_{-\pi}^{\pi} \sin kx \sin nx \, dx = \int_{-\pi}^{\pi} \cos kx \cos nx \, dx = 0, \qquad k, n \in \mathbf{N}, k \neq n,$$

$$\int_{-\pi}^{\pi} \sin kx \cos mx \, dx = 0, \qquad k, m = 0, 1, 2, \ldots.$$

38.3   LEMMA.   *If $f \in PC(2\pi)$ and T_n is a trigonometric polynomial of degree n (that is, T_n has the form (38.4)), then*

$$(38.5) \quad \|f - T_n\|_2^{\,2} = \|f\|_2^{\,2} - \pi\Big\{\tfrac{1}{2}a_0^{\,2} + \sum_{k=1}^{n} (a_k^{\,2} + b_k^{\,2})\Big\}$$
$$+ \pi\Big\{\tfrac{1}{2}(\alpha_0 - a_0)^2 + \sum_{k=1}^{n} [(\alpha_k - a_k)^2 + (\beta_k - b_k)^2]\Big\}.$$

where a_k, b_k denote the Fourier coefficients of f.

PROOF.   We have

$$\|f - T_n\|_2^{\,2} = \int_{-\pi}^{\pi} [f(t) - T_n(t)]^2 \, dt$$
$$= \int_{-\pi}^{\pi} [f(t)]^2 \, dt - 2\int_{-\pi}^{\pi} f(t)T_n(t) \, dt + \int_{-\pi}^{\pi} [T_n(t)]^2 \, dt.$$

Now it is easily seen that

$$\int_{-\pi}^{\pi} f(t)T_n(t) \, dt = \tfrac{1}{2}\alpha_0 \int_{-\pi}^{\pi} f(t) \, dt + \sum_{k=1}^{n} \alpha_k \int_{-\pi}^{\pi} f(t) \cos kt \, dt$$
$$+ \sum_{k=1}^{n} \beta_k \int_{-\pi}^{\pi} f(t) \sin kt \, dt$$
$$= \pi\Big\{\tfrac{1}{2}\alpha_0 a_0 + \sum_{k=1}^{n} (\alpha_k a_k + \beta_k b_k)\Big\}.$$

Moreover, using the relations cited above it is seen that

$$\int_{-\pi}^{\pi} [T_n(t)]^2 \, dt = \pi\Big\{\tfrac{1}{2}\alpha_0^{\,2} + \sum_{k=1}^{n} (\alpha_k^{\,2} + \beta_k^{\,2})\Big\}.$$

If we insert these two relations into the first formula and add and subtract $\pi\{\tfrac{1}{2}a_0^{\,2} + \sum_{k=1}^{n} (a_k^{\,2} + b_k^{\,2})\}$, we obtain formula (38.5).   Q.E.D.

Lemma 38.3 has the following important "geometrical" interpretation: among all trigonometric polynomials T_n of degree n, the one which minimizes the expression $\|f - T_n\|_2^{\,2}$ is uniquely determined and is obtained by choosing the coefficients α_k, β_k to be the Fourier coefficients a_k, b_k of f, $k = 0, 1, \ldots, n$. If we denote this (unique) minimizing trigonometrical polynomial by $S_n(f)$, then

$$(38.6) \qquad S_n(f)(x) = \tfrac{1}{2}a_0 + \sum_{k=1}^{n} (a_k \cos kx + b_k \sin kx)$$

is the nth partial sum of the Fourier series for f and formula (38.5) implies

$$(38.7) \qquad \|f - S_n(f)\|_2^{\,2} = \|f\|_2^{\,2} - \pi\Big\{\tfrac{1}{2}a_0^{\,2} + \sum_{k=1}^{n} (a_k^{\,2} + b_k^{\,2})\Big\}.$$

If we make use of Exercise 26.F we can show that

$$(38.8) \qquad \lim_n \|f - S_n(f)\|_2 = 0$$

for each continuous function with period 2π. However, since that exercise is the result of considerable analysis, we prefer to derive this result more directly. In doing so we shall need the following two results.

38.4 BESSEL'S INEQUALITY. *If $f \in PC(2\pi)$, then*

$$(38.9) \qquad \tfrac{1}{2}a_0{}^2 + \sum_{k=1}^{\infty} (a_k{}^2 + b_k{}^2) \le \frac{1}{\pi} \|f\|_2{}^2.$$

PROOF. If $n \in N$ is arbitrary, then it follows from (38.7) that

$$\tfrac{1}{2}a_0{}^2 + \sum_{k=1}^{n} (a_k{}^2 + b_k{}^2) \le \frac{1}{\pi} \|f\|_2{}^2.$$

Hence the partial sums of the series on the left side of (38.9) are bounded above. Since the terms are all positive, this series is convergent and (38.9) holds. Q.E.D.

The next result is a special case of what is usually called the Riemann-Lebesgue Lemma.

38.5 RIEMANN-LEBESGUE LEMMA. *If $g \in PC(2\pi)$, then*

$$\lim_n \int_{-\pi}^{\pi} g(t) \sin(n + \tfrac{1}{2})t \, dt = 0.$$

PROOF. Since $\sin(n + \tfrac{1}{2})t = \sin nt \cos \tfrac{1}{2}t + \cos nt \sin \tfrac{1}{2}t$, we have

$$\int_{-\pi}^{\pi} g(t) \sin(n + \tfrac{1}{2})t \, dt = \frac{1}{\pi} \int_{-\pi}^{\pi} [\pi g(t) \cos \tfrac{1}{2}t] \sin nt \, dt$$
$$+ \frac{1}{\pi} \int_{-\pi}^{\pi} [\pi g(t) \sin \tfrac{1}{2}t] \cos nt \, dt.$$

Since $g \in PC(2\pi)$, it follows that the functions defined for $t \in (-\pi, \pi]$, by

$$g_1(t) = \pi g(t) \cos \tfrac{1}{2}t, \qquad g_2(t) = \pi g(t) \sin \tfrac{1}{2}t,$$

have extensions to R which belong to $PC(2\pi)$. Therefore the integrals in the right side of the above formula give Fourier coefficients for g_1 and g_2; hence, by Bessel's Inequality, these integrals converge to 0 as $n \to \infty$. Q.E.D.

38.6 LEMMA. *If $f \in PC(2\pi)$, then the partial sum $S_n(f)$ of its Fourier series is given by*

$$(38.10) \qquad S_n(f)(x) = \frac{1}{\pi} \int_{-\pi}^{\pi} f(x + t)D_n(t) \, dt$$

where D_n is the nth Dirichlet kernel, defined by

$$D_n(t) = \tfrac{1}{2} + \sum_{k=1}^{n} \cos kt = \begin{cases} \dfrac{\sin (n+\tfrac{1}{2})t}{2 \sin \tfrac{1}{2}t}, & 0 < |t| \le \pi, \\ n+\tfrac{1}{2}, & t = 0. \end{cases}$$

PROOF. It follows from the formulas (38.2) and (38.6) that

$$S_n(f)(x) = \frac{1}{2\pi} \int_{-\pi}^{\pi} f(t)\, dt + \frac{1}{\pi} \int_{-\pi}^{\pi} \sum_{k=1}^{n} f(t)\{\cos kx \cos kt + \sin kx \sin kt\}\, dt$$

$$= \frac{1}{\pi} \int_{-\pi}^{\pi} f(t)\Big\{\tfrac{1}{2} + \sum_{k=1}^{n} \cos k(x-t)\Big\}\, dt.$$

If we let $t = x + s$ and use the fact that the cosine is an even function and that the integrand has period 2π, we have

$$S_n(f)(x) = \frac{1}{\pi} \int_{-\pi-x}^{\pi-x} f(x+s)\Big\{\tfrac{1}{2} + \sum_{k=1}^{n} \cos ks\Big\}\, ds$$

$$= \frac{1}{\pi} \int_{-\pi}^{\pi} f(x+s)\Big\{\tfrac{1}{2} + \sum_{k=1}^{n} \cos ks\Big\}\, ds.$$

We now apply formula (36.6) to obtain (38.10). Q.E.D.

Before we proceed, we recall (see Exercise 27.Q) that, by the **right-hand derivative** of a function $f : \mathbf{R} \to \mathbf{R}$ at a point $c \in \mathbf{R}$ where f has a right-hand limit $f(c+)$, we mean the limit

$$f'_+(c) = \lim_{\substack{t \to 0 \\ t > 0}} \frac{f(c+t) - f(c+)}{t}$$

whenever this limit exists. Similarly, the **left-hand derivative** of f at c is the limit

$$f'_-(c) = \lim_{\substack{t \to 0 \\ t < 0}} \frac{f(c+t) - f(c-)}{t}$$

38.7 POINTWISE CONVERGENCE THEOREM. *Suppose that $f \in PC(2\pi)$ and that f has right- and left-hand derivatives at c. Then, the Fourier series for f converges to $\tfrac{1}{2}\{f(c-) + f(c+)\}$ at the point c. In symbols,*

$$(38.11) \qquad \tfrac{1}{2}\{f(c-) + f(c+)\} = \tfrac{1}{2}a_0 + \sum_{n=1}^{\infty} (a_n \cos nc + b_n \sin nc).$$

PROOF. It follows from (36.6) that if $\sin \tfrac{1}{2}t \ne 0$, then

$$\tfrac{1}{2} + \sum_{k=1}^{n} \cos kt = \frac{\sin (n+\tfrac{1}{2})t}{2 \sin \tfrac{1}{2}t}.$$

Multiply by $(1/\pi)f(c+)$ and integrate with respect to t over $[0, \pi]$. Since $\int_0^\pi \cos kt\, dt = 0$ for $k \in \mathbf{N}$, we obtain

$$\tfrac{1}{2}f(c-) = \frac{1}{\pi}\int_0^\pi f(c+)\frac{\sin(n+\tfrac{1}{2})t}{2\sin\tfrac{1}{2}t}\, dt.$$

Similarly, if we multiply the above expression by $(1/\pi)f(c-)$ and integrate with respect to t over $[-\pi, 0]$, we obtain

$$\tfrac{1}{2}f(c+) = \frac{1}{\pi}\int_{-\pi}^0 f(c-)\frac{\sin(n+\tfrac{1}{2})t}{2\sin\tfrac{1}{2}t}\, dt.$$

If we subtract these expressions from formula (38.10), we get

$$(*) \quad S_n(f)(c) - \tfrac{1}{2}\{f(c-)+f(c+)\} = \frac{1}{\pi}\int_{-\pi}^0 \frac{f(c+t)-f(c-)}{2\sin\tfrac{1}{2}t}\sin(n+\tfrac{1}{2})t\, dt$$
$$+ \frac{1}{\pi}\int_0^\pi \frac{f(c+t)-f(c+)}{2\sin\tfrac{1}{2}t}\sin(n+\tfrac{1}{2})t\, dt.$$

Now since we have

$$\lim_{\substack{t\to 0 \\ t>0}} \frac{f(c+t)-f(c+)}{2\sin\tfrac{1}{2}t} = \lim_{\substack{t\to 0 \\ t>0}}\left\{\frac{f(c+t)-f(c+)}{t}\cdot\frac{t}{2\sin\tfrac{1}{2}t}\right\}$$

$$= f'_+(c)\cdot 1 = f'_+(c),$$

it follows that the function

$$F_+(t) = \frac{f(c+t)-f(c+)}{2\sin\tfrac{1}{2}t} \qquad \text{for} \qquad t\in(0, \pi],$$

$$= f'_+(c) \qquad\qquad \text{for} \qquad t=0,$$

$$= 0 \qquad\qquad\quad \text{for} \qquad t\in(-\pi, 0),$$

is piecewise continuous on $(-\pi, \pi]$. Hence the second integral in $(*)$ converges to 0 as $n \to \infty$.

Similarly, the first integral in $(*)$ converges to 0 as $n \to \infty$. Therefore the stated conclusion follows.   Q.E.D.

38.8 Examples. (a) The function f_1 in Example 38.2(a) is in $PC(2\pi)$, with $f(c-)=f(c)=f(c+)$ for $c\in[-\pi, \pi]$, $c\neq -\pi, 0, +\pi$, where we have $f(-\pi-)=+1$, $f(-\pi+)=-1$, $f(0-)=-1$, $f(0+)=1$, $f(\pi-)=1$, $f(\pi+)=-1$. Since one-sided derivatives exist everywhere (and equal 0), it follows from the Pointwise Convergence Theorem 38.7 that the Fourier series for f_1 converges to $f_1(c)$ provided $c\in[-\pi, \pi]$, $c\neq -\pi, 0, \pi$, and that at these three points the Fourier series for f_1 converges to 0.

(b) The function f_2 in Example 38.2(b) is continuous, has period 2π and has one-sided derivatives everywhere. Therefore the Fourier series for f_2 converges at every point to f_2 and, as we have seen, the convergence is uniform. We note that the (two-sided) derivative of f_2 exists in $[-\pi, \pi]$ except at the points 0, $\pm\pi$, and that f_2' agrees with the piecewise continuous function f_1 for $x \notin \{n\pi : n \in \mathbf{Z}\}$.

We remark that it follows from the Mean Value Theorem (see Exercise 27.N) that if $f' \in PC(2\pi)$, then the left- and right-hand derivatives of f exist at the points of discontinuity of f'. We now show that for a function f with period 2π and such that $f' \in PC(2\pi)$, the Fourier series for f is uniformly convergent to f.

38.9 UNIFORM CONVERGENCE THEOREM. *Let f be continuous, have period 2π, and suppose that $f' \in PC(2\pi)$. Then the Fourier series for f converges uniformly to f on $\mathbf{R}$.*

PROOF. Since f is continuous and the one-sided derivatives of f exists at every point, it follows from the Pointwise Convergence Theorem 38.7 that the Fourier series for f converges to f at every point. It remains to show that the convergence is uniform. In view of the inequality

$$\left| \sum_{k=1}^{\infty} (a_k \cos kx + b_k \sin kx) \right| \le \sum_{k=1}^{\infty} (|a_k| + |b_k|),$$

it is enough to establish the convergence of the latter series. In fact, if we apply Bessel's Inequality to f', we know that the series $\sum (|a_k'|^2 + |b_k'|^2)$ is convergent. But, as we have seen in Example 38.2(e), $a_k = -b_k'/k$ and $b_k = a_k'/k$. If we apply the Schwarz Inequality we have

$$\sum_{k=1}^{m} |a_k| = \sum_{k=1}^{m} \frac{1}{k} |b_k'| \le \left(\sum_{k=1}^{m} \frac{1}{k^2} \right)^{1/2} \left(\sum_{k=1}^{m} |b_k'|^2 \right)^{1/2}.$$

Since a similar inequality holds for $\sum |b_k|$, the desired assertion follows.

Q.E.D.

We now show that the partial sums of the Fourier series for *any* function f in $PC(2\pi)$ converges to f in the norm $\|\cdot\|_2$. While this does not guarantee that we can recover the value of f at any particular preassigned point, it can be interpreted as giving f in a certain "statistical" sense. For some applications this type of convergence is as useful as pointwise convergence, and there is the advantage that we do not have to impose differentiability restrictions.

38.10 NORM CONVERGENCE THEOREM. *If $f \in PC(2\pi)$ and if $(S_n(f))$ is the sequence of partial sums of the Fourier series for f, then*

$$\lim_{n \to \infty} \|f - S_n(f)\|_2 = 0.$$

PROOF. Let $f \in PC(2\pi)$ and let $\varepsilon > 0$ be given. It is an exercise to show that there exists a continuous function f_1 with period 2π such that

$\|f - f_1\|_2 < \varepsilon/7$. By Theorem 24.5 there is a continuous, piecewise linear function f_2 which can be chosen to have period 2π, such that $\|f_1 - f_2\|_\infty < \varepsilon/7$. It follows from the Uniform Convergence Theorem 38.9 that if n is sufficiently large, then $\|f_2 - S_n(f_2)\|_\infty < \varepsilon/7$. From formula (38.1) we have $\|g\|_2 \le \sqrt{2\pi}\|g\|_\infty \le 3\|g\|_\infty$ for any $g \in PC(2\pi)$; hence we deduce that

$$\|f - S_n(f_2)\|_2 < \|f - f_1\|_2 + \|f_1 - f_2\|_2 + \|f_2 - S_n(f_2)\|_2$$

$$\le \frac{\varepsilon}{7} + \frac{3\varepsilon}{7} + \frac{3\varepsilon}{7} = \varepsilon.$$

Now $S_n(f_2)$ is a trigonometric polynomial of degree n approximating f within ε (with respect to $\|\cdot\|_2$). Since it was established in Lemma 38.3 that the partial sum $S_n(f)$ is the trigonometric polynomial of degree n that gives the best such approximation, we infer that $\|f - S_n(f)\|_2 < \varepsilon$. Since $\varepsilon > 0$ is arbitrary, we conclude that $\lim \|f - S_n(f)\|_2 = 0$.   Q.E.D.

As a corollary of this result and Lemma 38.3 we obtain the following strengthening of Bessel's Inequality for $f \in PC(2\pi)$.

38.11 PARSEVAL'S EQUALITY.   *If $f \in PC(2\pi)$, then*

(38.12) $$\frac{1}{\pi}\|f\|_2^2 = \tfrac{1}{2}a_0^2 + \sum_{k=1}^\infty (a_k^2 + b_k^2),$$

where the a_k, b_k are the Fourier coefficients of f.

We shall end this section with a proof of the theorem of Fejér[†] on the Cesàro summability of the Fourier series of a continuous function. If $S_n(f)$, $n = 0, 1, 2, \ldots$, denote the partial sums of the Fourier series corresponding to f, let $\Gamma_n(f)$ denote the Cesàro means:

$$\Gamma_n(f) = \frac{1}{n}[S_0(f) + S_1(f) + \cdots + S_{n-1}(f)].$$

Now let D_n, $n = 0, 1, 2, \ldots$, be as in Lemma 38.6. If we make use of the elementary formula

$$2\sin(k - \tfrac{1}{2})t \sin \tfrac{1}{2}t = \cos(k-1)t - \cos kt, \quad k = 0, 1, 2, \ldots,$$

we can show that

(38.13) $\dfrac{1}{n}[D_0(t) + D_1(t) + \cdots + D_{n-1}(t)] = \begin{cases} \dfrac{1}{2n}\left(\dfrac{\sin\frac{1}{2}nt}{\sin\frac{1}{2}t}\right)^2, & 0 < |t| \le \pi, \\ \tfrac{1}{2}n, & t = 0, \end{cases}$

[†] LEOPOLD FEJÉR (1880–1959) studied and taught in Budapest. He made many interesting contributions to various areas of real and complex analysis.

and we let K_n be this function which is called the nth Fejér kernel. Clearly $K_n(t) \geq 0$ and since

$$\frac{1}{\pi} \int_{-\pi}^{\pi} D_k(t)\, dt = 1$$

for $k = 0, 1, 2, \ldots$, it follows that

$$(38.14) \qquad \frac{1}{\pi} \int_{-\pi}^{\pi} K_n(t)\, dt = 1.$$

Also, if $0 < \delta < \pi$, it follows from the fact that $\sin \theta \geq 2\theta/\pi$ for $0 \leq \theta \leq \pi/2$ that we have

$$(38.15) \qquad 0 \leq K_n(t) \leq \frac{1}{2n}\left(\frac{\pi}{2\delta}\right)^2 \qquad \text{for } \delta \leq |t| \leq \pi.$$

Finally, we note that it follows from Lemma 38.6 that we can express the Cesàro means by the formula

$$(38.16) \qquad \Gamma_n(f)(x) = \frac{1}{\pi} \int_{-\pi}^{\pi} f(x+t) K_n(t)\, dt.$$

We are now prepared to prove Fejér's Theorem.

38.12 FEJÉR'S THEOREM. *If f is continuous and has period 2π, then the Cesàro means of the Fourier series for f converge uniformly to f on $\mathbf{R}$.*

PROOF. It follows from (38.14) that

$$f(x) = \frac{1}{\pi} \int_{-\pi}^{\pi} f(x) K_n(t)\, dt.$$

Subtracting this from (38.16) we obtain

$$\Gamma_n(f)(x) - f(x) = \frac{1}{\pi} \int_{-\pi}^{\pi} \{f(x+t) - f(x)\} K_n(t)\, dt.$$

Since $K_n(t) \geq 0$ for all t, we have

$$|\Gamma_n(f)(x) - f(x)| \leq \frac{1}{\pi} \int_{-\pi}^{\pi} |f(x+t) - f(x)|\, K_n(t)\, dt.$$

Let $\varepsilon > 0$ be given; since f is uniformly continuous on $\mathbf{R}$, there exists a number δ with $0 < \delta < \pi$ such that if $|t| \leq \delta$, then

$$|f(x+t) - f(x)| < \varepsilon \qquad \text{for all } x \in [-\pi, \pi].$$

Hence we have

$$\frac{1}{\pi} \int_{-\delta}^{\delta} |f(x+t) - f(x)|\, K_n(t)\, dt \leq \frac{\varepsilon}{\pi} \int_{-\delta}^{\delta} K_n(t)\, dt \leq \frac{\varepsilon}{\pi} \int_{-\pi}^{\pi} K_n(t) = \frac{\varepsilon}{\pi}.$$

On the other hand, in view of (38.15) we have

$$\frac{1}{\pi} \int_\delta^\pi |f(x+t) - f(x)|\, K_n(t)\, dt \le \frac{\pi - \delta}{\pi} (2\,\|f\|_\infty) \left(\frac{1}{8n}\frac{\pi^2}{\delta^2}\right) \le \frac{1}{n}\left(\frac{\pi^2\,\|f\|_\infty}{4\delta^2}\right),$$

which can be made less than ε by taking n sufficiently large. Since a similar estimate holds for the integral over $[-\pi, -\delta]$, it follows that

$$\|\Gamma_n(f) - f\|_\infty < \left(2 + \frac{1}{\pi}\right)\varepsilon$$

for n sufficiently large.   Q.E.D.

Since the function $\Gamma_n(f)$ is readily seen to be a trigonometric polynomial (of degree $n-1$), we have another proof of the following theorem of Weierstrass.

38.13 WEIERSTRASS APPROXIMATION THEOREM. *If f is continuous and has period 2π, then it can be uniformly approximated by trigonometric polynomials.*

Exercises

38.A. Let g be a real-valued function defined on a cell J in R with end points $a < b$. We say that g is **piecewise continuous** on J if (i) g has a right-hand limit at a, (ii) g has a left-hand limit at b, and (iii) g is continuous at all interior points of J except, possibly, for a finite number of points at which g has right- and left-hand limits.

(a) Show that if g is piecewise continuous on $(-\pi, \pi]$, then there exists a unique function G in $PC(2\pi)$ such that $G(x) = g(x)$ for all $x \in (-\pi, \pi]$.

(b) The function g has a left-hand (respectively, right-hand, two-sided) derivative at $c \in (-\pi, \pi)$ if and only if G does.

(c) The function g has a right-hand derivative at $-\pi$ (respectively, a left-hand derivative at π) if and only if G does.

(d) The one-sided derivatives $g'_+(-\pi)$, $g'_-(\pi)$ exist and are equal if and only if G has a derivative at $\pm\pi$.

38.B. (a) If $f \in PC(2\pi)$ and the derivative $f'(x)$ exists for all $x \in R$, then f' has period 2π.

(b) If $f \in PC(2\pi)$ and $c \in R$, define $F : R \to R$ by $F(x) = \int_c^x f(t)\, dt$, so that F is continuous. Show that F has period 2π if and only if the **mean** of f is zero; that is,

$$\frac{1}{2} a_0 = \frac{1}{2\pi} \int_{-\pi}^\pi f(t)\, dt = 0.$$

38.C. (a) Let $f \in PC(2\pi)$ be odd. Then $f(\pm\pi) = 0$. If f is continuous at 0, then $f(0) = 0$.

(b) Let $g \in PC(2\pi)$ be even, then $g(0+) = g(0-)$. If the derivative $g'(x)$ exists for all $x \in R$, then (see Exercise 27.P) g' is odd, has period 2π, and $g'(0) = g'(\pm\pi) = 0$.

38.D. Let F and f belong to $PC(2\pi)$ and have Fourier coefficients A_n, B_n and a_n, b_n, respectively. If $\alpha, \beta \in \mathbf{R}$ and if $h = \alpha F + \beta f$, show that h belongs to $PC(2\pi)$ and has Fourier coefficients $\alpha A_n + \beta a_n$, $\alpha B_n + \beta b_n$. (Hence the Fourier coefficients of a function depend linearly on the function.)

38.E. (a) Let f_1 be the function in Example 38.2(a). Calculate the Fourier series for f_1, and show that this Fourier series does not converge uniformly on $[-\pi, \pi]$.

(b) Let f_2 be the function in Example 38.2(b). Calculate the Fourier series for f_2, and show that the term-by-term derivative of the Fourier series for f_2 coincides with the Fourier series for f_1.

(c) Using the fact that the Fourier series for f_2 converges to f_2, deduce that

$$\frac{\pi^2}{8} = \frac{1}{1^2} + \frac{1}{3^2} + \frac{1}{5^2} + \cdots.$$

(d) Let $f_3(x) = \frac{1}{2}\pi - f_2(x)$ so that $f_3(x) = \frac{1}{2}\pi - |x|$ for $x \in (-\pi, \pi)$. Use Exercise 38.D to show that the Fourier series for f_3 is given by

$$f_3(x) \sim \frac{4}{\pi}\left[\frac{\cos x}{1^2} + \frac{\cos 3x}{3^2} + \frac{\cos 5x}{5^2} + \cdots\right].$$

38.F. (a) Let $g_1 \in PC(2\pi)$ be such that $g_1(x) = x$ for $x \in (-\pi, \pi)$ and $g_1(\pi) = 0$. Show that g_1 is an odd function and that its Fourier series is given by

$$2\left[\frac{\sin x}{1} - \frac{\sin 2x}{2} + \frac{\sin 3x}{3} - \cdots\right].$$

Note that this Fourier series converges to 0 at $x = \pm\pi$. Use the Pointwise Convergence Theorem 38.7 to show that this Fourier series converges to $g_1(x)$ for every point $x \in [-\pi, \pi]$.

(b) Let $g_2 \in PC(2\pi)$ be such that $g_2(x) = x^2$ for $x \in (-\pi, \pi]$. Show that g_2 is an even function and that its Fourier series is given by

$$\frac{\pi^2}{3} - 4\left[\frac{\cos x}{1^2} - \frac{\cos 2x}{2^2} + \frac{\cos 3x}{3^2} - \cdots\right].$$

Show that this Fourier series converges uniformly to g_2 on $[-\pi, \pi]$, and that its term-by-term derivative is twice the Fourier series for g_1.

(c) Show that

$$\frac{\pi^2}{12} = \frac{1}{1^2} - \frac{1}{2^2} + \frac{1}{3^2} - \cdots.$$

(d) Let $h(x) = \frac{1}{3}\pi^2 - g_2(x)$ so that $h(x) = \frac{1}{3}\pi^2 - x^2$ for $x \in (-\pi, \pi]$. Then the Fourier series for h is given by

$$4\left[\frac{\cos x}{1^2} - \frac{\cos 2x}{2^2} + \frac{\cos 3x}{3^2} - \cdots\right].$$

38.G. (a) Let $k(x) = x^3$ for all $x \in \mathbf{R}$. Show that k is continuous and odd on $\mathbf{R}$.

However, the function k_1 in $PC(2\pi)$ which coincides with k on $(-\pi, \pi]$ is not continuous.

(b) Let $h(x) = x^3 - \pi^2 x$ so that h is continuous and odd on $\boldsymbol{R}$. Let h_1 be the function in $PC(2\pi)$ which coincides with h on $(-\pi, \pi]$. Show that h_1 is continuous on $\boldsymbol{R}$ and that $h_1'(x) = 3x^2 - \pi^2$ for $x \in (-\pi, \pi]$.

(c) Use Exercise 27.P, Example 38.2(e), and Exercise 38.C(d) to show that the Fourier series for h_1 is given by

$$-12\left[\frac{\sin x}{1^3} - \frac{\sin 2x}{2^3} + \frac{\sin 3x}{3^2} - \cdots\right].$$

38.H. Let $f:[0, \pi] \to \boldsymbol{R}$ be piecewise continuous and let $f_e \in PC(2\pi)$ be defined by

$$f_e(x) = f(x) \qquad \text{for } x \in [0, \pi],$$
$$= f(-x) \qquad \text{for } x \in [-\pi, 0).$$

(a) Show that f_e is an even function; it is called the **even extension** of f with period 2π.

(b) The Fourier series of f_e is called the (Fourier) **cosine series** of f. Show that it is given by

$$\tfrac{1}{2}a_0 + \sum_{n=1}^{\infty} a_n \cos nx.$$

where

$$a_n = \frac{2}{\pi} \int_0^{\pi} f(t) \cos nt \, dt, \qquad n = 0, 1, 2, \ldots,$$

(c) Show that if $c \in (0, \pi)$ and f has left- and right-hand derivatives at c, then the cosine series for f converges to $\frac{1}{2}[f(c-) + f(c+)]$. Also if f has a right-hand derivative at 0, then the cosine series for f converges to $f(0+)$. If f has a left-hand derivative at π, then the cosine series for f converges to $f(\pi-)$.

38.I. For each of the following functions defined on $[0, \pi]$, calculate the cosine series and determine the limit of this series at each point.

(a) $f(x) = x$;   (b) $f(x) = \sin x$;

(c) $f(x) = 1$ for $0 \le x \le \frac{1}{2}\pi$,   (d) $f(x) = \frac{1}{2}\pi - x$ for $0 \le x \le \frac{1}{2}\pi$,
$\quad\;\; = 0$ for $\frac{1}{2}\pi < x \le \pi$. $\qquad\;\; = 0$ for $\frac{1}{2}\pi < x \le \pi$.

(e) $f(x) = x(\pi - x)$.

38.J. Let $f:[0, \pi] \to \boldsymbol{R}$ be piecewise continuous and let $f_0 \in PC(2\pi)$ be defined by

$$f_0 = f(x) \qquad \text{for } x \in (0, \pi],$$
$$= 0 \qquad \text{for } x = 0,$$
$$= -f(-x) \qquad \text{for } x \in (-\pi, 0).$$

(a) Show that f_o is an odd function; it is called the **odd extension** of f with period 2π.

(b) The Fourier series of f_o is called the (Fourier) **sine series** of f. Show that it is given by

$$\sum_{n=1}^{\infty} b_n \sin nx,$$

where

$$b_2 = \frac{2}{\pi} \int_0^{\pi} f(t) \sin nt \, dt, \qquad n = 1, 2, \ldots.$$

(c) Show that if $c \in (0, \pi)$ and if f has left- and right-hand derivatives at c, then the sine series for f converges to $\frac{1}{2}[f(c-) + f(c+)]$. In any case, the sine series for f converges to 0 at $x = 0, \pi$.

38.K. For each of the following functions defined on $[0, \pi]$, calculate the sine series and determine the limit of this series at each point.

(a) $f(x) = 1$;

(b) $f(x) = \cos x$;

(c) $f(x) = 1 \qquad$ for $0 \le x \le \frac{1}{2}\pi$,

$= 0 \qquad$ for $\frac{1}{2}\pi < x \le \pi$;

(d) $f(x) = \pi - x$;

(e) $f(x) = x(\pi - x)$.

38.L. Let $f_n \in PC(2\pi)$ be the function such that $f_n(x) = n^{1/4}$ for $0 \le x \le 1/n$ and $= 0$ for other $x \in (-\pi, \pi]$. Show that $\|f_n\|_2 = 1/n^{1/4}$ so that the sequence (f_n) converges to the zero function in the norm $\|\cdot\|_2$ but, since it is unbounded, the convergence is not uniform.

38.M. If $f \in PC(2\pi)$ and if $\varepsilon > 0$, show that there exists a continuous function f_1 with period 2π such that $\|f - f_1\|_2 < \varepsilon$.

38.N. Use Parseval's Equality 38.11 to establish the following formulas.

(a) $\dfrac{\pi^2}{6} = \displaystyle\sum_{n=1}^{\infty} \dfrac{1}{n^2}$,

(b) $\dfrac{\pi^2}{8} = \displaystyle\sum_{n=1}^{\infty} \dfrac{1}{(2n-1)^2}$,

(c) $\dfrac{\pi^4}{90} = \displaystyle\sum_{n=1}^{\infty} \dfrac{1}{n^4}$,

(d) $\dfrac{\pi^6}{945} = \displaystyle\sum_{n=1}^{\infty} \dfrac{1}{n^6}$.

38.O. If f and F belong to $PC(2\pi)$ and have Fourier coefficients a_n, b_n and A_n, B_n, respectively, show that

$$\frac{1}{\pi} \int_{-\pi}^{\pi} f(t)F(t) \, dt = \frac{1}{2}a_0 A_0 + \sum_{n=1}^{\infty} (a_n A_n + b_n B_n).$$

(Hint: apply Parseval's Equality to $f + F$.)

38.P. Use Dirichlet's Test 36.2 and Example 36.8 to show that the trigonometrical series

$$\sum_{n=1}^{\infty} \frac{\sin nx}{n^{1/2}}$$

converges for all x. Show, however, that this series cannot be the Fourier series of any function in $PC(2\pi)$.

38.Q. Let $L > 0$ and let $PC(2L)$ be the vector space of all functions $f : \mathbf{R} \to \mathbf{R}$ which have period $2L$ and are piecewise continuous.

(a) If we define $f \cdot g = \int_{-L}^{L} f(t)g(t)\, dt$ for f, $g \in PC(2L)$, show that the map $(f, g) \mapsto f \cdot g$ is an inner product (in the sense of Definition 8.3) on $PC(2L)$. Moreover the norm induced by this inner product (see 8.7) is

$$\|f\|_2 = \left[\int_{-L}^{L} |f(t)|^2 \, dt \right]^{1/2}.$$

(b) We let C_0, C_n, S_n, $n \in \mathbf{N}$, be the functions in $PC(2L)$ given by

$$C_0(x) = \frac{1}{\sqrt{2L}}, \qquad C_n(x) = \frac{1}{\sqrt{L}} \cos \frac{n\pi x}{L}, \qquad S_n(x) = \frac{1}{\sqrt{L}} \sin \frac{n\pi x}{L}.$$

Show that this set of functions is **orthonormal** in the sense that

$$C_n \cdot S_m = 0, \qquad C_n \cdot C_m = \delta_{nm}, \qquad S_n \cdot S_m = \delta_{nm},$$

where $\delta_{nm} = 1$ if $n = m$ and $\delta_{nm} = 0$ if $n \neq m$. (Hint: if $L = \pi$, these are the relations given before 38.3.)

(c) If $f \in PC(2L)$, we define the **Fourier series** of f on $[-L, L]$ to be the series

$$\frac{1}{2} a_0 + \sum_{n=1}^{\infty} \left(a_n \cos \frac{n\pi x}{L} + b_n \sin \frac{n\pi x}{L} \right),$$

where we have

$$a_0 = \frac{1}{L} \int_{-L}^{L} f(t)\, dt, \qquad a_n = \frac{1}{L} \int_{-L}^{L} f(t) \cos \frac{n\pi t}{L} \, dt,$$

$$b_n = \frac{1}{L} \int_{-L}^{L} f(t) \sin \frac{n\pi t}{L} \, dt,$$

for $n = 1, 2, \ldots$.

(d) Reformulate the Convergence Theorems 38.7, 38.9, and 38.10 for Fourier series of functions in $PC(2L)$. (Hint: make a change of variable.)

(e) If $f \in PC(2L)$, then Parseval's Equality becomes

$$\frac{1}{L} \|f\|_2^2 = \frac{1}{2} a_0^2 + \sum_{n=1}^{\infty} (a_n^2 + b_n^2),$$

where the norm of f is as in part (a) and the Fourier coefficients are as in part (c).

38.R. For each of the following functions on the specified interval, calculate the Fourier series on this interval and determine the limit of this series at each point.

(a) $f(x) = x$ on $(-2, 2]$;

(b) $f(x) = 0$ for $-4 < x < 0$,

       $= x$ for $0 \leq x \leq 4$;

(c) $f(x) = 0$ for $-3 < x < 0$,

       $= 1$ for $0 \leq x \leq 1$,

       $= 0$ for $1 < x \leq 3$.

38.S. Let f be continuous and have period 2π. Show that if the Fourier series for f converges at $c \in [-\pi, \pi]$ to some number, then it converges to $f(c)$.

38.T. Let f belong to $PC(2\pi)$ and suppose that $c \in [-\pi, \pi]$. If $\Gamma_n(f)$ denotes the nth Fejér mean, defined in (38.16), show that

$$\lim \Gamma_n(f)(c) = \frac{1}{2}[f(c-) + f(c+)].$$

38.U. Suppose that f and f' are continuous with period 2π and that $f'' \in PC(2\pi)$. (a) Show that the Fourier coefficients a_n, b_n of f are such that the series

$$\sum_{n=1}^{\infty} n^2(|a_n| + |b_n|)$$

is convergent. Hence, there exists a constant $M > 0$ such that $|a_n| \leq M/n^2$ and $|b_n| \leq M/n^2$ for all $n \in N$.

(b) Show that the Fourier series for f' is the term-by-term derivative of the Fourier series for f.

38.V. (a) If $k \in PC(2\pi)$ and if x_0, $x \in [-\pi, \pi]$, use the Schwarz Inequality to show that

$$\left| \int_{x_0}^{x} k(t)\, dt \right| \leq \|k\|_2 |x - x_0|^{1/2} \leq \|k\|_2 \sqrt{2\pi}.$$

(b) Use part (a) and the Norm Convergence Theorem 38.10 to show that if $f \in PC(2\pi)$ and $x_0 \in [-\pi, \pi]$, then the Fourier series for f can be integrated term-by-term:

$$\int_{x_0}^{x} f(t)\, dt = \frac{1}{2} a_0(x - x_0) + \sum_{n=1}^{\infty} \int_{x_0}^{x} (a_n \cos nt + b_n \sin nt)\, dt,$$

and the resulting series is uniformly convergent for $x \in [-\pi, \pi]$.

38.W. (a) Suppose that $\alpha > 0$ is not an integer. Show that

$$\cos \alpha x = \frac{2\alpha \sin \alpha \pi}{\pi} \left[\frac{1}{2\alpha^2} - \frac{\cos x}{\alpha^2 - 1^2} + \frac{\cos 2x}{\alpha^2 - 2^2} - \frac{\cos 3x}{\alpha^2 - 3^2} + \cdots \right]$$

for all $x \in [-\pi, \pi]$.

(b) Use part (a) to show that if $x \notin Z$, then

$$\cot \pi x = \frac{1}{\pi x} + \frac{2x}{\pi} \sum_{n=1}^{\infty} \frac{1}{x^2 - n^2},$$

$$\csc \pi x = \frac{1}{\pi x} + \frac{2x}{\pi} \sum_{n=1}^{\infty} \frac{(-1)^n}{x^2 - n^2}.$$

(c) Differentiate the first series in (b) term-by-term (justify this) to show that if $x \notin Z$, then

$$\frac{\pi^2}{(\sin \pi x)^2} = \lim_m \sum_{n=-m}^{m} \frac{1}{(x - n)^2}.$$

(d) Integrate the first series in (b) term-by-term (justify this) to show that if $x \notin Z$, then

$$\frac{\sin \pi x}{\pi x} = \lim_m \left[\left(1 - \frac{x^2}{1^2}\right)\left(1 - \frac{x^2}{2^2}\right) \cdots \left(1 - \frac{x^2}{m^2}\right) \right].$$

VII
DIFFERENTIATION
IN R^p

In this chapter we shall present the theory of differentiable functions in R^p where $p > 1$. Although the theory is parallel to that presented in Sections 27 and 28, there are several complications and new features that arise. Some of these complications are due purely to the inevitable notational complexity, but most arise because it is possible to approach a point $c \in R^p$ from "many directions," so some new phenomena can occur.

In Section 27 we defined the derivative of a function $f : R \to R$ at a point $c \in R$ in the traditional manner; namely, as the number $L \in R$ such that

$$L = \lim_{x \to c} \frac{f(x) - f(c)}{x - c},$$

when this limit exists. Equivalently, we could have defined this derivative to be the number L such that

$$\lim_{x \to c} \frac{|f(x) - f(c) - L(x - c)|}{|x - c|} = 0.$$

This limiting relation can be regarded as making precise the sense in which we approximate the values $f(x)$, for x sufficiently near c, by the values of the affine† map

$$x \mapsto f(c) + L(x - c),$$

whose graph yields the line tangent to the graph of f at the point $(c, f(c))$.

It is this approach to the derivative that we shall use for functions on R^p to R^q. Thus the derivative of a function f defined on a neighborhood of a

---

† In elementary courses, such a map is called "linear." However, to be consistent with the more restricted use of the term "linear" introduced in Section 21, we shall use the term "affine" to refer to a function obtained by adding a constant to a linear function.

point $c \in \mathbf{R}^p$ with values in $\mathbf{R}^q$ will be a *linear map* $L : \mathbf{R}^p \to \mathbf{R}^q$ such that

$$\lim_{x \to c} \frac{\|f(x) - f(c) - L(x - c)\|}{\|x - c\|} = 0.$$

Hence we are approximating $f(x)$, for x sufficiently near c, by the affine map

$$x \mapsto f(c) + L(x - c)$$

of $\mathbf{R}^p$ into $\mathbf{R}^q$. [The reader should note that if $p = 1$, then the notation $L(x - c)$ denotes the product of the real numbers L and $x - c$; however, if $p > 1$, then $L(x - c)$ denotes the value of the linear map L at the vector $x - c$.]

Section 39 presents the definition and relates the derivative with the various "partial" derivatives. In Section 40 we obtain the Chain Rule and the Mean Value Theorem which are of central importance. Section 41 gives a penetrating analysis of the mapping properties of differentiable functions, leading to the important Inversion and Implicit Function Theorems, and culminating with the Parametrization and Rank Theorems. The final section deals with extremum properties of real valued functions on $\mathbf{R}^p$.

Section 39 The Derivative in $\mathbf{R}^p$

Section 27 considered the derivative of a function with domain and range in $\mathbf{R}$. In this section we shall consider a function defined on a subset of $\mathbf{R}^p$ and with values in $\mathbf{R}^q$ from a similar point of view.

If the reader will review Definition 27.1, he will note that it applies equally well to a function defined on an interval J in $\mathbf{R}$ and with values in the Cartesian space $\mathbf{R}^q$. Of course, in this case L is a vector in $\mathbf{R}^q$. The only change required for this extension is to replace the absolute value in equation (27.1) by the norm in the space $\mathbf{R}^q$. Except for this, Definition 27.1 applies *verbatim* to this more general situation. That this situation is worthy of study should be clear when it is realized that a function f on J to $\mathbf{R}^q$ can be regarded as being a *curve* in the space $\mathbf{R}^q$ and that the derivative (when it exists) of this function at the point $x = c$ yields a *tangent vector* to the curve at the point $f(c)$. Alternatively, if we think of x as denoting time, then the function f is the trajectory of a point in $\mathbf{R}^q$ and the derivative $f'(c)$ denotes the *velocity vector* of the point at time $x = c$.

A fuller investigation of these lines of thought would take us farther into differential geometry and dynamics than is desirable at present. Our aims are more modest: we wish to organize the analytical machinery that would make a satisfactory investigation possible and to remove the restriction that the domain is in a one-dimensional space and allow the domain to belong to the Cartesian space $\mathbf{R}^p$. We shall now proceed to do this.

An analysis of Definition 27.1 shows that the only place where it is necessary for the domain to consist of a subset of $\mathbf{R}$ is in equation (27.1), where a quotient appears. Since we have no meaning for the quotient of a vector in $\mathbf{R}^q$ by a vector in $\mathbf{R}^p$, we cannot interpret equation (27.1) as it stands. We are led, therefore, to find reformulations of this equation. One possibility which is of considerable interest is to take one-dimensional "slices" passing through the point c in the domain. For simplicity it will be supposed that c is an interior point of the domain D of the function; then for any u in $\mathbf{R}^p$, the point $c + tu$ belongs to D for sufficiently small real numbers t.

39.1 DEFINITION. Let f be defined on a subset A of $\mathbf{R}^p$ and have values in $\mathbf{R}^q$, let c be an interior point of A, and let u be any point in $\mathbf{R}^p$. A vector $L_u \in \mathbf{R}^q$ is said to be the **partial derivative of f at c with respect to** u if for each number $\varepsilon > 0$ there is a $\delta(\varepsilon) > 0$ such that for all $t \in \mathbf{R}$ satisfying $0 < |t| < \delta(\varepsilon)$, we have

$$(39.1) \qquad \left\| \frac{1}{t} \{ f(c + tu) - f(c) \} - L_u \right\| < \varepsilon.$$

It is readily seen that the partial derivative L_u defined in (39.1) is uniquely determined when it exists. Alternatively, we can define L_u as the limit

$$\lim_{t \to 0} \frac{1}{t} \{ f(c + tu) - f(c) \},$$

or as the derivative at $t = 0$ of the function F defined by $F(t) = f(c + tu)$ for $|t|$ sufficiently small, and having values in $\mathbf{R}^q$.

We shall write $D_u f(c)$ or $f_u(c)$ for the partial derivative L_u of f at c with respect to u. The first notation is greatly to be preferred when, as is often the case, the symbol denoting the function has a subscript. We denote the function $c \mapsto D_u f(c) = f_u(c)$ by $D_u f$ or f_u; it is defined for those interior points c of A for which the required limit exists, and has values in $\mathbf{R}^q$.

It is clear that if f is real valued (so that $q = 1$) and if u is the vector $e_1 = (1, 0, \ldots, 0)$ in $\mathbf{R}^p$, then the partial derivative of f with respect to e_1 coincides with what is usually called the **partial derivative of f with respect to its first variable,** which is often denoted by

$$D_1 f, \qquad f_{x_1}, \qquad \text{or} \qquad \frac{\partial f}{\partial x_1}.$$

In the same way, taking $e_2 = (0, 1, \ldots, 0), \ldots, e_p = (0, 0, \ldots, 1)$, we obtain the **partial derivatives of f with respect to the other variables:**

$$D_2 f = f_{x_2} = \frac{\partial f}{\partial x_2}, \ldots, D_p f = f_{x_p} = \frac{\partial f}{\partial x_p}.$$

In case the symbol denoting the function has a subscript, we shall sometimes insert a comma to indicate a partial derivative; thus, $D_j f_2 = f_{2,j}$.

It should be observed that the partial derivative of a function at a point with respect to one vector may exist, yet the partial derivative with respect to another vector need not exist (see Exercise 39.A). It is also plain that, under appropriate hypotheses, there are algebraic relations between partial derivatives of sums and products of functions, etc. We shall not bother to obtain these relations since they are either special cases of what we shall do below, or can be proved similarly.

A word about terminology is in order. If u is a unit vector in $\mathbf{R}^p$, then the partial derivative $D_u f(c) = f_u(c)$ is often called the **directional derivative of f at c in the direction of u.**

The Derivative

The main drawback of the partial derivative of a function f at a point c with respect to a vector u is that it only gives a picture of the behavior of f near c on the one-dimensional set $\{c + tu : t \in \mathbf{R}\}$. In order to obtain more complete information about f in a neighborhood of $c \in \mathbf{R}^p$, we shall introduce the notion of the derivative of f at c, which is a *linear map* from $\mathbf{R}^p$ to $\mathbf{R}^q$.

39.2 DEFINITION. Let f have domain A in $\mathbf{R}^p$ and range in $\mathbf{R}^q$, and let c be an interior point of A. We say that f is **differentiable** *at* c if there exists a linear function $L : \mathbf{R}^p \to \mathbf{R}^q$ such that for every $\varepsilon > 0$ there exists $\delta(\varepsilon) > 0$ such that if $x \in \mathbf{R}^p$ is any vector satisfying $\|x - c\| \le \delta(\varepsilon)$, then $x \in A$ and

$$(39.2) \qquad \|f(x) - f(c) - L(x - c)\| \le \varepsilon \|x - c\|.$$

Alternatively, (39.2) can be rephrased by requiring that for any $\varepsilon > 0$ there exists $\delta(\varepsilon) > 0$ such that if $u \in \mathbf{R}^p$ and $\|u\| \le \delta(\varepsilon)$, then

$$(39.3) \qquad \|f(c + u) - f(c) - L(u)\| \le \varepsilon \|u\|,$$

which, in turn, can be expressed more compactly by writing

$$(39.4) \qquad \lim_{\|u\| \to 0} \frac{\|f(c + u) - f(c) - L(u)\|}{\|u\|} = 0.$$

We will see below that such a linear function L is uniquely determined when it exists. It is called† the **derivative of f at c** and we shall often denote it by $Df(c)$ instead of L. We shall often write $Df(c)(u)$ for $L(u)$, and $Df(c)(x - c)$ for $L(x - c)$.

---

† The reader is warned that L is sometimes called the **Fréchet derivative**, or the **differential**, of f at c, and is sometimes denoted by $df(c)$ or $f'(c)$, etc.

From an analytic point of view, the existence of the derivative of f at c reflects the possibility of approximating the mapping $x \mapsto f(x)$ by the mapping $x \mapsto f(c) + L(x - c)$. Inequality (39.2) gives a measure of the closeness of this approximation when x is near to c. Because of the linearity of L, we have

$$f(c) + L(x - c) = (f(c) - L(c)) + L(x).$$

Hence we are approximating $x \mapsto f(x)$ by a function of the form $x \mapsto y_0 + L(x)$, where y_0 is fixed. Such functions are called **affine mappings** of $\mathbf{R}^p$ into $\mathbf{R}^q$; they are merely translations of linear mappings and so have a very simple character.

From a geometric point of view, the existence of the derivative of f at c reflects the existence of a **tangent plane** to the surface $\{(x, f(x)) : x \in A\}$ in $\mathbf{R}^p \times \mathbf{R}^q$ at the point $(c, f(c))$; namely, the plane given by the graph

$$(39.5) \qquad \{(x, f(c) + L(x - c)) : x \in \mathbf{R}^p\}.$$

We shall now establish the uniqueness of the derivative.

39.3 LEMMA. *The function f has at most one derivative at a point.*

PROOF. Suppose L_1, L_2 are linear functions from $\mathbf{R}^p$ to $\mathbf{R}^q$ and satisfy (39.3) for $\|u\| \le \delta(\varepsilon)$. Then we have

$$0 \le \|L_1(u) - L_2(u)\|$$
$$\le \|f(c + u) - f(c) - L_1(u)\| + \|f(c + u) - f(c) - L_2(u)\|$$
$$\le 2\varepsilon \|u\|.$$

Therefore we have $0 \le \|L_1(u) - L_2(u)\| \le 2\varepsilon \|u\|$ for all $u \in \mathbf{R}^p$ with $\|u\| \le \delta(\varepsilon)$. If $L_1 \ne L_2$, there exists $z \in \mathbf{R}^p$ with $L_1(z) \ne L_2(z)$, whence $z \ne 0$. Now let $z_0 = (\delta(\varepsilon)/\|z\|)z$ so that we have $\|z_0\| = \delta(\varepsilon)$ and hence $0 \le \|L_1(z_0) - L_2(z_0)\| \le 2\varepsilon \|z_0\|$. Hence $\|L_1(z) - L_2(z)\| \le 2\varepsilon \|z\|$ for all $\varepsilon > 0$, so $L_1(z) = L_2(z)$, a contradiction. Therefore $L_1 = L_2$.   Q.E.D.

39.4 EXAMPLES. (a) Let $A \subseteq \mathbf{R}^p$, let $y_0 \in \mathbf{R}^q$, and let $f_0 : A \to \mathbf{R}^q$ be the "constant function" defined by $f_0(x) = y_0$ for $x \in A$. If c is an interior point of A and $x \in A$, then $f_0(x) - f_0(c) = 0$. It follows that f_0 is differentiable at c and that the derivative $Df_0(c) = 0$, the "zero linear function" that maps every element of $\mathbf{R}^p$ into the zero element of $\mathbf{R}^q$. *Hence the derivative at any point of a constant function is the zero linear function.*

(b) Let $A = \mathbf{R}^p$ and let $f_1 : A \to \mathbf{R}^q$ be a linear function. If $c \in A$ and $x \in A$, then $f_1(x) - f_1(c) - f_1(x - c) = 0$. It follows from this that f_1 is differentiable at c and that $Df_1(c) = f_1$. *Hence the derivative at any point of a linear function is the linear function itself.*

39.5 LEMMA. *If $f: A \to \mathbf{R}^q$ is differentiable at $c \in A$, then there exist strictly positive numbers δ, K such that if $\|x - c\| \le \delta$, then*

$$(39.6) \qquad \|f(x) - f(c)\| \le K \|x - c\|.$$

In particular it follows that f is continuous at $x = c$.

PROOF. By Definition 39.2 it follows that there exists $\delta > 0$ such that if $0 < \|x - c\| \le \delta$, then (39.2) holds with $\varepsilon = 1$. If we use the Triangle Inequality, we have

$$\|f(x) - f(c)\| \le \|L(x - c)\| + \|x - c\|$$

for $0 < \|x - c\| \le \delta$. By Theorem 21.3 there exists $B > 0$ such that $\|L(x - c)\| \le B \|x - c\|$ for all $x \in \mathbf{R}^p$. Therefore, if $0 < \|x - c\| \le \delta$ we obtain

$$\|f(x) - f(c)\| \le (B + 1) \|x - c\|,$$

and this inequality remains true also for $x = c$.   Q.E.D.

We now show that the existence of the derivative at a point implies the existence of all of the partial derivatives at that point.

39.6 THEOREM. *If $A \subseteq \mathbf{R}^p$, if $f: A \to \mathbf{R}^q$ is differentiable at a point $c \in A$, and if u is any element of $\mathbf{R}^p$, then the partial derivative $D_u f(c)$ of f at c with respect to u exists. Moreover,*

$$(39.7) \qquad D_u f(c) = Df(c)(u).$$

PROOF. Since f is differentiable at c, given $\varepsilon > 0$ there exists $\delta(\varepsilon) > 0$ such that

$$\|f(c + tu) - f(c) - Df(c)(tu)\| \le \varepsilon \|tu\|$$

provided $\|tu\| \le \delta(\varepsilon)$. If $u = 0$, then the partial derivative with respect to 0 is readily seen to be $0 = Df(c)(0)$; hence we suppose $u \ne 0$. Thus if $0 < |t| \le \delta(\varepsilon)/\|u\|$, we have

$$\left\| \frac{f(c + tu) - f(c)}{t} - Df(c)(u) \right\| \le \varepsilon \|u\|.$$

This shows that $Df(c)(u)$ is the partial derivative of f at c with respect to u, as claimed.   Q.E.D.

39.7 COROLLARY. *Let $A \subseteq \mathbf{R}^p$, let $f: A \to \mathbf{R}$ and let c be an interior point of A. If the derivative $Df(c)$ exists, then each of the partial derivatives $D_1 f(c), \ldots, D_p f(c)$ exist in $\mathbf{R}$ and if $u = (u_1, \ldots, u_p) \in \mathbf{R}^p$, then*

$$(39.8) \qquad Df(c)(u) = u_1 D_1 f(c) + \cdots + u_p D_p f(c).$$

PROOF.   The theorem implies that for each of the vectors $e_1, \ldots, e_p$ the partial derivatives $D_1 f(c), \ldots, D_p f(c)$ exist and equal $Df(c)(e_1), \ldots, Df(c)(e_p)$. However, since $Df(c)$ is linear and $u = u_1 e_1 + \cdots + u_p e_p$, we deduce that

$$Df(c)(u) = \sum_{j=1}^{p} u_j Df(c)(e_j) = \sum_{j=1}^{p} u_j D_j f(c).$$ Q.E.D.

REMARKS.   (a) The converse of Corollary 39.7 is not always true, for the partial derivatives of f may exist without the derivative existing. For example, let $f : \mathbf{R}^2 \to \mathbf{R}$ be defined by

$$f(x, y) = 0 \qquad \text{for } (x, y) = (0, 0),$$

$$= \frac{xy^2}{x^2 + y^2} \qquad \text{for } (x, y) \neq (0, 0).$$

It is an exercise to show that the partial derivative of f with respect to the vector (a, b) at $(0, 0)$ is given by

$$(39.9) \qquad\qquad D_{(a,b)} f(0, 0) = \frac{ab^2}{a^2 + b^2}, \qquad (a, b) \neq (0, 0).$$

In particular, $D_1 f(0, 0) = 0$ and $D_2 f(0, 0) = 0$. If the derivative Df exists at $(0, 0)$, Corollary 39.7 would imply that

$$D_{(a,b)} f(0, 0) = Df(0, 0)(a, b) = a \cdot 0 + b \cdot 0 = 0,$$

contrary to (39.9).

(b) We shall see below that if $A \subseteq \mathbf{R}^p$ and if the partial derivatives of $f : A \to \mathbf{R}^q$ are continuous at c, then $Df(c)$ exists.

39.8 EXAMPLES.   (a) Let $A \subseteq \mathbf{R}$ and let $f : A \to \mathbf{R}$. Then f is differentiable at an interior point c of A in the sense of Definition 39.2 if and only if the ordinary derivative

$$\lim_{\substack{t \to 0 \\ t \neq 0}} \frac{f(c + t) - f(c)}{t} = f'(c)$$

exists. In this case the derivative $Df(c)$ is the linear function of $\mathbf{R}$ into $\mathbf{R}$ defined by

$$u \mapsto f'(c)u.$$

Thus $Df(c)$ maps $u \in \mathbf{R}$ into the product of $f'(c)$ and u. (In matrix terminology, the derivative $Df(c)$ is the linear mapping represented by the 1×1 matrix whose only element is $f'(c)$.)

Traditionally, instead of writing u for the real number on which the linear function of $Df(c)$ acts, one writes the somewhat peculiar symbol dx (here the "d" plays the role of a prefix and has no other significance). When this is done and the

Leibniz† notation for the derivative is used, the formula $Df(c)(u) = f'(c)u$ becomes

$$Df(c)(dx) = \frac{df}{dx}(c)\, dx.$$

(b) Let $A \subseteq \mathbf{R}$ and let $f: A \to \mathbf{R}^q$ $(q > 1)$. Hence f can be represented by the "coordinate functions"

$$f(x) = (f_1(x), \ldots, f_q(x)), \qquad x \in A.$$

As an exercise the reader should prove that f is differentiable at an interior point c of A if and only if each of the real-valued functions $f_1, \ldots, f_q$ has a derivative at c. In this case, the derivative $Df(c)$ is the linear function of $\mathbf{R}$ into $\mathbf{R}^q$ given by

$$u \mapsto u(f_1'(c), \ldots, f_q'(c)), \qquad u \in \mathbf{R}.$$

Hence $Df(c)$ maps a real number u into the product of u and a fixed vector $f'(c) = (f_1'(c), \ldots, f_q'(c))$. When f is thought of as being a "curve," this vector is called the "tangent vector" to f at the point $f(c)$.

(c) Let $A \subseteq \mathbf{R}^p$ $(p > 1)$ and let $f: A \to \mathbf{R}$. Then it follows from Corollary 39.7 that if the derivative $Df(c)$ exists at a point c interior to A, then each of the partial derivatives $D_1f(c), \ldots, D_pf(c)$ must exist and that $Df(c)$ is the linear mapping of $u = (u_1, \ldots, u_p) \in \mathbf{R}^p$ into $\mathbf{R}$ given by

$$Df(c)(u) = u_1 D_1 f(c) + \cdots + u_p D_p f(c).$$

Although the mere existence of these partial derivatives does *not* imply the existence of the derivative, we shall see below that their continuity at c does guarantee its existence.

Sometimes, instead of $u = (u_1, \ldots, u_p)$ we write $dx = (dx_1, \ldots, dx_p)$ for the point in $\mathbf{R}^p$ on which the derivative is to act. When this and Leibniz' notation is employed for the partial derivatives, the above formula becomes

$$Df(c)\,(dx) = \frac{\partial f}{\partial x_1}(c)\, dx_1 + \cdots + \frac{\partial f}{\partial x_p}(c)\, dx_p.$$

(d) Now let $A \subseteq \mathbf{R}^p$ and $f: A \to \mathbf{R}^q$ where both $p > 1$, $q > 1$. In this case we can represent $y = f(x)$ by a system

$$y_1 = f_1(x_1, \ldots, x_p)$$
$$\cdots\cdots\cdots\cdots$$
$$y_q = f_q(x_1, \ldots, x_p),$$

†GOTTFRIED WILHELM LEIBNIZ (1646–1716) is, with ISAAC NEWTON (1642–1727), one of the coinventors of calculus. Leibniz spent most of his life serving the dukes of Hanover and was a universal genius. He contributed greatly to mathematics, law, philosophy, theology, linguistics, and history.

of q functions of p arguments. If f is differentiable at a point $c = (c_1, \ldots, c_p)$ in A, then it is an exercise to show that each of the partial derivatives $D_j f_i(c)(= f_{i,j}(c))$ must exist at c. (Again this latter condition is not sufficient, in general, for the differentiability of f at c.) When $Df(c)$ exists, it is the linear function which maps the point $u = (u_1, \ldots, u_p)$ of $\mathbf{R}^p$ into the point $w = (w_1, \ldots, w_q)$ of $\mathbf{R}^q$ given by

$$
\begin{aligned}
w_1 &= D_1 f_1(c) u_1 + D_2 f_1(c) u_2 + \cdots + D_p f_1(c) u_p, \\
&\,\cdot\;\cdot\;\cdot\;\cdot\;\cdot\;\cdot\;\cdot\;\cdot\;\cdot\;\cdot\;\cdot\;\cdot\;\cdot\;\cdot\;\cdot\;\cdot \\
w_q &= D_1 f_q(c) u_1 + D_2 f_q(c) u_2 + \cdots + D_p f_q(c) u_p.
\end{aligned}
$$
(39.10)

The derivative $Df(c)$ is the linear mapping of $\mathbf{R}^p$ into $\mathbf{R}^q$ determined by the $q \times p$ matrix whose elements are

(39.11)
$$
\begin{bmatrix}
D_1 f_1(c) & D_2 f_1(c) & \cdots & D_p f_1(c) \\
D_1 f_2(c) & D_2 f_2(c) & \cdots & D_p f_2(c) \\
\cdot\;\cdot & \cdot\;\cdot & \cdots & \cdot\;\cdot \\
D_1 f_q(c) & D_2 f_q(c) & \cdots & D_p f_q(c)
\end{bmatrix}
$$
$$
= \begin{bmatrix}
f_{1,1}(c) & f_{1,2}(c) & \cdots & f_{1,p}(c) \\
f_{2,1}(c) & f_{2,2}(c) & \cdots & f_{2,p}(c) \\
\cdot\;\cdot & \cdot\;\cdot & \cdots & \cdot\;\cdot \\
f_{q,1}(c) & f_{q,2}(c) & \cdots & f_{q,p}(c)
\end{bmatrix}
$$

We have already remarked (see Theorem 21.2) that such an array of real numbers determines a linear function on $\mathbf{R}^p$ to $\mathbf{R}^q$. The matrix (39.11) is called the **Jacobian matrix** of the system (39.9) at point c. When $p = q$, the determinant of the matrix (39.11) is called the **Jacobian determinant**, or simply the **Jacobian** of the system (39.10) at the point c. Frequently, this Jacobian† determinant is denoted by

$$
\frac{\partial(f_1, f_2, \ldots, f_p)}{\partial(x_1, x_2, \ldots, x_p)}\bigg|_{x=c}
\qquad \text{or} \qquad J_f(c).
$$

Existence of the Derivative

It was proved in Theorem 39.6 that the existence of the derivative at a point implies the existence of all the partial derivatives at that point. It was seen in the remark after Corollary 39.7 that the mere existence of the partial derivatives does *not* imply the existence of the derivative even when $p = 2$, $q = 1$. We shall now show that the *continuity* of the partial derivatives at c is sufficient for the existence of the derivative at c.

---

† CARL (G. J.) JACOBI (1804–1851) was professor at Königsberg and Berlin. His main work was concerned with elliptic functions, but he is also known for his work in determinants and dynamics.

39.9   THEOREM. *Let $A \subseteq \mathbf{R}^p$, let $f : A \to \mathbf{R}^q$, and let c be an interior point of A. If the partial derivatives $D_j f_i$ $(i = 1, \ldots, q, j = 1, \ldots, p)$ exist in a neighborhood of c and are continuous at c, then f is differentiable at c. Moreover $Df(c)$ is represented by the $q \times p$ matrix (39.11).*

PROOF.   We shall treat the case $q = 1$ in detail. If $\varepsilon > 0$ let $\delta(\varepsilon) > 0$ be such that if $\|y - c\| \le \delta(\varepsilon)$ and $j = 1, 2, \ldots, p$, then

$$(39.12) \qquad |D_j f(y) - D_j f(c)| < \varepsilon.$$

If $x = (x_1, x_2, \ldots, x_p)$ and $c = (c_1, c_2, \ldots, c_p)$, let $z_1, z_2, \ldots, z_{p-1}$ denote the points

$$z_1 = (c_1, x_2, \ldots, x_p), \qquad z_2 = (c_1, c_2, x_3, \ldots, x_p),$$

$$\ldots, z_{p-1} = (c_1, c_2, \ldots, c_{p-1}, x_p)$$

and let $z_0 = x$ and $z_p = c$. If $\|x - c\| \le \delta(\varepsilon)$, then it is easily seen that $\|z_j - c\| \le \delta(\varepsilon)$ for $j = 0, 1, \ldots, p$. We write the difference $f(x) - f(c)$ as a telescoping sum:

$$f(x) - f(c) = \sum_{j=1}^{p} \{f(z_{j-1}) - f(z_j)\}.$$

If we apply the Mean Value Theorem 27.6 to the jth term of this sum, we obtain a point $\bar{z}_j$, lying on the line segment joining z_{j-1} and z_j, such that

$$f(z_{j-1}) - f(z_j) = (x_j - c_j) D_j f(\bar{z}_j).$$

Therefore, we obtain

$$f(x) - f(c) - \sum_{j=1}^{p} (x_j - c_j) D_j f(c) = \sum_{j=1}^{p} (x_j - c_j)\{D_j f(\bar{z}_j) - D_j f(c)\}.$$

In view of the inequality (39.12), each quantity appearing in braces in the last formula is dominated by ε. Applying the Schwarz Inequality to this last sum, we obtain the estimate

$$\left\| f(x) - f(c) - \sum_{j=1}^{p} (x_j - c_j) D_j f(c) \right\| \le (\varepsilon \sqrt{p}) \|x - c\|,$$

whenever $\|x - c\| \le \delta(\varepsilon)$.

We have proved that f is differentiable at c and that its derivative $Df(c)$ is the linear function from $\mathbf{R}^p$ to $\mathbf{R}$ given by

$$u = (u_1, \ldots, u_p) \mapsto Df(c)(u) = \sum_{j=1}^{p} u_j D_j f(c).$$

In the case where f takes values in $\mathbf{R}^q$ with $q > 1$, we apply the same argument to each of the real-valued functions f_i, $i = 1, 2, \ldots, q$, which occur in the coordinate representation of the mapping f. We shall leave the details of this argument as an exercise.   Q.E.D.

Exercises

39.A. Let $f: \boldsymbol{R}^2 \to \boldsymbol{R}$ be defined by

$$f(x, y) = \frac{x}{y} \qquad \text{for } y \neq 0,$$
$$= 0 \qquad \text{for } y = 0.$$

Show that the partial derivatives $D_1 f(0, 0)$, $D_2 f(0, 0)$ exist and equal 0. However, the derivative of f at $(0, 0)$ with respect to a vector $u = (a, b)$ does not exist if $ab \neq 0$. Show that f is not continuous at $(0, 0)$; indeed, f is not even bounded on a neighborhood of $(0, 0)$.

39.B. Let $g: \boldsymbol{R}^2 \to \boldsymbol{R}$ be defined by

$$g(x, y) = 0 \qquad \text{for } xy = 0,$$
$$= 1 \qquad \text{for } xy \neq 0.$$

Show that the partial derivatives $D_1 g(0, 0)$, $D_2 g(0, 0)$ exist and equal 0. However, the derivative of g at $(0, 0)$ with respect to a vector $u = (a, b)$ does not exist if $ab \neq 0$. Show that g is not continuous at $(0, 0)$; however, g is bounded on a neighborhood of $(0, 0)$.

39.C. Let $h: \boldsymbol{R}^2 \to \boldsymbol{R}$ be defined by

$$h(x, y) = 0 \qquad \text{for } (x, y) = (0, 0),$$
$$= \frac{xy}{x^2 + y^2} \qquad \text{for } (x, y) \neq (0, 0).$$

Show that the partial derivatives $D_1 h(0, 0)$, $D_2 h(0, 0)$ exist and equal 0. However, the derivative of h at $(0, 0)$ with respect to a vector $u = (a, b)$ does not exist if $ab \neq 0$. Show that h is not continuous at $(0, 0)$.

39.D. Let $k: \boldsymbol{R}^2 \to \boldsymbol{R}$ be defined by

$$k(x, y) = 0 \qquad \text{for } (x, y) = (0, 0),$$
$$= \frac{xy^2}{x^2 + y^4} \qquad \text{for } (x, y) \neq (0, 0).$$

Show that the partial derivative of k at $(0, 0)$ with respect to any vector $u = (a, b)$ exists and that

$$D_u k(0, 0) = \frac{b^2}{a} \qquad \text{if } a \neq 0.$$

Show that k is not continuous and hence not differentiable at $(0, 0)$.

39.E. Let $f: \boldsymbol{R}^2 \to \boldsymbol{R}$ be defined by

$$f(x, y) = 0 \qquad \text{for } (x, y) = (0, 0),$$
$$= \frac{xy^2}{x^2 + y^2} \qquad \text{for } (x, y) \neq (0, 0).$$

Show that the partial derivative of f at $(0, 0)$ with respect to any vector $u = (a, b)$ exists and that

$$D_u f(0, 0) = \frac{ab^2}{a^2 + b^2} \qquad \text{if } (a, b) \neq (0, 0).$$

Show that f is continuous but not differentiable at $(0, 0)$.

39.F. Let $F : \mathbf{R}^2 \to \mathbf{R}$ be defined by

$$F(x, y) = x^2 + y^2 \qquad \text{if both } x, y \text{ are rational,}$$
$$= 0 \qquad \text{otherwise.}$$

Show that F is continuous only at the point $(0, 0)$ and that it is differentiable there.

39.G. Let $G : \mathbf{R}^2 \to \mathbf{R}$ be defined by

$$G(x, y) = (x^2 + y^2) \sin 1/(x^2 + y^2) \qquad \text{for } (x, y) \neq (0, 0),$$
$$= 0 \qquad \text{for } (x, y) = (0, 0).$$

Show that G is differentiable at every point of $\mathbf{R}^2$ but the partial derivatives $D_1 G$, $D_2 G$ are not bounded (and hence not continuous) on a neighborhood of $(0, 0)$.

39.H. Let $H : \mathbf{R}^2 \to \mathbf{R}^2$ be defined by

$$H(x, y) = \left(x^2 + x^2 \sin \frac{1}{x}, y \right) \qquad \text{for } x \neq 0,$$
$$= (0, y) \qquad \text{for } x = 0.$$

Show that $D_1 H$ exists at every point and that $D_2 H$ exists and is continuous on a neighborhood of $(0, 0)$. Show that H is differentiable at $(0, 0)$.

39.I. Let $A \subseteq \mathbf{R}^p$, let $f : A \to \mathbf{R}^q$ be differentiable at a point c interior to A, and let $v \in \mathbf{R}^q$. If we define $g : A \to \mathbf{R}$ by $g(x) = f(x) \cdot v$ for all $x \in A$, show that g is differentiable at c and that

$$Dg(c)(u) = (Df(c)(u)) \cdot v \qquad \text{for } u \in \mathbf{R}^p.$$

39.J. Let c be an interior point of $A \subseteq \mathbf{R}^p$ and let $f : A \to \mathbf{R}$.

(a) If f is differentiable at c, show that there exists a unique vector $v_c \in \mathbf{R}^p$ such that

$$D_u f(c) = Df(c)(u) = v_c \cdot u \qquad \text{for all } u \in \mathbf{R}^p.$$

The vector v_c is called the **gradient** of f at c and is denoted by $\nabla_c f$, or by $\operatorname{grad} f(c)$. Show that

$$\nabla_c f = (D_1 f(c), \ldots, D_p f(c)).$$

(b) Use the Schwarz Inequality to show that if $u \in \mathbf{R}^p$ and $\|u\| = 1$, then the function $u \mapsto D_u f(c)$ has a maximum value when u is a positive multiple of $\nabla_c f$. Hence the direction in which the directional derivative of f at c is maximum is that of the gradient of f at c.

39.K. Let c be an interior point of $A \subseteq \mathbf{R}^p$, let $f, g : A \to \mathbf{R}$ be differentiable at c, and let $\alpha \in \mathbf{R}$. Show that

$$\nabla_c(\alpha f) = \alpha \nabla_c f, \qquad \nabla_c(f+g) = \nabla_c f + \nabla_c g,$$
$$\nabla_c(fg) = f(c)\,\nabla_c g + g(c)\,\nabla_c f.$$

39.L. Find the gradients of the following functions at an arbitrary point in $\mathbf{R}^3$.
(a) $f_1(x, y, z) = x^2 + y^2 + z^2$;
(b) $f_2(x, y, z) = x^2 - yz + z^2$;
(c) $f_3(x, y, z) = xyz$.

39.M. Find the directional derivatives of each of the functions in 39.L at the point $(0, 1, 2)$ in the direction toward the point $(0, 2, 3)$.

39.N. Let $A \subseteq \mathbf{R}^2$ and let a function $f : A \to \mathbf{R}$ represent a surface S_f in $\mathbf{R}^3$ explicitly as its graph:

$$S_f = \{(x, y, f(x, y)) : (x, y) \in A\}.$$

If f is differentiable at an interior point (x_0, y_0) of A, then the **tangent plane** to S_f at the point $(x_0, y_0, f(x_0, y_0))$ is given by the graph of the affine map $A_{(x_0, y_0)} : \mathbf{R}^2 \to \mathbf{R}$ defined by

$$A_{(x_0, y_0)}(x, y) = f(x_0, y_0) + Df(x_0, y_0)(x - x_0, y - y_0).$$

Show that the tangent plane to S_f at this point is

$$\{(x, y, z) \in \mathbf{R}^3 : z = f(x_0, y_0) + D_1 f(x_0, y_0)(x - x_0) + D_2 f(x_0, y_0)(y - y_0)\}.$$

39.O. Find the tangent planes to the surfaces in $\mathbf{R}^3$ represented as graphs of the following functions of the points specified. Draw a sketch.
(a) $f_1(x, y) = x^2 + y^2$ at $(0, 0)$ and at $(1, 2)$.
(b) $f_2(x, y) = xy$ at $(0, 0)$ and at $(1, 2)$.
(c) $f_3(x, y) = (4 - (x^2 + y^2))^{1/2}$ at $(0, 0)$ and at $(1, 1)$.

39.P. Let $J \subseteq \mathbf{R}$ be an interval and let $g : J \to \mathbf{R}^3$ represent a curve C_g in $\mathbf{R}^3$ parametrically:

$$C_g = \{(g_1(t), g_2(t), g_3(t)) : t \in J\}.$$

If g is differentiable at an interior point t_0 of J, then the **tangent space** to C_g at the point $g(t_0) = (g_1(t_0), g_2(t_0), g_3(t_0)) \in \mathbf{R}^3$ is given parametrically by the affine map $A_{t_0} : \mathbf{R} \to \mathbf{R}^3$ defined by

$$A_{t_0}(t) = g(t_0) + Dg(t_0)(t - t_0).$$

Show that the tangent space to C_g at this point is

$$\{(x, y, z) \in \mathbf{R}^3 : x = g_1(t_0) + g_1'(t_0)(t - t_0),$$
$$y = g_2(t_0) + g_2'(t_0)(t - t_0), \qquad z = g_3(t_0) + g_3'(t_0)(t - t_0)\}.$$

If $g_1'(t_0), g_2'(t_0), g_3'(t_0)$ are not all zero, then this tangent space is a line in $\mathbf{R}^3$ and is called the **tangent line**.

39.Q. Find parametric equations for the tangent lines to the following curves in

$\mathbf{R}^3$ at the specified points:

(a) $g : t \mapsto (x, y, z) = (t, t^2, t^3)$

at the points corresponding to $t = 0$ and $t = 1$.

(b) $g : t \mapsto (x, y, z) = (t - 1, t^2, 2)$

at the points corresponding to $t = 0$ and $t = 1$.

(c) $g : t \mapsto (x, y, z) = (2 \cos t, 2 \sin t, t)$

at the points corresponding to $t = \pi/2$ and $t = \pi$.

39.R. Let $A \subseteq \mathbf{R}^2$ and let $h : A \to \mathbf{R}^3$ represent a surface S_h in $\mathbf{R}^3$ parametrically:

$$S_h = \{(h_1(s, t), h_2(s, t), h_3(s, t)) : (s, t) \in A\}.$$

If h is differentiable at an interior point (s_0, t_0) of A, then the **tangent space** to S_h at the point $h(s_0, t_0) = (h_1(s_0, t_0), h_2(s_0, t_0), h_3(s_0, t_0)) \in \mathbf{R}^3$ is given parametrically by the affine map $A_{(s_0, t_0)} : \mathbf{R} \to \mathbf{R}^3$ defined by

$$A_{(s_0, t_0)}(s, t) = h(s_0, t_0) + Dh(s_0, t_0)(s - s_0, t - t_0).$$

Show that the tangent space to S_h at this point is

$$\{(x, y, z) \in \mathbf{R}^3 : x = h_1(s_0, t_0) + D_1h_1(s_0, t_0)(s - s_0) + D_2h_1(s_0, t_0)(t - t_0),$$
$$y = h_2(s_0, t_0) + D_1h_2(s_0, t_0)(s - s_1) + D_2h_2(s_0, t_0)(t - t_0),$$
$$z = h_3(s_0, t_0) + D_1h_3(s_0, t_0)(s - s_0) + D_2h_3(s_0, t_0)(t - t_0)\}.$$

If the vectors $(D_1h_1(s_0, t_0), D_1h_2(s_0, t_0), D_1h_3(s_0, t_0))$ and $(D_2h_1(s_0, t_0), D_2h_2(s_0, t_0), D_2h_3(s_0, t_0))$ in $\mathbf{R}^3$ are not multiples of each other, then this tangent space is a plane in $\mathbf{R}^3$ and is called the **tangent plane.**

39.S. Find parametric equations for the tangent planes to the following surfaces in $\mathbf{R}^3$ at the specified points.

(a) $h : (s, t) \mapsto (x, y, z) = (s, t, s^2 + t^2)$ at the points corresponding to $(s, t) = (0, 0)$ and $(1, 1)$.

(b) $h : (s, t) \mapsto (x, y, z) = (s + t, s - t, s^2 - t^2)$ at the points corresponding to $(s, t) = (0, 0)$ and $(1, 2)$.

(c) $h : (s, t) \mapsto (x, y, z) = (s \cos t, s \sin t, t)$ at the points corresponding to $(s, t) = (1, 0)$ and $(2, \pi/2)$.

(d) $h : (s, t) \mapsto (x, y, z) = (\cos s \sin t, \sin s \sin t, \cos t)$ at the points corresponding to $(s, t) = (0, 0)$, $(0, \pi/2)$ and $(\pi/4, \pi/4)$.

39.T. If $A \subseteq \mathbf{R}^p$ and $f : A \to \mathbf{R}$ is such that the partial derivatives $D_1 f, \ldots, D_p f$ exist and are bounded on some neighborhood of $c \in A$, then f is continuous at c. (Hint: argue as in the proof of Theorem 39.9.)

39.U. Let f be defined on a neighborhood of a point $c \in \mathbf{R}^2$ with values in $\mathbf{R}$. Suppose that $D_1 f$ exists and is continuous on a neighborhood of c and that $D_2 f$ exists at c. Show that f is differentiable at c.

39.V. Let $A \subseteq \mathbf{R}^p$ and let $f : A \to \mathbf{R}^q$ and $g : A \to \mathbf{R}^r$ be given. If $F : A \to \mathbf{R}^q \times \mathbf{R}^r = \mathbf{R}^{q+r}$ is defined by $F(x) = (f(x), g(x))$ for $x \in A$, show that F is differentiable at an interior point $c \in A$ if and only if f and g are differentiable at c. In this case we have

$$DF(c)(u) = (Df(c)(u), Dg(c)(u)) \qquad \text{for } u \in \mathbf{R}^p.$$

39.W. Let $A \subseteq \mathbf{R}^p$ and $B \subseteq \mathbf{R}^q$ and let $G : A \times B \to \mathbf{R}'$ be differentiable at a point (a, b) in $A \times B$. We define $g_1 : A \to \mathbf{R}'$ and $g_2 : B \to \mathbf{R}'$ to be the "partial maps" at (a, b) given by

$$g_1(x) = G(x, b), \qquad g_2(y) = G(a, y)$$

for all $x \in A$, $y \in B$. Show that g_1 and g_2 are differentiable at a and b, respectively, and that

$$Dg_1(a)(u) = DG(a, b)(u, 0), \qquad Dg_2(b)(v) = DG(a, b)(0, v),$$

for all $u \in \mathbf{R}^p$, $v \in \mathbf{R}^q$. Moreover, we have

$$DG(a, b)(u, v) = Dg_1(a)(u) + Dg_2(b)(v).$$

[Sometimes $Dg_1(a) \in \mathcal{L}(\mathbf{R}^p, \mathbf{R}')$ and $Dg_2(b) \in \mathcal{L}(\mathbf{R}^q, \mathbf{R}')$ are called the "block partial derivatives" of G at (a, b) and are denoted by $D_{(1)}G(a, b)$ and $D_{(2)}G(a, b)$.]

Section 40   The Chain Rule and Mean Value Theorems

We shall first establish the basic algebraic relations concerning the derivative. These properties, which are the same as those for real-valued functions of one variable, will be used frequently in the following.

40.1   THEOREM.   *Let $A \subseteq \mathbf{R}^p$ and let c be an interior point of A.*

(a) *If f and g are defined on A to $\mathbf{R}^q$ and are differentiable at c, and if α, $\beta \in \mathbf{R}$, then the function $h = \alpha f + \beta g$ is differentiable at c and*

$$Dh(c) = \alpha Df(c) + \beta Dg(c).$$

(b) *If $\varphi : A \to \mathbf{R}$ and $f : A \to \mathbf{R}^q$ are differentiable at c, then the product function $k = \varphi f : A \to \mathbf{R}^q$ is differentiable at c and*

$$Dk(c)(u) = \{D\varphi(c)(u)\}f(c) + \varphi(c)\{Df(c)(u)\} \qquad for \ u \in \mathbf{R}^p.$$

PROOF.   (a) If $\varepsilon > 0$, then there exist $\delta_1(\varepsilon) > 0$ and $\delta_2(\varepsilon) > 0$ such that if $\|x - c\| \le \inf\{\delta_1(\varepsilon), \delta_2(\varepsilon)\}$, then

$$\|f(x) - f(c) - Df(c)(x - c)\| \le \varepsilon \|x - c\|,$$
$$\|g(x) - g(c) - Dg(c)(x - c)\| \le \varepsilon \|x - c\|.$$

Thus, if $\|x - c\| \le \inf\{\delta_1(\varepsilon), \delta_2(\varepsilon)\}$, then

$$\|h(x) - h(c) - \{\alpha Df(c)(x - c) + \beta Dg(c)(x - c)\}\| \le (|\alpha| + |\beta|)\varepsilon \|x - c\|.$$

Since $\alpha Df(c) + \beta Dg(c)$ is a linear function of $\mathbf{R}^p$ into $\mathbf{R}^q$, it follows that h is differentiable at c and that $Dh(c) = \alpha Df(c) + \beta Dg(c)$.

(b) A simple calculation shows that

$$k(x) - k(c) - \{D\varphi(c)(x-c)f(c) + \varphi(c)Df(c)(x-c)\}$$
$$= \{\varphi(x) - \varphi(c) - D\varphi(c)(x-c)\}f(x)$$
$$+ D\varphi(c)(x-c)\{f(x) - f(c)\} + \varphi(c)\{f(x) - f(c) - Df(c)(x-c)\}.$$

Since $Df(c)$ exists, we infer from Lemma 39.5 that f is continuous at c; hence there exists a constant M such that $\|f(x)\| < M$ for $\|x - c\| \le \delta$. From this it can be seen that all the terms on the right side of the last equation can be made arbitrarily small by choosing $\|x - c\|$ small enough. This establishes (b).   Q.E.D.

The next result, which is very important, asserts that the derivative of the composition of two differentiable functions is the composition of their derivatives.

40.2   CHAIN RULE.   *Let f have domain $A \subseteq \mathbf{R}^p$ and range in $\mathbf{R}^q$, and let g have domain $B \subseteq \mathbf{R}^q$ and range in $\mathbf{R}^r$. Suppose that f is differentiable at c and that g is differentiable at $b = f(c)$. Then the composition $h = g \circ f$ is differentiable at c and*

(40.1) $$Dh(c) = Dg(b) \circ Df(c).$$

Alternatively, we write

(40.2) $$D(g \circ f)(c) = Dg(f(c)) \circ Df(c).$$

PROOF.   The hypothesis implies that c is an interior point of the domain of $h = g \circ f$. (Why?) Let $\varepsilon > 0$ and let $\delta(\varepsilon, f)$ and $\delta(\varepsilon, g)$ be as in Definition 39.2. It follows from Lemma 39.5 that there exist strictly positive numbers γ, K such that if $\|x - c\| \le \gamma$, then $f(x) \in B$ and

(40.3) $$\|f(x) - f(c)\| \le K \|x - c\|.$$

For simplicity, we write $L_f = Df(c)$ and $L_g = Dg(b)$. By Theorem 21.3 there is a constant M such that

(40.4) $$\|L_g(u)\| \le M \|u\|, \qquad \text{for } u \in \mathbf{R}^q.$$

If $\|x - c\| \le \inf\{\gamma, (1/K)\,\delta(\varepsilon, g)\}$, then (40.3) implies that $\|f(x) - f(c)\| \le \delta(\varepsilon, g)$, which means that

(40.5) $$\|g(f(x)) - g(f(c)) - L_g(f(x) - f(c))\| \le \varepsilon \|f(x) - f(c)\| \le \varepsilon K \|x - c\|.$$

If we also require that $\|x - c\| \le \delta(\varepsilon, f)$, then we infer from (40.4) that

$$\|L_g\{f(x) - f(c) - L_f(x - c)\}\| \le \varepsilon M \|x - c\|.$$

If we combine this last relation with (40.5), we infer that if $\delta_1 = \inf\{\gamma, (1/K)\,\delta(\varepsilon, g), \delta(\varepsilon, f)\}$ and if $x \in A$ and $\|x - c\| \leq \delta_1$, then

$$\|g(f(x)) - g(f(c)) - L_g(L_f(x - c))\| \leq \varepsilon(K + M)\,\|x - c\|,$$

which means that

$$\|g \circ f(x) - g \circ f(c) - L_g \circ L_f(x - c)\| \leq \varepsilon(K + M)\,\|x - c\|.$$

We conclude that $Dh(c) = L_g \circ L_f$.   Q.E.D.

Maintaining the notation of the proof of the theorem, $L_f = Df(c)$ is a linear function of $\mathbf{R}^p$ into $\mathbf{R}^q$ and $L_g = Dg(b)$ is a linear function of $\mathbf{R}^q$ into $\mathbf{R}'$. The composition $L_g \circ L_f$ is a linear function of $\mathbf{R}^p$ into $\mathbf{R}'$, as is required, since $h = g \circ f$ is a function defined on part of $\mathbf{R}^p$ with values in $\mathbf{R}'$. We now consider some examples of this result.

40.3 EXAMPLES. (a) Let $p = q = r = 1$; then the derivative $Df(c)$ is the linear function which takes the real number u into $f'(c)u$, and similarly for $Dg(b)$. It follows that the derivative of $g \circ f$ sends the real number u into $g'(b)f'(c)u$.

(b) Let $p > 1$, $q = r = 1$. According to Example 39.8(c), the derivative of f at c takes the point $w = (w_1, \ldots, w_p)$ of $\mathbf{R}^p$ into the real number

$$D_1f(c)w_1 + \cdots + D_pf(c)w_p$$

and so the derivative of $g \circ f$ at c takes this point of $\mathbf{R}^p$ into the real number

$$g'(b)[D_1f(c)w_1 + \cdots + D_pf(c)w_p].$$

(c) Let $q > 1$, $p = r = 1$. According to Examples 39.8(b), (c) the derivative $Df(c)$ takes the real number u into the point

$$Df(c)(u) = uf'(c) = (f_1'(c)u, \ldots, f_q'(c)u) \qquad \text{in } \mathbf{R}^q,$$

and the derivative $Dg(b)$ takes the point $w = (w_1, \ldots, w_q)$ in $\mathbf{R}^q$ into the real number

$$D_1g(b)w_1 + \cdots + D_qg(b)w_q.$$

It follows that the derivative of $h = g \circ f$ takes the real number u into the real number

$$Dh(c)u = \{D_1g(b)f_1'(c) + \cdots + D_qg(b)f_q'(c)\}u = u\{Dg(b)(f'(c))\}.$$

The quantity in the braces, which is $h'(c) = (g \circ f)'(c)$ is sometimes denoted by the less precise symbolism

$$\frac{\partial g}{\partial y_1}\frac{df_1}{dx} + \cdots + \frac{\partial g}{\partial y_q}\frac{df_q}{dx}.$$

In this connection, it must be understood that the derivatives are to be evaluated at appropriate points.

(d) We consider the case where $p = q = 2$ and $r = 3$. For simplicity in notation, we denote the coordinate variables in $\mathbf{R}^p$ by (x, y), in $\mathbf{R}^q$ by (w, z), and in $\mathbf{R}^r$ by (r, s, t). Then a function f on $\mathbf{R}^p$ to $\mathbf{R}^q$ can be expressed in the form

$$w = W(x, y), \qquad z = Z(x, y)$$

and a function g on $\mathbf{R}^q$ to $\mathbf{R}^r$ can be expressed in the form

$$r = R(w, z), \qquad s = S(w, z), \qquad t = T(w, z).$$

The derivative $Df(c)$ sends (ξ, η) into (ω, ζ) according to the formulas

(40.6)
$$\omega = W_x(c)\xi + W_y(c)\eta,$$
$$\zeta = Z_x(c)\xi + Z_y(c)\eta.$$

Here we write W_x for $D_1 W = D_x W$, etc. Also the derivative $Dg(b)$ sends (ω, ζ) into (ρ, σ, τ) according to the relations

(40.7)
$$\rho = R_w(b)\omega + R_z(b)\zeta,$$
$$\sigma = S_w(b)\omega + S_z(b)\zeta,$$
$$\tau = T_w(b)\omega + T_z(b)\zeta.$$

A routine calculation shows that the derivative of $g \circ f$ sends (ξ, η) into (ρ, σ, τ) by

(40.8)
$$\rho = \{R_w(b) W_x(c) + R_z(b) Z_x(c)\}\xi + \{R_w(b) W_y(c) + R_z(b) Z_y(c)\}\eta,$$
$$\sigma = \{S_w(b) W_x(c) + S_z(b) Z_x(c)\}\xi + \{S_w(b) W_y(c) + S_z(b) Z_y(c)\}\eta,$$
$$\tau = \{T_w(b) W_x(c) + T_z(b) Z_x(c)\}\xi + \{T_w(b) W_y(c) + T_z(b) Z_y(c)\}\eta.$$

A more classical notation would be to write dx, dy instead of ξ, η; dw, dz instead of ω, ζ; and dr, ds, dt instead of ρ, σ, τ. If we denote the values of the partial derivative W_x at the point c by $\partial w/\partial x$, etc., then (40.6) becomes

$$dw = \frac{\partial w}{\partial x} dx + \frac{\partial w}{\partial y} dy,$$

$$dz = \frac{\partial z}{\partial x} dx + \frac{\partial z}{\partial y} dy;$$

similarly, (40.7) becomes

$$dr = \frac{\partial r}{\partial w} dw + \frac{\partial r}{\partial z} dz,$$

$$ds = \frac{\partial s}{\partial w} dw + \frac{\partial s}{\partial z} dz,$$

$$dt = \frac{\partial t}{\partial w} dw + \frac{\partial t}{\partial z} dz;$$

and (40.8) is written in the form

$$dr = \left(\frac{\partial r}{\partial w}\frac{\partial w}{\partial x} + \frac{\partial r}{\partial z}\frac{\partial z}{\partial x}\right)dx + \left(\frac{\partial r}{\partial w}\frac{\partial w}{\partial y} + \frac{\partial r}{\partial z}\frac{\partial z}{\partial y}\right)dy,$$

$$ds = \left(\frac{\partial s}{\partial w}\frac{\partial w}{\partial x} + \frac{\partial s}{\partial z}\frac{\partial z}{\partial x}\right)dx + \left(\frac{\partial s}{\partial w}\frac{\partial w}{\partial y} + \frac{\partial s}{\partial z}\frac{\partial z}{\partial y}\right)dy,$$

$$dt = \left(\frac{\partial t}{\partial w}\frac{\partial w}{\partial x} + \frac{\partial t}{\partial z}\frac{\partial z}{\partial x}\right)dx + \left(\frac{\partial t}{\partial w}\frac{\partial w}{\partial y} + \frac{\partial t}{\partial z}\frac{\partial z}{\partial y}\right)dy.$$

In these last three sets of formulas it is important to realize that all of the indicated partial derivatives are to be evaluated at appropriate points. Hence the coefficients of dx, dy, etc., turn out to be real numbers.

We can express equation (40.6) in matrix terminology by saying that the mapping $Df(c)$ of (ξ, η) into (ω, ζ) is given by the 2×2 matrix

(40.9)
$$\begin{bmatrix} W_x(c) & W_y(c) \\ Z_x(c) & Z_y(c) \end{bmatrix} = \begin{bmatrix} \dfrac{\partial w}{\partial x}(c) & \dfrac{\partial w}{\partial y}(c) \\ \dfrac{\partial z}{\partial x}(c) & \dfrac{\partial z}{\partial y}(c) \end{bmatrix}.$$

Similarly, (40.7) asserts that the mapping $Dg(b)$ of (ω, ζ) into (ρ, σ, τ) is given by the 3×2 matrix

(40.10)
$$\begin{bmatrix} R_w(b) & R_z(b) \\ S_w(b) & S_z(b) \\ T_w(b) & T_z(b) \end{bmatrix} = \begin{bmatrix} \dfrac{\partial r}{\partial w}(b) & \dfrac{\partial r}{\partial z}(b) \\ \dfrac{\partial s}{\partial w}(b) & \dfrac{\partial s}{\partial z}(b) \\ \dfrac{\partial t}{\partial w}(b) & \dfrac{\partial t}{\partial z}(b) \end{bmatrix}.$$

Finally, relation (40.8) asserts that the mapping $D(g \circ f)(c)$ of (ξ, η) into (ρ, σ, τ) is given by the 3×2 matrix

$$\begin{bmatrix} R_w(b)W_x(c) + R_z(b)Z_x(c) & R_w(b)W_y(c) + R_z(b)Z_y(c) \\ S_w(b)W_x(c) + S_z(b)Z_x(c) & S_w(b)W_y(c) + S_z(b)Z_y(c) \\ T_w(b)W_x(c) + T_z(b)Z_x(c) & T_w(t)W_y(c) + T_z(b)Z_y(c) \end{bmatrix}$$

which is the product of the matrix in (40.10) with the matrix in (40.9) in that order.

Mean Value Theorem

We now turn to the problem of obtaining a generalization of the Mean Value Theorem 27.6 for differentiable functions on $\mathbf{R}^p$ to $\mathbf{R}^q$. It will be seen that the direct analog of Theorem 27.6 does not hold when $q > 1$. It might be expected that if f is differentiable at every point of $\mathbf{R}^p$ with values in $\mathbf{R}^q$, and if a, b belong to $\mathbf{R}^p$, then there exists a point c (lying between a, b) such that

$$(40.11) \qquad f(b) - f(a) = Df(c)(b - a).$$

This conclusion fails even when $p = 1$ and $q = 2$ as is seen by the function f defined on $\mathbf{R}$ to $\mathbf{R}^2$ by the formula

$$f(x) = (x - x^2, x - x^3).$$

Here $Df(c)$ is the linear function on $\mathbf{R}$ to $\mathbf{R}^2$ which sends the real number u into the element

$$Df(c)(u) = ((1 - 2c)u, (1 - 3c^2)u).$$

Now $f(0) = (0, 0)$ and $f(1) = (0, 0)$, but there is no point c such that $Df(c)(u) = (0, 0)$ for any non-zero u in $\mathbf{R}$. Hence the formula (40.11) cannot hold in general when $q > 1$, even when $p = 1$. However, for many applications it is sufficient to consider the case where $q = 1$ and here it is easy to extend the Mean Value Theorem.

40.4 MEAN VALUE THEOREM. *Let f defined on an open subset Ω of $\mathbf{R}^p$ and have values in $\mathbf{R}$. Suppose that the set Ω contains the points a, b and the line segment S joining them, and that f is differentiable at every point of this segment. Then there exists a point c on S such that*

$$(40.11) \qquad f(b) - f(a) = Df(c)(b - a).$$

PROOF. Let $\varphi : \mathbf{R} \to \mathbf{R}^p$ be defined by

$$\varphi(t) = (1 - t)a + tb = a + t(b - a),$$

so that $\varphi(0) = a$, $\varphi(1) = b$, and $\varphi(t) \in S \subseteq \Omega$ for $t \in [0, 1]$. Since Ω is open and φ is continuous, there is a number $\gamma > 0$ such that φ maps the interval $(-\gamma, 1 + \gamma)$ into Ω. Now let $F : (-\gamma, 1 + \gamma) \to \mathbf{R}$ be defined by

$$F(t) = f \circ \varphi(t) = f((1 - t)a + tb).$$

By the Chain Rule [see 40.3(c) and 40.P] it follows that

$$F'(t) = Df((1 - t)a + tb)(\varphi'(t))$$
$$= Df((1 - t)a + tb)(b - c).$$

If we apply the Mean Value Theorem 27.6 to F, we infer that there exists $t_0 \in (0, 1)$ such that $F(1) - F(0) = F'(t_0)$. Letting $c = \varphi(t_0) \in S$, we obtain

$$f(b) - f(a) = F(1) - F(0)$$
$$= F'(t_0) = Df(c)(b - a). \qquad \text{Q.E.D.}$$

Although the most natural extension of the Mean Value Theorem does not hold when the range space is $\mathbf{R}^q$, $q > 1$, there are some extensions which are available. One of the most useful is based on an inequality rather than an equality.

40.5 MEAN VALUE THEOREM. *Let $\Omega \subseteq \mathbf{R}^p$ be an open set and let $f : \Omega \to \mathbf{R}^q$. Suppose that Ω contains the points a, b and the line segment S joining these points, and that f is differentiable at every point of S. Then there exists a point c on S such that*

$$(40.12) \qquad \|f(b) - f(a)\| \leq \|Df(c)(b - a)\|.$$

PROOF. If $y_0 = f(b) - f(a)$ is the zero vector in $\mathbf{R}^q$, then the result is trivial. If $y_0 \neq 0$, let $y_1 = y_0 / \|y_0\|$ and use the inner product in $\mathbf{R}^q$ to define $H : \Omega \to \mathbf{R}$ by

$$H(x) = f(x) \cdot y_1 \qquad \text{for } x \in \Omega.$$

Evidently we have

$$H(b) - H(a) = \{f(b) - f(a)\} \cdot y_1 = \|f(b) - f(a)\|$$

and it is easily seen (cf. Exercise 40.H) that

$$DH(x)(u) = \{Df(x)(u)\} \cdot y_1$$

for $x \in S$, $u \in \mathbf{R}^p$. It follows from the Mean Value Theorem 40.4 that there is a point c on S such that

$$H(b) - H(a) = DH(c)(b - a)$$
$$= \{Df(c)(b - a)\} \cdot y_1.$$

If we use the Schwarz Inequality and the fact that $\|y_1\| = 1$, we have

$$\|f(b) - f(a)\| = \{Df(c)(b - a)\} \cdot y_1 \leq \|Df(c)(b - a)\|,$$

which is the desired result.   Q.E.D.

Since the exact value of the point c is usually not known, the theorem is often applied by using the following result, whose statement uses the notion of the norm of a linear map L from $\mathbf{R}^p$ to $\mathbf{R}^q$ that was introduced in Exercise 21.L. It is only necessary to recall that $\|L(u)\| \leq M \|u\|$ for all $u \in \mathbf{R}^p$, if and only if the norm $\|L\|_{pq} \leq M$.

40.6 COROLLARY. *Suppose the hypotheses of Theorem 40.5 are satisfied and that there exists $M > 0$ such that $\|Df(x)\|_{pq} \leq M$ for all $x \in S$. Then we have*

$$\|f(b) - f(a)\| \leq M \|b - a\|.$$

PROOF. Since $\|Df(c)(b-a)\| \le \|Df(c)\|_{pq} \|b-a\|$, and since $c \in S$, we have

$$\|f(b)-f(a)\| \le \|Df(c)(b-a)\| \le \|Df(c)\|_{pq} \|b-a\| \le M \|b-a\|.$$

Q.E.D.

Interchange of the Order of Differentiation

If f is a function with domain in $\mathbf{R}^p$ and range in $\mathbf{R}$, then f may have p (first) partial derivatives, which we denote by

$$D_i f \qquad \text{or} \qquad \frac{\partial f}{\partial x_i}, \qquad i = 1, 2, \dots, p.$$

Each of the partial derivatives is a function with domain in $\mathbf{R}^p$ and range in $\mathbf{R}$ and so each of these p functions may have p partial derivatives. Following the accepted American notation, we shall refer to the resulting p^2 functions (or to such ones that exist) as the **second partial derivatives** of f and we shall denote them by

$$D_{ji} f \qquad \text{or} \qquad \frac{\partial^2 f}{\partial x_j \, \partial x_i}, \qquad i, j = 1, 2, \dots, p.$$

It should be observed that the partial derivative intended by either of the latter symbols is the partial derivative with respect to x_j of the partial derivative of f with respect to x_i. (In other words: first x_i, then x_j).

In like manner, we can inquire into the existence of the third partial derivatives and those of still higher order. In principle, a function on $\mathbf{R}^p$ to $\mathbf{R}$ can have as many as p^n nth partial derivatives. However, it is a considerable convenience that if the resulting derivatives are continuous, then the order of differentiation is not significant. In addition to decreasing the number of (potentially distinct) higher partial derivatives, this result largely removes the danger from the rather subtle notational distinction employed for different orders of differentiation.

It is enough to consider the interchange of order for second derivatives. By holding all the other coordinates constant, we see that it is no loss of generality to consider a function on $\mathbf{R}^2$ to $\mathbf{R}$. In order to simplify our notation we let (x, y) denote a point in $\mathbf{R}^2$ and we shall show that if $D_x f$, $D_y f$, and $D_{yx} f$ exist and if $D_{yx} f$ is continuous at a point, then the partial derivative $D_{xy} f$ exists at this point and equals $D_{yx} f$. It will be seen in Exercise 40.U that it is possible that both $D_{yx} f$ and $D_{xy} f$ exists at a point and yet are not equal.

The device that will be used in this proof is to show that both of these mixed partial derivatives at the point $(0, 0)$ are the limit of the quotient

$$\frac{f(h, k) - f(h, 0) - f(0, k) + f(0, 0)}{hk},$$

as (h, k) approaches $(0, 0)$.

40.7　LEMMA.　*Suppose that f is defined on a neighborhood U of the origin in $\mathbf{R}^2$ with values in $\mathbf{R}$, that the partial derivatives $D_x f$ and $D_{yx} f$ exist in U, and that $D_{yx} f$ is continuous at $(0, 0)$. If A is the mixed difference*

(40.13)　　　　　$A(h, k) = f(h, k) - f(h, 0) - f(0, k) + f(0, 0),$

then we have

$$D_{yx} f(0, 0) = \lim_{(h, k) \to (0, 0)} \frac{A(h, k)}{hk}.$$

PROOF.　Let $\varepsilon > 0$ and let $\delta > 0$ be so small that if $|h| < \delta$ and $|k| < \delta$, then the point (h, k) belongs to U and

(40.14)　　　　　　　$|D_{yx} f(h, k) - D_{yx} f(0, 0)| < \varepsilon.$

If $|k| < \delta$, we define B for $|h| < \delta$ by

$$B(h) = f(h, k) - f(h, 0),$$

from which it follows that $A(h, k) = B(h) - B(0)$. By hypothesis, the partial derivative $D_x f$ exists in U and hence B has a derivative. Applying the Mean Value Theorem 27.6 to B, there exists a number h_0 with $0 < |h_0| < |h|$ such that

(40.15)　　　　　$A(h, k) = B(h) - B(0) = hB'(h_0).$

(It is noted that the value of h_0 depends on the value of k, but this will not cause any difficulty.) Referring to the definition of B, we have

$$B'(h_0) = D_x f(h_0, k) - D_x f(h_0, 0).$$

Applying the Mean Value Theorem to the right-hand side of the last equation, there exists a number k_0 with $0 < |k_0| < |k|$ such that

(40.16)　　　　　　　$B'(h_0) = k\{D_{yx} f(h_0, k_0)\}.$

Combining equations (40.15) and (40.16), we conclude that if $0 < |h| < \delta$ and $0 < |k| < \delta$, then

$$\frac{A(h, k)}{hk} = D_{yx} f(h_0, k_0),$$

where $0 < |h_0| < |h|$, $0 < |k_0| < |k|$. It follows from inequality (40.14) and the preceding expression that

$$\left|\frac{A(h, k)}{hk} - D_{yx}f(0, 0)\right| < \varepsilon$$

whenever $0 < |h| < \delta$ and $0 < |k| < \delta$.   Q.E.D.

We can now obtain a useful condition (due to H. A. Schwarz) for the equality of the mixed partials.

40.8 THEOREM. *Suppose that f is defined on a neighborhood U of a point (x, y) in $\mathbf{R}^2$ with values in $\mathbf{R}$. Suppose that the partial derivatives $D_x f$, $D_y f$, and $D_{yx}f$ exist in U and that $D_{yx}f$ is continuous at (x, y). Then the partial derivative $D_{xy}f$ exists at (x, y) and $D_{xy}f(x, y) = D_{yx}f(x, y)$.*

PROOF. It is no loss of generality to suppose that $(x, y) = (0, 0)$ and we shall do so. If A is the function defined in the preceding lemma, then it was seen that

(40.17) $$D_{yx}f(0, 0) = \lim_{(h, k) \to (0,)} \frac{A(h, k)}{hk},$$

the existence of this double limit being part of the conclusion. By hypothesis $D_y f$ exists in U, so that

(40.18) $$\lim_{k \to 0} \frac{A(h, k)}{hk} = \frac{1}{h}\{D_y f(h, 0) - D_y f(0, 0)\}, \qquad h \neq 0.$$

If $\varepsilon > 0$, there exists a number $\delta(\varepsilon) > 0$ such that if $0 < |h| < \delta(\varepsilon)$ and $0 < |k| < \delta(\varepsilon)$, then

$$\left|\frac{A(h, k)}{hk} - D_{yx}f(0, 0)\right| < \varepsilon.$$

By taking the limit in this inequality with respect to k and using (40.18), we obtain

$$\left|\frac{1}{h}\{D_y f(h, 0) - D_y f(0, 0)\} - D_{yx}f(0, 0)\right| \leq \varepsilon,$$

for all h satisfying $0 < |h| < \delta(\varepsilon)$. Therefore, $D_{xy}f(0, 0)$ exists and equals $D_{yx}f(0, 0)$.   Q.E.D.

Higher Derivatives

If f is a function with domain in $\mathbf{R}^p$ and range in $\mathbf{R}$, then the derivative $Df(c)$ of f at c is the linear function on $\mathbf{R}^p$ to $\mathbf{R}$ such that

$$\|f(c + z) - f(c) - Df(c)(z)\| \leq \varepsilon \|z\|,$$

for sufficiently small z. This means that $Df(c)$ is the linear function which most closely approximates the difference $f(c + z) - f(c)$ when z is small. Any other linear function would lead to a less exact approximation for small z. From this defining property, it is seen that if $Df(c)$ exists, then it is necessarily given by the formula

$$Df(c)(z) = D_1 f(c) z_1 + \cdots + D_p f(c) z_p,$$

where $z = (z_1, \ldots, z_p)$ in R^p.

Although linear approximations are particularly simple and are sufficiently exact for many purposes, it is sometimes desirable to obtain a finer degree of approximation than is possible by using linear functions. In such cases it is natural to turn to quadratic functions, cubic functions, etc., to effect closer approximations. Since our functions are to have their domains in R^p, we would be led into the study of multilinear functions on R^p to R for a thorough discussion of such functions. Although such a study is not particularly difficult, it would take us rather far afield in view of the limited applications we have in mind.

For this reason we shall define the **second derivative** $D^2 f(c)$ of f at c to be the function on $R^p \times R^p$ to R such that if (y, z) belongs to this product and $y = (y_1, \ldots, y_p)$ and $z = (z_1, \ldots, z_p)$, then

$$D^2 f(c)(y, z) = \sum_{i, j=1}^{p} D_{ij} f(c) y_i z_j.$$

In discussing the second derivative, we shall assume in the following that the second partial derivatives of f exist and are continuous on a neighborhood of c. Similarly, we define the **third derivative** $D^3 f(c)$ of f at c to be the function of (y, z, w) in $R^p \times R^p \times R^p$ given by

$$D^3 f(c)(y, z, w) = \sum_{i, j, k=1}^{p} D_{kji} f(c) y_i z_j w_k.$$

In discussing the third derivative, we shall assume that all of the third partial derivatives of f exist and are continuous in a neighborhood of c.

By now the method of formation of the higher derivatives should be clear. (In view of our preceding remarks concerning the interchange of order in differentiation, if the resulting mixed partial derivatives are continuous, then they are independent of the order of differentiation.)

One further notational device: we write

$$
\begin{array}{lll}
D^2 f(c)(w)^2 & \text{for} & D^2 f(c)(w, w), \\
D^3 f(c)(w)^3 & \text{for} & D^3 f(c)(w, w, w), \\
\cdots \cdots \cdots \cdots \cdots \cdots \cdots \\
D^n f(c)(w)^n & \text{for} & D^n f(c)(w, w, \ldots, w).
\end{array}
$$

If $p = 2$ and if we denote an element of $\mathbf{R}^2$ by (x, y) and $w = (h, k)$, then $D^2 f(c)(w)^2$ equals the expression

$$D_{xx}f(c)h^2 + 2D_{xy}f(c)hk + D_{yy}f(c)k^2;$$

similarly, $D^3 f(c)(w)^3$ equals

$$D_{xxx}f(c)h^3 + 3D_{xxy}f(c)h^2 k + 3D_{xyy}f(c)hk^2 + D_{yyy}f(c)k^3,$$

and $D^n f(c)(w)^n$ equals the expression

$$D_{x\cdots x}f(c)h^n + \binom{n}{1}D_{x\cdots xy}f(c)h^{n-1}k + \binom{n}{2}D_{x\cdots xyy}f(c)h^{n-2}k^2$$

$$+ \cdots + D_{y\cdots y}f(c)k^n.$$

Now that we have introduced this notation we shall establish an important generalization of Taylor's Theorem for functions on $\mathbf{R}^p$ to $\mathbf{R}$.

40.9 TAYLOR'S THEOREM. *Suppose that f is a function with open domain Ω in $\mathbf{R}^p$ and range in $\mathbf{R}$, and suppose that f has continuous partial derivatives of order n in a neighborhood of every point on a line segment S joining two points a, $b = a + u$ in Ω. Then there exists a point c on S such that*

$$f(a + u) = f(a) + \frac{1}{1!}Df(a)(u) + \frac{1}{2!}D^2 f(a)(u)^2$$

$$+ \cdots + \frac{1}{(n-1)!}D^{n-1}f(a)(u)^{n-1} + \frac{1}{n!}D^n f(c)(u)^n.$$

PROOF.   Let F be defined for t in $\mathbf{I}$ to $\mathbf{R}$ by

$$F(t) = f(a + tu).$$

In view of the assumed existence of the partial derivatives of f, it follows that

$$F'(t) = Df(a + tu)(u),$$
$$F''(t) = D^2 f(a + tu)(u)^2,$$
$$\cdots \cdots \cdots \cdots$$
$$F^{(n)}(t) = D^n f(a + tu)(u)^n.$$

If we apply the one-dimensional version of Taylor's Theorem 28.6 to the function F on $\mathbf{I}$, we infer that there exists a real number t_0 in $\mathbf{I}$ such that

$$F(1) = F(0) + \frac{1}{1!}F'(0) + \cdots + \frac{1}{(n-1)!}F^{(n-1)}(0) + \frac{1}{n!}F^{(n)}(t_0).$$

If we set $c = a + t_0 u$, then the result follows.   Q.E.D.

Exercises

40.A. If $f(x, y) = x^2 + y^2$ and $g(t) = (3t + 1, 2t - 3)$, let $F(t) = f \circ g(t)$. Evaluate $F'(t)$ both directly and by using the Chain Rule.

40.B. If $f(x, y) = xy$ and $g(s, t) = (2s + 3t, 4s + t)$, let $F(s, t) = f \circ g(s, t)$. Evaluate D_1F and D_2F both directly and by using the Chain Rule.

40.C. If $f(x, y, z) = xyz$ and $g(s, t) = (3s + st, s, t)$, let $F(s, t) = f \circ g(s, t)$. Evaluate D_1F and D_2F both directly and by using the Chain Rule.

40.D. If $f(x, y, z) = xy + yz + zx$ and $g(s, t) = (\cos s, \sin s \cos t, \sin t)$, let $F(s, t) = f \circ g(s, t)$. Evaluate D_1F and D_2F both directly and by using the Chain Rule.

40.E. If Cartesian axes are rotated in the plane by the angle θ, then the new coordinates u, v of a point are related to the original coordinates x, y by

$$x = u \cos \theta - v \sin \theta, \qquad y = u \sin \theta + v \cos \theta.$$

Let $f : \mathbf{R}^2 \to \mathbf{R}$ be differentiable on $\mathbf{R}^2$ and let $F(u, v) = f(x, y)$ for all x, y. Show that

$$[D_1F(u, v)]^2 + [D_2F(u, v)]^2 = [D_1f(x, y)]^2 + [D_2f(x, y)]^2.$$

40.F. Let $f : \mathbf{R}^2 \to \mathbf{R}$ be differentiable on $\mathbf{R}^2$, let $g : (0, +\infty) \times \mathbf{R} \to \mathbf{R}$ be defined by $g(r, \theta) = (r \cos \theta, r \sin \theta)$, and let $F = f \circ g$. Calculate D_1F and D_2F and show that

$$[D_1F(r, \theta)]^2 + \frac{1}{r^2} [D_2F(r, \theta)]^2 = [D_1f(r \cos \theta, r \sin \theta)]^2$$

$$+ [D_2f(r \cos \theta, r \sin \theta)]^2.$$

40.G. Let $f : \mathbf{R} \to \mathbf{R}$ be differentiable on $\mathbf{R}$.

(a) If $F(x, y) = f(xy)$, then $xD_1F(x, y) = yD_2F(x, y)$ for all (x, y).

(b) If $F(x, y) = f(ax + by)$ where $a, b \in \mathbf{R}$, then $bD_1F(x, y) = aD_2F(x, y)$ for all (x, y).

(c) If $F(x, y) = f(x^2 + y^2)$, then $yD_1F(x, y) = xD_2F(x, y)$ for all (x, y).

(d) If $F(x, y) = f(x^2 - y^2)$, then $yD_1F(x, y) + xD_2F(x, y) = 0$ for all (x, y).

40.H. Let $A \subseteq \mathbf{R}^p$ and let c be an interior point of A. Suppose that f, g are defined on A to $\mathbf{R}^q$ and are differentiable at c. If $h : A \to \mathbf{R}$ is defined by $h(x) = f(x) \cdot g(x)$ for all $x \in A$, show that h is differentiable at c and that if $u \in \mathbf{R}^p$, then

$$Dh(c)(u) = (Df(c)(u)) \cdot g(c) + f(c) \cdot (Dg(c)(u)).$$

40.I. Express the result of Exercise 40.H in terms of the coordinate functions.

40.J. Let $A \subseteq \mathbf{R}$ and let c be an interior point of A. Suppose that $f : A \to \mathbf{R}^p$ is differentiable at c and such that $\|f(x)\| = 1$ for $x \in A$. Show that $f(c) \cdot \nabla_c f = 0$, where $\nabla_c f$ denotes the gradient of f at c (see Exercise 39.J). Interpret this conclusion geometrically.

40.K. Let $f : \mathbf{R}^p \to \mathbf{R}$ be (positively) **homogeneous of degree** k in the sense that

$$f(tx) = t^k f(x) \qquad \text{for} \qquad x \in \mathbf{R}^p, t > 0.$$

(a) If f is differentiable on $\mathbf{R}^p$, show that it satisfies **Euler's†** **Relation**:

$$kf(x) = x_1 D_1 f(x) + \cdots + x_p D_p f(x)$$

for all $x = \{x_1, \ldots, x_p\}$ in $\mathbf{R}^p$ with $x \neq 0$.

(b) Conversely, let f satisfy Euler's Relation and let $c \in \mathbf{R}^p$, $c \neq 0$. Let $g(t) = f(tc)$ for $t > 0$ and show that $tg'(t) = kg(t)$ for $t > 0$. Use this to prove that f is homogeneous of degree k.

40.L. Let $A \subseteq \mathbf{R}^p$, $f : A \to \mathbf{R}^p$, and let the function $g : f(A) \to \mathbf{R}^p$ be inverse to f in the sense that

$$f \circ g(x) = x, \qquad g \circ f(y) = y$$

for all $x \in A$ and $y \in f(A)$. If f is differentiable at a point $a \in A$ and if g is differentiable at $b = f(a)$, show that the linear functions $Df(a)$ and $Dg(b)$ are inverse to each other; that is, $Df(a) \circ Dg(b)$ and $Dg(b) \circ Df(a)$ are the identity on $\mathbf{R}^p$.

40.M. Let $B : \mathbf{R}^p \times \mathbf{R}^p = \mathbf{R}^{2p} \to \mathbf{R}^q$ be **bilinear** in the sense that

$$B(ax + bx', y) = aB(x, y) + bB(x', y),$$

$$B(x, ay + by') = aB(x, y) + bB(x, y')$$

for all a, $b \in \mathbf{R}$ and all x, x', y, y' in $\mathbf{R}^p$. It can be proved that there exists $M > 0$ such that $\|B(x, y)\| \le M \|x\| \|y\|$ for all x, y in $\mathbf{R}^p$. Assuming this, prove that B is differentiable at every point $(x, y) \in \mathbf{R}^p \times \mathbf{R}^p = \mathbf{R}^{2p}$ and that

$$DB(x, y)(u, v) = B(x, v) + B(u, y)$$

for all (u, v) in $\mathbf{R}^p \times \mathbf{R}^p = \mathbf{R}^{2p}$.

40.N. Let $B : \mathbf{R}^p \times \mathbf{R}^p \to \mathbf{R}^q$ be bilinear in the sense of the preceding exercise and let $g(x) = B(x, x)$ for all $x \in \mathbf{R}^p$. Show that if x, $u \in \mathbf{R}^p$, then

(i) $g(tx) = t^2 g(x)$ for all $t \in \mathbf{R}$;

(ii) $Dg(x)(u) = B(x, u) + B(u, x) = Dg(u)(x)$;

(iii) $g(x + u) = g(x) + Dg(x)(u) + g(u)$.

Moreover, if B is symmetric in the sense that $B(x, y) = B(y, x)$, then

(iv) $Dg(x)(u) = 2B(x, u)$.

40.O. Give a proof of Exercise 40.H using 40.M.

40.P. Let $\Omega \subseteq \mathbf{R}^p$ be open and let $f : \Omega \to \mathbf{R}^q$ be differentiable on Ω. Let $I = (a, b)$ be an open interval in $\mathbf{R}$ and let $g : I \to \mathbf{R}^p$ be differentiable on I and such that $g(I) \subseteq \Omega$. If $h = f \circ g : I \to \mathbf{R}^q$ show that

$$h'(c) = Df(g(c))(g'(c)).$$

40.Q. Let $\Omega \subseteq \mathbf{R}^p$ be open and let $f : \Omega \to \mathbf{R}^q$. Suppose that Ω contains the points a, b and the line segment S joining these points, and that f is differentiable at every point of S. Show that there exists a linear mapping $L : \mathbf{R}^p \to \mathbf{R}^q$ such that $f(b) - f(a) = L(b - a)$.

† LEONARD EULER (1707–1783), a native of Basel, studied with Johann Bernoulli. He resided many years at the court in St. Petersburg, but this stay was interrupted by twenty-five years in Berlin. Despite the fact that he was the father of thirteen children and became totally blind, he was still able to write over eight hundred papers and books and make fundamental contributions to all branches of mathematics.

40.R. Let $\Omega \subseteq \mathbf{R}^p$ be an open connected set and let $f : \Omega \to \mathbf{R}^q$ be differentiable on Ω. If $Df(x) = 0$ for all $x \in \Omega$, show that $f(x) = f(y)$ for all $x, y \in \Omega$. Show that this conclusion may fail if Ω is not connected.

40.S. Let $J \subseteq \mathbf{R}^p$ be an open cell and suppose that $f : J \to \mathbf{R}$ is differentiable on J. Show that if the partial derivative $D_1 f(x) = 0$ for all $x \in J$, then f does not depend on the first variable in the sense that

$$f(x_1, x_2, \ldots, x_p) = f(x_1', x_2, \ldots, x_p)$$

for any two points in J whose second, $\ldots$, pth coordinates are the same.

40.T. Show that the conclusion of the preceding exercise may fail if J is not assumed to be a cell.

40.U. Let $f : \mathbf{R}^2 \to \mathbf{R}$ be defined by

$$f(x, y) = \frac{xy(x^2 - y^2)}{x^2 + y^2} \qquad \text{for } (x, y) \neq (0, 0),$$

$$= 0 \qquad \text{for } (x, y) = (0, 0).$$

Show that the second partial derivatives $D_{xy}f$ and $D_{yx}f$ exist at $(0, 0)$ but are not equal.

40.V. Use the Mean Value Theorem to determine approximately the distance from the point $(3.2, 4.1)$ to the origin. Give error bounds for your estimate.

40.W. Let $\Omega \subseteq \mathbf{R}^p$ be open and let $f : \Omega \to \mathbf{R}^q$. Suppose that Ω contains the points a, b and the line segment S joining the points, and that f has continuous partial derivatives on S. Show that

$$f(b) - f(a) = \int_0^1 Df(a + t(b - a))(b - a) \, dt.$$

40.X. Let f, $g : \mathbf{R} \to \mathbf{R}$ have continuous second derivatives on $\mathbf{R}$.

(a) If $c \in \mathbf{R}$ and $u(x, y) = f(x + cy) + g(x - cy)$, show that $u : \mathbf{R}^2 \to \mathbf{R}$ satisfies the "wave equation"

$$c^2 D_{xx} u(x, y) = D_{yy} u(y, y)$$

for all (x, y).

(b) If $v(x, y) = f(3x + 2y) + g(x - 2y)$, show that $v : \mathbf{R}^2 \to \mathbf{R}$ satisfies the equation

$$4 D_{xx} v(x, y) - 4 D_{xy} v(x, y) - 3 D_{yy} v(x, y) = 0$$

for all (x, y).

40.Y. If $f : \mathbf{R}^2 \to \mathbf{R}$ has continuous second partial derivatives and if $F(r, \theta) = f(r \cos \theta, r \sin \theta)$ for $r > 0$, $\theta \in \mathbf{R}$, show that

$$D_{xx} f(x, y) + D_{yy} f(x, y) = D_{rr} F(r, \theta) + \frac{1}{r} D_r F(r, \theta) + \frac{1}{r^2} D_{\theta\theta} F(r, \theta)$$

$$= \frac{1}{r} D_r (r D_r F(r, \theta)) + \frac{1}{r^2} D_{\theta\theta} F(r, \theta),$$

where $x = r \cos \theta$, $y = r \sin \theta$.

Project

40.α. (This project is a modification of the classical **Newton's Method** for the location of roots when a sufficiently close approximation is known.) Let f be defined and continuous on an open set containing the closed ball $B_r(x_0) = \{x \in \mathbf{R}^p : \|x - x_0\| \le r\}$ with values in $\mathbf{R}^q$. Suppose that f is differentiable at every point of $B_r(x_0)$ and that there exists a number C, with $0 < C < 1$, and an injective linear map $\Gamma : \mathbf{R}^q \to \mathbf{R}^p$ such that $\|\Gamma \circ f(x_0)\| \le (1 - C)r$ and such that

$$\|I - \Gamma \circ Df(x)\|_{pp} \le C \qquad \text{for } x \in B_r(x_0).$$

(a) Let $g : B_r(x_0) \to \mathbf{R}^p$ be defined by $g(x) = x - \Gamma \circ f(x)$ for $x \in B_r(x_0)$. Show that g is differentiable at every point of $B_r(x_0)$ and that g is a contraction with constant $C < 1$ (see 23.4) on $B_r(x_0)$.

(b) Define $x_1 = g(x_0)$ and $x_{n+1} = g(x_n)$ for $n \in \mathbf{N}$. Show that $\|x_{n+1} - x_n\| \le C^n \|x_1 - x_0\|$, whence it follows that $\|x_{n+1} - x_m\| \le C^m r$ for $n \ge m \ge 0$. Hence $\|x_k - x_0\| < r$ for $k = 0, 1, 2, \ldots$.

(c) Show that (x_k) is a Cauchy sequence and hence converges to an element $\bar{x} \in B_r(x_0)$, which is such that $g(\bar{x}) = \bar{x}$. Moreover, we have the estimate $\|x_k - \bar{x}\| \le C^k r$.

(d) Show that $f(\bar{x}) = 0$ and $\bar{x}$ is the only element in $B_r(x_0)$ where f vanishes.

Section 41 Mapping Theorems and Implicit Functions

Let Ω be an open set in $\mathbf{R}^p$ and let f be a function with domain Ω and range in $\mathbf{R}^q$; unless there is specific mention, we do *not* assume that $p = q$. It will be shown that, under assumptions that will be stated, the "local character" of the mapping f at a point $c \in \Omega$ is indicated by the linear mapping $Df(c)$. Somewhat more precisely:

(i) if $p \le q$ and $Df(c)$ is injective ($=$ one-one), then f is injective on small neighborhoods of c;

(ii) if $p \ge q$ and $Df(c)$ is surjective ($=$ maps $\mathbf{R}^p$ onto $\mathbf{R}^q$), then the image under f of a small neighborhood of c is a neighborhood of $f(c)$; and

(iii) if $p = q$ and $Df(c)$ is bijective ($=$ one-one and onto $=$ invertible), then f maps a neighborhood U of c in a one-one fashion onto a neighborhood V of $f(c)$. In case (iii), there is a function defined on V which is inverse to the restriction of f to U.

As a consequence of these mapping theorems we shall obtain the Implicit Function Theorem, which is one of the fundamental theorems in analysis and geometry. We also present a useful Parametrization Theorem and the important Rank Theorem.

The Class $C^1(\Omega)$

The mere existence of the derivative is not enough for our purposes; we need also the continuity of the derivative. We recall that if $f:\Omega \to \mathbf{R}^q$ is differentiable at every point of $\Omega \subseteq \mathbf{R}^p$, then the function $x \mapsto Df(x)$ is a map of Ω into the collection $\mathscr{L}(\mathbf{R}^p, \mathbf{R}^q)$ of all linear functions from $\mathbf{R}^p$ to $\mathbf{R}^q$. It was noted in Section 21 that this set $\mathscr{L}(\mathbf{R}^p, \mathbf{R}^q)$ is a vector space and, in Exercise 21.L, that this space is a normed space under the norm

$$(41.1) \qquad \|L\|_{pq} = \sup \{\|L(x)\| : x \in \mathbf{R}^p, \|x\| \le 1\}.$$

41.1 Definition. If Ω is open in $\mathbf{R}^p$ and $f:\Omega \to \mathbf{R}^q$, we say that f belongs to **Class** $C^1(\Omega)$ if the derivative $Df(x)$ exists for all $x \in \Omega$ and the mapping $x \mapsto Df(x)$ of Ω into $\mathscr{L}(\mathbf{R}^p, \mathbf{R}^q)$ is continuous under the norm (41.1).

We recall from Example 39.8(d), that for each $x \in \Omega$, the derivative $Df(x)$ can be represented by the $q \times p$ Jacobian matrix $[D_j f_i(x)]$. Hence $Df(x) - Df(y)$ is represented by the $q \times p$ matrix

$$[D_j f_i(x) - D_j f_i(y)].$$

Now it follows from the inequality (21.5), that

$$\|Df(x) - Df(y)\|_{pq} \le \left\{ \sum_{i=1}^{q} \sum_{j=1}^{p} |D_j f_i(x) - D_j f_i(y)|^2 \right\}^{1/2}.$$

Hence the continuity of each of the partial derivatives $D_j f_i$ on Ω implies continuity of $x \mapsto Df(x)$. We leave it to the reader to show that the converse is also true. Hence we have the following result.

41.2 Theorem. *If $\Omega \subseteq \mathbf{R}^p$ is open and $f:\Omega \to \mathbf{R}^q$ is differentiable at every point of Ω, then f belongs to Class $C^1(\Omega)$ if and only if the partial derivatives $D_j f_i$, $i = 1, \dots, q$, $j = 1, \dots, p$, of f are continuous on Ω.*

We shall need the next lemma, which is a variant of the Mean Value Theorem.

41.3 Lemma. *Let $\Omega \subseteq \mathbf{R}^p$ be an open set and let $f:\Omega \to \mathbf{R}^q$ be differentiable on Ω. Suppose that Ω contains the points a, b and the line segment S joining these points, and let $x_0 \in \Omega$. Then we have*

$$\|f(b) - f(a) - Df(x_0)(b - a)\| \le \|b - a\| \sup_{x \in S} \{\|Df(x) - Df(x_0)\|_{pq}\}.$$

PROOF. Let $g:\Omega \to \mathbf{R}^q$ be defined for $x \in \Omega$ by

$$g(x) = f(x) - Df(x_0)(x).$$

Since $Df(x_0)$ is linear, it follows that $Dg(x) = Df(x) - Df(x_0)$ for $x \in \Omega$. If we apply the Mean Value Theorem 40.5, we infer that there exists a point $c \in S$ such that

$$\|f(b) - f(a) - Df(x_0)(b - a)\| = \|g(b) - g(a)\|$$
$$\leq \|Dg(c)(b - a)\| = \|(Df(c) - Df(x_0))(b - a)\|$$
$$\leq \|b - a\| \sup_{x \in S} \{\|Df(x) - Df(x_0)\|_{pq}\}. \qquad \text{Q.E.D.}$$

The next result is the key lemma to the mapping theorems.

41.4 APPROXIMATION LEMMA. *Let $\Omega \subseteq \mathbf{R}^p$ be open and let $f : \Omega \to \mathbf{R}^q$ belong to Class $C^1(\Omega)$. If $x_0 \in \Omega$ and $\varepsilon > 0$, then there exists $\delta(\varepsilon) > 0$ such that if $\|x_k - x_0\| \leq \delta(\varepsilon)$, $k = 1, 2$, then $x_k \in \Omega$ and*

$$(41.2) \qquad \|f(x_1) - f(x_2) - Df(x_0)(x_1 - x_2)\| \leq \varepsilon \|x_1 - x_2\|.$$

PROOF. Since $x \mapsto Df(x)$ is continuous on Ω to $\mathcal{L}(\mathbf{R}^p, \mathbf{R}^q)$, then given $\varepsilon > 0$ there exists $\delta(\varepsilon) > 0$ such that if $\|x - x_0\| < \delta(\varepsilon)$ then $x \in \Omega$ and $\|Df(x) - Df(x_0)\|_{pq} \leq \varepsilon$. Now let x_1, x_2 satisfy $\|x_k - x_0\| \leq \delta(\varepsilon)$, whence the line segment joining x_1 and x_2 lies inside of the closed ball with center x_0 and radius $\delta(\varepsilon)$, and hence inside Ω. Now apply Lemma 41.3 to obtain the stated conclusion. Q.E.D.

The Injective Mapping Theorem

We shall now show that if f belongs to Class $C^1(\Omega)$ and if $Df(c)$ is injective, then the restriction of f to a suitable neighborhood of c is an injection.

A reader familiar with the notion of the "rank" of a linear transformation, will recall that $L : \mathbf{R}^p \to \mathbf{R}^q$ is injective if and only if rank $(L) = p \leq q$.

41.5 INJECTIVE MAPPING THEOREM. *Suppose that $\Omega \subseteq \mathbf{R}^p$ is open, that $f : \Omega \to \mathbf{R}^q$ belongs to Class $C^1(\Omega)$, and that $L = Df(c)$ is an injection. Then there exists a number $\delta > 0$ such that the restriction of f to $B_\delta = \{x \in \mathbf{R}^p : \|x - c\| \leq \delta\}$ is an injection. Moreover, the inverse of the restriction $f \mid B_\delta$ is a continuous function on $f(B_\delta) \subseteq \mathbf{R}^q$ to $B_\delta \subseteq \mathbf{R}^p$.*

PROOF Since the linear function $L = Df(c)$ is an injection, it follows from Corollary 22.7 that there exists an $r > 0$ such that

$$(41.3) \qquad r \|u\| \leq \|Df(c)(u)\| \qquad \text{for } u \in \mathbf{R}^p.$$

We now apply the Approximation Lemma 41.4 with $\varepsilon = \frac{1}{2}r$ to obtain a number $\delta > 0$ such that if $\|x_k - c\| \leq \delta$, $k = 1, 2$, then

$$\|f(x_1) - f(x_2) - L(x_1 - x_2)\| \leq \tfrac{1}{2}r \|x_1 - x_2\|.$$

If we apply the Triangle Inequality to the left side of this inequality, we obtain

$$\|L(x_1 - x_2)\| - \|f(x_1) - f(x_2)\| \le \tfrac{1}{2}r \, \|x_1 - x_2\|.$$

If we combine this and (41.3) with $u = x_1 - x_2$, we obtain

(41.4) $$\tfrac{1}{2}r \, \|x_1 - x_2\| \le \|f(x_1) - f(x_2)\|$$

for $x_k \in B_\delta$. This proves that the restriction of f to B_δ is an injection; hence this restriction has an inverse function which we shall denote by g. If $y_k \in f(B_\delta)$, then there exist unique points $x_k = g(y_k)$ in B_δ such that $y_k = f(x_k)$. It follows from (41.4) that

$$\|g(y_1) - g(y_2)\| \le (2/r) \, \|y_1 - y_2\|,$$

whence it follows that $g = (f \mid B_\delta)^{-1}$ is uniformly continuous on $f(B_\delta)$ to $\mathbf{R}^p$.

Q.E.D.

We note that g need not be defined on a neighborhood of $f(c)$; that is, $f(c)$ need not be an interior point of $f(B_\delta)$. For that reason we can make no assertion about the differentiability of g. A stronger inversion theorem will be established below under additional hypotheses.

The Surjective Mapping Theorem

The next result is a companion to the Injective Mapping Theorem. This theorem, which is due to L. M. Graves,† asserts that if f is in Class $C^1(\Omega)$ and if for some $c \in \Omega$, the linear map $Df(c)$ is a surjection of $\mathbf{R}^p$ onto $\mathbf{R}^q$, then f maps a suitable neighborhood of c to a neighborhood of $f(c)$. Thus every point of $\mathbf{R}^q$ which is sufficiently close to $f(c)$ is the image under f of a point near c.

A reader familiar with the notion of the "rank" of a linear transformation will recall that $L : \mathbf{R}^p \to \mathbf{R}^q$ is surjective if and only if rank $(L) = q \le p$.

41.6 SURJECTIVE MAPPING THEOREM. *Let $\Omega \subseteq \mathbf{R}^p$ be open and let $f : \Omega \to \mathbf{R}^q$ belong to Class $C^1(\Omega)$. Suppose that for some $c \in \Omega$, the linear function $L = Df(c)$ is a surjection of $\mathbf{R}^p$ onto $\mathbf{R}^q$. Then there exist numbers $m > 0$ and $\alpha > 0$ such that if $y \in \mathbf{R}^q$ and $\|y - f(c)\| \le \alpha/2m$, then there exists an $x \in \Omega$ such that $\|x - c\| \le \alpha$ and $f(x) = y$.*

PROOF. Since L is a surjection, each of the standard basic vectors

$$e_1 = (1, 0, \ldots, 0), \qquad e_2 = (0, 1, \ldots, 0), \ldots, e_q = (0, 0, \ldots, 1)$$

---

† LAWRENCE M. GRAVES (1896–1973) was born in Kansas, but was associated with the University of Chicago for many years as student and professor. He is best known for his contributions to functional analysis and the calculus of variations.

in $\mathbf{R}^q$ is the image under L of some vector in $\mathbf{R}^p$, say $u_1, u_2, \ldots, u_q$. Now let $M : \mathbf{R}^q \to \mathbf{R}^p$ be the linear function mapping e_j into u_j for $j = 1, 2, \ldots, q$; that is,

$$M\left(\sum_{i=1}^{q} a_i e_i\right) = \sum_{i=1}^{q} a_i u_i.$$

It follows that $L \circ M$ is the identity mapping on $\mathbf{R}^q$; that is, $L \circ M(y) = y$ for all $y \in \mathbf{R}^q$. If we let

$$m = \left\{\sum_{i=1}^{q} \|u_j\|^2\right\}^{1/2},$$

then an application of the Triangle and Schwarz Inequalities implies that if $y = \sum_{i=1}^{q} a_i e_i$, then

$$\begin{aligned}\|M(y)\| &\le \sum_{i=1}^{q} |a_i|\,\|u_i\| \\ &\le \left\{\sum_{i=1}^{q} |a_i|^2\right\}^{1/2}\left\{\sum_{i=1}^{q} \|u_i\|^2\right\}^{1/2} \\ &= m\,\|y\|.\end{aligned}$$

By the Approximation Lemma 41.4 there exists a number $\alpha > 0$ such that if $\|x_k - c\| \le \alpha$, $k = 1, 2$, then $x_k \in \Omega$ and

$$(41.5) \qquad \|f(x_1) - f(x_2) - L(x_1 - x_2)\| \le \frac{1}{2m}\|x_1 - x_2\|.$$

Now let $B_\alpha = \{x \in \mathbf{R}^p : \|x - c\| \le \alpha\}$ and suppose that $y \in \mathbf{R}^q$ is such that $\|y - f(c)\| \le \alpha/2m$. We will show that there exists a vector x with $x \in B_\alpha$ such that $y = f(x)$.

Let $x_0 = c$ and let $x_1 = x_0 + M(y - f(c))$ so that $\|x_1 - x_0\| \le m\,\|y - f(c)\| \le \frac{1}{2}\alpha$, whence

$$\|x_1 - x_0\| \le \frac{\alpha}{2} \qquad \text{and} \qquad \|x_1 - c\| \le \left(1 - \frac{1}{2}\right)\alpha.$$

Suppose that $c = x_0, x_1, \ldots, x_n$ have been chosen inductively in $\mathbf{R}^p$ such that

$$(41.6) \qquad \|x_k - x_{k-1}\| \le \alpha/2^k, \qquad \|x_k - c\| \le (1 - 1/2^k)\alpha,$$

for $k = 1, \ldots, n$. We now define x_{n+1} $(n \ge 1)$ by

$$(41.7) \qquad x_{n+1} = x_n - M[f(x_n) - f(x_{n-1}) - L(x_n - x_{n-1})].$$

It follows from (41.5) that

$$\begin{aligned}\|x_{n+1} - x_n\| &\le m\,\|f(x_n) - f(x_{n-1}) - L(x_n - x_{n-1})\| \\ &\le \frac{1}{2}\|x_n - x_{n-1}\|,\end{aligned}$$

whence it follows that $\|x_{n+1} - x_n\| \le \frac{1}{2}(\alpha/2^n) = \alpha/2^{n+1}$ and

$$\|x_{n+1} - c\| \le \|x_{n+1} - x_n\| + \|x_n - c\|$$

$$\le (\alpha/2^{n+1}) + (1 - 1/2^n)\alpha$$

$$= (1 - 1/2^{n+1})\alpha.$$

Hence (41.6) is also established for $k = n+1$. Therefore, we can construct a sequence (x_n) in B_α in this way. If $m \ge n$, then we have

$$\|x_n - x_m\| \le \|x_n - x_{n+1}\| + \|x_{n+1} - x_{n+2}\| + \cdots + \|x_{m-1} - x_m\|$$

$$\le \frac{\alpha}{2^{n+1}} + \frac{\alpha}{2^{n+2}} + \cdots + \frac{\alpha}{2^m} \le \frac{\alpha}{2^n}.$$

It follows that (x_n) is a Cauchy sequence in $\mathbf{R}^p$ and therefore converges to some element x. Since $\|x_n - c\| \le (1 - 1/2^n)\alpha$, it follows that $\|x - c\| \le \alpha$ so that $x \in B_\alpha$.

Since $x_1 - x_0 = M(y - f(c))$, it follows that

$$L(x_1 - x_0) = L \circ M(y - f(c)) = y - f(x_0).$$

Moreover, by (41.7) we have

$$L(x_{n+1} - x_n) = -L \circ M[f(x_n) - f(x_{n-1}) - L(x_n - x_{n-1})]$$

$$= -\{f(x_n) - f(x_{n-1}) - L(x_n - x_{n-1})\}$$

$$= L(x_n - x_{n-1}) - [f(x_n) - f(x_{n-1})]$$

By induction we find that

$$L(x_{n+1} - x_n) = y - f(x_n),$$

whence it follows that $y = \lim f(x_n) = f(x)$. Hence every point y satisfying $\|y - f(c)\| \le \alpha/2m$ is the image under f of a point $x \in \Omega$ with $\|x - c\| \le \alpha$.

Q.E.D.

41.7 OPEN MAPPING THEOREM. *Let* $\Omega \subseteq \mathbf{R}^p$ *be open and let* $f: \Omega \to \mathbf{R}^q$ *belong to Class* $C^1(\Omega)$. *If for each* $x \in \Omega$ *the derivative* $Df(x)$ *is a surjection, and if* $G \subseteq \Omega$ *is open, then* $f(G)$ *is open in* $\mathbf{R}^q$.

PROOF. If $b \in f(G)$, then there exists a point $c \in G$ such that $f(c) = b$. It follows from the Surjective Mapping Theorem 41.6 applied to $f \mid G$ that there exists $\beta > 0$ such that if $\|y - b\| \le \beta$ then there exists an $x \in G$ such that $y = f(x)$. Hence $f(G)$ is open in $\mathbf{R}^q$.

Q.E.D.

The Inversion Theorem

We shall now combine our two mapping theorems in the case $p = q$. Here the derivative $Df(c)$ is assumed to be a bijection. This is the case if and only if the derivative $Df(c)$ has an inverse which, in turn, is true if and only if the Jacobian determinant

$$J_f(c) = \det [D_j f_i(c)] = \det [f_{i,j}(c)]$$

is different from zero.

A reader familiar with the notion of the "rank" of a linear transformation will recall that $L : \mathbf{R}^p \to \mathbf{R}^q$ is bijective if and only if rank $(L) = p = q$.

It follows from the continuity of the partial derivatives and of the determinant that if $Df(a)$ is invertible, then $Df(x)$ is invertible for x sufficiently close to c.

41.8 INVERSION THEOREM. *Let $\Omega \subseteq \mathbf{R}^p$ be open and suppose that $f : \Omega \to \mathbf{R}^p$ belongs to Class $C^1(\Omega)$. If $c \in \Omega$ is such that $Df(c)$ is a bijection, then there exists an open neighborhood U of c such that $V = f(U)$ is an open neighborhood of $f(c)$ and the restriction of f to U is a bijection onto V with continuous inverse g. Moreover g belongs to Class $C^1(V)$ and*

$$Dg(y) = [Df(g(y))]^{-1} \qquad for \ y \in V.$$

PROOF. By hypothesis $L = Df(c)$ is injective; hence Corollary 22.7 implies that there exists $r > 0$ such that

$$2r \|z\| \le \|Df(c)(z)\| \qquad for \ z \in \mathbf{R}^p.$$

Since f is in Class $C^1(\Omega)$, there is a neighborhood of c on which $Df(x)$ is invertible and satisfies

$$(41.8) \qquad r \|z\| \le \|Df(x)(z)\| \qquad for \ z \in \mathbf{R}^p.$$

We further restrict our attention to a neighborhood U of c on which f is injective and which is contained in a ball with center c and radius α (as in the Surjective Mapping Theorem 41.6). Then $V = f(U)$ is a neighborhood of $f(c)$, and we infer from the preceding mapping theorems that the restriction $f \mid U$ has a continuous inverse function $g : V \to \mathbf{R}^p$.

It remains to show that g is differentiable at an arbitrary point $y_1 \in V$. Let $x_1 = g(y_1) \in U$; since f is differentiable at x_1, it follows that if $x \in U$, then

$$f(x) - f(x_1) - Df(x_1)(x - x_1) = \|x - x_1\| u(x),$$

where $\|u(x)\| \to 0$ as $x \to x_1$. If we let M_1 be the inverse of the linear function $Df(x_1)$, then

$$x - x_1 = M_1[Df(x_1)(x - x_1)]$$
$$= M_1[f(x) - f(x_1) - \|x - x_1\| u(x)].$$

If $x \in U$, then $x = g(y)$ for some $y = f(x) \in V$; moreover $y_1 = f(x_1)$, so this equation can be written in the form

$$g(y) - g(y_1) - M_1(y - y_1) = -\|x - x_1\| M_1(u(x)).$$

Since $Df(x_1)$ is injective, it follows as in the proof of the Injective Mapping Theorem 41.5 that

$$\|y - y_1\| = \|f(x) - f(x_1)\| \geq \tfrac{1}{2}r \|x - x_1\|$$

provided y is sufficiently close to y_1. Moreover, it follows from (41.8) that $\|M_1(u)\| \leq (1/r) \|u\|$ for all $u \in \mathbf{R}^p$. Therefore we have

$$\|g(y) - g(y_1) - M_1(y - y_1)\| \leq (2/r^2) \|u(x)\| \|y - y_1\|.$$

Now as $y \to y_1$, then $x = g(y) \to g(y_1) = x_1$ and so $\|u(x)\| \to 0$. We conclude, therefore, that $Dg(y_1)$ exists and equals $M_1 = (Df(x_1))^{-1}$.

The fact that g belongs to Class $C^1(V)$ follows from the relation $Dg(y) = [Df(g(y))]^{-1}$ for $y \in V$, and the continuity of the mappings

$$y \mapsto g(y), \qquad x \mapsto Df(x), \qquad L \mapsto L^{-1}$$

of $V \to U$, $U \to \mathscr{L}(\mathbf{R}^p, \mathbf{R}^p)$, and $\mathscr{L}(\mathbf{R}^p, \mathbf{R}^p) \to \mathscr{L}(\mathbf{R}^p, \mathbf{R}^p)$, respectively. (See Exercise 41.L.)   Q.E.D.

Implicit Functions

Suppose that F is a function that is defined on a subset of $\mathbf{R}^p \times \mathbf{R}^q$ into $\mathbf{R}^q$. (If we make the obvious identification of $\mathbf{R}^p \times \mathbf{R}^q$ with $\mathbf{R}^{p+q}$, then we do not need to define what it means to say that F is continuous, or is differentiable at a point, or is in Class C^1 on a set.) Suppose that F takes the point (a, b) into the zero vector of $\mathbf{R}^q$. The problem of implicit functions is to solve the equation

$$F(x, y) = 0$$

for one argument (say, y) in terms of the other in the sense that we find a function φ defined on a subset of $\mathbf{R}^p$ with values in $\mathbf{R}^q$ such that $b = \varphi(a)$ and

$$F(x, \varphi(x)) = 0$$

for all x in the domain of φ. We assume that F is continuous on a neighborhood of (a, b) and we hope to conclude that the "solution function" φ is continuous on a neighborhood of a. It will probably be no surprise to the reader that we shall assume that F belongs to Class C^1 on a neighborhood of (a, b); however, even this hypothesis is not enough to guarantee the existence and uniqueness of a continuous solution function φ defined on a neighborhood of a.

Indeed, if $p = q = 1$, then the function given by $F(x, y) = x^2 - y^2$ has two continuous solution functions $\varphi_1(x) = x$ and $\varphi_2(x) = -x$ corresponding to the point $(0, 0)$. It also has discontinuous solutions, such as

$$\varphi_3(c) = x, \qquad x \text{ rational,}$$
$$= -x, \qquad x \text{ irrational.}$$

The function $G(x, y) = x - y^2$ has two continuous solution functions corresponding to $(0, 0)$, but neither of them is defined on a neighborhood of the point $x = 0$. To give a more exotic example, the function $H : \mathbf{R}^2 \to \mathbf{R}$ defined by

$$H(x, y) = x, \qquad\qquad y = 0,$$
$$= x - y^3 \sin\left(\frac{1}{y}\right), \qquad y \neq 0,$$

belongs to Class C^1 on a neighborhood of $(0, 0)$, but there is no continuous solution function defined on a neighborhood of $x = 0$.

In all three of these examples the partial derivative with respect to y vanishes at the point under consideration. In the case $p = q = 1$, the additional assertion needed to guarantee the existence and uniqueness of the solution function is that this partial derivative be non-zero. In the general case, we observe that $DF(a, b)$ is a continuous linear function on $\mathbf{R}^p \times \mathbf{R}^q$ to $\mathbf{R}^q$ and induces a continuous linear function $L_2 : \mathbf{R}^q \to \mathbf{R}^q$ defined by

$$L_2(v) = DF(a, b)(0, v)$$

for $v \in \mathbf{R}^q$. In a very reasonable sense, L_2 is the "partial derivative" of F with respect to $y \in \mathbf{R}^q$ at the point (a, b). The additional assumption that we shall impose is that L_2 be invertible.

We now wish to interpret this problem in terms of coordinates. If $x = (x_1, \ldots, x_p)$ and $y = (y_1, \ldots, y_q)$, then the equation $F(x, y) = 0$ takes the form of q equations in the $p + q$ arguments $x_1, \ldots, x_p, y_1, \ldots, y_q$ given by

$$f_1(x_1, \ldots, x_p, y_1, \ldots, y_q) = 0,$$
(41.9) $\qquad \cdot \quad \cdot \quad \cdot \quad \cdot \quad \cdot \quad \cdot \quad \cdot \quad \cdot \quad \cdot \quad \cdot \quad \cdot \quad \cdot \quad \cdot$
$$f_q(x_1, \ldots, x_p, y_1, \ldots, y_q) = 0.$$

For the sake of convenience, suppose that $a = 0$ and $b = 0$ so that this system is satisfied for $x_1 = 0, \ldots, x_p = 0, y_1 = 0, \ldots, y_q = 0$, and it is desired to solve for the y_i in terms of the x_i at least when the $|x_i|$ are sufficiently small. If the functions f_i are linear, then the condition for solvability is that the determinant of coefficients of the y_i should be non-zero. If the functions f_i are not linear, then the condition is that the Jacobian determinant

$$\frac{\partial(f_1, \ldots, f_q)}{\partial(y_1, \ldots, y_q)}(a, b) \neq 0.$$

When this is the case, there are functions φ_j, $j = 1, \ldots, q$, defined and continuous near $a = 0$ such that if we substitute

$$y_1 = \varphi_1(x_1, \ldots, x_p),$$
$$\cdots \cdots \cdots \cdots$$
$$y_q = \varphi_q(x_1, \ldots, x_p),$$

into the system (41.9), then we obtain an identity in the x_i.

41.9 IMPLICIT FUNCTION THEOREM. *Let $\Omega \subseteq \mathbf{R}^p \times \mathbf{R}^q$ be open and let $(a, b) \in \Omega$. Suppose that $F : \Omega \to \mathbf{R}^q$ belongs to Class $C^1(\Omega)$, that $F(a, b) = 0$, and that the linear map defined by*

$$L_2(v) = DF(a, b)(0, v), \qquad v \in \mathbf{R}^q,$$

is a bijection of $\mathbf{R}^q$ onto $\mathbf{R}^q$.

(a) *Then there exists an open neighborhood W of $a \in \mathbf{R}^p$ and a unique function $\varphi : W \to \mathbf{R}^q$ belonging to Class $C^1(W)$ such that $b = \varphi(a)$ and*

$$F(x, \varphi(x)) = 0 \qquad \text{for all } x \in W.$$

(b) *There exists an open neighborhood U of (a, b) in $\mathbf{R}^p \times \mathbf{R}^q$ such that the pair $(x, y) \in U$ satisfies $F(x, y) = 0$ if and only if $y = \varphi(x)$ for $x \in W$.*

PROOF.   It is no loss of generality to assume that $a = 0$ and $b = 0$. Let $H : \Omega \to \mathbf{R}^p \times \mathbf{R}^q$ be defined by

$$H(x, y) = (x, F(x, y)) \qquad \text{for } (x, y) \in \Omega.$$

It follows readily (see Exercise 39.V) that H belongs to Class $C^1(\Omega)$ and that

$$DH(x, y)(u, v) = (u, DF(x, y)(u, v))$$

for $(x, y) \in \Omega$ and $(u, v) \in \mathbf{R}^p \times \mathbf{R}^q$. We now claim that $DH(0, 0)$ is invertible on $\mathbf{R}^q \times \mathbf{R}^q$. Indeed, if we let $L_1 \in \mathscr{L}(\mathbf{R}^p, \mathbf{R}^q)$ be defined by

$$L_1(u) = DF(0, 0)(u, 0) \qquad \text{for } u \in \mathbf{R}^p;$$

then the fact that $DF(0, 0)(u, v) = L_1(u) + L_2(v)$ shows that the inverse of $DH(0, 0)$ is the linear mapping K on $\mathbf{R}^p \times \mathbf{R}^q$ defined by

$$K(x, z) = (x, L_2^{-1}[z - L_1(x)]).$$

Hence it follows from the Inversion Theorem 41.8 that there is an open neighborhood U of $(0, 0) \in \mathbf{R}^p \times \mathbf{R}^q$ such that $V = H(U)$ is an open neighborhood of $(0, 0) \in \mathbf{R}^p \times \mathbf{R}^q$ and the restriction of H to U is a bijection onto V with a continuous inverse $\Phi : V \to U$ which belongs to Class $C^1(V)$ and with $\Phi(0, 0) = (0, 0)$. Now Φ has the form

$$\Phi(x, z) = (\varphi_1(x, z), \varphi_2(x, z)) \qquad \text{for } (x, z) \in V$$

where $\varphi_1: V \to \mathbf{R}^p$ and $\varphi_2: V \to \mathbf{R}^q$. Since

$$(x, z) = H \circ \Phi(x, z) = H[\varphi_1(x, z), \varphi_2(x, z)]$$
$$= [\varphi_1(x, z), F(\varphi_1(x, z), \varphi_2(x, z))],$$

we infer that $\varphi_1(x, z) = x$ for all $(x, z) \in V$. Hence Φ takes the simpler form

$$\Phi(x, z) = (x, \varphi_2(x, z)) \qquad \text{for } (x, z) \in V.$$

Now if $P: \mathbf{R}^p \times \mathbf{R}^q \to \mathbf{R}^q$ is defined by $P(x, z) = z$, then P is linear and continuous and $\varphi_2 = P \circ \Phi$; therefore φ_2 belongs to Class $C^1(V)$ and we have

$$z = F(x, \varphi_2(x, z)) \qquad \text{for } (x, z) \in V.$$

Now let $W = \{x \in \mathbf{R}^p : (x, 0) \in V\}$ so that W is an open neighborhood of 0 in $\mathbf{R}^p$, and define $\varphi(x) = \varphi_2(x, 0)$ for $x \in W$. Evidently $\varphi(0) = 0$, and it follows from the preceding formula that

$$F(x, \varphi(x)) = 0 \qquad \text{for } x \in W.$$

Moreover $D\varphi(x)(u) = D\varphi_2(x, 0)(u, 0)$ for $x \in W$, $u \in \mathbf{R}^p$, whence we conclude that φ belongs to Class $C^1(W)$. This proves part (a).

To complete the proof of part (b), suppose that $(x, y) \in U$ satisfies $F(x, y) = 0$. Then $H(x, y) = (x, F(x, y)) = (x, 0) \in V$ whence it follows that $x \in W$. Moreover $(x, y) = \Phi(x, 0) = (x, \varphi_2(x, 0)) = (x, \varphi(x))$ so that $y = \varphi(x)$.   Q.E.D.

It is sometimes useful to have an explicit formula for the derivative of φ. In order to give this it is convenient to introduce the notion of the **block partial derivatives** of F. If $(x, y) \in \Omega$, the block partial derivative $D_{(1)}F(x, y)$ is the linear function mapping $\mathbf{R}^p \to \mathbf{R}^q$ given by

$$D_{(1)}F(x, y)(u) = Df(x, y)(u, 0) \qquad \text{for } u \in \mathbf{R}^p,$$

and the block partial derivative $D_{(2)}F(x, y)$ is the linear function mapping $\mathbf{R}^q \to \mathbf{R}^q$ given by

$$D_{(2)}F(x, y)(v) = DF(x, y)(0, v) \qquad \text{for } v \in \mathbf{R}^q.$$

Since $(u, v) = (u, 0) + (0, v)$ it is clear that

(41.10) $$DF(x, y)(u, v) = D_{(1)}F(x, y)(u) + D_{(2)}F(x, y)(v).$$

Note that the maps L_1 and L_2 that entered in the preceding proof are $D_{(1)}F(0, 0)$ and $D_{(2)}F(0, 0)$, respectively.

41.10 COROLLARY. *With the hypotheses of the theorem, there exists a $\gamma > 0$ such that if $\|x - a\| < \gamma$, then the derivative of φ at x is the element of $\mathscr{L}(\mathbf{R}^p, \mathbf{R}^q)$ given by*

(41.11) $$D\varphi(x) = -[D_{(2)}F(x, \varphi(x))]^{-1} \circ [D_{(1)}F(x, \varphi(x))].$$

PROOF.   Let $K: W \to \mathbf{R}^p \times \mathbf{R}^q$ be defined by

$$K(x) = (x, \varphi(x)) \qquad \text{for } x \in W.$$

Then since $F \circ K(x) = F(x, \varphi(x)) = 0$, we have $F \circ K: W \to \mathbf{R}^q$ is a constant function.   Moreover, since it is readily seen that

$$DK(x)(u) = (u, D\varphi(x)(u)) \qquad \text{for } u \in \mathbf{R}^p,$$

it follows from the Chain Rule 40.2 applied to the constant function $F \circ K$ that

$$0 = D(F \circ K)(x) = DF(K(x)) \circ DK(x).$$

If we use (41.10), we have

$$DF(x, \varphi(x))(u, v) = D_{(1)}F(x, \varphi(x))(u) + D_{(2)}F(x, \varphi(x))(v).$$

It follows from this that if $u \in \mathbf{R}^p$, then

$$0 = DF(x, \varphi(x))(u) = D_{(1)}F(x, \varphi(x))(u) + D_{(2)}F(x, \varphi(x))(D\varphi(x)(u))$$
$$= D_{(1)}F(x, \varphi(x))(u) + [D_{(2)}F(x, \varphi(x)) \circ D\varphi(x)](u).$$

Hence we have

$$0 = D_{(1)}F(x, \varphi(x)) + D_{(2)}F(x, \varphi(x)) \circ D\varphi(x)$$

for all $x \in W$.   By hypothesis, $L_2 = D_{(2)}F(a, b)$ is invertible.   Since φ and F are continuous, there is a $\gamma > 0$ such that if $\|x - a\| < \gamma$, then $D_{(2)}F(x, \varphi(x))$ is also invertible.   Hence equation (41.11) follows from the preceding equation.   Q.E.D.

It may be useful to interpret formula (41.11) in terms of matrices. Suppose that we have the system of q equations in $p + q$ arguments given by (41.9).   As we have remarked, the hypothesis of the Implicit Function Theorem requires that the matrix

$$\begin{bmatrix} f_{1,p+1} & \cdots & f_{1,p+q} \\ \cdot & & \cdot \\ \cdot & & \cdot \\ \cdot & & \cdot \\ f_{q,p+1} & \cdots & f_{q,p+q} \end{bmatrix}$$

is invertible at the point (a, b).   (Recall that $f_{i,j}$ denotes the partial derivative of f_i with respect to the jth argument.)   In this case the derivative of the solution function φ at a point x is given by

$$- \begin{bmatrix} f_{1,p+1} & \cdots & f_{1,p+q} \\ \cdot & & \cdot \\ \cdot & & \cdot \\ \cdot & & \cdot \\ f_{q,p+1} & \cdots & f_{q,p+q} \end{bmatrix}^{-1} \begin{bmatrix} f_{1,1} & \cdots & f_{1,p} \\ \cdot & & \cdot \\ \cdot & & \cdot \\ \cdot & & \cdot \\ f_{q,1} & \cdots & f_{q,p} \end{bmatrix},$$

where it is understood that both matrices are evaluated at the point $(x, \varphi(x))$ near (a, b).

The Parametrization and Rank Theorems

The Implicit Function Theorem 41.9 can be regarded as giving conditions under which the "level curve"

$$C = \{(x, y) \in \mathbf{R}^p \times \mathbf{R}^q : F(x, y) = 0\}$$

passing through the point (a, b), can be parametrized at least locally as the *graph* in $\mathbf{R}^p \times \mathbf{R}^q$ of some function defined on a neighborhood W of $a \in \mathbf{R}^p$ to $\mathbf{R}^q$; that is,

$$C = \{(x, \varphi(x)) : x \in W\}.$$

We shall now present another theorem which gives conditions under which the image of a function mapping an open subset of $\mathbf{R}^p$ into $\mathbf{R}^q$ can be parametrized by means of a function φ defined on an open set in a space of lower dimension.

In presenting this theorem, we will need to use some elementary, but important, facts from linear algebra which may be familiar to the reader.† We recall that if $L : \mathbf{R}^p \to \mathbf{R}^q$ is a linear transformation then the **range** (or the **image**) R_L of L is the subspace of $\mathbf{R}^q$ given by

$$R_L = \{L(x) : x \in \mathbf{R}^p\},$$

and the **null space** (or the **kernel**) N_L of L is the subspace of $\mathbf{R}^p$ given by

$$N_L = \{x \in \mathbf{R}^p : L(x) = 0\}.$$

The dimension $r(L)$ of R_L is called the **rank** of L, and the dimension $n(L)$ of N_L is called the **nullity** of L. (Thus the rank of L is the number of linearly independent vectors in $\mathbf{R}^q$ needed to span the range R_L, and the nullity of L is the number of linearly independent vectors in $\mathbf{R}^p$ needed to span the null space N_L.) It is an exercise to prove that if $\{u_1, \ldots, u_n\}$ (where $n = n(L)$) is a linearly independent set of vectors in $\mathbf{R}^p$ spanning N_L to which we adjoin $p - n$ vectors $u_{n+1}, \ldots, u_p$ to get a basis for $\mathbf{R}^p$, then the set $\{L(u_{n+1}), \ldots, L(u_p)\}$ is a linearly independent set of vectors in $\mathbf{R}^q$ spanning R_L. Therefore it follows that $p = n(L) + r(L)$; hence: *the dimension of the domain of L is equal to the sum of the nullity and the rank of L.*

If we represent L by a $q \times p$ matrix as in (23.1), then it can be shown that the rank of L is the largest number r such that there is at least one $r \times r$ submatrix with non-zero determinant.

---

† For more detail, consult the books of Hoffman and Kunze or Finkbeiner listed in the References.

The Parametrization Theorem asserts that if f is a C^1 mapping of an open set $\Omega \subseteq \mathbf{R}^p$ into $\mathbf{R}^q$ such that $Df(x)$ has rank equal to r for all $x \in \Omega$ and if $f(a) = b \in \mathbf{R}^q$ for some $a \in \Omega$, then there is a neighborhood V of a such that the restriction of f to V can be given as a C^1 mapping φ defined on a neighborhood in $\mathbf{R}^r$.

41.11   PARAMETRIZATION THEOREM. *Let $\Omega \subseteq \mathbf{R}^p$ be open and let $f : \Omega \to \mathbf{R}^q$ belong to Class $C^1(\Omega)$. Let $Df(x)$ have rank r for all $x \in \Omega$ and let $f(a) = b \in \mathbf{R}^q$ for some $a \in \Omega$. Then*

(i) *there exists an open neighborhood $V \subseteq \Omega$ of a and a function $\alpha : V \to \mathbf{R}^r$ in Class $C^1(V)$, and*

(ii) *there exists an open set $W \subseteq \mathbf{R}^r$ and functions $\beta : W \to \mathbf{R}^p$ and $\varphi : W \to \mathbf{R}^q$, such that*

(iii) *$f(x) = \varphi \circ \alpha(x)$ for all $x \in V$, and $\varphi(t) = f \circ \beta(t)$ for all $t \in W$.*

PROOF.   Without any loss of generality we may assume that $a = 0 \in \mathbf{R}^p$ and $b = 0 \in \mathbf{R}^q$.

Let $L = Df(0)$ so that $L : \mathbf{R}^p \to \mathbf{R}^q$ has rank r, and let $\{x_1, \ldots, x_p\}$ be a basis in $\mathbf{R}^p$ such that $\{x_{r+1}, \ldots, x_p\}$ spans the null space of L. We let X_1 be the span of $\{x_1, \ldots, x_r\}$ and $X_2 = N_L$ be the span of $\{x_{r+1}, \ldots, x_p\}$. As mentioned above, it follows that $Y_1 = R_L$ is spanned by $\{y_1 = L(x_1), \ldots, y_r = L(x_r)\}$. We let $\{y_{r+1}, \ldots, y_q\}$ be chosen such that $\{y_1, \ldots, y_q\}$ is a basis for $\mathbf{R}^q$ and let Y_2 be the span of $\{y_{r+1}, \ldots, y_q\}$.

It follows that every vector $x \in \mathbf{R}^p$ has a unique representation in the form $x = c_1 x_1 + \cdots + c_p x_p$. We let P_1 and P_2 be the linear transformations in $\mathbf{R}^p$ defined by

$$P_1(x) = \sum_{j=1}^{r} c_j x_j, \qquad P_2(x) = \sum_{j=r+1}^{p} c_j x_j.$$

Clearly the range of P_j is equal to X_j, $j = 1, 2$. Similarly, we let Q_1 and Q_2 be the linear transformations in $\mathbf{R}^q$ defined for $y = c_1 y_1 + \cdots + c_q y_q$ by

$$Q_1(y) = \sum_{j=1}^{r} c_j y_j, \qquad Q_2(y) = \sum_{j=r+1}^{q} c_j y_j.$$

Clearly the range of Q_j is Y_j, $j = 1, 2$.

If L_1 is the restriction of L to X_1, then L_1 is a bijection of X_1 onto Y_1; we let $A : Y_1 \to X_1$ be the inverse of L_1. We note that $A \circ L(x) = x$ for all $x \in X_1$ and $L \circ A(y) = y$ for all $y \in Y_1$. We now define u on $\Omega \subseteq \mathbf{R}^p$ to $\mathbf{R}^p$ by

(41.12) $u(x) = A \circ Q_1 \circ f(x) + P_2(x), \qquad x \in \Omega,$

so that $u(0) = 0$, u maps $X_1 \cap \Omega$ into X_1, and that

$$Du(x) = A \circ Q_1 \circ Df(x) + P_2, \qquad x \in \Omega;$$

hence u belongs to Class $C^1(\Omega)$. Since it is readily seen that $Du(0)$ is the identity map on $\mathbf{R}^p$, then it follows from the Inversion Theorem 41.8 that there exists an open neighborhood U of $a = 0$ such that $U' = u(U)$ is an open neighborhood of 0, and that the restriction of u to U is a bijection onto U' with inverse $w = u^{-1} : U' \to \mathbf{R}^p$ which belongs to Class $C^1(U')$. Further, by replacing U and U' by smaller sets, we may also suppose that U' is convex (that is, contains the line segment joining any two of its points).

We now let $g : U' \to \mathbf{R}^q$ be defined by

$$g(z) = f(w(z)), \qquad z \in U' \subseteq \mathbf{R}^p.$$

Clearly g belongs to Class $C^1(U')$ and

$$Dg(z) = Df(w(z)) \circ Dw(z), \qquad z \in U'.$$

Since $Df(x)$ has rank r for all $x \in \Omega$ and $Dw(z)$ is invertible for $x \in U'$, then it follows from a theorem in linear algebra that $Dg(z)$ has rank r for all $z \in U'$. On the other hand,

$$\begin{aligned} g(z) &= (Q_1 + Q_2) \circ f(w(z)) \\ &= Q_1 \circ f(w(z)) + Q_2 \circ f(w(z)). \end{aligned}$$

Since $w = u^{-1}$, it follows from (41.12) that

$$z = u(w(z)) = A \circ Q_1 \circ f(w(z)) + P_2(w(z)), \qquad z \in U'.$$

But since $L \circ A \circ Q_1 = Q_1$ on $\mathbf{R}^q$ and $L \circ P_2 = 0$ on $\mathbf{R}^p$, we have

$$(41.13) \qquad L(z) = Q_1 \circ f(w(z)) = Q_1 \circ g(z),$$

whence it follows that $L = Q_1 \circ Dg(z)$ for $z \in U'$. Therefore, if $z \in U'$, then the operator Q_1 maps the range of $Dg(z)$ (which has dimension r) onto the range of L (which also has dimension r). It follows that Q_1 is injective on the range of $Dg(z)$ for $z \in U'$; hence, if $z \in U'$ and $x \in \mathbf{R}^p$ is such that $L(x) = 0$, then $Dg(z)(x) = 0$. Consequently, if $z \in U'$ and $z_2 \in X_2 = N_L$, then we infer that $Dg(z)(z_2) = 0$.

We will now show that $g : U' \to \mathbf{R}^q$ depends only on $z_1 \in X_1$ in the sense that if $z \in U'$ and $z_2 \in X_2$ is such that $z + z_2 \in U'$, then $g(z + z_2) = g(z)$. To see this, we apply the Mean Value Theorem 40.5 to deduce that there exists a point z_0 on the line segment joining z and $z + z_2$ (and hence in U') such that

$$0 \le \|g(z + z_2) - g(z)\| \le \|Dg(z_0)(z_2)\| = 0;$$

hence $g(z + z_2) = g(z)$, as claimed.

We are now prepared to define the maps α, β, φ. Let $C : \mathbf{R}^r \to \mathbf{R}^p$ be the linear transformation which maps the standard basis elements $e_1, \ldots, e_r$ of $\mathbf{R}^r$ into the vectors $x_1, \ldots, x_r$ which form a basis for X_1. Hence C is a bijection of R' onto X_1 and so $C^{-1}: X_1 \to R'$ exists. Let $W = C^{-1}(U') = C^{-1}(U' \cap X_1)$, so that $W \subseteq R'$ is an open neighborhood of 0 in R' and let $V \subseteq U$ be an open neighborhood of $a = 0$ such that $P_1 \circ u(V) \subseteq U'$. We now define $\alpha: V \to R'$ and $\beta: W \to R^p$ by

$$(41.14) \qquad \alpha(x) = C^{-1} \circ P_1 \circ u(x), \qquad \beta(t) = w \circ C(t)$$

for $x \in V$ and $t \in W$. It is clear that α belongs to Class $C'(V)$ and $\alpha(V) \subseteq W$, and that β belongs to Class $C'(W)$ and $\beta(W) \subseteq U$. We now define $\varphi: W \to R^q$ for $t \in W$ by

$$\varphi(t) = g \circ C(t),$$

whence it follows that

$$\varphi(t) = (f \circ w) \circ C(t) = f \circ \beta(t).$$

Moreover, if $x \in V$, then

$$f(x) = f(w \circ u(x)) = (f \circ w) \circ u(x) = g \circ u(x);$$

however, we have seen that $g \circ u(x) = g \circ P_1 \circ u(x)$ so that

$$
\begin{aligned}
f(x) = g \circ u(x) &= g \circ (C \circ C^{-1}) \circ (P_1 \circ u)(x) \\
&= (g \circ C) \circ (C^{-1} \circ P_1 \circ u)(x) \\
&= \varphi \circ \alpha(x).
\end{aligned}
$$

Hence, $f(x) = \varphi \circ \alpha(x)$ for all $x \in V$. Q.E.D.

In the course of this construction, we have actually established a bit more information. In this corollary we make use of the notation developed in the proof of the theorem.

41.12 COROLLARY. (a) *The mapping* $\varphi: W \to R^q$ *has the form* $\varphi_1 + \varphi_2$, *where* φ_1 *is the restriction to W of the linear map of* $R' \to R^q$ *which takes* $e_j \in R'$ *into* $y_j = L(x_j), j = 1, \ldots, r$, *and where* $\varphi_2(W) \subseteq Y_2$.

(b) *If* $t \in W$, *then* $\alpha \circ \beta(t) = t$.

(c) *If* $x \in U \cap X_1$, *then* $x \in V$ *and* $\beta \circ \alpha(x) = x$.

PROOF. (a) Since $g = Q_1 \circ g + Q_2 \circ g$, it follows from (41.13) that $g = L + Q_2 \circ g$. Hence, from the definition of φ, we have $\varphi = L \circ C + Q_2 \circ g \circ C$, which has the form stated in (a).

(b) If $t \in W$, then $x = \beta(t) = w \circ C(t) \in U$ has the property that $u(x) = u \circ w \circ C(t) = C(t) \in U' \cap X_1$; hence $P_1 \circ u(x) = C(t) \in U'$ so that $x \in V$ and

$$\alpha(x) = C^{-1} \circ P_1 \circ u(x) = C^{-1} \circ C(t) = t,$$

which proves statement (b).

(c) If $x \in \Omega \cap X_1$, then it follows from (41.12) and the fact that $P_2(x) = 0$, that $u(x) \in X_1$. Hence if $x \in U \cap X_1$ it follows that $P_1 \circ u(x) = u(x) \in$

$U' \cap X_1$ so that $x \in V$.   Moreover,

$$\beta \circ \alpha(x) = (w \circ C) \circ (C^{-1} \circ P_1 \circ u)(x)$$
$$= w \circ C \circ C^{-1} \circ u(x) = w \circ u(x) = x. \qquad \text{Q.E.D.}$$

We can now use the result of the Parametrization Theorem to prove the Rank Theorem.

41.13   RANK THEOREM.   *Let* $\Omega \subseteq \mathbf{R}^p$ *be open and let* $f : \Omega \to \mathbf{R}^q$ *belong to Class* $C^1(\Omega)$.   *Let* $Df(x)$ *have rank* r *for all* $x \in \Omega$ *and let* $f(a) = b \in \mathbf{R}^q$ *for some* $a \in \Omega$.   *Then:*

(i) *there exist open neighborhoods* V *of* a *and* V' *of* 0 *in* $\mathbf{R}^p$, *and a function* $\sigma : V \to V'$ *in Class* $C'(V)$ *which has an inverse* $\sigma^{-1} : V' \to V$ *in Class* $C^1(V')$;

(ii) *there exist open neighborhoods* Z *of* b *and* Z' *of* 0 *in* $\mathbf{R}^q$, *and a function* $\tau : Z' \to Z$ *in Class* $C^1(Z')$ *which has an inverse* $\tau^{-1} : Z \to Z'$ *in Class* $C^1(Z)$;

(iii) *if* $x \in V$ *then* $f(x) = \tau \circ i_r \circ \sigma(x)$, *where* $i_r : \mathbf{R}^p \to \mathbf{R}^q$ *is the mapping defined by*

$$i_r(c_1, \ldots, c_r, c_{r+1}, \ldots, c_p) = (c_1, \ldots, c_r, 0, \ldots, 0) \in \mathbf{R}^q.$$

PROOF.   We assume that $a = 0$ and $b = 0$ and shall employ the notation and results established during the proof of The Parametrization Theorem. Let $B : \mathbf{R}^p \to \mathbf{R}^p$ be the linear function which maps the standard basis elements $e_1, \ldots, e_p$ of $\mathbf{R}^p$ into the vectors $x_1, \ldots, x_p$; hence B is a bijection of $\mathbf{R}^p$ onto $\mathbf{R}^p$ and so B^{-1} exists.   The map $\sigma : \Omega \to \mathbf{R}^p$ defined by $\sigma(x) = B^{-1} \circ u(x)$ belongs to Class $C^1(\Omega)$ and, since the restriction of u to U has an inverse $w : U' \to \mathbf{R}^p$ mapping onto U, it follows that the restriction of σ to U has an inverse $\sigma^{-1} = w \circ B$ mapping $B^{-1}(U')$ onto U.

Let $W \subseteq \mathbf{R}^r$ and $\varphi : W \to \mathbf{R}^q$ be as in the Parametrization Theorem and let $H : \mathbf{R}^q \to \mathbf{R}^q$ be the linear function which maps the standard basis elements $e_1, \ldots, e_q$ of $\mathbf{R}^q$ into the vectors $y_1, \ldots, y_q$; hence H is a bijection of $\mathbf{R}^q$ onto $\mathbf{R}^q$ and so H^{-1} exists.   We define

$$W' = \{(c_1, \ldots, c_q) \in \mathbf{R}^q : (c_1, \ldots, c_r) \in W\}$$

and let $\tau : W' \to \mathbf{R}^q$ be defined by

$$\tau(c_1, \ldots, c_q) = \varphi(c_1, \ldots, c_r) + H(0, \ldots, 0, c_{r+1}, \ldots, c_q).$$

It follows from Corollary 41.12(a) that $D\tau(0) = H$; hence the inversion Theorem 41.8 implies that the restriction of τ to some neighborhood Z' of 0 is a bijection onto some neighborhood Z of $\tau(0) = 0$.

By further restricting V if necessary, we may assume that $f(V) \subseteq Z$. Now let $x \in V$ and consider $\sigma(x) = B^{-1} \circ u(x)$.   If i_r is as defined above, then $i_r \circ \sigma(x) = (C^{-1} \circ P_1 \circ u(x), 0) = (\alpha(x), 0)$.   Hence $\tau \circ i_r \circ \sigma(x) = \varphi \circ \sigma(x) = f(x)$ for all $x \in V$. \qquad\qquad Q.E.D.

Exercises

41.A. Let $\Omega \subseteq \mathbf{R}^p$ be open and $f : \Omega \to \mathbf{R}^q$. If $Df(x)$ exists for all $x \in \Omega$ and if $i = 1, \ldots, q$, $j = 1, \ldots, p$, then show that $|D_j f_i(x) - D_j f_i(y)| \leq \|Df(x) - Df(y)\|_{pq}$. Hence, if f belongs to Class $C^1(\Omega)$ then each of the partial derivatives $D_j f_i$ is continuous on Ω.

41.B. Let $\Omega \subseteq \mathbf{R}^p$ be open and $f : \Omega \to \mathbf{R}^q$. If f belongs to Class $C^1(\Omega)$ and $K \subseteq \Omega$ is compact, show that $x \mapsto Df(x)$ is uniformly continuous in the sense that for every $\varepsilon > 0$ there exists $\delta > 0$ such that if $x, y \in K$ and $\|x - y\| < \delta$ then $\|Df(x) - Df(y)\|_{pq} < \varepsilon$.

41.C. Let $\Omega \subseteq \mathbf{R}^p$ and $\Omega_1 \subseteq \mathbf{R}^q$ be open and let $f : \Omega \to \mathbf{R}^q$ belong to Class $C^1(\Omega)$ and $g : \Omega_1 \to \mathbf{R}^r$ belong to Class $C^1(\Omega_1)$. If $f(\Omega) \subseteq \Omega_1$, show that $g \circ f$ belongs to Class $C^1(\Omega)$.

41.D. Let $f : \mathbf{R} \to \mathbf{R}$ be defined by $f(x) = x^3$. Show that f belongs to Class $C^1(\mathbf{R})$ and that it is a bijection of $\mathbf{R}$ onto $\mathbf{R}$ with inverse $g(x) = x^{1/3}$ for all $x \in \mathbf{R}$. However $Df(0)$ is neither injective or surjective. Does g belong to Class $C^1(\mathbf{R})$?

41.E. Let $g : \mathbf{R} \to \mathbf{R}$ be such that $g'(x) \neq 0$ for all $x \in \mathbf{R}$. Show that g is a bijection of $\mathbf{R}$ onto $g(\mathbf{R})$.

41.F. Let $A \subseteq \mathbf{R}^p$, let $f : A \to \mathbf{R}^p$, and let $g : f(A) \to \mathbf{R}^p$ be inverse to f. Suppose that f is differentiable at $a \in A$ and g is differentiable at $b = f(a)$. If $Df(a)$ is not invertible, then show that $Dg(b)$ is not invertible.

41.G. Let $f : \mathbf{R}^2 \to \mathbf{R}^2$ be given by

$$f(x, y) = (x + y, 2x + ay).$$

(a) Calculate $Df(x, y)$ and show that $Df(x, y)$ is invertible if and only if $a \neq 2$.

(b) Examine the image of the unit square $\{(x, y) : x, y \in [0, 1]\}$ when $a = 1, 2, 3$.

41.H. Let f be the mapping of $\mathbf{R}^2$ into $\mathbf{R}^2$ which sends the point (x, y) into the point (u, v) given by

$$u = x, \qquad v = xy.$$

Draw some curves $u = $ constant, $v = $ constant in the (x, y)-plane and some curves $x = $ constant, $y = $ constant in the (u, v)-plane. Is this mapping one-one? Does f map onto all of $\mathbf{R}^2$? Show that if $x \neq 0$, then f maps some neighborhood of (x, y) in a one-one fashion onto a neighborhood of (x, xy). Into what region in the (u, v)-plane does f map the rectangle $\{(x, y) : 1 \leq x \leq 2, 0 \leq y \leq 2\}$? What points in the (x, y)-plane map under f into the rectangle $\{(u, v) : 1 \leq u \leq 2, 0 \leq v \leq 2\}$?

41.I. Let f be the mapping of $\mathbf{R}^2$ into $\mathbf{R}^2$ which sends the point (x, y) into the point (u, v) given by

$$u = x^2 - y^2, \qquad v = 2xy.$$

What curves in the (x, y)-plane map under f into the lines $u = $ constant, $v = $ constant? Into what curves in the (u, v)-plane do the lines $x = $ constant, $y = $ constant map? Show that each non-zero point (u, v) is the image under f of two points. Into what region does f map the square $\{(x, y) : 0 \leq x \leq 1, 0 \leq y \leq 1\}$? What region is mapped by f into the square $\{(u, v) : 0 \leq u \leq 1, 0 \leq v \leq 1\}$?

41.J. Let $h : \mathbf{R} \to \mathbf{R}$ be defined by

$$h(x) = x + 2x^2 \sin \frac{1}{x} \qquad \text{for } x \neq 0,$$

$$= 0 \qquad \text{for } x = 0.$$

Show that h does not belong to Class $C^1(\mathbf{R})$ and that h is not injective on a neighborhood of 0. However, it is surjective on a neighborhood of 0 and $Dh(0)$ is invertible.

41.K. Let $f : \mathbf{R}^2 \to \mathbf{R}^2$ be defined by $f(x, y) = (y, x + y^2)$ for $(x, y) \in \mathbf{R}^2$. Show that f belongs to Class $C^1(\mathbf{R}^2)$ and that f is invertible on some neighborhood of an arbitrary point of $\mathbf{R}^2$. Draw the image under f of the lines $x = 0, \pm 1, \pm 2$ and $y = 0$, $\pm 1, \pm 2$. Find the inverse $g = f^{-1} : \mathbf{R}^2 \to \mathbf{R}^2$ and show that $Dg(f(x_0, y_0)) = Df(x_0, y_0)^{-1}$.

41.L. (This exercise assumes familiarity with the notion of the determinant of a square matrix.) Let $L \in \mathscr{L}(\mathbf{R}^p, \mathbf{R}^p)$ and let $[c_{ij}]$ be the matrix representation of L with respect to the standard basis in $\mathbf{R}^p$. It is shown in linear algebra that L is invertible if and only if $\Delta = \det[c_{ij}]$ is not zero. Furthermore, if $\Delta \neq 0$, then the matrix of L^{-1} has the form $[p_{ij}/\Delta]$, where the p_{ij} are polynomials in the c_{ij}.

(a) Show that if L_0 is invertible and if $\|L - L_0\|_{pp}$ is sufficiently small, then L is invertible.

(b) Show that if L_0 is invertible, then the map $L \mapsto L^{-1}$ is continuous on a neighborhood of L_0 with respect to the norm in $\mathscr{L}(\mathbf{R}^p, \mathbf{R}^p)$.

(c) Let $\Omega \subseteq \mathbf{R}^p$ be open and $f : \Omega \to \mathbf{R}^p$ belong to Class $C^1(\Omega)$. If $Df(c)$ is invertible for some $c \in \Omega$, then $Df(x)$ is invertible on some neighborhood of c.

41.M. Let $F : \mathbf{R}^2 \to \mathbf{R}$ be defined by $F(x, y) = y^2 - x$. Show that F belongs to Class $C^1(\mathbf{R}^2)$ but that $D_2 F(0, 0) = 0$. Show that there does not exist a function φ defined on a neighborhood W of 0 such that $F(x, \varphi(x)) = 0$ for all $x \in W$.

41.N. Let $f : \mathbf{R}^3 \to \mathbf{R}^2$ be defined by

$$f(x, y, z) = (x + y + z, x - y - 2xz),$$

so that $f(0, 0, 0) = (0, 0)$ and $Df(0, 0, 0)$ is given by

$$\begin{bmatrix} 1 & 1 & 1 \\ 1 & -1 & 0 \end{bmatrix}.$$

(a) Show that we can solve for $(x, y) = \varphi(z)$ near $z = 0$ and that

$$D\varphi(0) = \begin{bmatrix} -\frac{1}{2} \\ -\frac{1}{2} \end{bmatrix}.$$

(b) Carry out the explicit solution of $(x, y) = \varphi(z)$ to obtain

$$\varphi(z) = \left(\frac{z}{2(z-1)}, \frac{2 - 2z^2}{2(z-1)} \right) \qquad \text{for } z < 1.$$

Check the result of part (a).

(c) Show that we can solve for $(y, z) = \psi(x)$ near $x = 0$ and that

$$D\psi(0) = \begin{bmatrix} 1 \\ -2 \end{bmatrix}.$$

(d) Carry out the explicit solution of $(y, z) = \psi(x)$ to obtain

$$\psi(x) = \left(\frac{2x^2 + x}{1 - 2x}, \frac{2x}{2x - 1}\right) \qquad \text{for } x < \tfrac{1}{2}.$$

Check the result of part (c).

41.O. Let $F: \mathbf{R}^5 \to \mathbf{R}^2$ be defined by

$$F(u, v, w, x, y) = (uy + vx + w + x^2, \, uvw + x + y + 1),$$

and note that $F(2, 1, 0, -1, 0) = (0, 0)$.

(a) Show that we can solve $F(u, v, w, x, y) = (0, 0)$ for (x, y) in terms of (u, v, w) near $(2, 1, 0)$.

(b) If $(x, y) = \varphi(u, v, w)$ is the solution in part (a), show that $D\varphi(2, 1, 0)$ is given by the matrix

$$-\begin{bmatrix} -1 & 2 \\ 1 & 1 \end{bmatrix}^{-1} \begin{bmatrix} 0 & -1 & 1 \\ 0 & 0 & 2 \end{bmatrix} = \frac{1}{3}\begin{bmatrix} 0 & -1 & -3 \\ 0 & 1 & -3 \end{bmatrix}.$$

41.P. Let $A \subseteq \mathbf{R}^3$ and let $F: A \to \mathbf{R}$ represent a surface S_F in $\mathbf{R}^3$ implicitly as the "level surface"

$$S_F = \{(x, y, z) \in A : F(x, y, z) = 0\}.$$

If F is differentiable at a point $(x_0, y_0, z_0) \in S_F$ which is interior to A, then the **tangent space** to S_F at this point is the set of points

$$\{(x, y, z) \in \mathbf{R}^3 : A_{(x_0, y_0, z_0)}(x, y, z) = 0\},$$

where $A_{(x_0, y_0, z_0)}$ is the affine map of $\mathbf{R}^3 \to \mathbf{R}$ defined by

$$A_{(x_0, y_0, z_0)}(x, y, z) = F(x_0, y_0, z_0) + DF(x_0, y_0, z_0)(x - x_0, y - y_0, z - z_0)$$
$$= DF(x_0, y_0, z_0)(x - x_0, y - y_0, z - z_0).$$

(a) Show that the tangent space at (x_0, y_0, z_0) is given by

$$\{(x, y, z) : D_1 F(x_0, y_0, z_0)(x - x_0) + D_2 F(x_0, y_0, z_0)(y - y_0) + D_3 F(x_0, y_0, z_0)(z - z_0) = 0\}.$$

Hence the tangent space to S_F is a plane if at least one of the numbers $D_1 F(x_0, y_0, z_0)$, $D_2 F(x_0, y_0, z_0)$, $D_3 F(x_0, y_0, z_0)$ is different from 0. In this case the tangent space to S_F is called the **tangent plane** to S_F at (x_0, y_0, z_0).

41.Q. Let $F: \mathbf{R}^3 \to \mathbf{R}$, given below, represent a surface S_F in $\mathbf{R}^3$ implicitly as the level surface

$$S_F = \{(x, y, z) \in \mathbf{R}^3 : F(x, y, z) = 0\}.$$

In each of the following cases, determine the tangent space to S_F at the indicated points.

(a) Let $F(x, y, z) = x^2 + y^2 - z$ at the points $(1, 1, 2)$ and $(0, 2, 4)$.

(b) Let $F(x, y, z) = x^2 + y^2 + z^2 - 25$ at the points $(3, 4, 0)$ and $(3, 3, \sqrt{7})$.

(c) Let $F(x, y, z) = z - xy$ at the points $(1, 1, 1)$ and $(4, \tfrac{1}{2}, 2)$.

41.R. (a) Suppose that, in addition to the hypotheses of the Inversion Theorem 41.8, it is known that the function f has continuous partial derivatives of order $m > 1$. Show that the inverse function $g: V \to \mathbf{R}^p$ has continuous partial derivatives of order m.

(b) Prove the analogous result for the Implicit Function Theorem 41.9.

41.S. Let $f: \mathbf{R}^2 \to \mathbf{R}$ belong to Class $C^1(\mathbf{R}^2)$. Show that f is not injective; indeed, the restriction of f to any open set of $\mathbf{R}^2$ is not injective.

41.T. Let $g: \mathbf{R} \to \mathbf{R}^2$ belong to Class $C^1(\mathbf{R})$. Show that if $c \in \mathbf{R}$, then the restriction of g to any neighborhood of c is not a surjective map onto a neighborhood of $g(c)$.

41.U. Let $L \in \mathcal{L}(\mathbf{R}^p, \mathbf{R}^q)$ be injective and let $r > 0$ be such that $r \|x\| \leq \|L(x)\|$ for all $x \in \mathbf{R}^p$. Show that if $L_1 \in \mathcal{L}(\mathbf{R}^p, \mathbf{R}^q)$ is such that $\|L_1 - L\|_{pq} < r$, then L_1 is injective. (Hence, the set of injective maps is open in $\mathcal{L}(\mathbf{R}^p, \mathbf{R}^q)$.)

41.V. Let $L \in \mathcal{L}(\mathbf{R}^p, \mathbf{R}^q)$ be surjective and let $m > 0$ be as in the proof of 41.6. Show that if $L_1 \in \mathcal{L}(\mathbf{R}^p, \mathbf{R}^q)$ is such that $\|L_1 - L\|_{pq} < m/2$, then L_1 is surjective. (Hence, the set of surjective maps is open in $\mathcal{L}(\mathbf{R}^p, \mathbf{R}^q)$).

41.W. Let $g: \mathbf{R}^p \to \mathbf{R}^p$ belong to Class $C^1(\mathbf{R}^p)$ and satisfy $\|Dg(x)\|_{pp} \leq a < 1$ for all $x \in \mathbf{R}^p$. If $f(x) = x + g(x)$ for $x \in \mathbf{R}^p$, show that f satisfies

$$\|f(x_1) - f(x_2) - (x_1 - x_2)\| \leq a \|x_1 - x_2\|$$

for all x_1, x_2 in $\mathbf{R}^p$ and that f is a bijection of $\mathbf{R}^p$ onto $\mathbf{R}^p$.

Projects

41.α. (This project gives a direct and elementary proof of the Implicit Function Theorem.) Let $\Omega \subseteq \mathbf{R}^2$ be open and let $F: \Omega \to \mathbf{R}$ belong to Class $C^1(\Omega)$. Suppose that $(a, b) \in \Omega$, that $F(a, b) = 0$, and that $D_2 F(a, b) > 0$.

(a) Show that there exists a closed cell $Q = [a_1, a_2] \times [b_1, b_2]$ with center (a, b) such that $D_2 F(x, y) > 0$ for all $(x, y) \in Q$, and such that $F(x, b_1) < 0$ and $F(x, b_2) > 0$ for all $x \in [a_1, a_2]$.

(b) If $x \in [a_1, a_2]$, then the function $F_x : [b_1, b_2] \to \mathbf{R}$ defined by $F_x(y) = F(x, y)$ for $y \in [b_1, b_2]$ is such that $F_x(b_1) < 0 < F_x(b_2)$ and $F_x'(y) > 0$ for $y \in [b_1, b_2]$.

(c) There exists a function φ mapping $[a_1, a_2]$ into $[b_1, b_2]$ such that $F(x, \varphi(x)) = 0$ for all $x \in [a_1, a_2]$.

(d) If $x \in (a_1, a_2)$ and $|h|$ is sufficiently small, show that there exists h_1 with $0 < |h_1| < |h|$ such that

$$0 = F[x + h, \varphi(x + h)] - F[x, \varphi(x)]$$
$$= D_1 F[x + h_1, \varphi(x + h_1)]h + D_2 F[x + h_1, \varphi(x + h_1)][\varphi(x + h) - \varphi(x)].$$

(e) Show that φ is differentiable on (a_1, a_2) and that $\varphi'(x) = -D_1 F[x, \varphi(x)]/D_2 F[x, \varphi(x)]$.

(f) Modify the preceding argument for a function F defined on an open set $\Omega \subseteq \mathbf{R}^p$.

(g) Let $\Omega \subseteq \mathbf{R}^p \times \mathbf{R}^2$ be open and let $F, G: \Omega \to \mathbf{R}$ belong to Class $C^1(\Omega)$ and suppose that for some point $(a, b) \in \mathbf{R}^p \times \mathbf{R}^2$ we have $F(a, b) = 0$, $G(a, b) = 0$. Suppose that

$$\Delta = \det \begin{bmatrix} D_{p+1}F(a, b) & D_{p+2}F(a, b) \\ D_{p+1}G(a, b) & D_{p+2}G(a, b) \end{bmatrix} \neq 0,$$

then at least one of $D_{p+1}F(a, b)$ and $D_{p+2}F(a, b)$ does not vanish. Suppose that $D_{p+2}F(a, b) \neq 0$ and use (f) to obtain $x_{p+2} = \varphi(x_1, \ldots, x_{p+1})$ in a neighborhood of

$(a, b_1) \in \mathbf{R}^p \times \mathbf{R}$. Hence $F(x_1, \ldots, x_{p+1}, \varphi(x_1, \ldots, x_{p+1})) = 0$, on this neighborhood. Now put

$$H(x_1, \ldots, x_{p+1}) = G(x_1, \ldots, x_{p+1}, \varphi(x_1, \ldots, x_{p+1})).$$

By Chain Rule

$$D_{p+1}H = D_{p+1}G + (D_{p+2}G)(D_{p+1}\varphi),$$

where these functions are evaluated at the appropriate points. Since $D_{p+1}\varphi = -(D_{p+1}F)/(D_{p+2}F)$ we infer that $D_{p+1}H = -\Delta/D_{p+2}F$ which does not vanish at (a, b_1). Hence we can use (f) to obtain $x_{p+1} = \psi(x_1, \ldots, x_p)$ on a neighborhood of $a \in \mathbf{R}^p$. (This establishes the Implicit Function Theorem in the case where $q = 2$; extensions to the case of general q are obtained by induction.)

41.β. (This project is parallel to Project 40.α and gives a more direct proof of the first part of Inversion Theorem 41.8 then the one given in the text.) We assume that $\Omega \subseteq \mathbf{R}^p$ is open, that $f : \Omega \to \mathbf{R}^p$ belongs to Class $C^1(\Omega)$, and that for some $x_0 \in \Omega$ the linear mapping $Df(x_0)$ is a bijection. We let $\Gamma = Df(x_0)^{-1}$.

(a) Show that there exists $r > 0$ such that if $\|x - x_0\| \le r$, then $\|I - \Gamma \circ Df(x)\|_{pp} \le \frac{1}{2}$.

(b) Let $s \le \frac{1}{2}r \|\Gamma\|_{pp}^{-1}$ and for fixed y with $\|y - f(x_0)\| \le s$, we define $F_y(x) = f(x) - y$ for $\|x - x_0\| \le r$. Then F_y is differentiable, $\|\Gamma \circ F_y(x_0)\| \le \frac{1}{2}r$, and $\|I - \Gamma \circ DF_y(x)\|_{pp} \le \frac{1}{2}$ for $\|x - x_0\| \le r$.

(c) If $\|y - f(x_0)\| \le s$, let G_y be defined for $\|x - x_0\| \le r$ by $G_y(x) = x - \Gamma \circ F_y(x)$. Then G_y is a contraction with constant $\frac{1}{2}$ on this ball.

(d) If $\|y - f(x_0)\| \le s$, define $\varphi(y) = x_0$ and $\varphi_{n+1}(y) = G_y(\varphi_n(y))$ for $n = 0, 1, 2, \ldots$. Show that $\|\varphi_{n+1}(y) - \varphi_n(y)\| \le 2^{-n} \|\varphi_1(y) - \varphi_0(y)\| \le 2^{-n-1}r$, whence it follows that $\|\varphi_{n+1}(y) - \varphi_m(y)\| \le 2^{-m}r$ for $n \ge m \ge 0$. In particular, $\|\varphi_k(y) - x_0\| \le r$, so that this iteration is possible.

(e) Show that each of the functions φ_k is continuous for $\|y - f(x_0)\| \le s$ and that the sequence (φ_k) is uniformly convergent to a continuous function φ which is such that $G_y(\varphi(y)) = \varphi(y)$ for $\|y - f(x_0)\| \le s$, whence it follows that $f(\varphi(y)) = y$ for $\|y - f(x_0)\| \le s$. Hence the function φ is the inverse of f on the set $\{y : \|y - f(x_0)\| \le s\}$ and maps it into the set $\{x : \|x - x_0\| \le r\}$.

41.γ. (This project is parallel to Projects 40.α and 41.β and gives a direct proof of the Implicit Function Theorem.) Let $\Omega \subseteq \mathbf{R}^p \times \mathbf{R}^q$ be open and let $(x_0, y_0) \in \Omega$. Suppose that $F : \Omega \to \mathbf{R}^q$ belongs to Class $C^1(\Omega)$, that $F(x_0, y_0) = 0$, and that the linear map $L_2 : \mathbf{R}^q \to \mathbf{R}^q$ defined by

$$L_2(v) = DF(x_0, y_0)(0, v) \qquad \text{for } v \in \mathbf{R}^q$$

is a bijection of $\mathbf{R}^q$ onto $\mathbf{R}^q$. Let $\Gamma = L_2^{-1}$.

(a) Show that there exists $r > 0$ such that if $\|x - x_0\|^2 + \|y - y_0\|^2 \le r^2$, then

$$\|v - \Gamma \circ DF(x, y)(0, v)\| \le \frac{1}{2}\|v\| \qquad \text{for } v \in \mathbf{R}^q.$$

(b) Let $0 < s \le \frac{1}{2}r$ be such that if $\|x - x_0\| \le s$, then

$$\|F(x, y_0)\| \le \frac{1}{4}r \|\Gamma\|_{qq}^{-1}.$$

For each fixed x with $\|x - x_0\| \le s$, we define $G_x(y) = y - \Gamma \circ F(x, y)$ for $\|y - y_0\| \le \frac{1}{2}r$, with values in $\mathbf{R}^q$. Show that for each x with $\|x - x_0\| \le s$, we have

$$\|G_x(y_1) - G_x(y_2)\| \le \frac{1}{2}\|y_1 - y_2\|$$

for all y_1, y_2 satisfying $\|y_i - y_0\| \le \frac{1}{2}r$.

(c) If $\|x - x_0\| \le s$, define $\psi_0(x) = y_0$ and $\psi_{n+1}(x) = G_x(\psi_n(x))$ for $n = 0, 1, 2, \ldots$. Show that $\|\psi_{n+1}(x) - \psi_m(x)\| \le 2^{-m-1}r$ for $n \ge m \ge 0$. Hence $\|\psi_k(x) - y_0\| \le \frac{1}{2}r$, so that this iteration is possible.

(d) Show that each of the functions ψ_k is continuous for $\|x - x_0\| \le s$ and that the sequence (ψ_k) is uniformly convergent to a continuous function ψ such that

$$F(x, \psi(x)) = 0 \qquad \text{for all } \|x - x_0\| \le s.$$

(e) To show that ψ is differentiable for $\|x = x_0\| < s$ use Exercise 39.W and employ an argument similar to that in (d) and (e) of Project 41.α for each component of F.

Section 42 Extremum Problems

In Section 27 we briefly discussed the familiar process of locating interior points at which a real-valued differentiable function of one variable attains relative extreme values. The question as to whether a **critical point** (that is, a point at which the derivative vanishes) is actually an extreme point is not always discussed, but often can be handled by means of Taylor's Theorem 28.6. The analysis of extreme points which belong to the boundary of the domain often yields to an application of the Mean Value Theorem 27.6.

In the case of a function with domain in R^p $(p > 1)$ and range in R, the situation is often considerably more complicated, and each function needs to be examined in its own right. However, there are a few general theorems that are useful and which will be presented here.

Let $\Omega \subseteq R^p$ and let $f : \Omega \to R$. A point $c \in \Omega$ is said to be a **point of relative minimum** of f if there exists $\delta > 0$ such that $f(c) \le f(x)$ for all $x \in \Omega$ with $\|x - c\| < \delta$. A point $c \in \Omega$ is said to be a **point of relative strict minimum** of f if there exists $\delta > 0$ such that $f(c) < f(x)$ for all $x \in \Omega$ with $0 < \|x - c\| < \delta$. We define a **point of relative [strict] maximum** of f similarly. Moreover, if $c \in \Omega$ is a point of relative [strict] minimum or relative [strict] maximum of f, we say that c is a **point of relative [strict] extremum** of f, or that f has a **relative [strict] extremum** at c.

The next result is very often useful.

42.1 THEOREM. Let $\Omega \subseteq R^p$, and let $f : \Omega \to R$. If an interior point c of Ω is a point of relative extremum of f, and if the partial derivative $D_u f(c)$ of f with respect to a vector $u \in R^p$ exists, then $D_u f(c) = 0$.

PROOF. By hypothesis the restriction of f to the intersection of Ω with the line $\{c + tu : t \in R\}$ has a relative extremum at c. It therefore follows from Theorem 27.4 that $D_u f(c) = 0$. Q.E.D.

42.2 COROLLARY. Let $\Omega \subseteq R^p$, and let $f : \Omega \to R$. If an interior point c of Ω is a point of relative extremum of f, and if the derivative $Df(c)$ exists, then $Df(c) = 0$.

PROOF.   It follows from Corollary 39.7 that each of the partial deriva-
tives $D_j f(c)$, $j = 1, \ldots, p$, exist and that if $u = (u_1, \ldots, u_p) \in \mathbf{R}^p$, then

$$Df(c)(u) = \sum_{j=1}^{p} u_j D_j f(c).$$

By the preceding theorem, $D_j f(c) = 0$ for $j = 1, \ldots, p$, whence $Df(c)(u) = 0$
for all $u \in \mathbf{R}^p$.   Q.E.D.

It follows that if $\Omega \subseteq \mathbf{R}^p$, and if $f : \Omega \to \mathbf{R}$ has a relative extremum at
$c \in \Omega$ and if $Df(c)$ exists, then

(42.1) $D_1 f(c) = 0, \ldots, D_p f(c) = 0.$

An interior point c at which $Df(c) = 0$ is called a **critical point** of f. We
infer that if Ω is an open set in $\mathbf{R}^p$ on which f is differentiable, then the set
of critical points of f will contain all of the relative extreme points of f. Of
course, this set of critical points may also contain points at which f does not
have a relative extremum. (In addition, the function f may have a relative
extremum at an interior point c of Ω at which the derivative $Df(c)$ does not
exist, or f may have a relative extremum at a point $c \in \Omega$ which is not an
interior point Ω; in either case, the point c will not be a critical point of f.)

42.3 EXAMPLES.   (a) Let $f_1(x) = x^3$ for $x \in [-1, 1]$. Then $Df_1(0) = 0$;
however, f_1 does not have an extremum at $x = 0$. On the other hand, f_1
does have strict extrema at the points ± 1 (which are not interior points of
the domain and are not critical points).

(b) Let $f_2(x) = |x|$ for $x \in [-1, 1]$. Then $Df_2(0)$ does not exist; however,
f_2 has a relative strict minimum at the interior point 0. On the other hand,
f_2 does have relative strict extrema at the points ± 1.

(c) Let $f_3 : \mathbf{R}^2 \to \mathbf{R}$ be defined by $f_3(x, y) = xy$. Then $Df_3(0, 0) = 0$ so
that the origin $(0, 0)$ is a critical point of f_3; however, it is not a relative
extremum of f_3 since

$$f_3(0, 0) < f_3(x, y) \qquad \text{for } xy > 0,$$
$$f_3(0, 0) > f_3(x, y) \qquad \text{for } xy < 0.$$

We say that the origin $(0, 0)$ is a **saddle point** of f_3 meaning that every
neighborhood of $(0, 0)$ contains points at which f_3 is strictly greater than
$f_3(0, 0)$ and also contains points at which f_3 is strictly less than $f_3(0, 0)$.

(d) Let $f_4 : \mathbf{R}^2 \to \mathbf{R}$ be defined by $f_4(x, y) = (y - x^2)(y - 2x^2)$. Show that
$Df_4(0, 0) = 0$ and that the restriction of f_4 to every line passing through
$(0, 0)$ has a relative minimum at the origin. However, show that in
every neighborhood of $(0, 0)$ there are points where f_4 is strictly positive
and those where f_4 is strictly negative.

The Second Derivative Test

In view of the examples given above, it is convenient to have conditions which are necessary (or are sufficient) to guarantee that a critical point is an extremum or that it is a saddle point. The next results give conditions in terms of the second derivative of f that was introduced at the end of Section 40.

42.4 THEOREM. *Let $\Omega \subseteq R^p$ be open and let $f : \Omega \to R$ have continuous second partial derivatives on Ω. If $c \in \Omega$ is a point of relative minimum [respectively, maximum] of f, then*

$$(42.2) \qquad D^2f(c)(w)^2 = \sum_{i,j=1}^{p} D_{ij}f(c)w_iw_j \geq 0$$

[respectively, $D^2f(c)(w)^2 \leq 0$] for all $w \in R^p$.

PROOF. Let $w \in R^p$, $\|w\| = 1$. If c is a point of relative minimum, there exists $\delta > 0$ such that if $|t| < \delta$ then $f(c + tw) - f(c) \geq 0$. Since Ω is open, there exists $\delta_1 > 0$ with $\delta_1 \leq \delta$ such that $c + tw$ belongs to Ω for $0 \leq t \leq \delta_1$. By Taylor's Theorem 40.9 there exists t_1 with $0 \leq t_1 \leq t \leq \delta_1$ such that if $c_t = c + t_1w$, then

$$f(c + tw) = f(c) + Df(c)(tw) + \tfrac{1}{2}D^2f(c_t)(tw)^2.$$

Since c is a point of relative minimum it follows from Corollary 42.2 that $Df(c) = 0$; hence we have

$$\tfrac{1}{2}D^2f(c_t)(tw)^2 \geq 0$$

for $0 \leq t \leq \delta_1$. It follows that $D^2f(c_t)(w)^2 \geq 0$. Since $\|c_t - c\| = |t_1| \leq |t|$, it follows that $c_t \to c$ as $t \to 0$. Since the second partial derivatives of f are continuous, then $D^2f(c)(w)^2 \geq 0$ for all $w \in R^p$ with $\|w\| = 1$, from which the result follows.   Q.E.D.

The next result is a partial converse to Theorem 42.4. However, note that its hypothesis is somewhat stronger than the conclusion of 42.4.

42.5 THEOREM. *Let $\Omega \subseteq R^p$ be open, let $f : \Omega \to R$ have continuous second partial derivatives on Ω, and let $c \in \Omega$ be a critical point of f.*
  (a) *If $D^2f(c)(w)^2 > 0$ for all $w \in R^p$, $w \neq 0$, then f has a relative strict minimum at c.*
  (b) *If $D^2f(c)(w)^2 < 0$ for all $w \in R^p$, $w \neq 0$, then f has a relative strict maximum at c.*
  (c) *If $D^2f(c)(w)^2$ takes on both strictly positive and strictly negative values for $w \in R^p$, then f has a saddle point at c.*

PROOF. (a) By hypothesis $D^2f(c)(w)^2 > 0$ for w in the compact set $\{w \in R^p : \|w\| = 1\}$. Since the map $w \mapsto D^2f(c)(w)^2$ is continuous, there exists $m > 0$ such that

$$D^2f(c)(w)^2 \geq m \qquad \text{for } \|w\| = 1.$$

Since the second partial derivatives of f are continuous on Ω, there exists $\delta > 0$ such that if $\|x - c\| < \delta$ then

$$D^2f(x)(w)^2 \geq \tfrac{1}{2}m \qquad \text{for } \|w\| = 1.$$

By Taylor's Theorem 40.9, if $0 \leq t \leq 1$ there is a point c_t on the line segment joining c and $c + tw$ such that

$$f(c + tw) = f(c) + Df(c)(tw) + \tfrac{1}{2}D^2f(c_t)(tw)^2.$$

Since c is a critical point, it follows that if $\|w\| = 1$ and $0 < t < \delta$, then

$$f(c + tw) - f(c) = \tfrac{1}{2}t^2D^2f(c_t)(w)^2 \geq \tfrac{1}{4}mt^2 > 0,$$

Thus $f(c + u) > f(c)$ for $0 < \|u - c\| < \delta$, whence f has a relative strict minimum at c. Thus part (a) is proved and the proof of part (b) is similar.

(c) Let w_+, w_- be unit vectors in $\mathbf{R}^p$ such that

$$D^2f(c)(w_+)^2 > 0, \qquad D^2f(c)(w_-)^2 < 0.$$

It follows from Taylor's Theorem that for sufficiently small $t > 0$ we have

$$f(c + tw_+) > f(c), \qquad f(c + tw_-) < f(c).$$

Thus c is a saddle point of f.   Q.E.D.

On comparing Theorems 42.4 and 42.5, one is led to make the following conjectures: (i) if $c \in \Omega$ is a point of relative strict minimum, then $D^2f(c)(w)^2 > 0$ for all $w \in \mathbf{R}^p$, $w \neq 0$, (ii) if $c \in \Omega$ is a saddle point of f, then $D^2f(c)(w)^2$ takes on both strictly positive and strictly negative values, (iii) if $D^2f(c)(w)^2 \geq 0$ for all $w \in \mathbf{R}^p$, then c is a point of relative minimum. All of these conjectures are false, as may be seen by examples.

In order to implement Theorem 42.5 it is necessary to know whether the function $w \mapsto D^2f(c)(w)^2$ is of one sign. An important and well-known result of algebra (see the book of Hoffman and Kunze cited in the References) can be used to determine this. For each $j = 1, 2, \ldots, p$, let Δ_j be the determinant of the (symmetric) matrix

$$\begin{bmatrix} D_{11}f(c) & \cdots & D_{1j}f(c) \\ \cdot & & \cdot \\ \cdot & & \cdot \\ \cdot & & \cdot \\ D_{j1}f(c) & \cdots & D_{jj}f(c) \end{bmatrix}$$

If the numbers $\Delta_1, \Delta_2, \ldots, \Delta_p$ are all strictly positive, then $D^2f(c)(w)^2 > 0$ for all $w \neq 0$ and f has a relative strict minimum at c. If the numbers $\Delta_1, \Delta_2, \ldots, \Delta_p$ are alternately strictly negative and strictly positive, then

$D^2f(c)(w)^2 < 0$ for all $w \neq 0$ and f has a relative strict maximum at c. In other cases there can be extreme or saddle points.

In the important special case $p = 2$ a less elaborate formulation is more convenient and a bit more information can be derived. Here we need to examine the quadratic function

$$Q = Au^2 + 2Buv + Cv^2.$$

If $\Delta = AC - B^2 > 0$, then $A \neq 0$ (and $C \neq 0$) and we can complete the square and write

$$Q = \frac{1}{A}[(Au + Bv)^2 + (AC - B^2)v^2].$$

Hence the sign of Q is the same as that of A (or C). On the other hand, if $\Delta < 0$, then Q has both strictly positive and strictly negative values. This is obvious from the above equation if $A \neq 0$ and is also readily established if $A = 0$.

We collect these remarks in a formal statement.

42.6 COROLLARY. Let $\Omega \subseteq \mathbf{R}^2$ be open, let $f : \Omega \to \mathbf{R}$ have continuous second partial derivatives on Ω, let $c \in \Omega$ be a critical point of f, and let

$$(42.3) \qquad \Delta = D_{11}f(c)D_{22}f(c) - [D_{12}f(c)]^2.$$

(a) If $\Delta > 0$ and if $D_{11}f(c) > 0$, then f has a relative strict minimum at c.
(b) If $\Delta > 0$ and if $D_{11}f(c) < 0$, then f has a relative strict maximum at c.
(c) If $\Delta < 0$, then f has a saddle point at c.

Some information concerning the case where $\Delta = 0$ will be given in the exercises.

Extremum Problems with Constraints

Until now we have been discussing the case where the extrema of the function $f : \Omega \to \mathbf{R}$ belong to the interior of its domain $\Omega \subseteq \mathbf{R}^p$. None of our remarks applies to the location of the extrema on the boundary. However, if the function is defined on the boundary of Ω and if this boundary of Ω can be parametrized by a function φ, then the extremum problem is deduced to an examination of the extrema of the composition $f \circ \varphi$.

There is a related problem which leads to an interesting and elegant procedure. Suppose that S is a "surface" contained in the domain Ω of the real-valued function f. It is often desired to find the values of f that are maximum or minimum among all those attained on S. For example, if $\Omega = \mathbf{R}^p$ and $f(x) = \|x\|$, then the problem we have posed is concerned with finding the points on the surface S which are closest to, or farthest from, the origin. If the surface S is given parametrically, then we can treat this problem by considering the composition of f with the parametric representation of S. However, it frequently is not convenient to express S in this fashion and another procedure is often more desirable.

Suppose S can be given as the points x in Ω satisfying a relation of the form

$$g(x) = 0,$$

for a function g defined on Ω to R. We are attempting to find the relative extreme values of f for those points x in Ω satisfying the **constraint** (or **side condition**) $g(x) = 0$. If we assume that f and g are in Class $C^1(\Omega)$ and that $Dg(c) \neq 0$, then a necessary condition that c be an extreme point of f relative to points x satisfying $g(x) = 0$, is that the derivative $Dg(c)$ is a multiple of $Df(c)$. In terms of partial derivatives, this condition is that there exists a real number λ such that

(42.4)
$$D_1 f(c) = \lambda D_1 g(c),$$
$$\cdots \cdots \cdots$$
$$D_p f(c) = \lambda D_p g(c).$$

In practice we wish to determine the p coordinates of the point c satisfying this necessary condition. However, the real number λ, which is usually called the **Lagrange multiplier,** is not known either. The p equations given above, together with the equation

$$g(c) = 0$$

are then solved for the $p + 1$ unknown quantities, of which the coordinates of c are of primary interest.

42.7 Lagrange's Theorem. *Let $\Omega \subseteq R^p$ be open and suppose that f and g are real-valued functions in Class $C^1(\Omega)$. Suppose $c \in \Omega$ is such that $g(c) = 0$ and that there exists on neighborhood U of C such that*

$$f(x) \leq f(c) \qquad [or\ f(x) \geq f(c)]$$

for all points $x \in U$ which satisfy $g(x) = 0$. Then there exist real numbers μ, λ, not both zero, such that

(42.5)
$$\mu Df(c) = \lambda Dg(c).$$

Moreover, if $Dg(c) \neq 0$, we can take $\mu = 1$.

PROOF. Let $F : U \to R^2$ be defined by

$$F(x) = (f(x), g(x)) \qquad \text{for } x \in U,$$

so that F belongs to Class $C^1(U)$ and

$$DF(x)(v) = (Df(x)(v), Dg(x)(v)), \qquad x \in U,\ v \in R^p,$$

Moreover, a point $x \in U$ satisfies the constraint $g(x) = 0$ if and only if $F(x) = (f(x), 0)$.

If $f(x) \le f(c)$ for all $x \in U$ satisfying $g(x) = 0$, then the points $(r, 0)$ with $f(c) < r$ are not in the image $F(U)$; hence $DF(c)$ is not a surjection of $\mathbf{R}^p$ onto $\mathbf{R}^2$. But since the range of the linear map $DF(c)$ is a subspace of $\mathbf{R}^2$ and does not coincide with $\mathbf{R}^2$, the range of $DF(c)$ is contained in some line in $\mathbf{R}^2$ passing through $(0, 0)$. Therefore, there exists a point $(\lambda, \mu) \ne (0, 0)$ such that the range of $DF(c)$ is contained in the line through $(0, 0)$ and (λ, μ). Hence we have

$$(42.6) \qquad \mu Df(c)(v) = \lambda Dg(c)(v) \qquad \text{for all } v \in \mathbf{R}^p, v \in \mathbf{R}^p.$$

whence equation (42.5) follows.

Finally, suppose that $Dg(c) \ne 0$. If $\mu = 0$, then equation (42.5) implies that $\lambda = 0$, which contradicts the fact that $(\mu, \lambda) \ne (0, 0)$. Therefore in this case we must have $\mu \ne 0$ and can divide by μ and replace λ/μ by λ.   Q.E.D.

Since $U \subseteq \mathbf{R}^p$, equation (42.6) with $v = e_1, \ldots, e_p$ yields the system of p equations:

$$\mu D_1 f(c) = \lambda D_1 g(c),$$
$$\cdots \cdots \cdots \cdots$$
$$\mu D_p f(c) = \lambda D_p g(c).$$

If not all of the $D_i g(c)$, $i = 1, \ldots, p$, vanish, then we can take $\mu = 1$ to obtain the system (42.4).

It should be emphasized that Lagrange's Theorem yields a necessary condition only, and that the points obtained by solving the equations (which is often difficult to do!) may be relative maxima, relative minima, or neither. However, Corollary 42.13 below often can be used to test for relative maxima or minima. Furthermore, in many applications the determination of whether the points are actually extrema can be based on geometrical or physical considerations.

42.8 EXAMPLES. (a) We wish to find a point on the plane $\{(x, y, z): 2x + 3y - z = 5\}$ in $\mathbf{R}^3$ which is nearest to the origin. To attack this problem, we shall minimize the function which gives the square of the distance to the origin:

$$f(x, y, z) = x^2 + y^2 + z^2,$$

under the constraint

$$g(x, y, z) = 2x + 3y - z - 5 = 0.$$

Since $Dg(c) \ne 0$ for all $c \in \mathbf{R}^3$, Lagrange's Theorem leads to the system

$$2x = 2\lambda,$$
$$2y = 3\lambda,$$
$$2z = -\lambda,$$
$$2x + 3y - z - 5 = 0.$$

Hence, on eliminating x, y, z, we get

$$2\lambda + 3(\tfrac{3}{2}\lambda) - (-\tfrac{1}{2}\lambda) - 5 = 0,$$

or $14\lambda = 4\lambda + 9\lambda + \lambda = 10$, whence $\lambda = 5/7$. We are lead to the single point $(5/7, 15/14, -5/14)$. From geometrical considerations we deduce that this is the point on the plane nearest $(0, 0, 0)$.

(b) Find the dimensions of the rectangular box, open at the top, with maximum volume and given surface area A. Let x, y, z be the dimensions of the box, with z as the height. Then we wish to maximize the function

$$V(x, y, z) = xyz$$

subject to the constraint

$$g(x, y, z) = xy + 2xz + 2yz - A = 0.$$

Since the desired point will have strictly positive coordinates, Lagrange's Theorem leads to the system

$$yz = \lambda(y + 2z),$$
$$xz = \lambda(x + 2z),$$
$$xy = \lambda(2x + 2y),$$
$$xy + 2xz + 2yz - A = 0.$$

If we multiply the first three equations by x, y, and z, respectively, equate, and divide by λ (why is $\lambda \neq 0$?), we are led to

$$xy + 2xz = xy + 2yz = 2xz + 2yz.$$

The first equality implies $x = y$, and the second implies $y = 2z$. Hence the ratio of the sides are $2:2:1$ and it follows from the last equation that $4z^2 + 4z^2 + 4z^2 = A$ which implies that $z = \tfrac{1}{2}(A/3)^{1/2}$. Therefore the volume of this box is $\tfrac{1}{2}(A/3)^{3/2}$.

Frequently there is more than one constraint; in this case the following result is useful.

42.9 THEOREM. Let $\Omega \subseteq \mathbf{R}^p$ be open and suppose that f and $g_1, \ldots, g_k$ are real-valued functions in $C'(\Omega)$. Suppose that $c \in \Omega$ satisfies the constraints

$$g_1(x) = 0, \ldots, g_k(x) = 0,$$

and that there exists an open neighborhood U of a such that $f(x) \leq f(c)$ [or $f(x) \geq f(c)$] for all $x \in U$ satisfying these constraints. Then there exist real numbers $\mu, \lambda_1, \ldots, \lambda_k$ not all zero such that

(42.7) $\mu Df(c) = \lambda_1 Dg_1(c) + \cdots + \lambda Dg_k(c).$

PROOF. Let $F: U \to \mathbf{R}^{k+1}$ be defined by

$$F(x) = (f(x), g_1(x), \ldots, g_k(x)) \qquad \text{for } x \in U,$$

and argue as in the proof of Theorem 42.7. Q.E.D.

42.10 COROLLARY. *In addition to the hypotheses of Theorem 42.9, suppose that the rank of the matrix*

(42.8)
$$\begin{bmatrix} D_1 g_1(c) & \cdots & D_1 g_k(c) \\ \cdot & & \cdot \\ \cdot & & \cdot \\ \cdot & & \cdot \\ D_p g_1(c) & \cdots & D_p g_k(c) \end{bmatrix}$$

is equal to $k (\le p)$. Then there are real numbers $\lambda_1, \ldots, \lambda_k$ not all zero such that

(42.9)
$$D_1 f(c) = \lambda_1 D_1 g_1(c) + \cdots + \lambda_k D_1 g_k(c),$$
$$\cdot \quad \cdot \quad \cdot \quad \cdot \quad \cdot \quad \cdot \quad \cdot \quad \cdot \quad \cdot \quad \cdot \quad \cdot \quad \cdot \quad \cdot \quad \cdot$$
$$D_p f(c) = \lambda_1 D_p g_1(c) + \cdots + \lambda_k D_p g_k(c).$$

PROOF. If we apply the formula (42.7) to $e_1, \ldots, e_p \in \mathbf{R}^p$, we obtain a system of equations with the right-hand side of (42.9) and the left-hand side of (42.9) multiplied by μ. If $\mu = 0$, then the assumption that the rank equals k implies that $\lambda_1 = \cdots = \lambda_k = 0$, contrary to hypothesis. Hence $\mu \ne 0$ and we can normalize this system to obtain (42.9). Q.E.D.

42.11 EXAMPLE. Find the points on the intersection of the cylinder $\{(x, y, z) : x^2 + y^2 = 4\}$ and the plane $\{(x, y, z) : 6x + 3y + 2z = 6\}$ which are nearest to the origin and those which are farthest from the origin.

We shall search for relative extrema of the function

$$f(x, y, z) = x^2 + y^2 + z^2$$

subject to the constraints

$$g_1(x, y, z) = x^2 + y^2 - 4 = 0,$$
$$g_2(x, y, z) = 6x + 3y + 2z - 6 = 0.$$

The matrix corresponding to (42.8) in this case is

$$\begin{bmatrix} 2x & 6 \\ 2y & 3 \\ 0 & 2 \end{bmatrix}$$

which has rank 2 except at the point $(x, y) = (0, 0)$ which does not satisfy the constraints. Hence we can apply the corollary to obtain the system

$$2x = \lambda_1(2x) + \lambda_2(6),$$
$$2y = \lambda_1(2y) + \lambda_2(3),$$
$$2z = \qquad \lambda_2(2),$$
$$x^2 + y^2 = 4,$$
$$6x + 3y + 2z = 6,$$

of five equations in five variables. The third equation gives $\lambda_2 = z$, so we can eliminate λ_2 from the first two equations. To eliminate λ_1, we multiply the resulting first equation by y and the second by x and subtract, to get

$$0 = 6yz - 3xz = 3z(2y - x).$$

It follows that either $z = 0$ or $x = 2y$.

If $z = 0$, the fifth equation yields $2x + y = 2$. When combined with the fourth equation this gives

$$x^2 + (2 - 2x)^2 = x^2 + 4 - 8x + 4x^2 = 4,$$

whence $5x^2 - 8x = x(5x - 8) = 0$, and hence $x = 0$ or $x = 8/5$. This case leads to the two points $(0, 2, 0)$ and $(8/5, -6/5, 0)$ each of which have distance 2 from the origin.

On the other hand, if $x = 2y$, the fourth equation yields $5y^2 = 4$ so that $y = 2/\sqrt{5}$ (and $x = 4/\sqrt{5}$) or $y = -2/\sqrt{5}$ (and $x = -4/\sqrt{5}$). Substituting in the fifth equation we get $z = 3(1 - \sqrt{5})$ and $z = 3(1 + \sqrt{5})$, respectively. Therefore, this case leads to the two points $(4/\sqrt{5}, 2/\sqrt{5}, 3(1 - \sqrt{5}))$ and $(-4/\sqrt{5}, -2/\sqrt{5}, 3(1 + \sqrt{5}))$. The squares of the distances between these points and the origin are seen to be $58 - 18\sqrt{5}$ and $58 + 18\sqrt{5}$, respectively.

We deduce that both of the points $(0, 2, 0)$ and $(8/5, -6/5, 0)$ minimize the distance from the origin and this intersection, and that the point $(-4/\sqrt{5}, -2/\sqrt{5}, 3(1 + \sqrt{5}))$ maximizes this distance. From geometrical considerations we also see that the point $(4/\sqrt{5}, 2/\sqrt{5}, 3(1 - \sqrt{5}))$ gives a relative maximum among points of this intersection. (The reader should draw a diagram to help him visualize this situation.)

Inequality Constraints

In recent years, extremum problems involving constraints which are inequalities rather than equalities have become increasingly important. Thus we may wish to find a relative extremum of a function $f: \Omega \to \mathbf{R}$ among all points in $\Omega \subseteq \mathbf{R}^p$ satisfying the constraints

$$h_1(x) \ge 0, \ldots, h_k(x) \ge 0.$$

We will see that such problems can also be handled by Lagrange's method.

Sometimes an extremum problem may involve both equalities and inequalities, but since the equality $g(x) = 0$ is equivalent to the inequality $-(g(x))^2 \ge 0$, such problems can always be reduced to one involving only inequality constraints.

42.12 THEOREM. *Let $\Omega \subseteq \mathbf{R}^p$ be open and suppose that f and $h_1, \ldots, h_k$ are real-valued functions in $C^1(\Omega)$. Suppose that $c \in \Omega$ satisfies the inequality constraints*

$$h_1(x) \ge 0, \ldots, h_k(x) \ge 0,$$

and that there exists an open neighborhood U of c such that $f(x) \le f(c)$ [or $f(x) \ge f(c)$] for all $x \in U$ satisfying these constraints. Then there exist real *numbers $\mu, \lambda_1, \ldots, \lambda_k$ not all zero such that*

(42.10) $$\mu Df(c) = \lambda_1 Dh_1(c) + \cdots + \lambda_k Dh_k(c).$$

Furthermore, if $h_i(c) > 0$ for some i, then $\lambda_i = 0$.

PROOF. Let $F: U \to \mathbf{R}^{k+1}$ be defined by

$$F(x) = (f(x), h_1(x), \ldots, h_k(x)) \qquad \text{for } x \in U.$$

If c is a point in U where the constraints are satisfied and where f is either maximized or minimized; then $DF(c)$ cannot be surjective and so (42.10) must hold.

If $h_1(c) = 0, \ldots, h_r(c) = 0$, but $h_{r+1}(c) > 0, \ldots, h_k(c) > 0$, then let $U_1 \subseteq U$ be an open neighborhood of c on which $h_{r+1}, \ldots, h_k$ are strictly positive and apply the theorem to the constraints $h_1(x) \geq 0, \ldots, h_r(x) \geq 0$. Q.E.D.

42.13 COROLLARY. *In addition to the hypotheses of Theorem 42.12, suppose that the rank of the matrix*

(42.11)
$$\begin{bmatrix} D_1h_1(c) & \cdots & D_1h_r(c) \\ \cdot & & \cdot \\ \cdot & & \cdot \\ \cdot & & \cdot \\ D_ph_1(c) & \cdots & D_ph_r(c) \end{bmatrix}$$

corresponding to those h_i for which $h_i(c) = 0$, is equal to r. Then we may take $\mu = 1$ in (42.10). In addition, if $f(x) \leq f(c)$ [respectively, $f(x) \geq f(c)$] for all $x \in U$ satisfying the constraints and if we take $\mu = 1$ in (42.10), then $\lambda_i \leq 0$ [respectively, $\lambda_i \geq 0$] for $i = 1, \ldots, r$.

PROOF. The proof that we can take $\mu = 1$ in (42.10) is similar to that of Corollary 42.10. Suppose, then, that $\mu = 1$ and that $f(x) \leq f(c)$ for all $x \in U$ satisfying the constraints. Since the rank of the matrix (42.11) is $r \leq k$, then if $1 \leq j \leq r$ there exists a vector $v_j \in \mathbf{R}^p$ such that

$$Dh_i(c)(v_j) = \delta_{ij}.$$

Therefore, if $t > 0$ is sufficiently small, then there exists a point c_t on the segment joining c and $c + tv_j$ such that

$$0 \geq f(c + tv_j) - f(c) = Df(c_t)(tv_j) = tDf(c_t)(v_j).$$

Consequently we have

$$0 \geq \lim_{\substack{t \to 0 \\ t > 0}} \frac{f(c + tv_j) - f(c)}{t} = Df(c)(v_j) = \sum_{i=1}^{r} \lambda_i Dh_i(c)(v_j) = \lambda_j.$$

Hence $\lambda_j \leq 0$ for $j = 1, \ldots, r$. Q.E.D.

For an elementary, but very different proof of the theorem of Lagrange involving inequality constraints, see the article of E. J. McShane† listed in the References.

Exercises

42.A. Find the critical points of the following functions and determine the nature of these points.
  (a) $f(x, y) = x^2 + 4xy$,
  (b) $f(x, y) = x^4 + 2y^4 + 32x - y + 17$,
  (c) $f(x, y) = x^2 + 4y^2 - 12y^2 - 36y$,
  (d) $f(x, y) = x^4 - 4xy$,
  (e) $f(x, y) = x^2 + 4xy + 2y^2 - 2y$,
  (f) $f(x, y) = x^2 + 3y^4 - 4y^3 - 12y^2$.

42.B. Let $\Omega \subseteq R^p$ be open, let $f : \Omega \to R$ have continuous second partial derivatives on Ω, let $c \in \Omega$ be a critical point of f, and let $\delta > 0$.
  (a) Show that if $D^2 f(x)(w)^2 \geq 0$ for all $0 < \|x - c\| < \delta$ and $w \in R^p$, then c is a point of relative minimum of f.
  (b) Show that if $D^2 f(x)(w)^2 > 0$ for all $0 < \|x - c\| < \delta$ and $w \in R^p$, $w \neq 0$, then c is a point of relative strict minimum of f.

42.C. Let $\Omega \subseteq R^2$ be open, let $f : \Omega \to R$ have continuous second partial derivatives on Ω, let $c \in \Omega$ be a critical point of f, and let

$$\Delta(x) = D_{11} f(x) D_{22} f(x) - (D_{12} f(x))^2$$

for $x \in \Omega$. Suppose that for some $\delta > 0$, then $\Delta(x) \geq 0$ for all $\|x - c\| < \delta$.
  (a) If $D_{11} f(x) > 0$ (or if $D_{22} f(x) > 0$) for all x such that $0 < \|x - c\| < \delta$, show that c is a point of relative minimum of f.
  (b) If $D_{11} f(x) < 0$ (or if $D_{22} f(x) < 0$) for all x such that $0 < \|x - c\| < \delta$, show that c is a point of relative maximum of f.

42.D. Let $f : R^p \to R$ be differentiable on R^p and $f(x) = 0$ for all $x \in R^p$ with $\|x\| = 1$. Show that there exists a point $c \in R^p$ with $\|c\| < 1$ such that $Df(c) = 0$. (This is a version of Rolle's Theorem in R^p.)

42.E. Use the Surjective Mapping Theorem 41.6 to establish Corollary 42.2.

42.F. Show that each of the following functions has a critical point at the origin. Find which have relative extrema and which have saddle points at the origin.
  (a) $f(x, y) = x^2 y^2$,   (b) $f(x, y) = x^2 - y^3$,
  (c) $f(x, y) = x^3 - y^3$,   (d) $f(x, y) = x^4 - x^2 y^2 + y^4$,
  (e) $f(x, y) = x^3 y - xy^3$,   (f) $f(x, y) = x^4 + y^4$.

42.G. Show that the function $f(x, y) = 2x + 4y - x^2 y^4$ has a critical point but no relative extreme points.

42.H. Study the behavior of the function $f(x, y) = x^3 - 3xy^2$ in a neighborhood of the origin. The graph of this function is sometimes called a "monkey saddle." Why?

---

† E. J. McShane (1904–) received his doctor's degree from the University of Chicago. He has long been associated with the University of Virginia and is widely known for his contributions to integration theory, the calculus of variations, optimal control theory, and exterior ballistics.

42.I. Find the minimum distance from the point $(2, 1, -3)$ to the plane $2x + y - 2z = 4$.

42.J. Find the dimensions of the rectangular box, open at the top, with given volume and minimum surface area.

42.K. Find the minimum distance between the lines $L_1 = \{(x, y, z) : x = 2 - t, y = 3 + t, z = 1 - 2t\}$ and $L_2 = \{(x, y, z) : x = 1 - s, y = 2 - s, z = 3 + s\}$.

42.L. Give examples to show that each of the conjectures stated after Theorem 42.5 is false.

42.M. Suppose we are given n points (x_j, y_j), $j = 1, \ldots, n$ in $\mathbf{R}^2$ and wish to find the affine function $F : \mathbf{R} \to \mathbf{R}$ given by $F(x) = Ax + B$, such that the quantity

$$\sum_{j=1}^{n} (F(x_j) - y_j)^2$$

is minimized. Show that this leads to the equations

$$A \sum_{j=1}^{n} x_j^2 + B \sum_{j=1}^{n} x_j = \sum_{j=1}^{n} x_j y_j,$$

$$A \sum_{j=1}^{n} x_j + nB = \sum_{j=1}^{n} y_j,$$

for the numbers A, B. [This function F is said to be the affine function which "best fits the n points in the sense of **least squares**."]

42.N. Let $f : [0, 1] \to \mathbf{R}$ be continuous on $[0, 1]$. We wish to choose real numbers A, B, C in such a way as to minimize the quantity

$$\int_0^1 [f(x) - (Ax^2 + Bx + C)]^2 \, dx.$$

Show that we should choose A, B, C to satisfy the system

$$\tfrac{1}{5}A + \tfrac{1}{4}B + \tfrac{1}{3}C = \int_0^1 x^2 f(x) \, dx,$$

$$\tfrac{1}{4}A + \tfrac{1}{3}B + \tfrac{1}{2}C = \int_0^1 x f(x) \, dx,$$

$$\tfrac{1}{3}A + \tfrac{1}{2}B + C = \int_0^1 f(x) \, dx.$$

[The resulting function $x \mapsto Ax^2 + Bx + C$ is said to be the quadratic function which "best fits f on $[0, 1]$ in the sense of **least squares**."]

42.O. Use Lagrange's Theorem to locate points on the curve $y = x^5 + x - 2$ where the function $f(x, y) = x - y$ may have a relative extremum. Then sketch the curve and the level curves for f to show that the point(s) located are not point(s) of relative extrema for f.

42.P. Let $f : \mathbf{R}^2 \to \mathbf{R}$ be the quadratic function $f(x, y) = ax^2 + 2bxy + cy^2$ for $(x, y) \in \mathbf{R}^2$. We wish to find the relative extrema of f on the unit circle $\{(x, y) : x^2 + y^2 = 1\}$. Use Lagrange's Theorem to show that the points (x_0, y_0) at which these relative extrema are taken must satisfy the system

$$(a - \lambda)x_0 + by_0 = 0,$$

$$bx_0 + (c - \lambda)y_0 = 0,$$

where the Lagrange multiplier λ is a root of the equation

$$\lambda^2 - (a+c)\lambda + (ac - b^2) = 0.$$

Show that the corresponding value of the multiplier λ is equal to the extreme value of f at such a relative extremum.

42.Q. The sum of three real numbers is 9. Find these numbers if their product is to be maximized.

42.R. Show that the volume of the largest box that can be inscribed in the ellipsoidal region

$$\left\{(x, y, z) : \frac{x^2}{a^2} + \frac{y^2}{b^2} + \frac{z^2}{c^2} \le 1\right\}$$

(where a, b, c are strictly positive numbers) is equal to $8\,abc/3\sqrt{3}$.

42.S. For each of the following functions, find the maximum and minimum values on the given set. (When appropriate, consider the signs of the multipliers.)

(a) $f(x, y) = x^4 - y^4$, $S = \{(x, y) : x^2 + y^2 \le 1\}$.

(b) $f(x, y) = x^2 + 2x + y^2$, $S = \{(x, y) : x^2 + y^2 \le 1\}$.

(c) $f(x, y) = x^2 + 2x + y^2$, $S = \{(x, y), |x| \le 1, |y| \le 1\}$.

(d) $f(x, y) = (1 - x^2) \sin y$, $S = \{(x, y), |x| \le 1, |y| \le \pi\}$.

42.T. Let f be defined for $x > 0$, $y > 0$ to $\mathbf{R}$ by $f(x, y) = 1/x + cxy + 1/y$.

(a) Locate the critical points of f and determine their nature.

(b) If $c > 0$, let $S = \{(x, y) : 0 < x, \ 0 < y, \ x + y \le c\}$. Determine the relative maximum and minimum values of f on S.

42.U. Find the extreme values of $f(x, y, z) = x^3 + y^3 + z^3$ subject to the constraints $x^2 + y^2 + z^2 = 1$ and $x + y + z = 1$.

42.V. Let f have continuous second partial derivatives on an open set containing the ball $\{x \in \mathbf{R}^p : \|x\| \le r\}$ to $\mathbf{R}$, and suppose there exists c with $\|c\| < r$ such that

$$M = f(c) > \sup\{f(x) : \|x\| = r\} = m.$$

Let g be defined by

$$g(x) = f(x) + \frac{M - m}{4r^2} \|x - c\|^2.$$

Show that $g(c) = M$, while $g(x) < M$ for $\|x\| = r$. Hence g attains a relative maximum at some point c_1 with $\|c_1\| < r$, where we have

$$\sum_{j=1}^{p} D_{jj}f(c_1) \le -\frac{p}{2r^2}(M - m) < 0.$$

42.W. Let $\Omega \subseteq \mathbf{R}^p$ be a bounded open set, let $b(\Omega)$ be the set of boundary points of Ω (see Definition 9.7), and let $\Omega^- = \Omega \cup b(\Omega)$ be the closure of Ω. A function $f : \Omega^- \to \mathbf{R}$ is said to be **harmonic** on Ω if it is continuous on Ω^- and satisfies the Laplace equation

$$\sum_{j=1}^{p} D_{jj}f(x) = 0$$

for all $x \in \Omega$.

(a) Use the argument of the preceding exercise to show that a harmonic function on Ω attains its supremum and infimum on $b(\Omega)$.

(b) If f and g are harmonic on Ω and if $f(x) = g(x)$ for $x \in b(\Omega)$, then $f(x) = g(x)$ for all $x \in \Omega^-$.

(c) If f and g are harmonic on Ω and if $f(x) = \varphi(x)$, $g(x) = \psi(x)$ for $x \in b(\Omega)$, then

$$\sup\{|f(x) - g(x)| : x \in \Omega\} = \sup\{|\varphi(x) - \psi(x)| : x \in b(\Omega)\}.$$

(This conclusion can be stated by saying that "the solutions of the Dirichlet problem for Ω depend continuously on the boundary data.")

42.X. Show that the maximum of $f(x_1, \ldots, x_p) = (x_1 \cdots x_p)^2$ subject to the constraint $x_1^2 + \cdots + x_p^2 = 1$ is equal to $1/p^p$. Use this to obtain the inequality:

$$|y_1 \cdots y_p| \le \frac{\|y\|^p}{p^{p/2}} \qquad \text{for } y \in \mathbf{R}^p.$$

42.Y. Show that the geometric mean of a collection of positive real numbers $\{a_1, \ldots, a_p\}$ does not exceed their arithmetic mean; that is,

$$(a_1 \cdots a_p)^{1/p} \le \frac{1}{p}(a_1 + \cdots + a_p).$$

42.Z. (a) Let $p > 1$, $q > 1$, $1/p + 1/q = 1$. Show that the minimum of $f(x, y) = (1/p)x^p + (1/q)y^q$ $(x > 0, y > 0)$ subject to the constraint $xy = 1$ is equal to 1.

(b) Use part (a) to show that if $a > 0$, $b > 0$, then

$$ab \le \frac{1}{p} a^p + \frac{1}{q} b^q.$$

(c) Let $\{a_i\}$, $\{b_i\}$, $i = 1, \ldots, n$, be positive real numbers. Prove Hölder's Inequality:

$$\sum_{i=1}^n a_i b_i \le \left(\sum_{i=1}^n a_i^p\right)^{1/p} \left(\sum_{i=1}^n b_i^q\right)^{1/q}$$

by letting $A = (\sum a_i^p)^{1/p}$, $B = (\sum b_i^q)^{1/q}$, and applying part (c) to $a = a_i/A$, $b = b_i/B$.

(d) Use Hölder's Inequality in (c) to obtain Minkowski's Inequality:

$$\left(\sum_{i=1}^n |a_i + b_i|^p\right)^{1/p} \le \left(\sum_{i=1}^n |a_i|^p\right)^{1/p} + \left(\sum_{i=1}^n |b_i|^{1/p}\right)^{1/p}$$

[Hint: $|a + b|^p = |a + b| \, |a + b|^{p/q} \le |a| \, |a + b|^{p/q} + |b| \, |a + b|^{p/q}.$]

VIII
INTEGRATION
IN R^p

In this chapter we shall present the theory of integration of real-valued functions on R^p where $p > 1$. The approach used here is the same as that initiated in Section 29 for the case $p = 1$, but we shall be concerned here only with the Riemann integral (and not the Riemann-Stieltjes integral).

It will be seen in Section 43 that, for bounded functions defined on a closed cell in R^p, the theory requires virtually no changes from that in R. However, in order to be able to integrate over more general sets in R^p it is necessary to develop, as we do in Section 44, a theory of "content" (as we shall call the p-dimensional notion of "area") for a suitable family of sets in R^p. We shall characterize the content function on this family of sets, and show how to express integrals in R^p as iterated integrals. The final section is devoted to developing important theorems on the transformations of sets and integrals under differentiable mappings. The theoretical difficulties are considerable, but we conclude with a very useful theorem justifying the change of variables even in cases when the transformation may possess a limited amount of "singular" behavior.

Section 43   The Integral in R^p

In Sections 29–31 we discussed the integral of bounded real-valued functions defined on a compact interval J in R. A reader for an eye for generalization will have noticed that a considerable part of what was done in those sections can be carried out when the *values* of the function lie in a Cartesian spaces R^q. Once this possibility has been recognized it is not difficult to carry out the modifications necessary to obtain an integration theory for functions on J to R^q.

It is also natural to ask whether we can obtain an integration theory for functions whose domain is a subset of the space $\mathbf{R}^p$, and the reader will recall that this was in fact done in calculus courses where one considered "double" and "triple" integrals. In this section we shall initiate a study of the Riemann integral of real-valued functions defined on a suitable bounded subset of $\mathbf{R}^p$. Although many of our results can be extended to permit the values to be in $\mathbf{R}^q$ for $q > 1$, we shall leave this extension to the reader.

Content Zero

We recall from Section 5 that a **cell** in $\mathbf{R}$ is a set having one of the four forms:

$$(a, b), \quad [a, b], \quad [a, b), \quad (a, b],$$

where $a \le b$. The numbers a, b are called the **end points** of these cells. A **cell** in $\mathbf{R}^p$ is the Cartesian product $J = J_1 \times \cdots \times J_p$ of p cells in $\mathbf{R}$. The cell J is said to be **closed** (respectively, **open**) if each of the cells $J_1, \ldots, J_p$ are closed (respectively, open) in $\mathbf{R}$. If the cells J_i have end points $a_i \le b_i$ $(i = 1, \ldots, p)$, we define the **content** of $J = J_1 \times \cdots \times J_p$ to be the product

$$c(J) = (b_1 - a_1) \cdots (b_p - a_p).$$

If $p = 1$, content is usually called "length"; if $p = 2$, the content is called "area"; if $p = 3$, the content is called "volume." We shall use the word "content" because it is free from special connotations that these other words may have.

Note that if $J = J_1 \times \cdots \times J_p$, and $K = K_1 \times \cdots \times K_p$ are cells in $\mathbf{R}^p$ such that the end points of J_i and of K_i are the same for each $i = 1, \ldots, p$, then $c(J) = c(K)$. Similarly, if $a_k = b_k$ for some $k = 1, \ldots, p$, then the cell J has content $c(J) = 0$; however, it is not necessary that $J = \emptyset$.

If J_i is a cell with end points $a_i \le b_i$ and if $b_1 - a_1 = \cdots = b_p - a_p > 0$, then we say that

$$J = J_1 \times \cdots \times J_p$$

is a **cube.** Cube may be closed, open, or neither. We call the number $b_1 - a_1 > 0$ the **side length** of the cube.

43.1 DEFINITION. A set $Z \subseteq \mathbf{R}^p$ has **content zero** if for each $\varepsilon > 0$ there exists a *finite* set $J_1, \ldots, J_n$ of cells whose union contains Z and such that

$$c(J_1) + \cdots + c(J_n) < \varepsilon.$$

It is important that the reader show that one can require the cells appearing in this definition to be closed, or to be open, or to be cubes, and the notion of content zero remains exactly the same.

43.2 **EXAMPLES.** (a) A point in $\boldsymbol{R}^p$ has content zero. (Why?) More generally, any *finite* subset of $\boldsymbol{R}^p$ has content zero.

(b) If $(z_n)_{n \in N}$ is a sequence in $\boldsymbol{R}^p$ which converges to $z_0 \in \boldsymbol{R}^p$, then the set $Z = \{z_n : n \ge 0\}$ has content zero. For, if $\varepsilon > 0$ let J_0 be an open cell containing z_0 such that $c(J_0) < \varepsilon$. Therefore there exists $k \in N$ such that $z_n \in J_0$ for all $n > k$, and we can take $J_i = \{z_i\}$ for $i = 1, \ldots, k$ to get $Z \subseteq J_0 \cup J_1 \cup \cdots \cup J_k$. Since

$$c(J_0) + c(J_1) + \cdots + c(J_k) < \varepsilon + 0 + \cdots + 0 = \varepsilon,$$

and since $\varepsilon > 0$ is arbitrary, it follows that Z has content zero.

(c) Any subset of a set with content zero has content zero. The union of a *finite* number of sets with content zero has content zero.

(d) In the space $\boldsymbol{R}^2$, the diamond-shaped set $S = \{(x, y) : |x| + |y| = 1\}$ has content zero. For, if $n \in N$, we introduce squares with diagonals along S and vertices at the points $x = y = \pm k/n$ ($k = 0, 1, \ldots, n$), then we see that we can enclose S in $4n$ closed squares each having content $1/n^2$. Hence the total content is $4/n$, which can be made arbitrarily small. (See Fig. 43.1.)

(e) The circle $S = \{(x, y) : x^2 + y^2 = 1\}$ in $\boldsymbol{R}^2$ has content zero. This can be proved by modifying the argument in (d).

(f) Let f be a continuous function on $J = [a, b]$ to $\boldsymbol{R}$. Then the graph

$$G = \{(x, f(x)) \in \boldsymbol{R}^2 : x \in J\}$$

has content zero. This can be proved by using the uniform continuity of f and modifying the argument in (d).

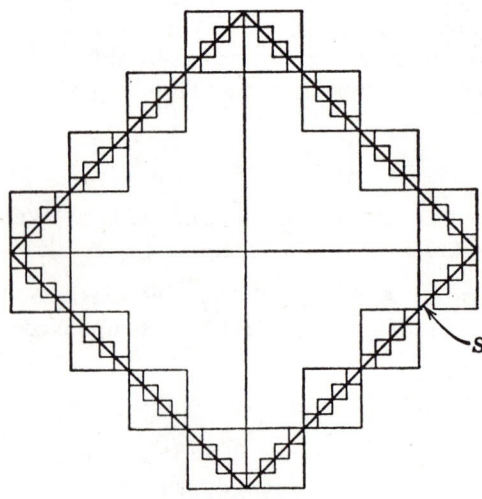

Figure 43.1

(g) The set $S \subseteq \mathbf{R}^2$ consisting of all points (x, y) where both x and y belong to $\mathbf{I} \cap \mathbf{Q}$ is a countable set but does not have content zero. Indeed, any finite union of cells containing S must also contain the cell $\mathbf{I} \times \mathbf{I}$, which has content 1.

In contrast to (f) we note that *there are "continuous curves" in* $\mathbf{R}^2$ *which have positive content.* Indeed, there are continuous functions f, g on $\mathbf{I} = [0, 1]$ to $\mathbf{R}$ such that the set

$$S = \{(f(t), g(t)); t \in \mathbf{I}\}$$

contains the cell $\mathbf{I} \times \mathbf{I}$ in $\mathbf{R}^2$. Such a curve is called a **space-filling curve,** or a **Peano curve.** (See Exercise 43.U.)

Definition of the Integral

We shall first define the integral for a bounded function f defined on a *closed cell* $I \subseteq \mathbf{R}^p$ and with values in $\mathbf{R}$. Let

$$I = [a_1, b_1] \times \cdots \times [a_p, b_p],$$

and, for each $k = 1, \ldots, p$, let P_k be a partition of $[a_k, b_k]$ into a finite number of closed cells in $\mathbf{R}$. This induces a **partition** P of I into a finite number of closed cells in $\mathbf{R}^p$. If P and Q are partitions of I, we say that P is a **refinement** of Q if each cell in P is contained in some cell in Q. (Alternatively, noting that a partition is determined by the vertices of its cells, P is a refinement of Q if and only if all of the vertices contained in Q are also in P.)

43.3 DEFINITION. A **Riemann sum** $S(P; f)$ corresponding to a partition $P = \{J_1, \ldots, J_n\}$ of I is given by

$$S(P; f) = \sum_{k=1}^{n} f(x_k) c(J_k),$$

where x_k is any "intermediate" point in J_k, $k = 1, \ldots, n$. A real number L is defined to be the **Riemann integral of** f **over** I if, for every $\varepsilon > 0$ there is a partition P_ε of I such that if P is any refinement of P_ε and $S(P; f)$ is any Riemann sum corresponding to P, then $|S(P; f) - L| \le \varepsilon$. In case this integral exists we say that f is **integrable over** I.

It is a routine exercise to show that the value L of the integral is uniquely determined when it exists; we shall generally write

$$L = \int_I f;$$

however, when $p = 2$ we occasionally denote the integral by

$$\iint_I f \quad \text{or} \quad \iint_I f(x, y)\, d(x, y),$$

and when $p = 3$ we occasionally write

$$\iiint_I f \quad \text{or} \quad \iiint_I f(x, y, z)\, d(x, y, z).$$

There is a convenient Cauchy Criterion for integrability. Since its proof is so similar to that of Theorem 29.4, we shall omit it.

43.4 CAUCHY CRITERION. *A bounded function $f : I \to \mathbf{R}$ is integrable on I if and only if for every $\varepsilon > 0$ there exists a partition Q_ε of I such that if P and Q are partitions of I which are refinements of Q_ε and $S(P; f)$ and $S(Q; f)$ are any corresponding Riemann sums, then*

$$|S(P; f) - S(Q; f)| \le \varepsilon.$$

We now wish to consider functions which are defined on bounded subsets of $\mathbf{R}^p$ more general than closed cells. Let $A \subseteq \mathbf{R}^p$ be a *bounded* set and let $f : A \to \mathbf{R}$ be a *bounded* function. Since A is bounded there exists a closed cell $I \subseteq \mathbf{R}^p$ such that $A \subseteq I$. We define $f_I : I \to \mathbf{R}$ by

$$f_I(x) = f(x) \qquad \text{for } x \in A,$$
$$= 0 \qquad \text{for } x \in I \setminus A.$$

If the function f_I is integrable on I in the sense of Definition 43.3, then it is an exercise (see 43.M) to show that the value $\int_I f_I$ does not depend on the choice of the closed cell I containing A. Because of this we shall say that f is **integrable on** A and define

$$\int_A f = \int_I f_I,$$

since the right-hand side depends only on f and A. (In subsequent arguments, we shall often denote f_I simply by f.)

Similarly, let A and B be bounded subsets of $\mathbf{R}^p$ and let $f : A \to \mathbf{R}$. Let I be a closed cell containing $A \cup B$ and define $f_1 : I \to \mathbf{R}$ by

$$f_1(x) = f(x) \qquad \text{for } x \in A \cap B,$$
$$= 0 \qquad \text{for } x \in I \setminus (A \cap B).$$

Note that f_1 is the extension to I of the restriction $f \mid A \cap B$. If f_1 is integrable over I, we say that f **is integrable on** B and define

$$\int_B f = \int_I f_1 \quad \left(= \int_{A \cap B} f \right).$$

Properties of the Integral

We shall now give some of the expected properties of the integral. Throughout, A will be a bounded subset of $\mathbf{R}^p$.

43.5 THEOREM. *Let f and g be functions on A to $\mathbf{R}$ which are integrable on A and let α, $\beta \in \mathbf{R}$. Then the function $\alpha f + \beta g$ is integrable on A and*

$$\int_A (\alpha f + \beta g) = \alpha \int_A f + \beta \int_A g.$$

PROOF. This result follows from the fact that the Riemann sums for a partition P of a cell $I \supseteq A$ satisfy

$$S(P; \alpha f + \beta g) = \alpha S(P; f) + \beta S(P; g),$$

when the same intermediate points x_k are used. Q.E.D.

43.6 THEOREM. *If $f : A \to \mathbf{R}$ is integrable on A and if $f(x) \geq 0$ for $x \in A$, then $\int_A f \geq 0$.*

PROOF. Note that $S(P; f) \geq 0$ for any Riemann sum. Q.E.D.

43.7 THEOREM. *Let $f : A \to \mathbf{R}$ be a bounded function and suppose that A has content zero. Then f is integrable on A and $\int_A f = 0$.*

PROOF. Let I be a closed cell containing A. If $\varepsilon > 0$ is given, let P_ε be a partition of I such that those cells in P_ε which contain points of A have total content less than ε. (Show there exists such a partition P_ε.) Now if P is a refinement of P_ε, then those cells in P which contain points of A will also have total content less than ε. Hence if $|f(x)| \leq M$ for $x \in A$, we have $|S(P; f)| \leq M\varepsilon$ for any Riemann sum corresponding to P. Since $\varepsilon > 0$ is arbitrary, this implies that $\int_A f = 0$. Q.E.D.

43.8 THEOREM. *Let $f, g : A \to \mathbf{R}$ be bounded functions and suppose that f is integrable on A. Let $E \subseteq A$ have content zero and suppose that $f(x) = g(x)$ for all $x \in A \setminus E$. Then g is integrable on A and*

$$\int_A f = \int_A g.$$

PROOF. Extend f and g to functions f_I, g_I defined on a closed cell I containing A. The hypotheses imply that $h_I = f_I - g_I$ is bounded and equals 0 except on E. By Theorem 43.7 we deduce that h_I is integrable on I and the value of its integral is 0. Applying Theorem 43.5, we infer that $g_I = f_I - h_I$ is integrable on I and

$$\int_A g = \int_I g_I = \int_I (f_I - h_I) = \int_I f_I = \int_A f.$$

Q.E.D.

Existence of the Integral

It is to be expected that if f is continuous on a closed cell I to $\mathbf{R}$, then f is integrable on I. We shall establish a stronger result which permits f to be discontinuous on the complement of a set with content zero.

43.9 INTEGRABILITY THEOREM. *Let $I \subseteq \mathbf{R}^p$ be a closed cell and let $f: I \to \mathbf{R}$ be bounded. If there exists a subset $E \subseteq I$ with content zero such that f is continuous on $I \setminus E$, then f is integrable on I.*

PROOF. Let $|f(x)| \leq M$ for all $x \in I$ and let $\varepsilon > 0$ be given. Then there exists (why?) a partition P_ε of I such that (i) the cells in P_ε which contain any points of E contain them in their interior, and (ii) these cells have total content less than ε. The union C of the closed cells in P_ε which do not contain points of E is a compact subset on which f is continuous. According to the Uniform Continuity Theorem 23.3, the restriction of f is uniformly continuous on C. Replacing P_ε by a refinement, if necessary, we may suppose that if J_k is a cell in P_ε which is contained in C, and if x, $y \in J_k$, then $|f(x) - f(y)| < \varepsilon$.

Now suppose that P and Q are refinements of P_ε. If $S'(P; f)$ and $S'(Q; f)$ denote the portions of the Riemann sums extended over the cells in C, then an argument similar to that in the second part of the proof of Theorem 30.1, yields

$$|S'(P; f) - S'(Q; f)| \leq |S'(P; f) - S'(P_\varepsilon; f)| + |S'(P_\varepsilon; f) - S'(Q; f)|$$
$$\leq 2\varepsilon c(I).$$

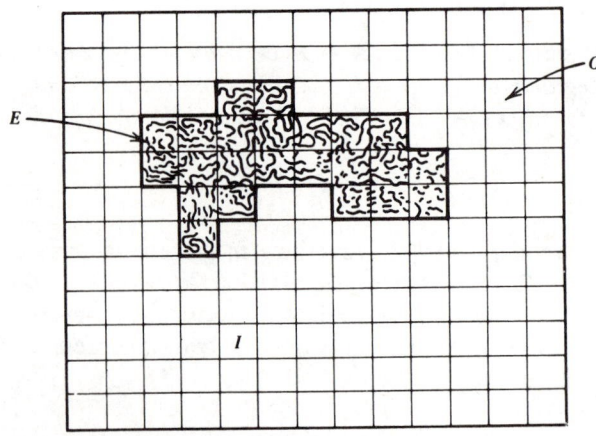

Figure 43.2

Similarly, if $S''(P; f)$ and $S''(Q; f)$ denote the remaining portion of the Riemann sums, then

$$|S''(P; f) - S''(Q; f)| \leq |S''(P; f)| + |S''(Q; f)| \leq 2M\varepsilon.$$

It therefore follows that

$$|S(P; f) - S(Q; f)| \leq \varepsilon(2c(I) + 2M);$$

since $\varepsilon > 0$ is arbitrary, we infer from the Cauchy Criterion that f is integrable on I. Q.E.D.

Necessary and sufficient conditions for integrability will be given in Exercise 43.P and Project 44.α.

Exercises

43.A. (a) Let $a = (a_1, \ldots, a_p) \in \mathbf{R}^p$ and let $J_1 = [a_1, a_1], \ldots, J_p = [a_p, a_p]$ be cells in $\mathbf{R}$. Show that $J = J_1 \times \cdots \times J_p$ has content zero in $\mathbf{R}^p$. Hence the set $\{a\}$ has content zero in $\mathbf{R}^p$.

(b) If we take $J_1' = (a_1, a_1)$, then the cell $J' = J_1' \times J_2 \times \cdots \times J_p$ is empty and has content zero.

43.B. Show that a set $Z \subseteq \mathbf{R}^p$ has content zero if and only if for each $\varepsilon > 0$ there exists a finite set $K_1, \ldots, K_n$ of cubes whose union contains Z and such that $c(K_1) + \cdots + c(K_n) < \varepsilon$.

43.C. Write out the details of the proof of the assertion, made in Example 43.2(f), that the graph $S \subseteq \mathbf{R}^2$ of a continuous function $f : [a, b] \to \mathbf{R}$ has content zero.

43.D. If J is a closed cell in $\mathbf{R}^2$ and $g : J \to \mathbf{R}$ is continuous, show that the graph $\{(x, y, g(x, y)) : (x, y) \in J)\} \subseteq \mathbf{R}^3$ of g has content zero.

43.E. Let $A \subseteq \mathbf{R}^2$ be the set consisting of all pairs $(i/p, j/p)$ where p is a prime number, and $i, j = 1, 2, \ldots, p - 1$. Show that each horizontal and each vertical line in $\mathbf{R}^2$ intersects A in a finite number (often zero) of points. Does A have content zero?

43.F. Let $I \subseteq \mathbf{R}^p$ be a closed cell and let $P = \{I_1, \ldots, I_n\}$ and $Q = \{J_1, \ldots, J_m\}$ be two partitions of I into closed cells. Show that $R = \{I_i \cap J_j : i = 1, \ldots, n; j = 1, \ldots, m\}$ is a partition of I and that R is a refinement of both P and Q. The partition R is called the **common refinement** of P and Q.

43.G. If $I \subseteq J$ are cells in $\mathbf{R}^p$ and if P is a partition of I, show that there exists a partition Q of J such that every cell in P belongs to Q.

43.H. Let $Z \subseteq \mathbf{R}^p$ be a set with content zero and let I be a closed cell containing Z. If $J_1, \ldots, J_n$ are cells contained in I whose union contains Z, show that there exists a partition P of I such that the closure of each J_k is the union of cells in P.

43.I. Let $Z \subseteq \mathbf{R}^p$ be a set with content zero and let I be a closed cell containing Z. If $\varepsilon > 0$, show that there exists a partition P_ε of I such that the cells in P_ε which contain points in Z have total content less than ε.

43.J. In the preceding exercise, show that we can choose P_ε to have the additional property that the cells in P_ε which contain any points of Z contain them in their interior.

43.K. In Exercise 43.I, if I is a cube, show that there exists a partition Q_ϵ of I into cubes such that the cubes in Q_ϵ which contain points in Z have total content less than ϵ.

43.L. Let $A \subseteq \mathbf{R}^p$ be bounded and let I, J be closed cells in $\mathbf{R}^p$ such that $A \subseteq I \subseteq J$. If $f : A \to \mathbf{R}$ is a bounded function, define $f_I : I \to \mathbf{R}$ (respectively, $f_J : J \to \mathbf{R}$) to be the function which agrees with f on A and vanishes on $I \setminus A$ (respectively, $J \setminus A$). Show that the integral of f_I over I exists if and only if the integral of f_J over J exists in which case these integrals are equal.

43.M. Let $A \subseteq \mathbf{R}^p$ be bounded and let I_1 and I_2 be closed cells in $\mathbf{R}^p$ such that $A \subseteq I_j$. Let $f : A \to \mathbf{R}$ be a bounded function, and, for $j = 1, 2$, define $f_j : I_j \to \mathbf{R}$ by $f_j(x) = f(x)$ for $x \in A$ and $f_j(x) = 0$ for $x \in I_j \setminus A$. Show that the integral of f_1 over I_1 exists if and only if the integral of f_2 over I_2 exists, in which case these integrals are equal.

43.N. Establish the uniqueness of the integral of a bounded function f defined on a closed cell $I \subseteq \mathbf{R}^p$.

43.O. Write out the details of the proof of the Cauchy Criterion 43.4.

43.P. Let $I \subseteq \mathbf{R}^p$ be a closed cell and let $f : I \to \mathbf{R}$ be bounded. Then f is integrable on I if and only if for every $\epsilon > 0$ there exists a partition P_ϵ of I such that if $P = \{J_1, \ldots, J_n\}$ is a refinement of P_ϵ, then

$$\sum_{j=1}^{n} (M_j - m_j) c(J_j) < \epsilon,$$

where $M_j = \sup \{f(x) : x \in J_j\}$ and $m_j = \inf \{f(x) : x \in J_j\}$ for $j = 1, \ldots, n$. This result is called the **Riemann Criterion for Integrability** (cf. Theorem 30.1).

43.Q. Let $I \subseteq \mathbf{R}^p$ be a closed cell and let $f : I \to \mathbf{R}$ be bounded by M. If f is integrable on I, show that the function $|f|$ is integrable on I and that $\int_I |f| \le Mc(I)$.

43.R. Let $I \subseteq \mathbf{R}^p$ be a closed cell and let f, $g : I \to \mathbf{R}$ be integrable on I. Show the product function fg is integrable on I.

43.S. Let $I \subseteq \mathbf{R}^p$ be a closed cell and let (f_n) be a sequence of functions which are integrable on I. If the sequence converges uniformly on I to f, show that f is integrable on I and that

$$\int_I f = \lim_n \left(\int_I f_n \right).$$

43.T. Let $K \subseteq \mathbf{R}^p$ be a closed cube and let f, $g : K \to \mathbf{R}$ be continuous. Show that if $\epsilon > 0$, then there exists a partition $P_\epsilon = \{K_1, \ldots, K_r\}$ of K into cubes such that if x_j, y_j are any points of K_j, $j = 1, \ldots, r$, then

$$\left| \int_K fg - \sum_{j=1}^{r} f(x_j)g(y_j)c(K_j) \right| < \epsilon c(K).$$

43.U. (This exercise gives an example due to I. J. Schoenberg[†] of a space-filling

---

[†] ISAAC J. SCHOENBERG (1903–   ) was born in Romania and educated there and in Germany. For many years at the University of Pennsylvania, he has worked in number theory, real and complex analysis, and the calculus of variations.

curve.) Let $\varphi : R \to R$ be continuous, even, with period 2, and such that

$$\varphi(t) = 0 \qquad \text{for} \qquad 0 \le t \le \tfrac{1}{3},$$
$$= 3t - 1 \qquad \text{for} \qquad \tfrac{1}{3} < t < \tfrac{2}{3},$$
$$= 1 \qquad \text{for} \qquad \tfrac{2}{3} \le t \le 1.$$

(a) Draw a sketch of the graph of φ. Note that $\|\varphi\|_R = 1$.

(b) If $t \in I$, define $f(t)$ and $g(t)$ by

$$f(t) = \frac{1}{2}\varphi(t) + \frac{1}{2^2}\varphi(3^2 t) + \frac{1}{2^3}\varphi(3^4 t) + \cdots,$$

$$g(t) = \frac{1}{2}\varphi(3t) + \frac{1}{2^2}\varphi(3^3 t) + \frac{1}{2^3}\varphi(3^5 t) + \cdots$$

Show that these series are uniformly convergent, so that f and g are continuous on I.

(c) Evaluate $f(t)$ and $g(t)$, where t has the ternary (base 3) expansion given by

$$0.2020, \qquad 0.0220, \qquad 0.0022, \qquad 0.2002.$$

(d) Let (x, y) belong to the graph $S = \{(f(t), g(t)) : t \in I\}$ and write x and y in their binary (base 2) expansions:

$$x = 0.\alpha_1\alpha_2\alpha_3 \ldots, \qquad y = 0.\beta_1\beta_2\beta_3 \ldots,$$

where α_m, β_m are either 0 or 1. Let t be corresponding real number whose ternary expansion is

$$t = 0.(2\alpha_1)(2\beta_1)(2\alpha_2)(2\beta_2) \ldots.$$

Show that $x = f(t)$ and $y = g(t)$. Hence every point in the cube $I \times I$ belongs to the graph S.

43.V. A set $Z \subseteq R^p$ has **measure zero** if for each $\varepsilon > 0$ there is a sequence (J_n) of cells whose union contains Z and such that $\sum c(J_n) < \varepsilon$.

(a) Since the empty set is a cell, show that a set which has content zero also has measure zero.

(b) Show that every countable set in R^p has measure zero. Hence the set in Example 43.2(g) has measure zero (but it does not have content zero).

(c) Show that, in the definition of "measure zero" given above, we can require the cells to be open, or to be cubes.

(d) Show that every compact set with measure zero also has content zero.

(e) The union of a countable family of sets with content zero has measure zero.

Projects

43.α. Let $I \subseteq R^p$ be a closed cell and let $f : I \to R$ be bounded. If $P = \{J_1, \ldots, J_n\}$ is a partition of I, let

$$m_j = \inf\{f(x) : x \in J_j\}, \qquad M_j = \sup\{f(x) : x \in J_j\}$$

for $j = 1, \ldots, n$ and define the **lower** and **upper sums** of f for P to be

$$L(P; f) = \sum_{j=1}^{n} m_j c(J_j), \qquad U(P; f) = \sum_{j=1}^{n} M_j c(J_j).$$

(a) If $S(P; f)$ is any Riemann sum corresponding to P, then $L(P; f) \le S(P; f) \le U(P; f)$. If $\varepsilon > 0$, then there exist Riemann sums $S_1(P; f)$ and $S_2(P; f)$ corresponding to P such that

$$S_1(P; f) \le L(P; f) + \varepsilon, \qquad U(P; f) - \varepsilon \le S_2(P; f).$$

(b) If P is a partition of I and Q is a refinement of P, then

$$L(P; f) \le L(Q; f) \le U(Q; f) \le U(P; f).$$

(c) If P_1 and P_2 are any partitions of I, then $L(P_1; f) \le U(P_2; f)$.

(d) Define the **lower** and the **upper integral** of f on I to be

$$L(f) = \sup\{L(P; f)\}, \qquad U(f) = \inf\{U(P; f)\},$$

respectively, where the supremum and the infimum are taken over all partitions of I. Show that $L(f) \le U(f)$.

(e) Show that f is integrable (in the sense of Definition 43.3) if and only if $L(f) = U(f)$, in which case $L(f) = U(f) = \int_I f$.

(f) Show that f is integrable if and only if for each $\varepsilon > 0$ there exists a partition P such that $U(P; f) - L(P; f) < \varepsilon$. (This condition is sometimes called Riemann's Condition; compare it with Exercise 43.P.)

43.β. This project develops the integral of functions on a closed cell $I \subseteq \mathbf{R}^p$ and with values in $\mathbf{R}^q$. If $P = \{J_1, \ldots, J_n\}$ is a partition of I, then a Riemann sum corresponding to P is a sum

$$S(P; f) = \sum_{k=1}^{n} c(J_k) f(x_k)$$

where $x_k \in J_k$. An element $L \in \mathbf{R}^q$ is the Riemann integral of f over I if, for every $\varepsilon > 0$ there exists a partition P_ε of I such that if P is any refinement of P_ε and $S(P; f)$ is any corresponding Riemann sum, then $\|S(P; f) - L\| < \varepsilon$. Examine which of the theorems in this section remain true for functions with values in $\mathbf{R}^q$. Show that $f: I \to \mathbf{R}^q$ is integrable if and only if each function $f_j = e_j \cdot f$, $j = 1, \ldots, q$, is integrable. (Here $e_1, \ldots, e_q$ are the standard basis vectors in $\mathbf{R}^q$.)

Section 44   Content and the Integral

In this section we shall introduce the collection of sets with content, and characterize the content function as a real-valued function defined on this collection of sets. Next we shall obtain some further properties of the integral over sets with content, and show how the integral can be evaluated as an "iterated integral."

44.1 DEFINITION. If $A \subseteq \mathbf{R}^p$, then we recall that a point $x \in \mathbf{R}^p$ is said to be a **boundary point** of A if every neighborhood of x contains both points in A and points in its complement $\mathscr{C}(A)$. The **boundary** of A is the subset of $\mathbf{R}^p$ consisting of all the boundary points of A; it will be denoted by $b(A)$.

If $A \subseteq \mathbf{R}^p$, we recall that a point of $\mathbf{R}^p$ is precisely one of the following: it is an interior point of A, it is a boundary point of A, or it is an exterior point of A. The **interior** A^0 consists of all of the interior points of A; it is an open set in $\mathbf{R}^p$. As noted above, the **boundary** $b(A)$ consists of all of the boundary points of A; it is a closed set in $\mathbf{R}^p$. The **closure** A^- is the union $A \cup b(A)$; it is a closed set in $\mathbf{R}^p$.

We ordinarily expect the boundary of a set to be small, but this is because we are accustomed to thinking about rectangles, circles, and other elementary figures. Example 43.2(g) shows that the boundary of a countable set in $\mathbf{R}^2$ can have boundary equal to $\mathbf{I} \times \mathbf{I}$.

Sets with Content

We shall now define the content of a subset of $\mathbf{R}^p$ whose boundary has zero content.

44.2 Definition. A bounded set $A \subseteq \mathbf{R}^p$ whose boundary $b(A)$ has content zero is said **to have content.** The collection of all subsets of $\mathbf{R}^p$ which have content will be denoted by $\mathcal{D}(\mathbf{R}^p)$. If $A \in \mathcal{D}(\mathbf{R}^p)$ and if I is a closed cell containing A, then the function g_I defined by

$$g_I(x) = 1 \qquad \text{for } x \in A,$$
$$= 0 \qquad \text{for } x \in I \setminus A,$$

is continuous on I except possibly at points of $b(A)$. Hence g_I is integrable on I and we define the **content** $c(A)$ *of* A to be equal to $\int_I g_I$. Thus

$$c(A) = \int_I g_I = \int_A 1.$$

Note that if $J \subseteq \mathbf{R}^p$ is a cell, then its boundary consists of the union of a finite number of "faces," which are cells each having content zero. [For example, if $J = [a, b] \times [c, d]$, then $b(J)$ is the union of the four cells

$$[a, b] \times [c, c], \qquad\qquad [a, b] \times [d, d],$$
$$[a, a] \times [c, d], \qquad\qquad [b, b] \times [c, d].$$

These same four cells are also the boundary of the cell $(a, b) \times (c, d)$.] It follows that a cell in $\mathbf{R}^p$ has content; moreover, one easily sees that if $J = [a_1, b_1] \times \cdots \times [a_p, b_p]$, then

$$c(J) = \int_J 1 = (b_1 - a_1) \cdots (b_p - a_p).$$

Hence the content of a cell, as given by Definition 44.2, is consistent with the definition of the content assigned to a closed cell in Section 43. Similar remarks apply to other cells in $\mathbf{R}^p$; in particular, it is seen that if $K = [a_1, b_1) \times \cdots \times [a_p, b_p)$, then

$$c(K) = \int_K 1 = (b_1 - a_1) \cdots (b_p - a_p).$$

We shall now show that the notion of content zero introduced in Definition 43.1 is consistent with the notion of content introduced in Definition 44.2.

44.3 LEMMA. *A set $A \subseteq \mathbf{R}^p$ has content zero (in the sense of Definition 43.1) if and only if it has content (in the sense of Definition 44.2) and $c(A) = 0$.*

PROOF. Suppose that $A \subseteq \mathbf{R}^p$ has content zero. Then, if $\varepsilon > 0$, we can enclose A in the union U of a finite number of closed cells with total content less than ε. Since this union U is a bounded set, then A is bounded; since U is closed, it also contains $b(A)$. Since $\varepsilon > 0$ is arbitrary, we infer that $b(A)$ has content zero; hence A has content and

$$c(A) = \int_A 1.$$

It now follows from Lemma 43.7 that $c(A) = 0$.

Conversely, suppose that $A \subseteq \mathbf{R}^p$ has content and that $c(A) = 0$. Hence there is a closed cell I containing A and such that the function

$$g_I(x) = 1 \qquad \text{for} \qquad x \in A,$$
$$= 0 \qquad \text{for} \qquad x \in I \setminus A,$$

is integrable on I. Let $\varepsilon > 0$ be given and let P_ε be a partition of I such that any Riemann sum corresponding to P_ε satisfies $0 \le S(P_\varepsilon; g_I) < \varepsilon$. If we take the intermediate points in $S(P_\varepsilon; g_I)$ to belong to A whenever possible, we infer that A is contained in the union of a finite number of cells in P_ε with total content less than ε. Thus A has content zero in the sense of Definition 43.1.   Q.E.D.

44.4 THEOREM. *Let A, B belong to $\mathcal{D}(\mathbf{R}^p)$ and let $x \in \mathbf{R}^p$.*
(a) *The sets $A \cap B$ and $A \cup B$ belong to $\mathcal{D}(\mathbf{R}^p)$ and*

$$c(A) + c(B) = c(A \cap B) + c(A \cup B).$$

(b) *The sets $A \setminus B$ and $B \setminus A$ belong to $\mathcal{D}(\mathbf{R}^p)$ and*

$$c(A \cup B) = c(A \setminus B) + c(A \cap B) + c(B \setminus A).$$

(c) *If $x + A = \{x + a : a \in A\}$, then $x + A$ belongs to $\mathcal{D}(\mathbf{R}^p)$ and*
$$c(x + A) = c(A).$$

PROOF. By hypothesis, the boundaries $b(A)$, $b(B)$ have content zero. We leave it as an exercise to the reader to show that the boundaries

$$b(A \cap B), \quad b(A \cup B), \quad b(A \setminus B), \quad b(B \setminus A)$$

are contained in $b(A) \cup b(B)$, It follows from this and Example 43.2(c) that $A \cap B$, $A \cup B$, $A \setminus B$, and $B \setminus A$ belong to $\mathcal{D}(\mathbf{R}^p)$.

Now let I be a closed cell containing $A \cup B$ and let f_a, f_b, f_i, f_u be the functions equal to 1 on A, B, $A \cap B$, $A \cup B$, respectively, and equal to 0 elsewhere on I. Since each of these functions are continuous except on sets with content zero, they are integrable on I. Since

$$f_a + f_b = f_i + f_u,$$

it follows from Theorem 43.5 and the definition of content that

$$c(A) + c(B) = \int_I f_a + \int_I f_b = \int_I (f_a + f_b)$$

$$= \int_I (f_i + f_u) = \int_I f_i + \int_I f_u$$

$$= c(A \cap B) + c(A \cup B).$$

This establishes the formula given in (a); the one in (b) can be proved similarly.

To prove (c), note that if $\varepsilon > 0$ is given and if $J_1, \ldots, J_n$ are cells with total content less than ε whose union contains $b(A)$, then $x + J_1, \ldots, x + J_n$ are cells with total content less than ε whose union contains $b(x + A)$. Since $\varepsilon > 0$ is arbitrary, the set $x + A$ belongs to $\mathcal{D}(\mathbf{R}^p)$. To show that $c(x + A) = c(A)$, let I be a closed cell containing A; hence $x + I$ is a closed cell containing $x + A$. Let $f_1 : I \to \mathbf{R}$ be such that $f_1(y) = 1$ for $y \in A$ and $f_1(y) = 0$ for $y \in I \setminus A$, and let $f_2 : x + I \to \mathbf{R}$ be such that $f_2(z) = 1$ for $z \in x + A$ and $f_2(z) = 0$ for $z \in (x + I) \setminus (x + A)$. Show that to each Riemann sum for f_1 there corresponds a Riemann sum for f_2 which is equal to it. Hence

$$c(A) = \int_I f_1 = \int_{x+I} f_2 = c(x + A). \qquad \text{Q.E.D.}$$

44.5 COROLLARY. *Let A and B belong to $\mathcal{D}(\mathbf{R}^p)$.*
(a) *If $A \cap B = \emptyset$, then $c(A \cup B) = c(A) + c(B)$.*
(b) *If $A \subseteq B$, then $c(B \setminus A) = c(B) - c(A)$.*

Characterization of the Content Function

We have seen that the content function $c : \mathcal{D}(\mathbf{R}^p) \to \mathbf{R}$ is positive, "additive," "invariant under translation," and assigns the value 1 to the "half-open" cube

$$K_0 = [0, 1) \times [0, 1) \times \cdots \times [0, 1).$$

We shall now show that these four properties characterize c.

44.6 THEOREM. *Let $\gamma : \mathcal{D}(\mathbf{R}^p) \to \mathbf{R}$ be a function with the following properties:*
 (i) *$\gamma(A) \ge 0$ for all $A \in \mathcal{D}(\mathbf{R}^p)$;*
 (ii) *if A, $B \in \mathcal{D}(\mathbf{R}^p)$ and $A \cap B = \emptyset$, then $\gamma(A \cup B) = \gamma(A) + \gamma(B)$;*
 (iii) *if $A \in \mathcal{D}(\mathbf{R}^p)$ and $x \in \mathbf{R}^p$, then $\gamma(A) = \gamma(x + A)$;*
 (iv) *$\gamma(K_0) = 1$.*
Then we have $\gamma(A) = c(A)$ for all $A \in \mathcal{D}(\mathbf{R}^p)$.

PROOF. If $n \in \mathbf{N}$, let K_n be the "half-open" cube

$$K_n = [0, 2^{-n}) \times [0, 2^{-n}) \times \cdots \times [0, 2^{-n}).$$

We note that K_0 is the union of 2^{np} disjoint translates of K_n; hence $1 = \gamma(K_0) = 2^{np}\gamma(K_n)$ and so $\gamma(K_n) = 1/2^{np} = c(K_n)$.

Let A, B belong to $\mathcal{D}(\mathbf{R}^p)$ and let $A \subseteq B$. Then we can write $B = A \cup (B \setminus A)$; since $A \cap (B \setminus A) = \emptyset$, it follows from (i) and (ii) that

$$\gamma(B) = \gamma(A) + \gamma(B \setminus A) \ge \gamma(A).$$

Hence γ is monotone in the sense that if $A \subseteq B$ then $\gamma(A) \le \gamma(B)$. Now let $A \in \mathcal{D}(\mathbf{R}^p)$. Since A is bounded, then for some $M \in \mathbf{N}$, the set A is contained in the interior of the closed cube I with half side length 2^M and center at 0. If $\varepsilon > 0$, there is a partition of I into small cubes of side length 2^{-n}, say, such that the content of the union of all the cubes $I_1, \ldots, I_r$ which are contained in A exceeds $c(A) - \varepsilon$, and such that the content of all the cells $I_1, \ldots, I_s$ $(r \le s)$ which contain points of A does not exceed $c(A) + \varepsilon$. (See Exercise 44.I.) Now each of these sets I_i differ from a translate $x_i + K_n$ by a set of content zero. Hence we have

$$c(A) - \varepsilon \le c\left(\bigcup_{i=1}^{r} (x_i + K_n)\right) \le c(A) \le c\left(\bigcup_{i=1}^{s} (x_i + K_n)\right) \le c(A) + \varepsilon.$$

Now c and γ are both invariant under translation of the set and agree on K_n. Moreover c and γ are additive over disjoint finite unions. Hence it follows that

$$c\left(\bigcup_{i=1}^{r} (x_i + K_n)\right) = \sum_{i=1}^{r} c(x_i + K_n) = \sum_{i=1}^{r} \gamma(x_i + K_n)$$

$$= \gamma\left(\bigcup_{i=1}^{r} (x_i + K_n)\right).$$

It follows from this and the fact that γ is monotone that

$$c(A) - \varepsilon \le \gamma\left(\bigcup_{i=1}^{r} (x_i + K_n)\right) \le \gamma(A) \le \gamma\left(\bigcup_{i=1}^{s} (x_i + K_n)\right) \le c(A) + \varepsilon,$$

whence $|\gamma(A) - c(A)| \leq \varepsilon$. Since $\varepsilon > 0$ is arbitrary, we infer that $\gamma(A) = c(A)$.
$\hfill$ Q.E.D.

44.7 COROLLARY. *Let $\mu : \mathscr{D}(\mathbf{R}^p) \to \mathbf{R}$ be a function satisfying properties* (i), (ii), *and* (iii). *Then there exists a constant $m \geq 0$ such that $\mu(A) = mc(A)$ for all $A \in \mathscr{D}(\mathbf{R}^p)$.*

PROOF. Since μ possesses properties (i) and (ii), it is easily seen that μ is monotone in the sense that $A \subseteq B$ implies that $\mu(A) \leq \mu(B)$. If $\mu(K_0) = 0$, then μ of any bounded set is 0, whence it follows that $\mu(A) = 0$ for all $A \in \mathscr{D}(\mathbf{R}^p)$, so we can take $m = 0$. If $\mu(K_0) \neq 0$, let

$$\gamma(A) = \frac{1}{\mu(K_0)} \mu(A) \qquad \text{for all } A \in \mathscr{D}(\mathbf{R}^p).$$

Since it is readily seen that γ has properties (i), (ii), (iii), and (iv) of the theorem, it follows that $\gamma = c$. Hence we take $m = \mu(K_0)$.
$\hfill$ Q.E.D.

Further Properties of the Integral

We shall now present some additional properties of the integral that are often useful.

44.8 THEOREM. *Let $A \in \mathscr{D}(\mathbf{R}^p)$ and let $f : A \to \mathbf{R}$ be bounded and continuous on A. Then f is integrable on A.*

PROOF. Let I be a closed cell with $A \subseteq I$ and let $f_I : I \to \mathbf{R}$ be equal to f on A and to 0 on $I \setminus A$. Since f_I is bounded on I and is continuous on $I \setminus b(A)$, it follows from the Integrability Theorem 43.9 that f_I is integrable on I. Therefore f is integrable on A.
$\hfill$ Q.E.D.

We shall now show that the integral is additive with respect to the set over which the integral is extended.

44.9 THEOREM. (a) *Let A_1 and A_2 belong to $\mathscr{D}(\mathbf{R}^p)$ and suppose that $A_1 \cap A_2$ has content zero. If $A = A_1 \cup A_2$ and if $f : A \to \mathbf{R}$ is integrable on A_1 and A_2, then f is integrable on A and*

$$(44.1) \qquad\qquad \int_A f = \int_{A_1} f + \int_{A_2} f.$$

(b) *Let A belong to $\mathscr{D}(\mathbf{R}^p)$ and let $A_1, A_2 \in \mathscr{D}(\mathbf{R}^p)$ be such that $A = A_1 \cup A_2$ and such that $A_1 \cap A_2$ has content zero. If $f : A \to \mathbf{R}$ is integrable on A, and if the restrictions of f to A_1 and A_2 are integrable, then (44.1) holds.*

PROOF. (a) Let I be a closed cell containing $A = A_1 \cup A_2$ and let $f_i : I \to \mathbf{R}$, $i = 1, 2$, be equal to f on A_i and equal to 0 elsewhere on I. By hypothesis, f_1 and f_2 are integrable on I and

$$\int_I f_i = \int_{A_i} f, \qquad i = 1, 2.$$

It follows from Theorem 43.5 that $f_1 + f_2$ is integrable on I and that

$$\int_I (f_1 + f_2) = \int_I f_1 + \int_I f_2.$$

Now since $f(x) = f_1(x) + f_2(x)$ provided $x \in A \setminus (A_1 \cap A_2)$, it follows from Lemma 43.8 that f is integrable on A and that (44.1) holds.

(b) We preserve the notation of the proof of (a). By hypothesis, f_i is integrable on I. Now, $f_I(x) = f_1(x) + f_2(x)$ except for x in $A_1 \cap A_2$, a set with content zero. It therefore follows from Theorem 43.5 and Lemma 43.8 that

$$\int_A f = \int_I f = \int_I (f_1 + f_2) = \int_I f_1 + \int_I f_2$$

$$= \int_{A_1} f + \int_{A_2} f. \qquad \text{Q.E.D.}$$

We remark that if $f : A \to \mathbf{R}$ is a bounded integrable function, then the assumption made in 44.9(b) that the restrictions of f to A_1 and A_2 are integrable is automatically satisfied. (See Exercise 44.J.)

The next result is often useful to estimate the magnitude of an integral.

44.10 THEOREM. *Let $A \in \mathcal{D}(\mathbf{R}^p)$ and let $f : A \to \mathbf{R}$ be integrable on A and such that $|f(x)| \le M$ for all $x \in A$. Then*

$$(44.2) \qquad \left| \int_A f \right| \le Mc(A).$$

More generally, if f is real valued and $m \le f(x) \le M$ for all $x \in A$, then

$$(44.3) \qquad mc(A) \le \int_A f \le Mc(A).$$

PROOF. Let f_I be the extension of f to a closed cell I containing A. If $\varepsilon > 0$ is given, then there exists a partition $P_\varepsilon = \{J_1, \ldots, J_h\}$ of I such that if $S(P_\varepsilon; f_I)$ is any corresponding Riemann sum, then

$$S(P_\varepsilon; f_I) - \varepsilon \le \int_I f_I \le S(P_\varepsilon; f) + \varepsilon.$$

We note that if the intermediate points of the Riemann sum are chosen outside of A whenever possible, we have

$$S(P_\varepsilon; f) = \sum{}' f(x_j) c(J_j),$$

where the sum is extended over those cells in P_ε entirely contained in A. Hence

$$S(P_\varepsilon; f) \le M \sum' c(J_k) \le Mc(A).$$

Therefore we have

$$\int_A f = \int_I f_I \le Mc(A) + \varepsilon,$$

and since $\varepsilon > 0$ is arbitrary we obtain the right side of inequality (44.3). The left side is established in a similar manner.   Q.E.D.

As a consequence of this result, we obtain the following theorem, which is an extension of the First Mean Value Theorem 30.6.

44.11 MEAN VALUE THEOREM. *Let $A \in \mathcal{D}(\mathbf{R}^p)$ be a connected set and let $f: A \to \mathbf{R}$ be bounded and continuous on A. Then there exists a point $p \in A$ such that*

$$(44.4) \qquad\qquad \int_A f = f(p)c(A).$$

PROOF. If $c(A) = 0$, the conclusion is trivial; hence we suppose that $c(A) \ne 0$. Let

$$m = \inf\{f(x): x \in A\}, \qquad M = \sup\{f(x): x \in A\};$$

it follows from the second part of the preceding theorem that

$$(44.5) \qquad\qquad m \le \frac{1}{c(A)} \int_A f \le M.$$

If both inequalities in (44.5) are strict, the result follows from Bolzano's Intermediate Value Theorem 22.4.

Now suppose that $\int_A f = Mc(A)$. If the supremum M is attained at $p \in A$, the conclusion also follows. Hence we assume that the supremum M is not attained on A. Since $c(A) \ne 0$, there exists a closed cell $K \subseteq A$ such that $c(K) \ne 0$ (see Exercise 44.G). Since K is compact and f is continuous on K, there exists $\varepsilon > 0$ such that $f(x) \le M - \varepsilon$ for all $x \in K$. Since $A = K \cup (A \setminus K)$ it follows from Theorem 44.9 and 44.10 that

$$Mc(A) = \int_A f = \int_K f + \int_{A \setminus K} f$$
$$\le (M - \varepsilon)c(K) + Mc(A \setminus K) < Mc(A),$$

a contradiction. If $\int_A f = mc(A)$, then a similar argument applies. Q.E.D.

The Integral as an Iterated Integral

It is desirable to know that if f is integrable on a closed cell $J = [a_1, b_1] \times \cdots \times [a_p, b_p]$ in $\mathbf{R}^p$ and has values in $\mathbf{R}$, then the integral $\int_J f$ can be calculated in terms of a p-fold "iterated integral":

$$\int_{a_p}^{b_p} \left\{ \cdots \left\{ \int_{a_2}^{b_2} \left\{ \int_{a_1}^{b_1} f(x_1, x_2, \ldots, x_p) \, dx_1 \right\} dx_2 \right\} \cdots \right\} dx_p.$$

This is the method of evaluating double and triple integrals by means of iterated integrals that is familiar to the reader from elementary calculus. We shall give a justification of this procedure; for simplicity we shall suppose that $p = 2$, but clearly the result extends to higher dimensions.

44.12 THEOREM. *If f is continuous on the closed cell $J = [a, b] \times [c, d]$ to $\mathbf{R}$, then*

$$\int_J f = \int_c^d \left\{ \int_a^b f(x, y) \, dx \right\} dy$$
$$= \int_a^b \left\{ \int_c^d f(x, y) \, dy \right\} dx.$$

PROOF. It was seen in the Interchange Theorem 31.9 that the two iterated integrals are equal. To show that the integral of f on J is given by the first iterated integral, let F be defined for $y \in [c, d]$ by

$$F(y) = \int_a^b f(x, y) \, dx.$$

Let $c = y_0 \le y_1 \le \cdots \le y_r = d$ be a partition of the interval $[c, d]$, let $a = x_0 \le x_1 \le \cdots \le x_s = b$ be a partition of $[a, b]$, and let P denote the partition of J obtained by using the cells $[x_{k-1}, x_k] \times [y_{j-1}, y_j]$. Let y_j^* be any point in $[y_{j-1}, y_j]$ and note that

$$F(y_j^*) = \int_a^b f(x, y_j^*) \, dx = \sum_{k=1}^s \int_{x_{k-1}}^{x_k} f(x, y_j^*) \, dx.$$

According to the First Mean Value Theorem 30.6, for each j and k there exists a point x_{jk}^* in $[x_{k-1}, x_k]$ such that

$$F(y_j^*) = \sum_{k=1}^s f(x_{jk}^*, y_j^*)(x_k - x_{k-1}).$$

We multiply by $(y_j - y_{j-1})$ and add to obtain

$$\sum_{j=1}^r F(y_j^*)(y_j - y_{j-1}) = \sum_{j=1}^r \sum_{k=1}^s f(x_{jk}^*, y_j^*)(x_k - x_{k-1})(y_j - y_{j-1}).$$

Now the expression on the left side of this formula is an arbitrary Riemann sum for the integral

$$\int_c^d F(y) \, dy = \int_c^d \left\{ \int_a^b f(x, y) \, dx \right\} dy.$$

We have shown that this Riemann sum is equal to a particular Riemann sum corresponding to the partition P. Since f is integrable on J, the existence of this iterated integral and its equality with the integral over J is established. Q.E.D.

A minor modification of the proof given for the preceding theorem yields the following, slightly stronger, assertion.

44.13 THEOREM. *Let f be integrable on the rectangle $J = [a, b] \times [c, d]$ to $\mathbf{R}$ and suppose that, for each $y \in [c, d]$, the function $x \mapsto f(x, y)$ of $[a, b]$ into $\mathbf{R}$ is continuous except possibly for a finite number of points, at which it has one-sided limits. Then*

$$\int_J f = \int_c^d \left\{ \int_a^b f(x, y) \, dx \right\} dy.$$

As a consequence of this theorem, we shall obtain a result which is often used in evaluating integrals of functions defined on a set which is bounded by continuous curves. For convenience, we shall state the result in the case where the set has horizontal line segments as its top and bottom boundaries, and continuous curves as its lateral boundaries. Clearly, a similar result holds if the lateral boundaries are vertical line segments and the top and bottom boundaries are curves. More complicated sets are handled by decomposing the sets into the union of subsets of these two types.

44.14 THEOREM. *Let $A \subseteq \mathbf{R}^2$ be given by*

$$A = \{(x, y) : \alpha(y) \le x \le \beta(y), \, c \le y \le d\},$$

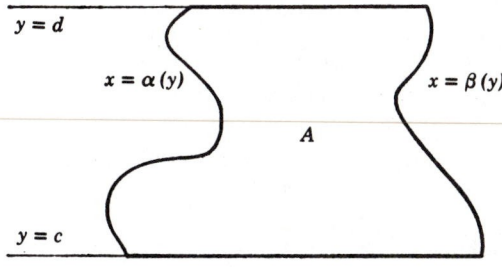

Figure 44.1

where α and β are continuous functions on $[c, d]$ with values in $[a, b]$. If f is continuous on $A \to \mathbf{R}$, then f is integrable on A and

$$\int_A f = \int_c^d \left\{ \int_{\alpha(y)}^{\beta(y)} f(x, y)\, dx \right\} dy.$$

PROOF. Let J be a closed cell containing A and let f_J be the extension of f to J. A variation of Example 43.2(f) shows that the boundary of A has content zero; hence f_J is integrable on J. Now for each $y \in [c, d]$ the function $x \mapsto f_J(x, y)$ is continuous except possibly at the two points $\alpha(y)$ and $\beta(y)$, at which it has one-sided limits. It follows from the preceding theorem that

$$\int_A f = \int_J f_J = \int_c^d \left\{ \int_a^b f_J(x, y)\, dx \right\} dy$$

$$= \int_c^d \left\{ \int_{\alpha(y)}^{\beta(y)} f(x, y)\, dx \right\} dy. \qquad \text{Q.E.D.}$$

Exercises

44.A. If $A \subseteq \mathbf{R}^p$, then a point is a boundary point of A if and only if it is a boundary point of the complement $\mathscr{C}(A)$ of A. Hence $b(A) = b(\mathscr{C}(A))$.

44.B. Let $A \subseteq \mathbf{R}^p$, and let $b(A)$ be the boundary of A.

(a) The set $b(A)$ is closed in $\mathbf{R}^p$.

(b) The interior $A^\circ = A \setminus b(A)$ is open in $\mathbf{R}^p$ and contains every open set G with $G \subseteq A$.

(c) The closure $A^- = A \cup b(A)$ is closed in $\mathbf{R}^p$ and is contained in every closed set F with $A \subseteq F$.

44.C. Let $A \subseteq \mathbf{R}^p$ and let $A^- = A \cup b(A)$ be the closure of A. Show that $b(A^-) \subseteq b(A)$. Give an example to show that equality can hold, and an example that equality can fail.

44.D. Let A, B be subsets of $\mathbf{R}^p$. Show that the boundary of each of the sets

$$A \cap B, \quad A \setminus B, \quad A \cup B$$

is contained in $b(A) \cup b(B)$. (Hint: $b(A) = A^- \cap (\mathscr{C}(A))^-$.)

44.E. A set $A \subseteq \mathbf{R}^p$ is closed in $\mathbf{R}^p$ if and only if $b(A) \subseteq A$. A set $B \subseteq \mathbf{R}^p$ is open in $\mathbf{R}^p$ if and only if $B \cap b(B) = \emptyset$.

44.F. If $A \in \mathscr{D}(\mathbf{R}^p)$, show that its interior $A^\circ = A \setminus b(A)$ and its closure $A^- = A \cup b(A)$ also belong to $\mathscr{D}(\mathbf{R}^p)$ and that $c(A^\circ) = c(A) = c(A^-)$.

44.G. If $A \in \mathscr{D}(\mathbf{R}^p)$ and $c(A) > 0$, prove that there exists a closed cell $K \subseteq A$ such that $c(K) \neq 0$.

44.H. If $A \subseteq \mathbf{R}^p$, we define the **inner** and **outer content** of A to be

$$c_*(A) = \sup c(U), \qquad c^*(A) = \inf c(V),$$

where the supremum is taken over the set of all finite unions of cells contained in A, and the infimum is taken over the set of all finite unions of cells containing points of A.

(a) Prove that $c_*(A) \le c^*(A)$ and that A has content if and only if $c_*(A) = c^*(A)$, in which case $c(A)$ is this common value.

(b) If A and B are disjoint subsets of $\mathbf{R}^p$, show that $c^*(A \cup B) \le c^*(A) + c^*(B)$.

(c) Give an example of disjoint sets A, B such that $0 \ne c^*(A) = c^*(B) = c^*(A \cup B)$.

44.I. Let $M \in \mathbf{N}$ and let $I_M \subseteq \mathbf{R}^p$ be the cube with half length 2^M and center 0. For $n \in \mathbf{N}$ we divide I_M into a **grid** $G_{M,n}$ of length 2^{-n} formed by the collection of all cubes in I_M with side length 2^{-n} and dyadic rational end points (that is, end points of the form $k/2^n$ where $k \in \mathbf{Z}$).

(a) If $J \subseteq I_M$ is a closed cell and $\varepsilon > 0$, show that there exists $n \in \mathbf{N}$ such that the union of all the cubes in $G_{M,n}$ which are contained in J has a total content more than $c(J) - \varepsilon$ and the union of all the cubes in $G_{M,n}$ which contain points in J has total content less than $c(J) + \varepsilon$.

(b) If $A \subseteq I_M$ has content and $\varepsilon > 0$, show that there exists $n \in \mathbf{N}$ such that the union of all cubes in $G_{M,n}$ which are contained in A has total content exceeding $c(A) - \varepsilon$ and the union of all cubes in $G_{M,n}$ which contain points in A has total content less than $c(A) + \varepsilon$.

44.J. Let $I \subseteq \mathbf{R}^p$ be a closed cell and let $f : I \to \mathbf{R}$ be integrable on I. If $A \subseteq I$ has content, then the restriction of f to A is integrable on A. (Hint: Use Exercise 43.P.)

44.K. Let $A \in \mathcal{D}(\mathbf{R}^p)$ and suppose that f and g are integrable on A and that $g(x) \ge 0$ for all $x \in A$. If $m = \inf f(A)$, $M = \sup f(A)$, then there exists a real number $\mu \in [m, M]$ such that

$$\int_A fg = \mu \int_A g.$$

44.L. If, in addition to the hypotheses of the previous exercise, we suppose that A is connected and f is continuous on A, then there exists a point $p \in A$ such that

$$\int_A fg = f(p) \int_A g.$$

44.M. Let $\{(x_n, y_n) : n \in \mathbf{N}\}$ be an enumeration of the points in $(0, 1) \times (0, 1)$ with rational coordinates. For each $n \in \mathbf{N}$, let I_n be an open cell in $(0, 1) \times (0, 1)$ containing (x_n, y_n), and let $G = \bigcup_{n \in \mathbf{N}} I_n$. Show that G is an open set in $\mathbf{R}^2$ whose boundary $b(G)$ is $(0, 1) \times (0, 1) \setminus G$. Show that if $\sum c(I_n) < 1$, then the open set G does not have content.

44.N. Using the terminology of Exercise 7.K, let $A \subseteq [0, 1]$ be a "Cantor-like" set with length $\frac{1}{2}$. If $K = A \times [0, 1]$, show that K is a compact subset of $\mathbf{R}^2$, that $b(K) = K$, and that K does not have content.

44.O. Let $a \le b$ and let $f : [a, b] \to \mathbf{R}$ be continuous and such that $f(x) \ge 0$ for all $x \in [a, b]$. Let $S_f = \{(x, y) : a \le x \le b, 0 \le y \le f(x)\}$ be called the **ordinate set** of f. By examining the boundary of S_f, show that it has content. Show that

$$c(S_f) = \int_a^b \left\{ \int_0^{y=f(x)} 1dy \right\} dx = \int_a^b f(x)\, dx.$$

44.P. Let $A \subseteq \mathbf{R}^2$ be the set in Exercise 43.E and let f be defined on $Q = [0, 1] \times [0, 1] \to \mathbf{R}$ by $f(x, y) = 1$ for $(x, y) \in A$ and $f(x, y) = 0$ otherwise. Show that A does not have content and that f is not integrable on Q. However, the iterated integrals exist and satisfy

$$\int_0^1 \left\{ \int_0^1 f(x, y)\, dx \right\} dy = \int_0^1 \left\{ \int_0^1 f(x, y)\, dy \right\} dx.$$

44.Q. Let $Q = [0, 1] \times [0, 1]$ and let $f : Q \to \mathbf{R}$ be defined by $f(x, y) = 0$ if either x or y is irrational and $f(x, y) = 1/n$ if y is rational and $x = m/n$ where m and $n > 0$ are relatively prime integers. Show that

$$\int_Q f = \int_0^1 \left\{ \int_0^1 f(x, y)\, dx \right\} dy = 0,$$

but that $\int_0^1 f(x, y)\, dy$ does not exist for rational x.

44.R. Let $J \subseteq \mathbf{R}^2$ be an open cell containing $(0, 0)$ and let $f : J \to \mathbf{R}$ be continuous on J. Define $F : J \to \mathbf{R}$ by the iterated integral:

$$F(x, y) = \int_0^x \left\{ \int_0^y f(s, t)\, dt \right\} ds.$$

Show that $D_2 D_1 F(x, y) = f(x, y) = D_1 D_2 F(x, y)$ for $(x, y) \in J$.

44.S. Let J be as in the previous exercise and let $G : J \to R$ be such that $D_2 D_1 G$ is continuous on J. Use this exercise to show that $D_1 D_2 G$ exists and equals $D_2 D_1 G$.

44.T. Let $J = [a_1, b_1] \times \cdots \times [a_p, b_p]$ and let $f : J \to \mathbf{R}$ be continuous. Let $J_{(1)} = [a_2, b_2] \times \cdots \times [a_p, b_p]$ in $\mathbf{R}^{p-1}$ and let $F_{(1)} : J_{(1)} \to \mathbf{R}$ be defined by

$$F_{(1)}(x_2, \ldots, x_p) = \int_{a_1}^{b_1} f(x_1, x_2, \ldots, x_p)\, dx_1.$$

(a) Show that $F_{(1)}$ is continuous on $J_{(1)}$.

(b) Given $(x_2^*, \ldots, x_p^*)$ in $J_{(1)}$ and any partition $a_1 = x_{1,0} < x_{1,1} < \cdots < x_{1,s} = b_1$ of $[a_1, b_1]$, show that there exists points $x_{1,k}^*$ in $[x_{1,k-1}, x_{1,k}]$ such that

$$F_{(1)}(x_2^*, \ldots, x_p^*) = \sum_{k=1}^{s} f(x_{1,k}^*, x_2^*, \ldots, x_p^*)(x_{1,k} - x_{1,k-1}).$$

(c) Prove that

$$\int_{J_{(1)}} F_{(1)}(x_2, \ldots, x_p)\, d(x_2, \ldots, x_p) = \int_{J_{(1)}} \left\{ \int_{a_1}^{b_1} f(x_1, x_2, \ldots, x_p)\, dx_1 \right\} d(x_2, \ldots, x_p)$$

$$= \int_J f(x_1, \ldots, x_p)\, d(x_1, \ldots, x_p).$$

(d) Extend the result to the case where for each point $(x_2, \ldots, x_p)$ in $J_{(1)}$ the function $x_1 \mapsto F(x_1, x_2, \ldots, x_p)$ of $[a_1, b_1] \to \mathbf{R}$ is continuous except possibly for a finite number of points at which it has one-sided limits.

44.U. (a) Let $\alpha, \beta : [a, b] \to \mathbf{R}$ be continuous with $\alpha(x) \le \beta(x)$ for all $x \in [a, b]$. Show that the set

$$B = \{(x, y) \in \mathbf{R}^2 : a \le x \le b, \alpha(x) \le y \le \beta(x)\}$$

is a compact set in $\mathbf{R}^2$ with content.

(b) Now let γ, $\delta : B \to \mathbf{R}$ be continuous functions with $\gamma(x, y) \le \delta(x, y)$ for all $(x, y) \in B$. Show that the set

$$D = \{(x, y, z) \in \mathbf{R}^3 : (x, y) \in B, \gamma(x, y) \le z \le \delta(x, y)\}$$

is a compact set in $\mathbf{R}^3$ with content.

(c) If $f : D \to \mathbf{R}$ is continuous, show that f is integrable on D and that

$$\int_D f = \int_a^b \left\{ \int_{\alpha(x)}^{\beta(x)} \left\{ \int_{\gamma(x, y)}^{\delta(x, y)} f(x, y, z) \, dz \right\} dy \right\} dx.$$

44.V. Let $I = [a_1, b_1] \times \cdots \times [a_p, b_p]$ and for each $j = 1, \ldots, p$, let $f_j : [a_j, b_j] \to \mathbf{R}$ be an integrable function.

If $\varphi : I \to \mathbf{R}$ is defined by $\varphi(x_1, \ldots, x_p) = f_1(x_1) \cdots f_p(x_p)$, show that φ is integrable on I and that

$$\int_I \varphi = \left\{ \int_{a_1}^{b_1} f_1 \right\} \cdots \left\{ \int_{a_p}^{b_p} f_p \right\}.$$

44.W. Use the Weierstrass Approximation Theorem to show that if $I = [a_1, b_1] \times \cdots \times [a_p, b_p]$ and if $g : I \to \mathbf{R}$ is continuous, then

$$\int_I g = \int_{a_1}^{b_1} \left\{ \int_{a_2}^{b_2} \cdots \left\{ \int_{a_p}^{b_p} g(x_1, x_2, \ldots, x_p) \, dx_p \right\} \cdots dx_2 \right\} dx_1.$$

44.X. Let $\varphi : [0, +\infty) \to \mathbf{R}$ with $\varphi(0) = 0$ be continuous, unbounded, and strictly increasing, and let ψ be its inverse function. Hence ψ is also continuous and strictly increasing on $[0, +\infty)$.

(a) If α, β are positive numbers, compare the area of the interval $[0, \alpha] \times [0, \beta]$ with the areas bounded by the coordinate axes and the graph of φ to obtain Young's Inequality:

$$\alpha\beta \le \int_0^\alpha \varphi + \int_0^\beta \psi.$$

(b) If $p \ge 1$ and $q \ge 1$ are such that $(1/p) + (1/q) = 1$, and if $\varphi(x) = x^{p/q}$ and $\psi(y) = y^{q/p}$, use Young's Inequality to establish the inequality

$$\alpha\beta \le \alpha^p/p + \beta^q/q.$$

(c) If a_i, b_i, $i = 1, \ldots, n$ are real numbers and if $A = (|a_1|^p + \cdots + |a_n|^p)^{1/p}$ and $B = (|b_1|^q + \cdots + |b_n|^q)^{1/q}$, use the above inequality to derive Hölder's Inequality

$$\sum_{i=1}^n |a_i b_i| \le AB,$$

which was obtained in Project 8.β(b).

Projects

44.α. Let $I \subseteq \mathbf{R}^p$ be a closed cell and let $f : I \to \mathbf{R}$ be bounded. For $\alpha > 0$, let $D_\alpha = \{x \in I : \omega_f(x) \ge \alpha\}$, where $\omega_f(x)$ denotes the oscillation of f at x (see Project 23.α).

(a) Suppose that D_α has content zero. Let $P_\alpha = \{I_1, \ldots, I_n\}$ be a partition of I such that (i) each point of D_α is contained in the interior of one of the cells $I_1, \ldots, I_r$ ($r \le n$), (ii) $c(I_1) + \cdots + c(I_r) \le \alpha/2 \|f\|_I$, and (iii) if $x, y \in K_j$ for $j = r+1, \ldots, n$, then $|f(x) - f(y)| < \alpha$. If P is a refinement of P_α, show that $|S(P; f) - S(P_\alpha; f)| \le \alpha(c(I) + 1)$.

(b) Deduce that if D_α has content zero for each $\alpha > 0$, then f is integrable on I.

(c) Suppose that for some $\alpha > 0$, the outer content $c^*(D_\alpha) > 0$. Show that for any partition $P = \{J_1, \ldots, J_n\}$ of I we have

$$\sum_{j=1}^n (M_j - m_j)c(J_j) \ge \alpha c^*(D_\alpha).$$

Deduce that f is not integrable on I.

(d) Conclude that f is integrable on I if and only if the set D_α has content zero for all $\alpha > 0$.

(e) Recall that $D = \bigcup_{n \in \mathbf{N}} D_{1/n}$ is the set of points where f is discontinuous. Show that D has measure zero (in the sense of Exercise 43.V) if and only if each set $D_{1/n}$ has content zero.

(f) Conclude that f is integrable on I if and only if its set D of points of discontinuity has measure zero. (This result is **Lebesgue's Criterion for Integrability**.)

44.β. This project considers lower and upper integrals (introduced in Project 43.α) and their iterations. Let $I \subseteq \mathbf{R}^r$ and $J \subseteq \mathbf{R}^s$ be closed cells, $p = r + s$, and let $K = I \times J \subseteq \mathbf{R}^p = \mathbf{R}^r \times \mathbf{R}^s$. Suppose that $f : K \to \mathbf{R}$ is bounded.

(a) For each $x \in I$, define $g_x : J \to \mathbf{R}$ by $g_x(y) = f(x, y)$ for $y \in J$. Let $\lambda : I \to \mathbf{R}$ be defined to be the lower integral $\lambda(x) = L(g_x)$ of g_x, and let $\mu : I \to \mathbf{R}$ be defined to be the upper integral $\mu(x) = U(g_x)$ of g_x. If R is any partition of I, and S is any partition of J, and $P = R \times S$ the resulting partition of K, then show that

$$L(P; f) \le L(R; \lambda) \le U(R; \lambda) \le U(R; \mu) \le U(P; f).$$

(b) Show that

$$L(f) \le L(\lambda) \le U(\lambda) \le U(f), \qquad L(f) \le L(\mu) \le U(\mu) \le U(f).$$

Hence, if f is integrable on K, then λ and μ are integrable on I and

$$\int_K f = \int_I \lambda = \int_I \mu.$$

(c) For each $y \in J$, define $h_y : I \to \mathbf{R}$ by $h_y(x) = f(x, y)$ for $x \in I$. Let $\lambda' : J \to \mathbf{R}$ and $\mu' : J \to \mathbf{R}$ be defined by

$$\lambda'(y) = L(h_y), \qquad \mu'(y) = U(h_y).$$

Show that if f is integrable on K, then λ' and μ' are integrable on J and

$$\int_K f = \int_J \lambda' = \int_J \mu'.$$

(d) If $g_x : J \to \mathbf{R}$ is integrable on J for each $x \in I$, then $\lambda = \mu$ and

$$\int_K f(x, y)\, d(x, y) = \int_K f = \int_J \left\{ \int_I f(x, y)\, dy \right\} dx.$$

Similarly, if $h_y : I \to R$ is integrable on I for each $y \in J$, then

$$\int_K f(x, y) \, d(x, y) = \int_K f = \int_J \left\{ \int_I f(x, y) \, dx \right\} dy.$$

44.γ. Let $\Omega \subseteq R^p$ be open and let $\mathscr{D}(\Omega)$ be the collection of all sets $A \in \mathscr{D}(R^p)$ with $A^- \subseteq \Omega$. In this project we shall introduce the notion of an "additive" function on $\mathscr{D}(\Omega)$ and of its "strong density." A function $G : \mathscr{D}(\Omega) \to R$ is said to be **additive** if

$$G(A \cup B) = G(A) + G(B)$$

whenever $A, B \in \mathscr{D}(\Omega)$ and $A \cap B = \emptyset$.

(a) If $f : \Omega \to R$ is integrable on every set in $\mathscr{D}(\Omega)$ and if we define $F : \mathscr{D}(\Omega) \to R$ by

$$F(A) = \int_A f,$$

then F is additive on $\mathscr{D}(\Omega)$.

(b) Let $G : \mathscr{D}(\Omega) \to R$ be an additive function and let $g : \Omega \to R$. We say that g is a **strong density** for G if, for every $\varepsilon > 0$ and every set $A \in \mathscr{D}(\Omega)$, there exists $\delta > 0$ such that if K is a closed cube with side length less than δ contained in Ω, and if $x \in A \cap K$, then

$$\left| \frac{G(K)}{c(K)} - g(x) \right| < \varepsilon.$$

(c) Let $\Omega = R^p$. Show that the content function $c : \mathscr{D}(R^p) \to R$ has strong density identically equal to 1 on R^p.

(d) Suppose that $\mu : \mathscr{D}(R^p) \to R$ is a positive additive function which is invariant under the translation of sets [that is, $\mu(x + A) = \mu(A)$ for all $x \in R^p$, $A \in \mathscr{D}(R^p)$]. Show that μ has a strong density on R^p which is a constant on R^p.

(e) If $f : \Omega \to R$ is continuous and if F is defined as in (a), show that F has strong density f on Ω.

(f) If $G : \mathscr{D}(\Omega) \to R$ is additive and has strong density $g : \Omega \to R$, show that g is continuous on Ω. Hence g is uniformly continuous on every $A \in \mathscr{D}(\Omega)$.

(g) Suppose that $G : \mathscr{D}(\Omega) \to R$ is additive and has strong density identically zero on Ω. Show that if K is a closed cube and if $\varepsilon > 0$, then there exists a partition of K into cubes $\{K_1, \ldots, K_r\}$ such that $|G(K_j)| \le \varepsilon c(K_j)$ for $j = 1, \ldots, r$, whence it follows that $|G(K)| \le \varepsilon c(K)$. Conclude that $G(K) = 0$ for all closed cubes $K \subseteq \Omega$.

(h) Suppose that F_1 and F_2 are additive functions on $\mathscr{D}(\Omega)$ such that for some $M > 0$ we have $|F_j(A)| \le M c(A)$ for all $A \in \mathscr{D}(\Omega)$, $j = 1, 2$. If $F_1(K) = F_2(K)$ for every cube $K \subseteq \Omega$, prove that $F_1(A) = F_2(A)$ for all $A \in \mathscr{D}(\Omega)$.

Section 45 Transformation of Sets and Integrals

It was noted in Section 43 that continuous mappings of an interval in R can cover a closed cube in R^2. We will show that this phenomenon cannot happen if the mapping is in Class C^1 and shall study the mapping of sets with content under C^1 maps. The case of a linear map is particularly important and the result is satisfyingly simple. In the case of a non-linear mapping, it will be seen that the Jacobian of the mapping indicates the extent of the "distortion" of the transformation.

These results will be used to establish a theorem concerned with the "change of variable" of an integral over a set in $\mathbf{R}^p$. The special cases of polar and spherical coordinates are briefly examined, and a stronger theorem is given that applies to many transformations that exhibit a mild amount of singularity.

45.1 Lemma. *The $\Omega \subseteq \mathbf{R}^p$ be open and let $\varphi : \Omega \to \mathbf{R}^p$ belong to Class $C^1(\Omega)$. Let A be a bounded set with $A^- \subseteq \Omega$. Then there exists a bounded open set Ω_1 with $A^- \subseteq \Omega_1 \subseteq \Omega_1^- \subseteq \Omega$ and a constant $M > 0$ such that if A is contained in the union of a finite number of closed cubes in Ω_1 with total content at most α, then $\varphi(A)$ is contained in the union of a finite number of closed cubes with total content at most $(\sqrt{p}\, M)^p \alpha$.*

proof. If $\Omega = \mathbf{R}^p$, let $\delta = 1$; otherwise let $\delta = \frac{1}{2} \inf \{\|a - x\| : a \in A^-, x \notin \Omega\}$. Since A^- is compact, it follows that $\delta > 0$. (Why?) Now let $\Omega_1 = \{y \in \mathbf{R}^p : \|y - a\| < \delta$ for some $a \in A\}$, so that Ω_1 is open and bounded and $A^- \subseteq \Omega_1$ and $\Omega_1^- \subseteq \Omega$. Since $\varphi \in C^1(\Omega)$ and Ω_1^- is compact, it follows that $M = \sup \{\|D\varphi(x)\|_{pp} : x \in \Omega_1\}$ is finite. If $A \subseteq I_1 \cup \cdots \cup I_n$, where the I_j are closed cubes contained in Ω_1, then it follows from Corollary 40.6 that for $x, y \in I_j$ we have

$$\|\varphi(x) - \varphi(y)\| \le M \|x - y\|.$$

Suppose the side length of I_j is $2r_j$ and take x to be the center of I_j; then if $y \in I_j$, we have $\|x - y\| \le \sqrt{p}\, r_j$ and so $\|\varphi(x) - \varphi(y)\| \le \sqrt{p}\, M r_j$. Thus $\varphi(I_j)$ is contained in a closed cube of side length $2\sqrt{p}\, M r_j$. Hence it follows that $\varphi(A)$ is contained in the union of a finite number of closed cubes with total content at most $(\sqrt{p}\, M)^p \alpha$. q.e.d.

45.2 Theorem. *Let $\Omega \subseteq \mathbf{R}^p$ be open and let $\varphi : \Omega \to \mathbf{R}^p$ belong to Class $C^1(\Omega)$. If $A \subset \Omega$ has content zero and if $A^- \subseteq \Omega$, then $\varphi(A)$ has content zero.*

proof. Apply the lemma for arbitrary $\alpha > 0$. q.e.d.

45.3 Corollary. *Let $r < p$, let $\Omega \subseteq \mathbf{R}^r$ be open, and let $\psi : \Omega \to \mathbf{R}^p$ belong to Class $C^1(\Omega)$. If $A \subseteq \Omega$ is a bounded set with $A^- \subseteq \Omega$, then $\psi(A)$ has content zero in $\mathbf{R}^p$.*

proof. Let $\Omega_0 = \Omega \times \mathbf{R}^{p-r}$ so that Ω_0 is open in $\mathbf{R}^p$, and define $\varphi : \Omega_0 \to \mathbf{R}^p$ by

$$\varphi(x_1, \ldots, x_r, x_{r+1}, \ldots, x_p) = \psi(x_1, \ldots, x_r).$$

Evidently $\varphi \in C^1(\Omega_0)$. Let $A_0 = A \times \{0, \ldots, 0\}$ so that $A_0^- \subseteq \Omega_0$ and A_0 has content zero in $\mathbf{R}^p$. It follows that $\psi(A) = \varphi(A_0)$ has content zero in $\mathbf{R}^p$.

Q.E.D.

We note that this corollary asserts that the C^1 image of any bounded set of "lower dimensionality" has content zero.

Since the boundary of a set A with content has content zero, it follows from Theorem 45.2 that if φ is in Class C^1 then $\varphi(b(A))$ has content zero. Unfortunately $\varphi(b(A))$ need have little relation, in general, to $b(\varphi(A))$. This observation enhances the interest of the next two results.

45.4 THEOREM. *Let $\Omega \subseteq \mathbf{R}^p$ be open and let $\varphi : \Omega \to \mathbf{R}^p$ belong to Class $C^1(\Omega)$. Suppose that A has content, $A^- \subseteq \Omega$, and $J_\varphi(x) \neq 0$ for all $x \in A^\circ$. Then $\varphi(A)$ has content.*

PROOF. Since A^- is compact and φ is continuous, then $\varphi(A) \subseteq \varphi(A^-)$ is bounded. To show that $\varphi(A)$ has content we shall show that $b(\varphi(A)) \subseteq \varphi(b(A))$ and that $\varphi(b(A))$ has content zero.

Since $\varphi(A^-)$ is compact, we have $b(\varphi(A)) \subseteq \varphi(A^-) = \varphi(A^\circ \cup b(A))$. Hence, if $y \in b(\varphi(A))$, there exists an $x \in A^\circ \cup b(A)$ such that $y = \varphi(x)$. If $x \in A^\circ$, then $J_\varphi(x) \neq 0$ and it follows from the Surjective Mapping Theorem 41.6 that $y = \varphi(x)$ is an interior point of $\varphi(A^\circ) \subseteq \varphi(A)$. But this contradicts the hypothesis that $y \in b(\varphi(A))$. Therefore we infer that $b(\varphi(A)) \subseteq \varphi(b(A))$.

Now, since A has content, its boundary $b(A) \subseteq \Omega$ is a closed set with content zero, whence it follows from Theorem 45.2 that $\varphi(b(A))$ has content zero.

Q.E.D.

45.5 COROLLARY. *Let $\Omega \subseteq \mathbf{R}^p$ be open, and let $\varphi : \Omega \to \mathbf{R}^p$ belong to Class $C^1(\Omega)$ and be injective on Ω. If A has content, $A^- \subseteq \Omega$, and $J_\varphi(x) \neq 0$ for $x \in A^\circ$, then $b(\varphi(A)) = \varphi(b(A))$.*

PROOF. It suffices to show that $\varphi(b(A)) \subseteq b(\varphi(A))$, since the reverse inclusion was established in the proof of the theorem. Let $x \in b(A)$, so that there exists a sequence (x_n) in A and a sequence (y_n) in $\Omega \setminus A$, both of which converge to x. Since φ is continuous, then $\varphi(x_n) \to \varphi(x)$ and $\varphi(y_n) \to \varphi(x)$. Since φ is injective on Ω, then $\varphi(y_n) \notin \varphi(A)$ and hence $\varphi(x) \in b(\varphi(A))$. Therefore $\varphi(b(A)) \subseteq b(\varphi(A))$.

Q.E.D.

Transformations by Linear Maps

We shall now see that sets with content are mapped by a *linear* map in $\mathbf{R}^p$ into sets whose content is a fixed multiple of the original content. Moreover, this multiple is the absolute value of the determinant corresponding to the linear map. (In this theorem we shall assume that the notion and elementary properties of the determinant of a linear map in $\mathbf{R}^p$ are familiar to the reader.)

45.6 THEOREM. *Let* $L \in \mathcal{L}(\mathbf{R}^p)$. *If* $A \in \mathcal{D}(\mathbf{R}^p)$, *then* $c(L(A)) = |\det L| \, c(A)$.

PROOF. If L is singular (that is, if $\det L = 0$), then L maps $\mathbf{R}^p$ into a proper linear subspace of $\mathbf{R}^p$. Since this subspace can also be obtained as the image of some $L' : \mathbf{R}^r \to \mathbf{R}^p$ with $r < p$, it follows from Corollary 45.3 that $c(L(A)) = 0$ for all $A \in \mathcal{D}(\mathbf{R}^p)$. Hence the statement is true for linear maps which are singular.

If L is not singular (that is, if $\det L \neq 0$), then Theorem 45.4 implies that if $A \in \mathcal{D}(\mathbf{R}^p)$, then $L(A) \in \mathcal{D}(\mathbf{R}^p)$. We now define $\lambda : \mathcal{D}(\mathbf{R}^p) \to \mathbf{R}$ by $\lambda(A) = c(L(A))$. (i) It is clear that $\lambda(A) \geq 0$ for all $A \in \mathcal{D}(\mathbf{R}^p)$. (ii) Suppose A, $B \in \mathcal{D}(\mathbf{R}^p)$ and $A \cap B = \emptyset$; then

$$\lambda(A \cup B) = c(L(A \cup B)) = c(L(A) \cup L(B)).$$

Since L is injective, then $L(A) \cap L(B) = \emptyset$ and hence

$$c(L(A) \cup L(B)) = c(L(A)) + c(L(B)) = \lambda(A) + \lambda(B).$$

(iii) Let $x \in \mathbf{R}^p$ and $A \in \mathcal{D}(\mathbf{R}^p)$; then

$$\lambda(x + A) = c(L(x + A)) = c(L(x) + L(A)) = c(L(A)) = \lambda(A).$$

Therefore it follows from Corollary 44.7 that there exists a constant $m_L \geq 0$ such that $\lambda(A) = m_L c(A)$ for all $A \in \mathcal{D}(\mathbf{R}^p)$.

We next examine how m_L depends on $L \in \mathcal{L}(\mathbf{R}^p)$. Let $M \in \mathcal{L}(\mathbf{R}^p)$ be non-singular; then if $A \in \mathcal{D}(\mathbf{R}^p)$, we have

$$m_{L \circ M} c(A) = c(L \circ M(A)) = c(L(M(A)))$$
$$= m_L c(M(A)) = m_L m_M c(A).$$

Hence we have $m_{L \circ M} = m_L m_M$ for all non-singular L, $M \in \mathcal{L}(\mathbf{R}^p)$.

It remains to show that $m_L = |\det L|$. To do this we shall use the fact from linear algebra that every non-singular $L \in \mathcal{L}(\mathbf{R}^p)$ is the composition of linear maps of the following three forms:

(a) $L_1(x_1, \ldots, x_p) = (\alpha x_1, x_2, \ldots, x_p)$ for some $\alpha \neq 0$;
(b) $L_2(x_1, \ldots, x_i, x_{i+1}, \ldots, x_p) = (x_1, \ldots, x_{i+1}, x_i, \ldots, x_p)$;
(c) $L_3(x_1, \ldots, x_p) = (x_1 + x_2, x_2, \ldots, x_p)$.

Note that if K_0 is the half-open cube $[0, 1) \times \cdots \times [0, 1)$ in $\mathbf{R}^p$ and if $\alpha > 0$, then $L_1(K_0) = [0, \alpha) \times [0, 1) \times \cdots \times [0, 1)$, whence it follows that

$$\alpha = c(L_1(K_0)) = m_{L_1} c(K_0) = m_{L_1}.$$

Similarly, if $\alpha < 0$, then $L_1(K_0) = (\alpha, 0] \times [0, 1) \times \cdots \times [0, 1)$, and

$$-\alpha = c(L_1(K_0)) = m_{L_1} c(K_0) = m_{L_1}.$$

Hence, in either case we have $m_{L_1} = |\alpha| = |\det L_1|$.

Since $L_2(K_0) = K_0$, it follows that $m_{L_2} = 1 = |\det L_2|$.

Finally, let Δ_1 and Δ_2 be the two sets

$$\Delta_1 = \{(x_1, \ldots, x_p) : 0 \le x_i < 1, x_1 < x_2\},$$

$$\Delta_2 = \{(x_1, \ldots, x_p) : 0 \le x_i < 1, x_2 \le x_1\}.$$

It is clear that $\Delta_1 \cap \Delta_2 = \emptyset$ and $K_0 = \Delta_1 \cup \Delta_2$. Since it can be seen that

$$L_3(K_0) = \Delta_2 \cup \{(1, 0, \ldots, 0) + \Delta_1\}$$

it follows that

$$c(L_3(K_0)) = c(\Delta_2) + c((1, 0, \ldots, 0) + \Delta_1) = c(\Delta_2) + c(\Delta_1)$$
$$= c(\Delta_1 \cup \Delta_2) = c(K_0).$$

Hence $m_{L_3} = 1 = |\det L_3|$.

Now let the non-singular linear map L be the composition of linear maps $L_1, L_2, \ldots, L_r$ having one of the three forms given above. Since

$$m_L = m_{L_1 \circ L_2 \circ \cdots \circ L_r} = m_{L_1} m_{L_2} \cdots m_{L_r}$$
$$= |\det L_1| \, |\det L_2| \cdots |\det L_r|$$
$$= |(\det L_1)(\det L_r) \cdots (\det L_r)|$$
$$= |\det (L_1 \circ L_2 \circ \cdots \circ L_r)| = |\det L|,$$

the theorem is proved.   Q.E.D.

Transformation by Non-linear Mappings

We shall now obtain an extension of Theorem 45.6 for C^1 mappings which are not linear. Of course, in this case the content of the image of an arbitrary set need not be a fixed multiple of the content of the given set, but may vary from point to point. The Jacobian Theorem implies that if K is a sufficiently small cube with center x, then $c(\varphi(K))$ is approximately equal to $|J_\varphi(x)| \, c(K)$. This result is crucial in order to establish the Change of Variables Theorem. It will be technically convenient to consider first the following special case.

45.7 LEMMA. *Let $K \subseteq \mathbf{R}^p$ be a closed cube with center 0. Let Ω be an open set containing K and let $\psi : \Omega \to \mathbf{R}^p$ belong to Class $C^1(\Omega)$ and be injective. Suppose further that $J_\psi(x) \ne 0$ for $x \in K$ and that*

$$(45.1) \qquad \|\psi(x) - x\| \le \alpha \, \|x\| \qquad \text{for } x \in K,$$

where α satisfies $0 < \alpha < 1/\sqrt{p}$.   *Then*

$$(1 - \alpha\sqrt{p})^p \leq \frac{c(\psi(K))}{c(K)} \leq (1 + \alpha\sqrt{p})^p.$$

PROOF.   It follows from Theorem 45.4 that $\psi(K)$ has content and from Corollary 45.5 that $b(\psi(K)) = \psi(b(K))$. If the side length of K is $2r$ and if $x \in b(K)$, then (by Theorem 8.10) we have $r \leq \|x\| \leq r\sqrt{p}$. The inequality (45.1) implies that $\psi(x)$ is within distance $\alpha r\sqrt{p}$ of $x \in b(K)$. Therefore the compact set $\psi(b(K)) = b(\psi(K))$ does not intersect an open cube C_i with center 0 and side length $2(1 - \alpha\sqrt{p})r$. If we let A (respectively, B) be the set of all interior (respectively, exterior) points of $\psi(K)$, then A and B are disjoint non-empty open sets with union $\mathbf{R}^p \setminus b(\psi(K))$. Since C_i is connected in $\mathbf{R}^p$, we must have either $C_i \subseteq A$ or $C_i \subseteq B$. But since $0 \in C_i \cap A$, we infer that $C_i \subseteq A \subseteq \psi(K)$. In an analogous fashion the reader can show that if C_o is the closed cube with center 0 and side length $2(1 + \alpha\sqrt{p})r$, then $\psi(K) \subseteq C_o$. The stated conclusion now follows from these inclusions.   Q.E.D.

45.8   THE JACOBIAN THEOREM.   *Let $\Omega \subseteq \mathbf{R}^p$ be open and suppose that $\varphi : \Omega \to \mathbf{R}^p$ belongs to Class $C^1(\Omega)$, is injective on Ω, and that $J_\varphi(x) \neq 0$ for $x \in \Omega$. Suppose that A has content and $A^- \subseteq \Omega$. If $\varepsilon > 0$ is given, then there exists $\gamma > 0$ such that if K is a closed cube with center $x \in A$ and side length less then 2γ, then*

$$(45.2) \qquad |J_\varphi(x)| (1 - \varepsilon)^p \leq \frac{c(\varphi(K))}{c(K)} \leq |J_\varphi(x)| (1 + \varepsilon)^p.$$

PROOF.   Construct $\delta > 0$ and Ω_1 as in the proof of Lemma 45.1. Since $\det D\varphi(x) = J_\varphi(x) \neq 0$ for all $x \in \Omega$, it follows that $L_x = (D\varphi(x))^{-1}$ exists; since $1 = \det (L_x \circ D\varphi(x)) = (\det L_x)(\det D\varphi(x))$, it follows that

$$\det L_x = 1/J_\varphi(x) \qquad \text{for } x \in \Omega.$$

Since the entries in the standard matrix representation for L_x are continuous functions, it follows from the compactness of Ω_1^- and (21.4) that there exists a constant $M > 0$ such that $\|L_x\|_{pp} \leq M$ for all $x \in \Omega_1$.

Now let ε, with $0 < \varepsilon < 1$, be given. Since the map $x \mapsto D\varphi(x)$ is uniformly continuous on Ω_1, there exists β with $0 < \beta < \delta$ such that if $x_1, x_2 \in \Omega_1$ and $\|x_1 - x_2\| \leq \beta$, then $\|D\varphi(x_1) - D\varphi(x_2)\|_{pp} \leq \varepsilon/M\sqrt{p}$. We now let $x \in A$ be given; hence if $\|z\| \leq \beta$, then x and $x + z$ belong to Ω_1. Hence it follows from Lemma 41.3 that

$$(45.3) \quad \|\varphi(x + z) - \varphi(x) - D\varphi(x)(z)\| \leq \|z\| \sup_{0 \leq t \leq 1} \|D\varphi(x + tz) - D\varphi(x)\|_{pp}$$

$$\leq \frac{\varepsilon}{M\sqrt{p}} \|z\|.$$

Let $x \in A$ and define $\psi(z)$ for $\|z\| \le \beta$ by

$$\psi(z) = L_x[\varphi(x+z) - \varphi(x)].$$

Since $L_x = (D\varphi(x))^{-1}$, the inequality (45.3) yields

$$\|\psi(z) - z\| \le \frac{\varepsilon}{\sqrt{p}}\|z\| \qquad \text{for } \|z\| \le \beta.$$

We now apply the preceeding lemma with $\alpha = \varepsilon/\sqrt{p}$ to infer that if K_1 is any closed cube with center 0 and contained in the open ball with radius β, then

$$(1-\varepsilon)^p \le \frac{c(\psi(K_1))}{c(K_1)} \le (1+\varepsilon)^p.$$

It follows from the definition of ψ and Theorem 45.6 that if $K = x + K_1$, then K is a closed cube with center x and that $c(K) = c(K_1)$ and

$$c(\psi(K_1)) = |\det L_x| \, c(\varphi(x + K_1) - \varphi(x))$$
$$= \frac{1}{|J_\varphi(x)|} c(\varphi(K)).$$

Hence, if K is a closed cube with center $x \in A$ and side length less than 2γ (where $\gamma = \beta/\sqrt{p}$), then inequality (45.2) holds. Q.E.D.

Change of Variables

We shall now apply the Jacobian Theorem to obtain an important theorem which is a generalization to $\mathbf{R}^p$ of the Change of Variables Theorem 30.12. The latter result asserts that if $\varphi : [\alpha, \beta] \to \mathbf{R}$ has a continuous derivative and if f is continuous on the range of φ, then

(45.4) $$\int_{\varphi(\alpha)}^{\varphi(\beta)} f = \int_\alpha^\beta (f \circ \varphi)\varphi'.$$

The result we shall establish concerns an injective mapping φ defined on an open subset $\Omega \subseteq \mathbf{R}^p$ with values in $\mathbf{R}^p$. We shall assume that $\varphi \in C^1(\Omega)$ and that its Jacobian determinant

$$J_\varphi(x) = \det[D_j\varphi_i(x)]$$

does not vanish on Ω. It will be shown that if A has content, if $A^- \subseteq \Omega$, and if f is bounded and continuous on $\varphi(A)$ to $\mathbf{R}$, then $\varphi(A)$ has content and

(45.5) $$\int_{\varphi(A)} f = \int_A (f \circ \varphi)|J_\varphi|.$$

It will be observed that the hypotheses are somewhat more restrictive than in the case $p = 1$. Indeed, in (45.4) we do not assume that φ is injective or that $\varphi'(x) \ne 0$

for $x \in [\alpha, \beta]$. If φ happens to be injective, we note that the exact analog for (45.5) in the case $p = 1$ is

$$\int_A^B f = \int_\alpha^\beta (f \circ \varphi) |\varphi'|,$$

where $A = \inf \{\varphi(\alpha), \varphi(\beta)\}$ and $B = \sup \{\varphi(\alpha), \varphi(\beta)\}$. Of course, if $\varphi'(x) > 0$ for $\alpha \le x \le \beta$, then formula (45.5) reduces to (45.4); while if $\varphi'(x) < 0$ for $\alpha \le x \le \beta$, then formula (45.5) reduces to

$$\int_{\varphi(\beta)}^{\varphi(\alpha)} f = \int_\alpha^\beta (f \circ \varphi)(-\varphi'),$$

whence (45.4) also follows. The explanation for this difference is that the integral over intervals in $\mathbf{R}$ is "oriented" in the sense that we define

$$\int_u^v f = -\int_v^u f$$

for any real numbers u, v. No such orientation has been defined for integrals over $\mathbf{R}^p$.

The proof given here is essentially due to J. T. Schwartz.[†] It is "elementary" in the sense that it does not make use of any results from measure theory. However, the argument is very delicate and makes use of a number of the deeper properties of continuous functions, compact and connected sets, and the properties of the integral. Even so, the theorem that will be proved is not quite sufficient for all the important cases that arise, and will be augmented below with a stronger form which permits J_φ to vanish and $f \circ \varphi$ to be discontinuous on a set with content zero.

45.9 CHANGE OF VARIABLES THEOREM. *Let $\Omega \subseteq \mathbf{R}^p$ be open and suppose that $\varphi : \Omega \to \mathbf{R}^p$ belongs to Class $C^1(\Omega)$, is injective on Ω, and $J_\varphi(x) \ne 0$ for $x \in \Omega$. Suppose that A has content, $A^- \subseteq \Omega$, and $f : \varphi(A) \to \mathbf{R}$ is bounded and continuous. Then*

(45.5) $$\int_{\varphi(A)} f = \int_A (f \circ \varphi) |J_\varphi|.$$

PROOF. It follows from Theorem 45.4 that $\varphi(A)$ has content. Since the integrands are continuous, it follows that the integrals in (45.5) exist; it remains to establish their equality. By letting $f = f^+ - f^-$, where $f^+ = \frac{1}{2}(f + |f|)$ and $f^- = \frac{1}{2}(|f| - f)$, and using the linearity of the integral, it is enough to suppose that $f(y) \ge 0$ for all $y \in \varphi(A)$.

[†] J. T. SCHWARTZ (1930–) was graduated from CCNY, received his doctorate at Yale University, and is a professor at the Courant Institute of New York University. Although he is best known for his work in functional analysis, he has also contributed to differential equations, geometry, computer languages, various aspects of mathematical physics, and mathematical economics.

Now let Ω_1 be as in Lemma 45.1 and let

$$M_\varphi = \sup\{\|D\varphi(x)\|_{pp} : x \in \Omega_1\},$$
$$M_f = \sup\{f(y) : y \in \varphi(A)\},$$
$$M_J = \sup\{|J_\varphi(x)| : x \in A\}.$$

Let $\varepsilon > 0$ be arbitrary except that $0 < \varepsilon < 1$, let I be a closed cell containing A, and let $\{K_i : i = 1, \ldots, M\}$ be a partition of I into non-overlapping closed cubes with side length less than 2γ, where γ is the constant in the Jacobian Theorem. Let those cubes that are completely contained in A be enumerated $K_1, \ldots, K_m$; let those that have points of both A and its complement be enumerated $K_{m+1}, \ldots, K_n$, and let those cubes completely contained in the complement of A be enumerated $K_{n+1}, \ldots, K_M$. Since A has content, we may assume that the partition has been chosen sufficiently fine that

(i) $$c(A) \le \sum_{i=1}^m c(K_i) + \varepsilon, \qquad \sum_{i=m+1}^n c(K_i) < \varepsilon.$$

We let $B = K_1 \cup \cdots \cup K_m$ so that $B \subseteq A$. Since $c(A \setminus B) = c(A) - c(B) < \varepsilon$, we have

(ii) $$\left| \int_A (f \circ \varphi) |J_\varphi| - \int_B (f \circ \varphi) |J_\varphi| \right|$$
$$= \left| \int_{A \setminus B} (f \circ \varphi) |J_\varphi| \right| \le M_f M_J c(A \setminus B) \le [M_f M_J]\varepsilon.$$

It follows from Lemma 45.1 that $c(\varphi(A \setminus B)) \le (\sqrt{p}\, M_\varphi)^p \varepsilon$, so that

(iii) $$\left| \int_{\varphi(A)} f - \int_{\varphi(B)} f \right| = \left| \int_{\varphi(A \setminus B)} f \right| \le [M_f (\sqrt{p}\, M_\varphi)^p]\varepsilon.$$

If x_i is the center of K_i, $i = 1, \ldots, m$, then it follows from the Jacobian Theorem that

$$|J_\varphi(x_i)| (1 - \varepsilon)^p \le \frac{c(\varphi(K_i))}{c(K_i)} \le |J_\varphi(x_i)| (1 + \varepsilon)^p.$$

Now since $0 < \varepsilon < 1$, it is seen that $1 - 2^p \varepsilon \le (1 - \varepsilon)^p$ and $(1 + \varepsilon)^p \le 1 + 2^p \varepsilon$, so we can write this inequality in the form

(iv) $$|c(\varphi(K_i)) - |J_\varphi(x_i)| c(K_i)| \le [c(K_i) M_J 2^p]\varepsilon.$$

Now because of the continuity of the functions in the integrand on the compact set B, it follows that we may assume that for any point $y_i \in K_i$, then

(v) $$\left| \int_B (f \circ \varphi) |J_\varphi| - \sum_{i=1}^m (f \circ \varphi)(y_i) |J_\varphi(x_i)| c(K_i) \right| < \varepsilon c(B).$$

(For, if necessary, we can divide the cubes $K_1, \ldots, K_m$ into small cubes; see Exercise 43.T.)

Since φ is injective, two sets from $\{\varphi(K_i) : i = 1, \ldots, m\}$ intersect at most in a set $\varphi(K_i \cap K_j)$ which has content 0 since $c(K_i \cap K_j) = 0$. Also, since $\varphi(K_i)$ has content, then f is integrable on $\varphi(K_i)$; hence it follows from Theorem 44.9(b) that

$$\int_{\varphi(B)} f = \sum_{i=1}^{m} \int_{\varphi(K_i)} f.$$

Now since K_i is connected, then $\varphi(K_i)$ is connected. Since f is bounded and continuous on $\varphi(K_i)$, it follows from the Mean Value Theorem 44.11 that there exists $p_i \in \varphi(K_i)$ such that

$$\int_{\varphi(K_i)} f = f(p_i) c(\varphi(K_i)), \qquad i = 1, \ldots, m.$$

Since $p_i \in \varphi(K_i)$, there exists a unique $y_i \in K_i$ with $p_i = \varphi(y_i)$, $i = 1, \ldots, m$. Hence we have

(vi) $$\int_{\varphi(B)} f = \sum_{i=1}^{m} (f \circ \varphi)(y_i) c(\varphi(K_i)).$$

But since $(f \circ \varphi)(y_i) \geq 0$, it follows from (iv) that

$$\left| \sum_{i=1}^{m} (f \circ \varphi)(y_i) c(\varphi(K_i)) - \sum_{i=1}^{m} (f \circ \varphi)(y_i) |J_\varphi(x_i)| \, c(K_i) \right|$$

$$\leq \left[M_J 2^p \sum_{i=1}^{m} (f \circ \varphi)(y_i) c(K_i) \right] \varepsilon$$

$$\leq \left[M_J M_f 2^p \sum_{i=1}^{m} c(K_i) \right] \varepsilon \leq [M_J M_f 2^p c(A)] \varepsilon.$$

If we combine this last relation with (v) and (vi), we get

(vii) $$\left| \int_{\varphi(B)} f - \int_B (f \circ \varphi) |J_\varphi| \right| \leq (1 + M_J M_f 2^p) c(A) \varepsilon.$$

Combining (vii) with (ii) and (iii), we obtain

$$\left| \int_{\varphi(A)} f - \int_A (f \circ \varphi) |J_\varphi| \right| \leq [M_f (\sqrt{p} M_\varphi)^p + (1 + M_J M_f 2^p) c(A) + M_f M_J] \varepsilon.$$

Since ε is an arbitrary number with $0 < \varepsilon < 1$, the equation (45.5) is established.   Q.E.D.

Applications

The *use* of the theorem on the change of variables when $p > 1$ is generally different from the application of the corresponding theorem when $p = 1$. For example, in evaluating

$$\int_0^1 x(1+x^2)^{1/2}\, dx$$

we usually note that if we introduce $\varphi(x) = 1 + x^2$ then $\varphi'(x) = 2x$; hence the integrand has the form $\frac{1}{2}(\varphi(x))^{1/2}\varphi'(x)$ and so

$$\int_0^1 x(1+x^2)^{1/2}\, dx = \frac{1}{2}\frac{2}{3}(\varphi(x))^{3/2}\Big|_{x=0}^{x=1}$$

$$= \frac{1}{3}(1+x^2)^{3/2}\Big|_{x=0}^{x=1} = \frac{1}{3}(2^{3/2}-1).$$

Thus the integration is performed by observing that the given integrand is a composition of some function and φ, multiplied by the derivative of φ. Similar applications to evaluate integrals in more than one variable are usually possible only when the Jacobian term is constant (or very simple). For example, an integral of the form

$$\iint_A f(x+2y, 2x-3y)\, d(x, y)$$

may be treated by introducing the linear transformation $\varphi(x, y) = (x+2y, 2x-3y)$. Here

$$J_\varphi(x, y) = \det \begin{bmatrix} 1 & 2 \\ 2 & -3 \end{bmatrix} = -3-4 = -7$$

and so we have

$$\iint_A f(x+2y, 2x-3y)\, d(x, y) = \frac{1}{7}\iint_{\varphi(A)} f(u, v)\, d(u, v).$$

This second integral may be simpler if $f(u, v)$ is simpler [for example, if $f(u, v) = g(u)h(v)$], or if $\varphi(A)$ is simple (for example, if it is a cell). Otherwise, the transformation may not simplify things very much.

A more typical use of the theorem is to evaluate a multiple integral $\int_D f$ by observing that the set D is the image of a simpler set A (for example, a cell) under a suitable map φ.

45.10 EXAMPLES. (a) Let D denote the rectangle with vertices $(0, 0)$, $(2, 2)$, $(1, 3)$, $(-1, 1)$; that is, the region bounded by the lines given by

$$y = x, \quad y = -x+4, \quad y = x+2, \quad y = -x.$$

If we let $u = y - x$ and $v = y + x$, these lines become

$$u = 0, \quad v = 4, \quad u = 2, \quad v = 0.$$

Hence, if φ is the map $\varphi(u, v) = (x, y)$, then φ maps the cell $A = [0, 2] \times [0, 4]$ into D. We leave it to the reader to show that

$$\iint_D f(x, y) \, d(x, y) = \iint_A f[\tfrac{1}{2}(v - u), \tfrac{1}{2}(u + v)]\tfrac{1}{2} \, d(u, v)$$

$$= \frac{1}{2} \int_0^4 \left\{ \int_0^2 f[\tfrac{1}{2}(v - u), \tfrac{1}{2}(u + v)] \, du \right\} dv.$$

(b) Let $D \subseteq \mathbf{R}^2$ be the set of points in $\mathbf{R}^2$ given by

$$D = \{(u, v) : 1 \leq u^2 - v^2 \leq 9, 1 \leq uv \leq 4\};$$

Hence D is bounded by four hyperbolas. If we define $\psi : (u, v) \mapsto (x, y)$ by

$$x = u^2 - v^2, \qquad y = uv,$$

then it is clear that ψ maps these hyperbolas in (u, v)-plane into the lines $x = 1$, $x = 9$, $y = 1$, $y = 4$ in the (x, y)-plane. Although ψ is not injective on all of $\mathbf{R}^2$, it is injective in the set $Q = \{(u, v) : u > 0, v > 0\}$ and $J_\psi(u, v) = 2(u^2 + v^2)$. Moreover, $\psi(Q) = \{(x, y) : x \in \mathbf{R}, y > 0\}$.

Hence we define φ on $\{(x, y) : x \in \mathbf{R}, y > 0\}$ to $Q \subseteq \mathbf{R}^2$ be the inverse of ψ. From the above it is clear that φ maps the lines $x = 1$, $x = 9$, $y = 1$, $y = 4$ into the hyperbolas

$$u^2 - v^2 = 1, \quad u^2 - v^2 = 9, \quad uv = 1, \quad uv = 4,$$

respectively, and that the set D is the image under φ of the cell $A = [1, 9] \times [1, 4]$. Direct calculation shows that φ has the form $\varphi(x, y) = (u, v)$ where

(45.6) $\qquad u = \left[\dfrac{x + (x^2 + 4y^2)^{1/2}}{2} \right]^{1/2}, \qquad v = \left[\dfrac{-x + (x^2 + 4y^2)^{1/2}}{2} \right]^{1/2}.$

It follows from this that $u^2 + v^2 = (x^2 + 4y^2)^{1/2}$ so that $J_\varphi(x, y) = \tfrac{1}{2}(x^2 + 4y^2)^{-1/2}$. [This fact also follows from the identity $(u^2 + v^2)^2 = (u^2 - v^2)^2 + 4u^2v^2 = x^2 + 4y^2$.] Thus we have

$$\iint_D f(u, v) \, d(u, v) = \iint_A \frac{f(u(x, y), v(x, y))}{2\sqrt{x^2 + 4y^2}} \, d(x, y),$$

where $A = [1, 9] \times [1, 4]$ and where $u(x, y)$ and $v(x, y)$ are given in (45.6).

Polar and Spherical Coordinates

It is often convenient to specify points in the plane $\mathbf{R}^2$ by giving their "polar coordinates." Usually we think of the plane as possessing both the Cartesian coordinates (given by vertical and horizontal lines) and the polar system (given by rays through the origin and circles centered at the origin). Alternatively, we can think of polar coordinates as a map of $(r, \theta) \in \mathbf{R}^2$ into $(x, y) \in \mathbf{R}^2$ given by

$$(45.7) \qquad (x, y) = \varphi(r, \theta) = (r \cos \theta, r \sin \theta).$$

Any pair of numbers $(r, \theta) \in \mathbf{R}^2$ such that $(x, y) = (r \cos \theta, r \sin \theta)$ is called a set of **polar coordinates** of the point (x, y). Usually one requires $r \geq 0$; even so, each point (x, y) in $\mathbf{R}^2$ has infinitely many sets of polar coordinates.

For example, if $(x, y) = (0, 0)$, then $(0, \theta)$ is a set of polar coordinates of $(0, 0)$ for all $\theta \in \mathbf{R}$; if $(x, y) \neq (0, 0)$ and (r, θ) is a set of polar coordinates for (x, y), then for each $n \in \mathbf{Z}$ the pair $(r, \theta + n2\pi)$ is also a set of polar coordinates for (x, y).

If $(x, y) \neq (0, 0)$, then the unique pair (r, θ) with $r > 0$, $0 \leq \theta < 2\pi$, is called the **principal set of polar coordinates of the point** (x, y). Thus the function φ gives rise to an injective map of $(0, +\infty) \times [0, 2\pi)$ onto $\mathbf{R}^2 \setminus \{(0, 0)\}$. It also gives a map of $[0, +\infty) \times [0, 2\pi)$ onto $\mathbf{R}^2$ but it is not injective, since it sends all the points $(0, \theta)$, $0 \leq \theta < 2\pi$, into $(0, 0)$. Note also that the Jacobian is given by

$$(45.8) \qquad J_\varphi(r, \theta) = \det \begin{bmatrix} \cos \theta & -r \sin \theta \\ \sin \theta & r \cos \theta \end{bmatrix}$$

$$= r(\cos \theta)^2 + r(\sin \theta)^2 = r,$$

which vanishes for $r = 0$.

It is clear that φ maps the cell $A = [0, 1] \times [0, 2\pi]$ in the (r, θ)-plane into the unit disk $D = \{(x, y) : x^2 + y^2 \leq 1\}$ but since φ is not injective on A and since J_φ vanishes for $r = 0$, we cannot apply the Change of Variables Theorem 45.9 to convert integration over D into integration over A.

We encounter analogous difficulties with spherical coordinates in $\mathbf{R}^3$. Recall that spherical coordinates are defined by the map $\Phi : \mathbf{R}^3 \to \mathbf{R}^3$ where

$$(45.9) \qquad \Phi(r, \theta, \phi) = (r \cos \theta \sin \phi, r \sin \theta \sin \phi, r \cos \phi).$$

Any triple of numbers $(r, \theta, \phi) \in \mathbf{R}^3$ such that $(x, y, z) = \Phi(r, \theta, \phi)$ is called a set of **spherical coordinates** of (x, y, z). Usually one requires $r \geq 0$, but even with this restriction each point in $\mathbf{R}^3$ has infinitely many sets of spherical coordinates.

For example, if $(x, y, z) = (0, 0, 0)$, then $(0, \theta, \phi)$ is a set of spherical coordinates for all $\theta \in \mathbf{R}$, $\phi \in \mathbf{R}$; if $(x, y, z) \neq (0, 0, 0)$ and (r, θ, ϕ) is a set of polar coordinates for (x, y, z), then for each m, $n \in \mathbf{Z}$, the triples $(r, \theta + 2m\pi, \phi + 2n\pi)$ and $(r, \theta + (2m + 1)\pi, \phi + (2n + 1)\pi)$ are sets of spherical coordinates for this point.

If (x, y, z) is such that $(x, y) \neq (0, 0)$, then the unique triple (r, θ, ϕ) with $r > 0$, $0 \leq \theta < 2\pi$, $0 < \phi < \pi$, is called the **principal set of spherical coordinates** of (x, y, z). Thus the function Φ yields an injective map of

$(0, +\infty) \times [0, 2\pi) \times (0, \pi)$ onto $\mathbf{R}^3 \setminus \{(0, 0, z) : z \in \mathbf{R}\}$. The restriction of Φ to $[0, +\infty) \times [0, 2\pi] \times [0, \pi]$ gives a map onto all of $\mathbf{R}^3$ but it is not injective, since it sends all points $(0, \theta, \phi)$ into $(0, 0, 0)$ and, if $\phi = 0$ or π, then all the points (r, θ, ϕ) map into $(0, 0, r \cos \phi)$. Note also that

$$(45.10) \qquad J_\Phi(r, \theta, \phi) = \det \begin{bmatrix} \cos \theta \sin \phi & -r \sin \theta \sin \phi & r \cos \theta \cos \phi \\ \sin \theta \sin \phi & r \cos \theta \sin \phi & r \sin \theta \cos \phi \\ \cos \phi & 0 & -r \sin \phi \end{bmatrix}$$

$$= -r^2 \sin \phi.$$

It is readily seen that Φ maps the cell $A = [0, 1] \times [0, 2\pi] \times [0, \pi]$ in the (r, θ, ϕ)-space into the unit ball $D = \{(x, y, z) : x^2 + y^2 + z^2 \le 1\}$, but since Φ is not injective on A and J_Φ vanishes when $r^2 \sin \phi = 0$, we cannot use the Change of Variables Theorem 45.9 to convert integration over D into integration over A.

We shall now present a theorem which enables us to handle the difficulties we have encountered in the use of polar and spherical coordinates and which is often useful in other "transformations with singularities." It will be noted that the theorem does not require φ to be injective on the set A, though it is injective on A^0.

45.11 CHANGE OF VARIABLES THEOREM (STRONG FORM). *Let $\Omega \subseteq \mathbf{R}^p$ be open and let $\varphi : \Omega \to \mathbf{R}^p$ belong to Class $C^1(\Omega)$. Let Ω_0 be an open set with content such that $\Omega_0^- \subseteq \Omega$ and such that φ is injective on Ω_0. Let $E \subseteq \Omega$ be a compact set with content zero such that $J_\varphi(x) \ne 0$ for $x \in \Omega_0 \setminus E$. Suppose that $A \subseteq \Omega$ has content, $A^- \subseteq \Omega_0^-$, that $f : \varphi(A) \to \mathbf{R}$ is bounded, and that f is continuous on $\varphi(A \setminus E)$. Then*

$$(45.5) \qquad \int_{\varphi(A)} f = \int_A (f \circ \varphi) |J_\varphi|.$$

PROOF. Since $b(A)$ and $b(\Omega_0)$ are compact and have content zero, we may assume that they are contained in E; therefore $A^0 \setminus E \subseteq \Omega_0 \setminus E$. Since A and E have content, the set $A \setminus E$ has content; moreover, since E is closed, then $(A \setminus E)^0 = A^0 \setminus E$ so that $J_\varphi(x) \ne 0$ for $x \in (A \setminus E)^0$. Therefore, by Theorem 45.4 applied to $A \setminus E$, we deduce that $\varphi(A \setminus E)$ has content. It follows from Theorem 45.2 that $\varphi(E)$ has content zero, and since $\varphi(A) = \varphi((A \setminus E) \cup (A \cap E)) = \varphi(A \setminus E) \cup \varphi(A \cap E)$, that $\varphi(A)$ has content. Since f is bounded on $\varphi(A)$ and continuous except on a subset of $\varphi(E)$, we deduce that f is integrable over $\varphi(A)$. Moreover, since $f \circ \varphi$ is continuous except on a subset of E, we deduce that $(f \circ \varphi) |J_\varphi|$ is integrable over A. It remains to show that these integrals are equal.

Now apply Lemma 45.1 to E to obtain a bounded open set Ω_1 with $E \subseteq \Omega_1 \subseteq \Omega_1^- \subseteq \Omega$ and a constant $M_1 > 0$, with the property that if E is contained in a finite union of closed cubes in Ω_1 with content at most $\alpha > 0$, then $\varphi(E)$ is contained in a corresponding finite union of closed cubes with content at most $(\sqrt{p}\,M_1)^p\alpha$.

We now let $\varepsilon>0$ be given and enclose E in a finite union U_ε of open cubes in Ω_1 with $c(U_\varepsilon)<\varepsilon$ and such that the union W_ε of the closures of the cubes in U_ε is still contained in Ω_1. Then $c(W_\varepsilon)<\varepsilon$ and it follows from Lemma 45.1 that $c(\varphi(U_\varepsilon))\le c(\varphi(W_\varepsilon))\le(\sqrt{p}\,M_1)^p\varepsilon$. We now let $B=A\setminus U_\varepsilon$ so that B has content. Now since U_ε is open and contains $b(\Omega_0)$ and E, we infer that $B^-\subseteq\Omega_0\setminus E$. We now apply the Change of Variables Theorem 45.9 to B, $\Omega_0\setminus E$ in place of A, Ω, to get

$$\int_{\varphi(B)} f = \int_B (f\circ\varphi)\,|J_\varphi|.$$

Now it is readily seen that $\varphi(A)\setminus\varphi(B)\subseteq\varphi(A\cap U_\varepsilon)$ whence

$$\left|\int_{\varphi(A)} f - \int_{\varphi(B)} f\right| \le \left|\int_{\varphi(A\cap U_\varepsilon)} f\right| \le M_f c(\varphi(A\cap U_\varepsilon))$$

$$\le (\sqrt{p}\,M_1)^p M_f \varepsilon.$$

Similarly we have that

$$\left|\int_A (f\circ\varphi)\,|J_\varphi| - \int_B (f\circ\varphi)\,|J_\varphi|\right| \le \int_{A\cap U_\varepsilon} (f\circ\varphi)\,|J_\varphi|$$

$$\le M_f M_J c(A\cap U_\varepsilon) \le M_f M_J \varepsilon.$$

It follows that

$$\left|\int_{\varphi(A)} f - \int_A (f\circ\varphi)\,|J_\varphi|\right| \le [(\sqrt{p}\,M_1)^p M_f + M_f M_J]\varepsilon.$$

Since $\varepsilon>0$ is arbitrary, the conclusion follows. \hfill Q.E.D.

For polar coordinates, we take Ω_0 to be an open set with content contained in $(0,+\infty)\times(0,2\pi)$. For spherical coordinates, we take Ω_0 to be an open set with content contained in $(0,+\infty)\times(0,2\pi)\times(0,\pi)$.

Exercises

45.A. Let $\Omega\subseteq\mathbf{R}^p$ be an open set and let $f:\Omega\to\mathbf{R}^q$ satisfy a Lipschitz condition on Ω; that is, for some $M>0$, $\|f(x)-f(y)\|\le M\|x-y\|$ for all $x,y\in\Omega$. If $K\subseteq\Omega$ is a cube with side length $s>0$, show that $f(K)$ is contained in a cube with side length $M\sqrt{p}s$. Show that if $A\subseteq\Omega$ is a compact set with content zero, then $f(A)$ has content zero, and if $B\subseteq\Omega$ is a compact set with content, then $f(B)$ has content.

45.B. Consider the polar coordinate map $(x,y)=\varphi(r,\theta)=(r\cos\theta, r\sin\theta)$ defined on $\mathbf{R}^2$, and its behavior on the set $A=[0,1]\times[0,2\pi]$. Use Theorem 45.4 to obtain the reassuring information that the image $D=\varphi(A)$, which is the unit disk $D=\{(x,y):x^2+y^2\le1\}$, has content. Investigate the manner in which φ maps the boundary of A. Show that the boundary of D is the image under φ of only one side of A, and that the other three sides of A get mapped into the interior of D.

45.C. Consider the map $(x, y) = \psi(u, v) = (\sin u, \sin v)$ defined on $\mathbf{R}^2$. Determine the image of the boundary of $B = [-\frac{3}{4}\pi, \frac{3}{4}\pi] \times [-\frac{3}{4}\pi, \frac{3}{4}\pi]$ under ψ, and the boundary of $\psi(B)$. Show that most, though not quite all, of the boundary points of $\psi(B)$ are images of interior points of B.

45.D. Given that the area of the circular disk $\{(x, y) : x^2 + y^2 \le 1\}$ is equal to π, find the areas of the elliptical disks given by:

(a) $\left\{(x, y) : \dfrac{x^2}{4} + \dfrac{y^2}{9} \le 1\right\}$;

(b) $\{(x, y) : 2x^2 + 2xy + 5y^2 \le 1\}$.

(Hint: $2x^2 + 2xy + 5y^2 = (x + 2y)^2 + (x - y)^2$.)

45.E. Let B be the set $\{(x, y) : 0 \le x, 0 \le y, 1 \le x + y \le 2\}$. Let $u = x + y$, $v = y$ so that B is the image under the map $(x, y) = \varphi(u, v) = (u - v, v)$ of the trapezoid $C = \{(u, v) : 1 \le u \le 2, 0 \le v \le u\}$. Show that φ is injective on all of $\mathbf{R}^2$ and that $J_\varphi(u, v) = 1$. Deduce that

$$\iint_B (x + y)\, d(x, y) = \iint_C u\, d(u, v) = \tfrac{7}{3}.$$

45.F. Let $B = \{(u, v) : 0 \le u + v \le 2, 0 \le v - u \le 2\}$. By using the transformation $(x, y) \mapsto (u, v) = (x - y, x + y)$, evaluate the integral

$$\iint_B (v^2 - u^2) e^{(u^2 + v^2)/2}\, d(u, v).$$

45.G. Evaluate the iterated integral

$$\int_1^3 \left\{ \int_{x^2}^{x^2+1} xy\, dy \right\} dx$$

directly. Then use the transformation $(x, y) \mapsto (u, v) = (x, y - x^2)$ to evaluate this integral.

45.H. Determine the area of the region bounded by the curves

$$xy = 1, \quad xy = 2, \quad y = x^2, \quad y = 2x^2$$

by introducing an appropriate change of variable.

45.I. Let $\psi : \mathbf{R}^2 \to \mathbf{R}^2$ be defined by $(u, v) = \psi(x, y) = (x^2 - y^2, x^2 + y^2)$. Note that the inverse image under ψ of the line $u = a > 0$ is a hyperbola, and the inverse image under ψ of the line $v = c > 0$ is a circle. Show that ψ is not injective on $\mathbf{R}^2$, but its restriction to $Q = \{(x, y) : x > 0, y > 0\}$ is an injective map onto $\{(u, v) : v > |u|\}$. Let φ be the inverse of the restriction $\psi \mid Q$ and show that if $0 < a < b < c < d$, then φ maps the rectangle $A = [a, b] \times [c, d]$ into the region

$$\varphi(A) = \{(x, y) : a \le x^2 - y^2 \le b, c \le x^2 + y^2 \le d\}.$$

Show that if $f : Q \to \mathbf{R}$ is continuous, then

$$\iint_{\varphi(A)} f(x, y)\, d(x, y) = \iint_A f\left(\left(\frac{u + v}{2}\right)^{1/2}, \left(\frac{v - u}{2}\right)^{1/2}\right) \frac{1}{4(v^2 - u^2)^{1/2}}\, d(u, v).$$

In particular, we have

$$\iint_{\varphi(A)} xy \, d(x, y) = \iint_A \tfrac{1}{8} \, d(u, v) = \tfrac{1}{8}(b-a)(d-c).$$

45.J. Let $\psi: \mathbf{R}^2 \to \mathbf{R}$ be as in the preceding exercise. Show that ψ maps the triangular region $\Delta = \{(x, y): 0 \le x \le 1, 0 \le y \le x\}$ into the triangular region

$$\Delta_1 = \psi(\Delta) = \{(u, v): 0 \le u \le 1, u \le v \le 2-u\}.$$

Here $J_\psi(x, y) = 8xy$. If $\Omega_0 = (0, 2) \times (0, 2)$, and if f is continuous on Δ_1, apply Theorem 45.11 to show that

$$\iint_{\Delta_1} f(u, v) \, d(u, v) = \iint_\Delta f \circ \psi(x, y) \, |J_\psi(x, y)| \, d(x, y).$$

In particular show that

$$\iint_\Delta (x^2 - y^2)(x^2 + y^2)^{1/2} xy \, d(x, y) = \frac{1}{8} \iint_{\Delta_1} uv^{1/2} \, d(u, v).$$

45.K. Let $\alpha < \beta$ belong to $[0, 2\pi]$ and let $h: [\alpha, \beta] \to \mathbf{R}$ be continuous and such that $h(\theta) \ge 0$ for $\theta \in [\alpha, \beta]$. Let $H = \{(\theta, r) \in \mathbf{R}^2 : \alpha \le \theta \le \beta, 0 \le r \le h(\theta)\}$ be the ordinate set of h (see Exercise 44.O), so that H has content. The **polar curve** generated by h is the curve in $\mathbf{R}^2$ defined by $\theta \mapsto (h(\theta) \cos \theta, h(\theta) \sin \theta)$, and the **polar ordinate set** of this curve is the set

$$H_1 = \{(r \cos \theta, r \sin \theta) \in \mathbf{R}^2 : \alpha \le \theta \le \beta, 0 \le r \le h(\theta)\}.$$

Note that H_1 is the image of H under the (reversed) polar map $\varphi_1(\theta, r) = (r \cos \theta, r \sin \theta)$ and use Theorem 45.11 to show that

$$c(H_1) = \frac{1}{2} \int_\alpha^\beta (f(\theta))^2 \, d\theta.$$

45.L. Let $a < b$ and let $f: [a, b] \to \mathbf{R}$ be continuous and such that $f(x) \ge 0$ for all $x \in [a, b]$. As in Exercise 44.O, let $S_f = \{(x, y): a \le x \le b, 0 \le y \le f(x)\}$ be the **ordinate set** of f. Let $\rho_x: \mathbf{R}^3 \to \mathbf{R}^3$ be defined by $\rho_x(x, y, \theta) = (x, y \cos \theta, y \sin \theta)$ and let X_f be the image of $S_f \times [0, 2\pi]$ under ρ_x. (The set X_f is called the "solid of revolution obtained by revolving the ordinate set S_f about the x-axis.") Use Theorem 45.11 to show that

$$c(X_f) = \pi \int_a^b (f(x))^2 \, dx.$$

45.M. Let $0 \le a < b$ and let $f: [a, b] \to \mathbf{R}$ and S_f be as in the preceding exercise. Let $\rho_y: \mathbf{R}^3 \to \mathbf{R}^3$ be defined by $\rho_y(x, y, \theta) = (x \cos \theta, y, x \sin \theta)$ and let Y_f be the image of $S_f \times [0, 2\pi]$ under ρ_y. (The set Y_f is called the "solid of revolution obtained by revolving the ordinate set S_f about the y-axis.") Use Theorem 45.11 to show that

$$c(Y_f) = 2\pi \int_a^b xf(x) \, dx.$$

45.N. (a) By changing to polar coordinates, show that

$$\iint_{C_R} e^{-(x^2+y^2)}\,d(x, y) = \frac{\pi}{4}(1 - e^{-R^2}),$$

where $C_R = \{(x, y): 0 \le x, 0 \le y, x^2 + y^2 \le R^2\}$.

(b) If $B_L = \{(x, y): 0 \le x \le L, 0 \le y \le L\}$, show that

$$\iint_{B_L} e^{-(x^2+y^2)}\,d(x, y) = \left(\int_0^L e^{-x^2}\,dx\right)^2.$$

(c) From the fact that $C_R \subseteq B_R \subseteq C_{R\sqrt{2}}$, show that

$$\lim_{R \to \infty} \left(\int_0^R e^{-x^2}\,dx\right)^2 = \frac{\pi}{4},$$

whence it follows that $\int_0^{+\infty} e^{-x^2}\,dx = \frac{1}{2}\sqrt{\pi}$.

45.O. Let $B = \{(x, y): 4x^2 + 9y^2 \le 4\}$. Use an appropriate change of variables to evaluate

$$\iint_B e^{-(4x^2+9y^2)}\,d(x, y) = \frac{\pi}{6}(1 - e^{-4}).$$

45.P. Observe that the set $\{(x, y, z): 0 \le x^2 + y^2 \le \frac{1}{2},\ (x^2+y^2)^{1/2} \le z \le (1 - x^2 - y^2)^{1/2}\}$ is a "conical sector cut out of the unit ball" in $\mathbf{R}^3$. Obtain this set as the image under the spherical coordinate map Φ of the cell $[0, 1] \times [0, 2\pi] \times [0, \frac{1}{4}\pi]$.

(a) Show that the content of this set in $\mathbf{R}^3$ equals $\pi(2 - \sqrt{2})/3$.

(b) Obtain the content of this set by using the cylindrical coordinate map $\Gamma: (r, \theta, z) \to (x, y, z) = (r \cos \theta, r \sin \theta, z)$.

45.Q. Let $a > 0$ and let A be the intersection of the sets

$$\{(x, y, z): x^2 + y^2 + z^2 \le 4a^2\} \quad \text{and} \quad \{(x, y, z): z \ge a\}.$$

(a) Use the spherical coordinate map to show that $c(A) = 5\pi a^3/3$.

(b) Use the cylindrical coordinate map to evaluate $c(A)$.

45.R. Let B be the intersection of the sets

$$\{(x, y, z): x^2 + y^2 + z^2 \le 2\} \quad \text{and} \quad \{(x, y, z): x^2 + y^2 + z^2 \le 2z\}.$$

(a) Use the spherical coordinate map to show that $c(B) = \pi(4\sqrt{2} - 3)/3$.

(b) Use the cylindrical coordinate map to evaluate $c(B)$.

45.S. Let $B_p(r) = \{x \in \mathbf{R}^p: \|x\| \le r\}$ be the ball with radius $r > 0$ in the space $\mathbf{R}^p$. We shall compute the content $\omega_p(r)$ of $B_p(r)$.

(a) Use a change of variables to show that $\omega_p(r) = r^p \omega_p(1)$.

(b) If $p \ge 3$, express the integral for $\omega_p(1)$ as an iterated integral and use part (a) to show that

$$\omega_p(1) = \omega_{p-2}(1) \int_0^{2\pi} \left\{ \int_0^1 (1 - r^2)^{(p/2)-1} r\,dr \right\} d\theta$$

$$= \omega_{p-2}(1) 2\pi/p.$$

(c) Conclude that if $p = 2k$ is even, then $\omega_p(1) = \pi^k/k!$ If $p = 2k - 1$ is odd, then $\omega_p(1) = 4^k \pi^{k-1} k!/(2k)!$ In terms of the Gamma function, we have $\omega_p(1) = \pi^{p/2}/\Gamma(\frac{1}{2}p + 1)$.

(d) Obtain the remarkable result that $\lim(\omega_p(1)) = 0$.

45.T. We shall obtain the result of the preceding exercise in a different way. Let $p \in N$ and let $\sigma : \mathbf{R}^p \to \mathbf{R}^p$ be defined by $\sigma(\theta) = \sigma(\theta_1, \ldots, \theta_p) = (\cos\theta_1, \sin\theta_1\cos\theta_2, \ldots, \sin\theta_1\sin\theta_2 \cdots \sin\theta_{p-1}\cos\theta_p)$.

(a) Show that $\|\sigma(\theta)\|^2 \le 1$ and that $\|\sigma(\theta)\| = 1$ only when $\theta_j = 0$ or $\theta_j = \pi$ for some value of $j = 1, \ldots, p$.

(b) Show that σ is an injective map of $(0, \pi)^p = (0, \pi) \times \cdots \times (0, \pi)(p$ times$)$ onto the interior $\{x \in \mathbf{R}^p : \|x\| < 1\}$ of the unit ball $B_p(1)$. Show also that σ maps $[0, \pi]^p$ onto the unit ball, but that it is not injective on the boundary.

(c) Evaluating the Jacobian of σ, we obtain

$$J_\sigma(\theta) = (-1)^p (\sin\theta_1)^p (\sin\theta_2)^{p-2} \cdots (\sin\theta_{p-1})^2 (\sin\theta_p).$$

Hence, $J_\sigma(\theta) \ne 0$ for $\theta \in (0, \pi)^p$.

(d) Using the Wallis product formulas for $\int_0^{\pi/2}(\sin\theta)^k\,d\theta$ obtained in Project 30.γ, derive the expressions for $\omega_p(1)$ given in the preceding exercise.

Project

45.α. This project is based on Project 44.γ and provides an alternative approach to the Change of Variables Theorem 45.9. Let $\Omega \subseteq \mathbf{R}^p$ be open, and let $\varphi : \Omega \to \mathbf{R}^p$ belong to Class $C^1(\Omega)$, be injective on Ω, and such that $J_\varphi(x) \ne 0$ for all $x \in \Omega$. For simplicity, we also suppose that there exists $M > 0$ such that $\|\varphi(x) - \varphi(y)\| \le M\|x - y\|$ for $x, y \in \Omega$.

(a) If $\Phi : \mathscr{D}(\Omega) \to \mathbf{R}$ is defined by

$$\Phi(A) = c(\varphi(A)) \qquad \text{for } A \in \mathscr{D}(\Omega),$$

then Φ is additive on $\mathscr{D}(\Omega)$ and has strong density equal to $|J_\varphi|$. Moreover, for some $M_1 > 0$, we have $\Phi(A) \le M_1 c(A)$ for all $A \in \mathscr{D}(\Omega)$.

(b) If f is a bounded function which is integrable on every set $\varphi(A)$, for $A \in \mathscr{D}(\Omega)$, and if $\Psi : \mathscr{D}(\Omega) \to \mathbf{R}$ is defined by

$$\Psi(A) = \int_{\varphi(A)} f,$$

then Ψ is additive on $\mathscr{D}(\Omega)$. Moreover, for some $M_2 > 0$, we have $|\Psi(A)| \le M_2 c(A)$ for all $A \in \mathscr{D}(\Omega)$.

(c) If f is a bounded and continuous function on $\varphi(\Omega)$, and if Ψ is defined as in (b), show that Ψ has strong density equal to $(f \circ \varphi)|J_\varphi|$.

(d) If f is as in (c), show that

$$\int_{\varphi(A)} f = \int_A (f \circ \varphi)|J_\varphi| \qquad \text{for } A \in \mathscr{D}(\Omega).$$

REFERENCES

This list includes books and articles that were cited in the text and some additional references that will be useful for further study.

Apostol, T. M., *Mathematical Analysis*, Second Edition, Addison-Wesley, Reading, Mass., 1974.

Bartle, R. G., *The Elements of Integration*, Wiley, New York, 1966.

Boas, R. P., Jr., *A Primer of Real Functions*, Carus Monograph Number 13, Math. Assn. of America, 1960.

Bruckner, A. M., "Differentiation of Integrals," *Amer. Math. Monthly*, Vol. 78, No. 9, Part II, 1–51 (1971). (H. E. Slaught Memorial Paper, Number 12.)

Burkill, J. C., and H. Burkill, *A Second Course in Mathematical Analysis*, Cambridge Univ. Press, Cambridge, 1970.

Cartan, H. P., *Cours de Mathématiques*, I. *Calcul Différentiel*; II. *Formes Différentielles*, Hermann, Paris, 1967. (English translation, Houghton-Mifflin, Boston, 1971.)

Cheney, E. L., *Introduction to Approximation Theory*, McGraw-Hill, New York, 1966.

Dieudonné, J., *Foundations of Modern Analysis*, Academic Press, New York, 1960.

Dunford, N., and J. T. Schwartz, *Linear Operators*, Part I, Wiley-Interscience, New York, 1958.

Finkbeiner, D. T., II, *Introduction to Matrices and Linear Transformations*, Second Edition, W. H. Freeman, San Francisco, 1966.

Gelbaum, B. R., and J. M. H. Olmsted, *Counterexamples in Analysis*, Holden-Day, San Francisco, 1964.

Halmos, P. R., *Naive Set Theory*, Van Nostrand, Princeton, 1960. (Republished by Springer-Verlag, New York, 1974.)

Hamilton, N. T., and J. Landin, *Set Theory*, Allyn-Bacon, Boston, 1961.

Hardy, G. H., J. E. Littlewood, and G. Pólya, *Inequalities*, Second Edition, Cambridge University Press, Cambridge, 1959.

Hewitt, E., and K. Stromberg, *Real and Abstract Analysis*, Springer-Verlag, New York, 1965.

Hoffman, K., and R. Kunze, *Linear Algebra*, Prentice-Hall, Englewood Cliffs, 1961.

Kelley, J. L., *General Topology*, Van Nostrand, New York, 1955.

Knopp, K., *Theory and Application of Infinite Series* (English translation), Hafner, New York, 1951.

Lefschetz, S., *Introduction to Topology*, Princeton University Press, Princeton, 1949.

Luxemberg, W. A. J., "Arzelà's Dominated Convergence Theorem for the Riemann Integral," *Amer. Math. Monthly*, Vol. 78, 970–979 (1971).

McShane, E. J., "A Theory of Limits," published in *MAA Studies in Mathematics*, Vol. 1, R. C. Buck, editor, Math. Assn. America, 1962.

———, "The Lagrange Multiplier Rule," *Amer. Math. Monthly*, Vol. 80, 922–925 (1973).

Royden, H. L., *Real Analysis*, Second Edition, Macmillan, New York, 1968.

Rudin, W., *Principles of Mathematical Analysis*, Second Edition, McGraw-Hill, New York, 1964.

Schwartz, J., "The Formula for Change of Variables in a Multiple Integral," *Amer. Math. Monthly*, Vol. 61, 81–85 (1954).

Simmons, G. F., *Introduction to Topology and Modern Analysis*, McGraw-Hill, New York, 1963.

Spivak, M., *Calculus on Manifolds*, W. A. Benjamin, New York, 1965.

Stone, M. H., "The Generalized Weierstrass Approximation Theorem," *Mathematics Magazine*, Vol. 21, 167–184, 237–254 (1947/48). (Reprinted in *MAA Studies in Mathematics*, Vol. 1, R. C. Buck, editor, Math. Assn. America, 1962.)

Suppes, P., *Axiomatic Set Theory*, Van Nostrand, Princeton, 1961.

Titchmarsh, E. C., *The Theory of Functions*, Second Edition, Oxford University Press, London, 1939.

Varberg, D. E., "Change of Variables in Multiple Integrals," *Amer. Math. Monthly*, Vol. 78, 42–45 (1971).

Woll, J. W., Jr., *Functions of Several Variables*, Harcourt, Brace and World, New York, 1966.

Wilder, R. L., *The Foundations of Mathematics*, Wiley, New York, 1952.

HINTS FOR SELECTED EXERCISES

The reader is urged not to look at these hints unless he is stymied. Many of the exercises call for proofs, and there is no single way that is correct; even if the reader has given a totally different argument, his may be entirely correct. However, in order to help the reader learn the material and to develop his technical skill, some hints and a few solutions are offered. It will be observed that more detail is presented for the early material.

Section 1

1.D. By definition $A \cap B \subseteq A$. If $A \subseteq B$, then $A \cap B \supseteq A$ so that $A \cap B = A$. Conversely, if $A \cap B = A$, then $A \cap B \supseteq A$ whence it follows that $B \supseteq A$.

1.E, F. The symmetric difference of A and B is the union of $\{x : x \in A \text{ and } x \notin B\}$ and $\{x : x \notin A \text{ and } x \in B\}$.

1.H. If x belongs to $E \cap \bigcup A_j$, then $x \in E$ and $x \in \bigcup A_j$. Therefore, $x \in E$ and $x \in A_j$ for at least one j. This implies that $x \in E \cap A_j$ for at least one j, so that

$$E \cap \bigcup A_j \subseteq \bigcup (E \cap A_j).$$

The opposite inclusion is proved by reversing these steps. The other equality is handled similarly.

1.L. If $x \in \mathscr{C}(\bigcap\{A_j : j \in J\})$, then $x \notin \bigcap\{A_j : j \in J\}$. This implies that there exists a $k \in J$ such that $x \notin A_k$. Therefore, $x \in \mathscr{C}(A_k)$, and hence $x \in \bigcup\{\mathscr{C}(A_j) : j \in J\}$. This proves that $\mathscr{C}(\bigcap A_j) \subseteq \bigcup \mathscr{C}(A_j)$. The opposite inclusion is proved by reversing these steps. The other equality is similar.

Section 2

2.A. If (a, c) and (a, c') belong to $g \circ f$, then there exist b, b' in B such that (a, b), (a, b') belong to f and (b, c), (b', c') belong to g. Since f is a function, $b = b'$; since g is a function, $c = c'$.

2.B. No. Both $(0, 1)$ and $(0, -1)$ belong to C.

2.D. Let $f(x) = 2x$, $g(x) = 3x$.

2.E. If (b, a), (b, a') belong to f^{-1}, then (a, b), (a', b) belong to f. Since f is injective, then $a = a'$. Hence f^{-1} is a function.

2.G. If $f(x_1) = f(x_2)$ then $x_1 = g \circ f(x_1) = g \circ f(x_2) = x_2$. Hence f is injective.

2.H. Apply Exercise 2.G twice.

Section 3

3.A. Let $f(n) = n/2$, $n \in E$.

3.B. Let $f(n) = (n + 1)/2$, $n \in O$.

3.C. Let $f(n) = n + 1$, $n \in N$.

3.E. Let $A_n = \{n\}$, $n \in N$. Then each set A_n has a single point, but $N = \bigcup \{A_n : n \in N\}$ is infinite.

3.F. If A is infinite and $B = \{b_n : n \in N\}$ is a subset of A, then the function defined by

$$f(x) = b_{n+1}, \qquad x = b_n \in B,$$
$$= x, \qquad x \in A \setminus B.$$

is one-one and maps A onto $A \setminus \{b_1\}$.

3.H. If f is a one-one map of A onto B and g is a one-one map of B onto C, then $g \circ f$ is a one-one map of A onto C.

Section 4

4.G. Consider three cases: $p = 3k$, $p = 3k + 1$, $p = 3k + 2$.

Section 5

5.A. Since $a^2 \geq 0$ and $b^2 \geq 0$, then $a^2 + b^2 = 0$ implies that $a^2 = b^2 = 0$.

5.D. If $c = 1 + a$ with $a > 0$, then $c^n = (1 + a)^n \geq 1 + na \geq 1 + a = c$.

5.G. Observe that $1 < 2^1 = 2$. If $k < 2^k$ for $k \geq 1$, then $k + 1 \leq 2k < 2 \cdot 2^k = 2^{k+1}$. Therefore, $n < 2^n$ for all $n \in N$.

5.H. Note that $b^n - a^n = (b - a)(b^{n-1} + \cdots + a^{n-1}) = (b - a)p$, where $p > 0$.

5.M. $\{(x, y) : y = \pm x\}$.

5.N. A square with vertices $(\pm 1, 0)$, $(0, \pm 1)$.

Section 6

6.A. If $A = \{x_1\}$, then $x_1 = \sup A$. If $A = \{x_1, \ldots, x_n, x_{n+1}\}$ and if $u = \sup \{x_1, \ldots, x_n\}$, show that $\sup \{u, x_{n+1}\}$ is the supremum of A.

6.C. Let $S = \{x \in Q : x^2 < 2\}$.

6.E. In fact, $\sup A \cup B = \sup \{\sup A, \sup B\}$.

6.H. If $S = \sup \{f(x, y) : x \in X, y \in Y\}$, then $f(x, y) \leq S$ for all $x \in X$, $y \in Y$, and so $f_1(x) \leq S$ for all $x \in X$. Hence $\sup \{f_1(x) : x \in X\} \leq S$. Conversely, if $\varepsilon > 0$ there exists (x_0, y_0) such that $S - \varepsilon < f(x_0, y_0)$. Hence $S - \varepsilon < f_1(x_0)$ and therefore $S - \varepsilon < \sup \{f_1(x) : x \in X\}$. Since $\varepsilon > 0$ is arbitrary, we infer that $S \leq \sup \{f_1(x) : x \in X\}$.

6.K. Since $f(x) \leq \sup \{f(z) : z \in X\}$, it follows that

$$f(x) + g(x) \leq \sup \{f(z) : z \in X\} + \sup \{g(z) : z \in X\}.$$

Therefore $\sup\{f(x)+g(x):x\in X\}$ is less than or equal to the right hand side. Similarly, if $x\in X$, then

$$\inf\{f(z):z\in X\}+g(x)\le f(x)+g(x).$$

If we use 6.J, we infer that

$$\inf\{f(z):z\in X\}+\sup\{g(x):x\in X\}\le \sup\{f(x)+g(x):x\in X\}.$$

The other assertions are proved in a similar way.

Section 7

7.B. Let $a\in A$; if $a\notin A'$ then $a\in B'$ and so $\xi<\xi'\le a$, a contradiction. Therefore $a\in A'$ and since $a\in A$ is arbitrary we have $A\subseteq A'$. Since $\xi<\xi'$, there exists $x\in \mathbf{R}$ with $\xi<x<\xi'$. Since $\xi<x$, we must have $x\in B$. But since $x\in A'$ we infer that $A\ne A'$.

7.C. Let $A=\{x:x<1\}$, $B=\{x:x\ge 1\}$ and $A'=\{x:x\le 1\}$, $B'=\{x:x>1\}$.

7.E. If $x\in I_n$ for all n, we have a contradiction to the Archimedean Property 6.6.

7.F. If $x\in J_n$ for all n, we have a contradiction to Corollary 6.7(b).

7.H. Every element in F_1 has a ternary expansion whose first digit is either 0 or 2. The points in the four subintervals of F_2 have ternary expansions beginning

$$0.00\ldots,0.02\ldots,0.20\ldots,0.22\ldots,$$

and so forth.

7.J. If n is sufficiently large, $1/3^n<b-a$.

7.K. As close to 1 as desired.

Section 8

8.E. Property 8.3(ii) is not satisfied.

8.H. The set S_1 is the interior of the square with vertices $(0,\pm 1)$, $(\pm 1,0)$ and S_2 is the interior of the square with vertices $(1,\pm 1)$, $(-1,\pm 1)$.

8.K. Take $a=1/\sqrt{p}$, $b=1$.

8.L. Take $a=1/p$, $b=1$.

8.M. We have $|x\cdot y|\le \sum |x_i||y_i|\le \{\sum |x_i|\}\sup|y_i|\le \|x\|_1\|y\|_1$. But $|x\cdot y|\le p\|x\|_\infty\|y\|_\infty$ and if $x=y=(1,1,\ldots,1)$, equality is attained.

8.N. The stated relation implies that

$$\|x\|^2+2(x\cdot y)+\|y\|^2=\|x+y\|^2=(\|x\|+\|y\|)^2$$
$$=\|x\|^2+2\|x\|\,\|y\|+\|y\|^2.$$

Hence $x\cdot y=\|x\|\,\|y\|$ and the condition for equality in Theorem 8.7 holds provided the vectors are non-zero.

8.P. Since $\|x+y\|^2=\|x\|^2+2(x\cdot y)+\|y\|^2$, the stated relation holds if and only if $x\cdot y=0$.

8.Q. A set K is convex if and only if it contains the line segment joining any two points in K. If $x,y\in K_1$, then $\|tx+(1-t)y\|\le t\|x\|+(1-t)\|y\|\le t+(1-t)=1$ so

$tx + (1 - t)y \in K_1$ for $0 \le t \le 1$. The points $(\pm 1, 0)$ belong to K_4, but their midpoint $(0, 0)$ does not belong to K_4.

8.R. If x, y belong to $\bigcap K_\alpha$, then x, $y \in K_\alpha$ for all α. Hence $tx + (1 - t)y \in K_\alpha$ for all α; whence it follows that $\bigcap K_\alpha$ is convex. Consider the union of two disjoint intervals.

Section 9

9.A. If $x \in G$, let $r = \inf\{x, 1 - x\}$. Then, if $|y - x| < r$, we have $x - r < y < x + r$ whence $0 \le x - r < y < x + r \le 1$, so that $y \in G$. If $z = 0$, then there does *not* exist a real number $r > 0$ such that every point y in $\mathbf{R}$ satisfying $|y| < r$ belongs to F. Similarly for $z = 1$.

9.B. If $x \in G$, take $r = 1 - \|x\|$. If $x \in H$, take $r = \inf\{\|x\|, 1 - \|x\|\}$. If $z = (1, 0)$, then for any $r > 0$, there is a point y in $\mathscr{C}(F)$ such that $\|y - z\| < r$.

9.G. Enumerate the points in the open set with all coordinates rational numbers. Then proceed as in the proof of Theorem 9.11 using open balls with center at these rational points.

9.H. Argue as in the preceding exercise, but this time use closed balls.

9.I. Take complements and apply 9.H.

9.J. The set A° is the union of the collection of all open sets in A. Hence any open set $G \subseteq A$ must be contained in A°. By its definition we must have $A^\circ \subseteq A$. It follows that $(A^\circ)^\circ \subseteq A^\circ$. Since A° is open and $(A^\circ)^\circ$ is the union of all open sets in A°, we must have $A^\circ \subseteq (A^\circ)^\circ$. Therefore $(A^\circ)^\circ = A^\circ$. Since $A^\circ \subseteq A$ and $B^\circ \subseteq B$ it follows that $A^\circ \cap B^\circ \subseteq A \cap B$; but since $A^\circ \cap B^\circ$ is open this implies that $(A^\circ \cap B^\circ) \subseteq (A \cap B)^\circ$. On the other hand $(A \cap B)^\circ$ is an open set and is contained in A and B; therefore $(A \cap B)^\circ \subseteq A^\circ$ and $(A \cap B)^\circ \subseteq B^\circ$, whence $(A \cap B)^\circ \subseteq A^\circ \cap B^\circ$. Consequently $A^\circ \cap B^\circ = (A \cap B)^\circ$. Since $\mathbf{R}^p$ is open, $(\mathbf{R}^p)^\circ = \mathbf{R}^p$. Let A be the set of all rational numbers in $(0, 1)$ and B be the set of all irrational numbers in $(0, 1)$. Then $A^\circ \cup B^\circ = \emptyset$, while $(A \cup B)^\circ = (0, 1)$.

9.L. Either argue as in 9.J, or take complements and use 9.J.

9.N. If $p = 1$, take $A = \mathbf{Q}$. In $\mathbf{R}^p$, take $\mathbf{Q}^p$.

9.O. Suppose A, B are open in $\mathbf{R}$. Let $(x, y) \in A \times B$, so that $x \in A$ and $y \in B$. There exist $r > 0$ such that if $|x' - x| < r$ then $x' \in A$ and $s > 0$ such that if $|y' - y| < s$ then $y' \in B$. Now let $t = \inf\{r, s\}$; the open ball with radius t is contained in $A \times B$. The converse is similar.

Section 10

10.C. If x is a cluster point of A in $\mathbf{R}^p$ and N is a neighborhood of x, then $N \cap \{y \in \mathbf{R}^p : \|y - x\| < 1\}$ contains a point $a_1 \in A$, $a_1 \neq x$. The set $N \cap \{y \in \mathbf{R}^p : \|y - x\| < \|a_1\|\}$ contains a point $a_2 \in A$, $a_2 \neq x$ and also $a_2 \neq a_1$. Continue this process.

10.F. Every neighborhood of x contains infinitely many points of $A \cup B$. Hence either A or B (or possibly both) must possess an infinite number of elements in this neighborhood.

Section 11

11.A. Let $G_n = \{(x, y) : x^2 + y^2 < 1 - 1/n\}$ for $n \in N$.

11.B. Let $G_n = \{(x, y) : x^2 + y^2 < n^2\}$ for $n \in N$.

11.C. Let $\mathscr{G} = \{G_\alpha\}$ be an open covering for F and let $G = \mathscr{C}(F)$, so that G is open in R^p. If $\mathscr{G}_1 = \mathscr{G} \cup \{G\}$, then $\mathscr{G}_1$ is an open covering for K; hence K has a finite subcovering $\{G, G_\alpha, G_\beta, \ldots, G_\omega\}$. Then $\{G_\alpha, G_\beta, \ldots, G_\omega\}$ forms a subcovering of $\mathscr{G}$ for the set F.

11.D. Observe that if G is open in R, then there exists an open subset G_1 of R^2 such that $G = G_1 \cap R$. Alternatively, use the Heine-Borel Theorem.

11.E. Let $\mathscr{G} = \{G_\alpha\}$ be an open covering of the closed unit interval J in R^2. Consider those real numbers x such that the square $[0, x] \times [0, x]$ is contained in the union of a finite number of sets in $\mathscr{G}$ and let x^* be their supremum.

11.G. Let $x_n \in F_n$, $n \in N$. If there are only a finite number of points in $\{x_n : n \in N\}$, then at least one of them occurs infinitely often and is a common point. If there are infinitely many points in the bounded set $\{x_n\}$, then there is a cluster point x. Since $x_m \in F_n$ for $m \geq n$ and since F_n is closed, then $x \in F_n$ for all $n \in N$.

11.H. If $d(x, F) = 0$, then x is a cluster point of the closed set F.

11.J. No. Let $F = \{y \in R^p : \|y - x\| = r\}$, then every point of F has the same distance to x.

11.K. Let G be an open set and let $x \in R^p$. If $H = \{y - x : y \in G\}$, then H is an open set in R^p.

11.M. Follow the argument in 11.7, except use open cells instead of open balls.

11.Q. Suppose that $Q = \bigcap \{G_n : n \in N\}$, where G_n is open in R. The complement F_n of G_n is a closed set which does not contain any non-empty open subset, by Theorem 6.10. Hence the set of irrationals is the union of a countable family of closed sets not one of which contains a non-empty open set; but this contradicts Exercise 11.P.

Section 12

12.B. Let A, B be a disconnection for $C' = C \cup \{x\}$. Then $A \cap C'$ and $B \cap C'$ are disjoint, non-empty, and have union C'. One of these sets must contain x; suppose it is B. Since B is an open set, it also contains points of C so $C \cap (B \setminus \{x\}) \neq \emptyset$. But then A, $B \setminus \{x\}$ form a disconnection of C.

12.E. Modify the proof of Theorem 12.4.

12.G. By Theorem 12.8, the sets C_1 and C_2 are intervals. It is easily seen that $C_1 \times C_2$ is convex so 12.E applies.

Section 13

13.A. Examine the geometrical position of $iz = (-y, x)$ in terms of $z = (x, y)$.

13.B. Note that $cz = (x \cos \theta - y \sin \theta, x \sin \theta + y \cos \theta)$, and this corresponds to a counter-clockwise rotation of θ radians about the origin.

13.C. The circle $|z - c| = r$ is mapped into the circle $|w - (ac + b)| = |a| r$. We can write $z = a^{-1} w - a^{-1} b$ and calculate $x = \operatorname{Re} z$, $y = \operatorname{Im} z$ in terms of $u = \operatorname{Re} w$, $v = \operatorname{Im} w$. Doing so we easily see that the equation $ax + by = c$ transforms into an equation of the form $Au + Bv = C$.

13.D. A circle is left fixed by g if and only if its center lies on the real axis. The only lines left fixed by g are the real and the imaginary axis.

13.E. Circles passing through the origin are sent into lines by h. All lines not passing through the origin are sent into circles passing through the origin; all lines passing through the origin are sent into lines passing through the origin.

13.F. Every point of C, except the origin, is the image under g of two elements of C. If Re $g(z) = k$, then $x^2 - y^2 = k$. If Im $g(z) = k$, then $2xy = k$. If $|g(z)| = k$, then $k \geq 0$ and $|z| = \sqrt{k}$.

Section 14

14.B. Note that $0 \leq \dfrac{1}{n} - \dfrac{1}{n+1} = \dfrac{1}{n(n+1)} \leq \dfrac{1}{n}$.

14.D. We have $0 \leq |\,\|x_n\| - \|x\|\,| \leq \|x_n - x\|$.

14.I. Let $r \in R$ be such that $\lim (x_{n+1}/x_n) < r < 1$. Since the interval $(-1, r)$ is a neighborhood of this limit, there exists $K \in N$ such that $0 < x_{n+1}/x_n < r$ for all $n \geq K$. Now show that $0 < x_n < Cr^n$ for some C and $n \geq K$.

14.K. Consider $(1/n)$ and (n).

14.L. Sequences (a), (b), (e), and (f) converge; sequences (c) and (d) diverge.

14.M. Let $r \in R$ be such that $\lim (x_n^{1/n}) < r < 1$. Since the interval $(-1, r)$ is a neighborhood of this limit, there exists $K \in N$ such that $0 < x_n^{1/n} < r$, whence $0 < x_n < r^n$ for all $n \geq K$.

Section 15

15.A. Consider $z_n = y_n - x_n$ and apply Example 15.5(c) and Theorem 15.6(a).

15.C. (a) Converges to 1. (b) Diverges. (f) Diverges.

15.D. Let $Y = -X$.

15.F. Consider two cases: $x = 0$ and $x > 0$.

15.G. Yes.

15.H. Use the hint in Exercise 15.F.

15.L. Observe that $b \leq x_n \leq b2^{1/n}$.

Section 16

16.A. By induction $1 < x_n < 2$ for $n \geq 2$. Since $x_{n+1} - x_n = (x_n - x_{n-1})/(x_n x_{n-1})$, the sequence is monotone.

16.C. The sequence is monotone and bounded. The limit is $(1 + (1 + 4a)^{1/2})/2$.

16.D. The sequence X is monotone decreasing and bounded.

16.E. An element x_k of $X = (x_n)$ is called a "peak" for X if $x_k \geq x_n$ for $n > k$.

(i) If there are infinitely many peaks with indices $k_1 < k_2 < \ldots$, then the sequence (x_{k_j}) of peaks is a decreasing subsequence of X.

(ii) If there are only a finite number of peaks with indices $k_1 < \cdots < k_r$, let $m_1 > k_r$. Since x_{m_1} is not a peak, there exists $m_2 > m_1$ such that $x_{m_1} < x_{m_2}$. Continuing in this way we obtain a strictly increasing subsequence (x_{m_i}) of X.

16.G. The sequence is increasing and $x_n \leq n/(n+1) < 1$.

16.K. There exists $K \in N$ such that if $n \geq K$, then $L - \varepsilon \leq x_{n+1}/x_n \leq L + \varepsilon$. Now use an argument similar to the one in Exercise 14.I.

16.M. (a) e, (b) $e^{1/2}$, (c) Hint: $(1 + 2/n) = (1 + 1/n)(1 + 1/(n+1))$, (d) e^3.

16.P. Let $y_n \in F$ be such that $\|x - y_n\| < d + 1/n$. If $y = \lim (y_n)$, then $\|x - y\| = d$.

Section 17

17.A. All.

17.C. If $x \in Z$, the limit is 1; if $x \notin Z$, the limit is 0.

17.E. If $x = 0$, the limit is 1; if $x \neq 0$, the limit is 0.

17.G. If $x > 0$ and $0 < \varepsilon < \pi/2$, then $\tan (\pi/2 - \varepsilon) > 0$. Therefore $nx \geq \tan (\pi/2 - \varepsilon)$ for all $n \geq n_*$, from which $\pi/2 - \varepsilon \leq \text{Arc} \tan nx \leq \pi/2$.

17.H. If $x > 0$, then $e^{-x} < 1$.   17.J. Not necessarily.

17.M. Consider the sequence $(1/n)$ or note that $\|f_n\|_D \geq \frac{1}{2}$.

17.P. Yes.   17.Q. Yes.

Section 18

18.A. (a) ± 1.   (b) 0.   (c) ± 1.   (d) ± 1.

18.E. Let m, $p \in N$, $p \leq m$. Then $v_m(X + Y) = \sup \{x_n + y_n : n \geq m\} \leq \sup \{x_n : n \geq m\} + \sup \{y_n : n \geq m\} = v_m(X) + v_m(Y) \leq v_p(X) + v_m(Y)$. Therefore

$$(x + y)^* = \inf \{v_m(X + Y) : m \in N\} \leq v_p(X) + y^*.$$

Since this is true for all $p \in N$, we infer that $(x + y)^* \leq x^* + y^*$.

18.G. (a) $\pm \infty$.   (c) 0, $+\infty$.

Section 19

19.E. If $j \leq n$, then $x_j \leq x_{n+1}$ and $x_j(1 + 1/n) \leq x_j + (1/n)x_{n+1}$. Now add.

19.I. If X is increasing and not convergent in R, then X is not bounded.

19.K. (a) None exist. (b, c) All three are equal. (d) The iterated limits are different and the double limit does not exist. (e) The double limit and one iterated limit are equal. (f) The iterated limits are equal, but the double limit does not exist.

19.L. Let $x_{mn} = n$ if $m = 1$ and $x_{mn} = 0$ if $m > 1$.

19.N. In (b, c, e).

19.O. Apply Corollary 19.7 to $x = \sup \{x_{mn} : m, n \in N\}$.

19.P. Let $x_{mn} = 0$ for $m < n$ and $x_{mn} = (-1)^m/n$ for $m \geq n$.

Section 20

20.A. If $a = 0$, take $\delta(\varepsilon) = \varepsilon^2$. If $a > 0$, use the estimate

$$|\sqrt{x} - \sqrt{a}| = \frac{|x - a|}{\sqrt{x} + \sqrt{a}} \leq \frac{|x - a|}{\sqrt{a}}.$$

20.B. Apply Example 20.5(b) and Theorem 20.6.

20.C. Apply Exercise 20.B and Theorem 20.6.

20.E. Show that $|f(x)-f(\tfrac{1}{2})|=|x-\tfrac{1}{2}|$.

20.F. Every real number is a limit of a sequence of rational numbers.

20.J. There exist sequences (x_n), (y_n) such that $\lim (h(x_n)) = 1$, $\lim (h(y_n)) = -1$.

20.L. Show that $f(a+h)-f(a) = f(h)-f(0)$. If f is monotone on $\mathbf{R}$, then it is continuous at some point.

20.M. Show that $f(0) = 0$ and $f(n) = nc$ for $n \in \mathbf{N}$. Also $f(n)+f(-n) = 0$, so $f(n) = nc$ for $n \in \mathbf{Z}$. Since $f(m/n) = mf(1/n)$, it follows on taking $m = n$ that $f(1/n) = c/n$, whence $f(m/n) = c(m/n)$. Now use the continuity of f.

20.N. Either $g(0) = 0$, in which case $g(x) = 0$ for all x in $\mathbf{R}$, or $g(0) = 1$, in which case $g(a+h)-g(a) = g(a)\{g(h)-g(0)\}$.

Section 21

21.C. $f(1, 1) = (3, 1, -1)$, $f(1, 3) = (5, 1, -3)$.

21.D. A vector (a, b, c) is in the range of f if and only if $a - 2b + c = 0$.

21.G. If $\Delta = 0$, then $f(-b, a) = (0, 0)$. If $\Delta \neq 0$, then the only solution of

$$ax + by = 0, \qquad cx + dy = 0$$

is $(x, y) = (0, 0)$.

21.I. Note that $g(x) = g(y)$, if and only if $g(x-y) = \theta$.

21.P. Note that $c_{ij} = e_i \cdot f(e_j)$ and apply the Schwarz Inequality.

Section 22

22.C. If $f(x_0) > 0$, then $V = \{y \in \mathbf{R} : y > 0\}$ is a neighborhood of $f(x_0)$.

22.H. Let $f(s, t) = 0$ if $st = 0$ and $f(s, t) = 1$ if $st \neq 0$.

22.L. Suppose the coefficient of the highest power is positive. Show that there exist $x_1 < 0 < x_2$ such that $f(x_1) < 0 < f(x_2)$.

22.M. Let $f(x) = x^n$. If $c > 1$, then $f(0) = 0 < c < f(c)$.

22.N. If $f(c) > 0$, there is a neighborhood of c on which f is positive, whence $c \neq \sup N$. Similarly if $f(c) < 0$.

22.O. Since f is strictly increasing, and $a < b$, then f maps the open interval (a, b) in a one-one fashion onto the open interval $(f(a), f(b))$, from which it follows that f^{-1} is continuous.

22.P. Yes. Let $a < b$ be fixed and suppose that $f(a) < f(b)$. If c is such that $a < c < b$, then either (i) $f(c) = f(a)$, (ii) $f(c) < f(a)$, or (iii) $f(a) < f(c)$. Case (i) is excluded by hypothesis. If (ii), then there exists a_1 in (c, b) such that $f(a_1) = f(a)$, a contradiction. Hence (iii) must hold. Similarly, $f(c) < f(b)$ and f is strictly increasing.

22.Q. Assume that g is continuous and let $c_1 < c_2$ be the two points in I where g attains its supremum. If $0 < c_1$, choose numbers a_1, a_2 such that $0 < a_1 < c_1 < a_2 < c_2$ and let k satisfy $g(a_i) < k < g(c_i)$. Then there exist three numbers b_i such that $a_1 < b_1 < c_1 < b_2 < a_2 < b_3 < c_2$ and where $k = g(b_i)$, a contradiction. Therefore, we must have $c_1 = 0$ and $c_2 = 1$. Now apply the same type of argument to the points where g attains its infimum to obtain a contradiction.

22.S. Note that $\varphi^{-1}(S)$ is not compact. Also φ^{-1} is not continuous at $(1, 0)$.

Section 23

23.A. The functions in Example 20.5(a, b, i) are uniformly continuous on $\mathbf{R}$.

23.G. The function g is bounded and uniformly continuous on $[0, p]$.

23.I. If (x_n) is a sequence in $(0, 1)$ with $x_n \to 0$, then $(f(x_n))$ is a Cauchy sequence and is therefore convergent in $\mathbf{R}$.

23.K. Take $f(x) = \sin x$, $g(x) = x$, for $x \in \mathbf{R}$.

Section 24

24.B. Take $((1/n)f)$, where f is as in Example 20.5(g).

24.C. Obtain the function in Example 20.5(h) in this way.

24.E. (a) The convergence is uniform on $[0, 1]$. (b) The convergence is uniform on any closed set not containing 1. (c) The convergence is uniform on $[0, 1]$ or on $[c, +\infty)$, $c > 1$.

24.J. It follows that f is monotone increasing. Since f is uniformly continuous, if $\varepsilon > 0$, let $0 = x_0 < x_1 < \cdots < x_h = 1$ be such that $f(x_j) - f(x_{j-1}) < \varepsilon$ and let n_j be such that if $n \geq n_j$; then $|f(x_j) - f_n(x_j)| < \varepsilon$. If $n \geq \sup\{n_0, n_1, \ldots, n_h\}$, show that $|f(x) - f_n(x)| < 3\varepsilon$ for all $x \in \mathbf{I}$.

24.S. Any polynomial (or uniform limit of a sequence of polynomials) is bounded on a bounded interval.

Section 25

25.G. (b) If $\varepsilon > 0$, there exists $\delta(\varepsilon) > 0$ such that if $c < x < c + \delta(\varepsilon)$, $x \in D(f)$, then $|f(x) - b| < \varepsilon$. (c) If (x_n) is any sequence in $D(f)$ such that $c < x_n$ and $c = \lim(x_n)$, then $b = \lim(f(x_n))$.

25.J. (a) If $M > 0$, there exists $m > 0$ such that if $x \geq m$ and $x \in D(f)$, then $f(x) \geq M$. (b) If $M < 0$, there exists $\delta > 0$ such that if $0 < |x - c| < \delta$, then $f(x) < M$.

25.L. (a) Let $\varphi(r) = \sup\{f(x) : x > r\}$ and set $L = \lim_{r \to +\infty} \varphi(r)$. Alternatively, if $\varepsilon > 0$ there exists $m(\varepsilon)$ such that if $x \geq m(\varepsilon)$, then $|\sup\{f(x) : x > r\} - L| < \varepsilon$.

25.M. Apply Lemma 25.12.

25.N. Consider the function $f(x) = -1/|x|$ for $x \neq 0$ and $f(0) = 0$.

25.P. Consider Example 20.5(h).

25.R. Not necessarily. Consider $f_n(x) = -x^n$ for $x \in \mathbf{I}$.

25.S. Yes.

Section 26

26.B. Show that the collection $\mathscr{A}$ of polynomials in $\cos x$ satisfies the hypotheses of the Stone-Weierstrass Theorem.

26.E. If $f(0) = f(\pi) = 0$, first approximate f by a function g vanishing on some intervals $[0, \delta]$ and $[\pi - \delta, \pi]$. Then consider $h(x) = g(x)/\sin x$ for $x \in (0, \pi)$, $h(x) = 0$ for $x = 0, \pi$.

26.I. Consider $f(x) = \sin(1/x)$ for $x \neq 0$.

26.K. Use the Heine-Borel Theorem or the Lebesgue Covering Theorem as in the proof of the Uniform Continuity Theorem.

26.Q. (a) Domain compact, sequence uniformly equicontinuous but not bounded. (b) Domain compact, sequence bounded but not uniformly equicontinuous. (c) Domain not compact, sequence bounded, and uniformly equicontinuous.

Section 27

27.D. Observe that $g'(0) = 0$ and that $g'(x) = 2x \sin(1/x) - \cos(1/x)$ for $x \neq 0$.
27.E. Yes.
27.L. We can write

$$\frac{f(x)-f(y)}{x-y} = \frac{x-c}{x-y} \cdot \frac{f(x)-f(c)}{x-c} - \frac{y-c}{x-y} \cdot \frac{f(y)-f(c)}{y-c}.$$

27.S. (b) If $b \neq 0$, then for $n \in N$ sufficiently large, given $x > n$, there is an $x_n > n$ such that

$$|(f(x)-f(n))/x| = |(x-n)/x|\,|f'(x_n)| \geq |(x-n)/x|\,|b|/2.$$

Section 28

28.F. Between consecutive roots of p' the polynomial is strictly monotone. If x_0 is a root of odd multiplicity of p', then x_0 is a point of strict extremum for p.
28.H. The function f has roots of multiplicity n at $x = \pm 1$; f' has roots of multiplicity $n-1$ at $x = \pm 1$, and a simple root inside $(-1, 1)$; etc.
28.O. Use Exercise 27.O.

Section 29

29.D. If $\varepsilon > 0$, then there are rational numbers $r_1, \ldots, r_m$ in I such that $0 \leq f(x) < \varepsilon$ if $x \neq r_k$. Let P be a partition such that each of the (at most $2m$) subintervals containing one of the $r_1, \ldots, r_m$ has length less than $\varepsilon/2m$. Show that $0 \leq S(P; f, g) \leq 2\varepsilon$.
29.J. If $f_1(x) = f(x)$ for $x \notin \{c_1, \ldots, c_m\}$ and $\varepsilon > 0$, let P be a partition such that each of the subintervals containing one of the $c_1, \ldots, c_m$ has length less than $\varepsilon/2mM$, where $M \geq \sup\{\|f\|_J, \|f_1\|_J\}$. Using the same intermediate points, we have $|S(P; f, g) - S(P; f_1, g)| < \varepsilon$, where $g(x) = x$ for $x \in J$.
29.N. Suppose $c \in (a, b)$; then f is g-integrable over $[a, c]$ and $[c, b]$. If g_1 is the restriction of g to $[a, c]$, it follows from 27.N than g_1' is continuous on $[a, c]$; similarly for the restriction g_2 of g to $[c, b]$. It follows from Theorem 29.8 that fg_1' is integrable over $[a, c]$ and that fg_2' is integrable over $[c, b]$ and that

$$\int_a^c f\,dg = \int_a^c fg_1', \qquad \int_c^b f\,dg = \int_c^b fg_2'.$$

Now let $(fg')(x) = f(x)g_1'(x)$ for $a \leq x \leq c$ and $(fg')(x) = f(x)g_2'(x)$ for $c < x \leq b$.
29.P. If $\|P\| < \delta$ and if Q is a refinement of P, then $\|Q\| < \delta$.
29.R. If $\varepsilon > 0$, let $P_\varepsilon = (x_0, x_1, \ldots, x_n)$ be a partition of J such that if $P \supseteq P_\varepsilon$, and $S(P; f)$ is any corresponding Riemann sum, then $|S(P; f) - \int_a^b f| < \varepsilon$. Let $M \geq \|f\|_J$

and let $\delta = \varepsilon/4nM$. If $Q = (y_0, y_1, \ldots, y_m)$ is a partition with norm $\|Q\| < \delta$, let $Q^* = Q \cup P_\varepsilon$ so that $Q^* \supseteq P_\varepsilon$ and has at most $n - 1$ more points than Q. Show that $S(Q^*; f) - S(Q; f)$ reduces to at most $2(n - 1)$ terms of the form $\pm\{f(\xi) - f(\eta)\}$ $(x_j - y_k)$ with $|x_j - y_k| < \delta$.

Section 30

30.C. If $\varepsilon > 0$ is given, let P_ε be as in the proof of 30.2. If P is any refinement of P_ε, then

$$|S(P_\varepsilon; f, g) - S(P; f, g)| \leq \sum |f(u_k) - f(v_k)| \, |g(x_k) - g(x_{k-1})|$$

where $|u_k - v_k| < \delta(\varepsilon)$ and hence this sum is dominated by εM. Now use the Cauchy Criterion.

30.E. Direct estimation yields

$$\left(\int_a^b (f(x))^n \, dx \right)^{1/n} \leq M(b - a)^{1/n}.$$

Conversely, $f(x) \geq M - \varepsilon$ on some subinterval of $[a, b]$.

30.H. If $m \leq f(x) \leq M$ for $\alpha \leq x \leq \beta$, there exists an A with $m \leq A \leq M$ such that

$$F(\beta) - F(\alpha) = \int_\alpha^\beta f \, dg = A\{g(\beta) - g(\alpha)\}.$$

30.I. Let $f(x) = -1$ for $x \in [-1, 0)$ and $f(x) = 1$ for $x \in [0, 1]$.

30.J. Apply the Mean Value Theorem 27.6 to obtain $F(b) - F(a)$ as a Riemann sum for the integral of f.

30.M. If $m \leq f(x) \leq M$ for $x \in J$, then

$$m \int_a^b p \leq \int_a^b fp \leq M \int_a^b p.$$

Now use Bolzano's Theorem 22.4.

30.P. The functions φ, φ^{-1} are one-one and continuous. The partitions of $[c, d]$ are in one-one correspondence with the partitions of $[a, b]$ and Riemann-Stieltjes sums of $f \circ g$ with respect to $g \circ \varphi$ are in one-one correspondence with those of f with respect to g.

30.V. (a) $\frac{3}{4}$. (c) 9. (e) $\pi/2$.

Section 31

31.K. Since $f_n(x) - f_n(c) = \int_c^x f_n'$, we can apply Theorem 31.2 to obtain $f(x) - f(c) = \int_c^x g$ for all $x \in J$. Show that $g = f'$.

31.S. Apply Theorem 30.9 to (31.2) with $h(t) = (b - t)^{n-1}$.

31.V. Prove that the functions G and H are continuous. The remainder of the proof is as in 31.9.

31.X. The function f is uniformly continuous on $J_1 \times J_2$.

31.Y. Let $g_2(0) = 0$, $g_2(x) = \frac{1}{2}$ for $0 < x < 1$, and $g_2(x) = 1$.

Section 32

32.D. (a), (b), (d), (e) are convergent.

32.E. (a) is convergent for p, $q > -1$. (b) is convergent for $p + q > -1$.

32.F. (a) and (c) are absolutely convergent. (b) is divergent.

32.G. (a) is absolutely convergent if $q > p + 1$. (b) is convergent if $q > 0$ and absolutely convergent if $q > 1$.

Section 33

33.A. If $0 \le t \le \beta$, then $x^t e^{-x} \le x^\beta e^{-x}$.

33.B. Apply the Dirichlet Test 33.4.

33.C. (a) is uniformly convergent for $|t| \ge a > 0$. (b) diverges if $t \le 0$ and is uniformly convergent if $t \ge c > 0$. (c) and (e) are uniformly convergent for all t.

33.F. $\sqrt{\pi}$.

Section 34

34.C. Group the terms in the series $\sum_{n=1}^{\infty} (-1)^n$ to produce convergence to -1 and to 0.

34.G. Consider $\sum ((-1)^n n^{-1/2})$. However, consider also the case where $a_n \ge 0$.

34.H. If a, $b \ge 0$, then $2(ab)^{1/2} \le a + b$.

34.I. Show that $b_1 + b_2 + \cdots + b_n \ge a_1(1 + \frac{1}{2} + \cdots + 1/n)$.

34.J. Use Exercise 34.F(a).

34.K. Show that $a_1 + a_2 + \cdots + a_{2^n}$ is bounded below by $\frac{1}{2}\{a_1 + 2a_2 + \cdots + 2^n a_{2^n}\}$ and above by $a_1 + 2a_2 + \cdots + 2^{n-1} a_{2^{n-1}} + a_{2^n}$.

34.O. Consider the partial sums s_k with $n/2 \le k \le n$ and apply the Cauchy Criterion.

Section 35

35.C. (a) and (e) are divergent. (b) is convergent.

35.D. (b), (c), and (e) are divergent.

35.G. (a) is convergent. (c) is divergent.

35.L. If $r < 1$, then $\log m < \log (m + 1) - r/m$ for sufficiently large $m \in \mathbf{N}$. Show that the sequence $(x_n n \log n)$ is increasing.

Section 36

36.A. Apply Dirichlet's Test.

36.D. (a) is convergent. (b) is divergent.

36.E. (c) If $\sum (a_n)$ is absolutely convergent, so is $\sum (b_n)$. If $a_n = 0$ except when $\sin n$ is near ± 1, we can obtain a counter-example. (d) Consider $a_n = 1/n(\log n)^2$.

36.I. If $m > n$, then $s_{mn} = +1$; if $m = n$, then $s_{mn} = 0$; if $m < n$, then $s_{mn} = -1$.

36.K. Note that $2mn \le m^2 + n^2$.

Section 37

37.A. (a) and (c) converge uniformly for all x. (b) converges for $x \neq 0$ and uniformly for x in the complement of any neighborhood of $x = 0$. (d) converges for $x > 1$ and uniformly for $x \geq a$, where $a > 1$.

37.C. If the series is uniformly convergent, then

$$|c_n \sin nx + \cdots + c_{2n} \sin 2nx| < \varepsilon,$$

provided n is sufficiently large. Now restrict attention to x in an interval such that $\sin kx > \frac{1}{2}$ for $n \leq k \leq 2n$.

37.H. (a) ∞, (c) $1/e$, (f) 1.

37.L. Apply the Uniqueness Theorem 37.17.

37.N. Show that if $n \in N$, then there exists a polynomial P_n such that if $x \neq 0$, then $f^{(n)}(x) = e^{-1/x^2} P_n(1/x)$.

37.T. The series $A(x) = \Sigma (a_n x^n)$, $B(x) = \Sigma (b_n x^n)$, and $C(x) = \Sigma (c_n x^n)$ converge to continuous functions on I. By the Multiplication Theorem 37.8, $C(x) = A(x)B(x)$ for $0 \leq x < 1$, and by continuity $C(1) = A(1)B(1)$.

37.U. The sequence of partial sums is increasing on the interval $[0, 1]$.

37.V. If $\varepsilon > 0$, then $|a_n| \leq \varepsilon p_n$ for $n > N$. Break the sum $\Sigma (a_n x^n)$ into a sum over $n = 1, \ldots, N$ and a sum over $n > N$.

Section 38

38.B. (b) If $a_n = 0$, then

$$F(x + 2\pi) = \int_c^{x+2\pi} f(t)\, dt = \int_c^x f(t)\, dt + \int_x^{x+2\pi} f(t)\, dt$$
$$= F(x) + 0 = F(x)$$

for all $x \in R$.

38.E. (c) Calculate $f_2(0)$ in two ways.

38.G. (a) If k_1 were continuous, then $k_1(-\pi) = -\pi^3$ but since k_1 has period 2π, then $k_1(-\pi) = k_1(\pi) = \pi^3$.

38.I. (b) $\dfrac{2}{\pi} - \dfrac{4}{\pi}\left[\dfrac{\cos 2x}{1 \cdot 3} + \dfrac{\cos 4x}{3 \cdot 5} + \dfrac{\cos 6x}{5 \cdot 7} + \cdots\right].$

(c) $\dfrac{1}{2} + \dfrac{2}{\pi}\left[\dfrac{\cos x}{1} - \dfrac{\cos 3x}{3} + \dfrac{\cos 5x}{5} - \cdots\right].$

(e) $\dfrac{\pi^2}{6} - \left[\dfrac{\cos 2x}{1^2} + \dfrac{\cos 4x}{2^2} + \dfrac{\cos 6x}{3^2} + \cdots\right].$

38.K. (b) $\dfrac{8}{\pi}\left[\dfrac{\sin 2x}{1 \cdot 3} + \dfrac{2 \sin 4x}{3 \cdot 5} + \dfrac{3 \sin 6x}{5 \cdot 7} + \cdots\right].$

(e) $\dfrac{8}{\pi}\left[\dfrac{\sin x}{1^3} + \dfrac{\sin 3x}{3^3} + \dfrac{\sin 5x}{5^3} + \cdots\right].$

38.N. (d) Use Exercise 38.G(b).

38.R. (a) $\dfrac{4}{\pi}\left[\dfrac{\sin\frac{1}{2}\pi x}{1}-\dfrac{\sin\pi x}{2}+\dfrac{\sin\frac{3}{2}\pi x}{3}-\cdots\right].$

(b) $1+\dfrac{4}{\pi}\displaystyle\sum_{n=1}^{\infty}\left[-\dfrac{1+(-1)^{n+1}}{n^2\pi}\cos\dfrac{n\pi x}{4}+\dfrac{(-1)^{n+1}}{n}\sin\dfrac{n\pi x}{4}\right].$

38.S. Use Fejér's Theorem 38.12 and Theorem 19.3.
38.T. Modify the proofs of 38.7 and 38.12.

Section 39

39.G. We have $|G(u,v)-G(0,0)|\le|u^2+v^2|=\|(u,v)\|^2$ so that $DG(0,0)(u,v)=0$. If $(x,y)\ne(0,0)$, then $D_1G(x,x)=2x\sin(2x^2)^{-1}-x^{-1}\cos(2x^2)^{-1}$, which is not bounded as $x\to0$.
39.L. (a) $\nabla_{(a,b,c)}f_1=(2a,2b,2c)$.

(c) $\nabla_{(a,b,c)}f_3=(bc,ac,ab)$.
39.M. (a) $3\sqrt{2}$. (c) 0.
39.O. (a) At $(1,2)$ we have $\{(x,y,z):z-5=2(x-1)+4(y-2)\}$.

(c) At $(1,1)$ we have $\{(x,y,z):z-\sqrt{2}=-(x+y-2)/\sqrt{2}\}$.
39.Q. (a) At $t=0$ we have $\{(x,y,z):x=t,y=0,z=0\}$; at $t=1$ we have $\{(x,y,z):x=1+s,y=1+2s,z=1+3s\}$.

(c) At $t=\frac{1}{2}\pi$ we have $\{(x,y,z):x=-2s,y=2,z=\frac{1}{2}\pi+s\}$.
39.S. (b) At the point $(3,-1,-3)$ corresponding to $(s,t)=(1,2)$ we have:

$$S_h=\{(x,y,z):x=3+(s-1)+(t-2),\ y=-1+(s-1)-(t-2),$$
$$z=-3+2(s-1)-4(t-2)\}.$$

(d) At the point $(1,0,0)$ corresponding to $(s,t)=(0,\frac{1}{2}\pi)$ we have: $S_h=\{(x,y,z):x=1,y=s,z=-(t-\frac{1}{2}\pi)\}$.
39.V. Note that if $y\in R^q$, $z\in R^r$, then $(y,z)\in R^q\times R^r=R^{q+r}$ is such that $\|(y,z)\|^2=\|y\|^2+\|z\|^2$.

Section 40

40.A. $F'(t)=2(3t+1)3+2(2t-3)2=26t-6$.
40.D. $D_1F(s,t)=(\sin s\cos t+\sin t)(-\sin s)+(\cos s+\sin t)(\cos s\cos t)+0$.
40.G. (a) $D_1F(x,y)=f'(xy)y,\quad D_2F(x,y)=f'(xy)x$.

(d) $D_1F(x,y)=f'(x^2-y^2)(2x),\quad D_2F(x,y)=f'(x^2-y^2)(-2y)$.
40.K. (b) Since $g'(t)=D_1f(tc)c_1+\cdots+D_pf(tc)c_p$, it follows from Euler's Relation that we have

$$tg'(t)=(tc_1)D_1f(tc)+\cdots+(tc_p)D_pf(tc)=kf(tc)=kg(t).$$

Therefore (why?) $g(t)=Ct^k$ for some constant C. Since $f(c)=g(1)=C$, we deduce that $f(tc)=g(t)=t^kf(c)$, whence f is homogeneous of degree k.
40.M. Since

$$\|B(x+u,y+v)-B(x,y)-(B(x,v)+B(u,y))\|$$
$$=\|B(u,v)\|\le M\|u\|\,\|v\|\le\tfrac{1}{2}M(\|u\|^2+\|v\|^2)=\tfrac{1}{2}M\|(u,v)\|^2,$$

it follows that $DB(x,y)(u,v)$ exists and equals $B(x,v)+B(u,y)$.

40.P. Since $Dg(c)(u) = (ug_1'(c), \ldots, ug_p'(c)) = ug'(c)$ for $u \in \mathbf{R}$, it follows from the Chain Rule that

$$Dh(c)(u) = Df(g(c))(Dg(c)(u)) = Df(g(c))(ug'(c)) = uDf(g(c))(g'(c))$$

whence $h'(c) = Df(g(c))(g'(c))$.

40.Q. If $f = (f_1, \ldots, f_q)$, there exist points $c_i \in S$ such that $f_i(b) - f_i(a) = Df_i(c_i)(b - a)$. Now let L have the matrix representation $[D_j f_i(c_i)]$.

40.R. By Theorem 12.7 any two points in Ω can be joined by a polygonal curve lying inside Ω. Apply the Mean Value Theorem to each segment of this curve.

40.U. Indeed $D_x f(x, y) = y(x^2 - y^2)(x^2 + y^2)^{-1} + 4x^2 y^3 (x^2 + y^2)^{-2}$ and $D_{yx} f(0, 0) = -1$, while $D_{xy} f(0, 0) = +1$.

40.W. If $\varphi : (-\varepsilon, 1+\varepsilon) \to \mathbf{R}^q$ is defined by $\varphi(t) = f(a + t(b - a))$, then $\varphi'(t) = Df(a + t(b - a))(b - a)$. Write $\varphi(t) = (\varphi_1(t), \ldots, \varphi_q(t))$ where $\varphi_i(t) = f_i(a + t(b - a))$ and note that $\varphi_i(1) - \varphi_i(0) = \int_0^1 \varphi_i'(t)$.

Section 41

41.A. Use Exercise 21.P.   41.D. Here $Df(0) = 0$. No.

41.E. Consider Exercises 27.H and 22.O.

41.F. Consider Exercise 40.L.

41.J. Find the relative extrema near 0.

41.Q. (a) At $(1, 1, 2)$ we have $S_F\{(x, y, z) : 2x + 2y - z = 2\}$.

(c) At $(4, \frac{1}{2}, 2)$ we have $S_F = \{(x, y, z) : x + 8y - 2z = 4\}$.

41.S. If $D_1 f$ vanishes on an open set, apply Exercise 40.S. If $D_1 f(x_0, y_0) \neq 0$, then consider $F(x, y) = (f(x, y), y)$ near (x_0, y_0).

41.T. If $D_1 g(c) \neq 0$, then consider $G(x, y) = g(x) + (0, y)$.

41.V. Show, as in the proof of 41.6, that if $\|y\| < m/2$, then there is a vector $x \in \mathbf{R}^p$ with $\|x\| \le 1$ such that $y = L_1(x)$.

41.W. If $y \in \mathbf{R}^p$, let $x_0 = 0$ and

$$x_{n+1} = x_n - (f(x_n) - f(x_{n-1}) - (x_n - x_{n-1})).$$

Show, as in the proof of 41.6, that $\bar{x} = \lim(x_n)$ exists and $f(\bar{x}) = y$.

Section 42

42.A. (a) Saddle point at $(0, 0)$.   (b) Relative strict minimum at $(-2, \frac{1}{2})$.

(c) Saddle point at $(0, -1)$; relative strict minimum at $(0, 3)$. (f) Saddle point at $(0, 0)$; relative strict minima at $(0, -1)$ and $(0, 2)$.

42.D. If f is not constant, then either the supremum or the infimum of f on $S = \{x \in \mathbf{R}^p : \|x\| \le 1\}$ is not 0. Since S is compact, this supremum (or infimum) is attained at a point $c \in S$. The hypothesis rules out the possibility that $\|c\| = 1$.

42.F. (a, d) Relative minimum at $(0, 0)$. (b, c, e) Saddle point at $(0, 0)$. (f) Relative strict minimum at $(0, 0)$.

42.G. Saddle point at $(1,1)$.

42.H. Monkeys have tails.

42.I. $\frac{7}{3}$.   42.K. $\frac{2}{7}$.

42.S. (a) Maximum value $= 1$, attained at $(\pm 1, 0)$; minimum value $= -1$, attained at $(0, \pm 1)$.   (b) Maximum value $= 3$, attained at $(1, 0)$; minimum value $= -1$, attained at $(-1, 0)$.   (c) Maximum value $= 4$, attained at $(1, \pm 1)$; minimum value $= -1$, attained at $(-1, 0)$.   (d) Maximum value $= 1$, attained at $(0, \pi/2)$; minimum value $= -1$; attained at $(0, -\pi/2)$.

42.U. Maximum value $= 1$, attained at $(1, 0, 0)$; minimum value $= \frac{5}{9}$, attained at $(-\frac{1}{3}, \frac{2}{3}, \frac{2}{3})$.

Section 43

43.B. If $p \in N$ is given, let $n > (2^{1/p} - 1)^{-1}$. If a cell I in R^p has side lengths $0 < a_1 \le a_2 \le \cdots \le a_p$, let $c = a_1/n$. Then I is contained in the union of $n([a_2/c] + 1) \cdots ([a_p/c] + 1)$ cubes with side length c, having total content less than $2(a_1 \cdots a_p) = 2c(I)$. Hence, if Z is contained in the union of cells with total content less than ε, it is contained in the union of cubes with total content less than 2ε.

43.E. No.

43.H. If the closure of J_j is $[a_{j1}, b_{j1}] \times \cdots \times [a_{jp}, b_{jp}]$, for $j = 1, \ldots, n$, and if $I = [a_1, b_1] \times \cdots \times [a_p, b_p]$, let P_1 be the partition of $[a_1, b_1]$ obtained by using the points $\{a_{j1}, b_{j1} : j = 1, \ldots, n\}, \ldots,$ and P_p be the partition of $[a_p, b_p]$ obtained by using $\{a_{jp}, b_{jp} : j = 1, \ldots, n\}$. The partitions $P_1, \ldots, P_p$ induce a partition of I.

43.I. Enclose Z in the union of a finite number of closed cells in I with total content less than ε. Now apply 43.H.

43.J. Enclose Z in the union of a finite number of open cells in I with total content less then ε. Now apply 43.H.

43.K. We form a sequence of partitions of I into cubes with side length $2^{-n}\delta$ by successive bisection of the sides of I. Given a cube $K \subseteq I$ with side length r, enclose K in the union of all cubes in the nth partition which have non-empty intersection with K. If n is so large that $(1 + \delta/2^{n-1}r)^p < 2$, then this union has total content less than $2c(K)$.

43.L. Use 43.G.   43.M. Use 43.L.

43.R. First treat the case $f = g$; then consider $(f + g)^2$.

43.T. Let $M > \|f\|_I, \|g\|_I$. Since f and g are uniformly continuous on K, if P_ε is sufficiently fine, then f and g vary less than $\varepsilon/2M$ on each K_j and such that for any $p_j \in K_j$ we have $|\int_K fg - \sum f(p_j)g(p_j)c(K_j)| \le (\varepsilon/2)c(K)$. Then we have

$$\left| \int_K fg - \sum f(x_j)g(y_j)c(K_j) \right| \le \left| \int_K fg - \sum f(x_j)g(x_j)c(K_j) \right|$$

$$+ \left| \sum f(x_j)[g(x_j) - g(y_j)]c(K_j) \right| \le \varepsilon c(K).$$

43.V. (d) If Z is compact and contained in the union of open cells $J_1, J_2, \ldots,$ then it is contained in the union of a finite number of these cells.

Section 44

44.B. (a) If $c \notin b(A)$, then either c is an interior point of A or it is an interior point of $\mathscr{C}(A)$. In either case, there is a neighborhood of c disjoint from $b(A)$, so $\mathscr{C}(b(A))$ is open.

44.C. In Example 43.2(g), we have $S^- = b(S) = I \times I$. However $b(I \times I) = I \times \{0, 1\} \cup \{0, 1\} \times I$.

44.D. Since $(A \cap B)^- \subseteq A^- \cap B^-$, it follows that

$$b(A \cap B) = (A \cap B)^- \cap (\mathscr{C}(A \cap B))^- \subseteq A^- \cap B^- \cap (\mathscr{C}(A) \cup \mathscr{C}(B))^-$$

$$= A^- \cap B^- \cap (\mathscr{C}(A)^- \cup (\mathscr{C}(B)^-)$$

$$= (B^- \cap b(A)) \cup (A^- \cap b(B)) \subseteq b(A) \cup b(B).$$

44.J. If $\varepsilon > 0$, let P_ε be a partition such as that in Exercise 43.P and such that the union of the cells in P_ε which contain points in $b(A)$ has total content less than $\varepsilon/2 \|f\|_I$. Now apply Exercise 43.P to the restriction of f to A.

44.K. Since $mg(x) \le f(x)g(x) \le Mg(x)$ for $x \in A$, it follows that $m\int_A g \le \int_A fg \le M \int_A g$. If $\int_A g \ne 0$, take $\mu = (\int_A fg)(\int_A g)^{-1}$.

44.N. The set $b(K) = K$ does not have content zero.

44.R. Note that $F(x, y) = \int_0^y \{\int_0^x f(s, t) \, ds\} \, dt$.

Section 45

45.A. Examine the proofs of 45.1–45.4.

45.D. (a) 6π.   45.F. $(e - 1)^2$.

45.H. Let $u = xy$, $v = y/x^2$. The area equals $(\log 2)/3$.

Index